CURRENT

A A R O N   S A C H S

Nineteenth-Century Exploration and the Roots of

American Environmentalism

PENGUIN BOOKS

PENGUIN BOOKS

Published by the Penguin Group

Penguin Group (USA) Inc., 375 Hudson Street, New York, New York 10014, U.S.A.

Penguin Group (Canada), 90 Eglinton Avenue East, Suite 700, Toronto,

Ontario, Canada M4P 2Y3 (a division of Pearson Penguin Canada Inc.)

Penguin Books Ltd, 80 Strand, London WC2R 0RL, England

Penguin Ireland, 25 St Stephen's Green, Dublin 2, Ireland (a division of Penguin Books Ltd)

Penguin Group (Australia), 250 Camberwell Road, Camberwell,

Victoria 3124, Australia (a division of Pearson Australia Group Pty Ltd)

Penguin Books India Pvt Ltd, 11 Community Centre, Panchsheel Park, New Delhi – 110 017, India

Penguin Group (NZ), 67 Apollo Drive, Rosedale, North Shore 0745, Auckland,

New Zealand (a division of Pearson New Zealand Ltd)

Penguin Books (South Africa) (Pty) Ltd, 24 Sturdee Avenue,

Rosebank, Johannesburg 2196, South Africa

Penguin Books Ltd, Registered Offices:
80 Strand, London WC2R 0RL, England

First published in the United States of America by Viking Penguin,
a member of Penguin Group (USA) Inc. 2006
Published in Penguin Books 2007

10 9 8 7 6 5 4 3 2 1

ISBN 0-670-03775-3 (hc.)
ISBN 978-0-14-311192-4 (pbk.)
CIP data available

Printed in the United States of America
Designed by Carla Bolte • Set in Scala

Praise for *The Humboldt Current* by Aaron Sachs

**Honorable Mention for the Frederick Jackson Turner Award,
given annually by the Organization of American Historians (OAH)
for the best first book in American history**

"Sachs creates a different relation between past and present that is quite un-Whiggish and quite liberating. . . . The book is . . . smart, closely observed, lively, and full of sharply etched characters, who carry his story."

—Richard White, *Raritan*

"Humboldt, the central figure in Aaron Sachs's ambitious first book, *The Humboldt Current*, was the toast of all Europe, as well as a subject of great admiration in the fledgling country across the sea. . . . *The Humboldt Current* is not lacking in resonant voices. Sachs's subjects are strong, and he describes them in extensive detail."

—Candice Millard, *The New York Times Book Review*

"Sachs is clearly smitten with his subject, and his enthusiasm bubbles over in the lively chapters he devotes to Humboldt's life. . . . Sachs has done something worthy of gratitude: he has reintroduced a nineteenth-century sage to a generation that sorely needs his wisdom."   —Judith Lewis, *Los Angeles Times*

"The portraits of these early environmentalists are compelling, particularly the surprising depiction of John Muir."   —Kathleen McGowan, *Audubon*

"Sachs picks his way carefully through the lives and writings of his Humboldtians. He wants an American environmentalism that doesn't separate people from nature, that is as attuned to social justice as it is to conservation and preserving the wild. And history is as good a place as any—and better than many—to go looking for what is lacking in the present and may be useful for making a better future."   —Jon Christensen, *San Francisco Chronicle*

"As a work of history, *The Humboldt Current* is impressive. It is smartly conceived and superbly written. Most important, it argues persuasively that the course of American empire was 'many-sided and intensely contested,' even by those whose explorations led the way."

—Gregory Summers, *History: Reviews of New Books*

"American history of the nineteenth century is dominated by the Civil War, the expansion to the Pacific, and the push to industrialization, but it is worth recalling the prominent interest in natural history in the U.S., a movement of which the tremendously popular Prussian scientist Alexander von Humboldt (1769–1859) was more or less the first practitioner. . . . This ambitious subject is admirably tackled in this complexly argued book."   —*Publishers Weekly*

"The book's greatest achievement lies in its deeply impressive scope, its integration not just of science and exploration, but also of the art, literature, and politics of the nineteenth century. In this, the author achieves a unity and harmony of vision not unlike that of Humboldt himself."

—*Kirkus Reviews* (starred review)

"*The Humboldt Current* is an astonishingly good piece of writing and research, and an essential piece of American naturalist history that has been too long in coming. Science and history buffs and the lay reader will equally enjoy this outstanding book."

—Science Book Reviews

"Alexander von Humboldt was one of my heroes, as were the explorer-scientists of the American West, and as were their contemporaries, poets and writers such as Whitman and Thoreau, precursors of cosmic consciousness and American environmentalism. But it never occurred to me to bring them all together in one all-encompassing, yet detailed, narrative. That is left to Aaron Sachs in a work of striking originality, meticulous scholarship, and deep humanist sympathy."

—Yi-Fu Tuan

"In this groundbreaking book, Aaron Sachs plucks from relative obscurity the nineteenth-century Prussian scientist Alexander von Humboldt and demonstrates his profound, lasting influence on many aspects of American culture, including literature, art, science, and environmentalism. Sachs's sweeping study argues convincingly that Humboldt was second only to Darwin among scientists in his impact on the history of American thought."

—David S. Reynolds

"*The Humboldt Current* is a dazzling debut performance by a young scholar-writer of extraordinary gifts. The book itself is a gift—of carefully researched, and beautifully expressed, and deeply humane, understanding. This is one of those rare works in which historical learning makes a lasting difference on our way of seeing both past and present worlds."

—John Demos

"Through the lives of Americans who followed or echoed Humboldt, this fascinating, insightful book gives us a brilliant new account of U.S. geography and ecology, exploration and eccentricity."

—Felipe Fernández-Armesto

"In this magnificent book, Aaron Sachs reintroduces us to a forgotten giant, Alexander von Humboldt, who cast an extraordinary spell over our Victorian ancestors and inspired some of the most heroic adventures in American science. Humboldt, Sachs reminds us, was a revolutionary figure whose bold vision of global ecology and human fellowship remains as urgent as ever."

—Mike Davis

PENGUIN BOOKS

THE HUMBOLDT CURRENT

Aaron Sachs is a professor of history and American Studies at Cornell University and an award-winning environmental journalist.

The
HUMBOLDT

For my mother and father

Strange mobility of the imagination of man,
eternal source of our enjoyments and our pains!

— Alexander von Humboldt —

— CONTENTS —

Part Four

— NORTH —

George Wallace Melville and John Muir in Extremis

— IMAGES —

THE HUMBOLDT CURRENT

Humboldt in America

1804–2004

It would have been easier to sail straight back to Europe. Politics and weather both favored the conservative course: to make the detour from Havana to Philadelphia, his ship would be forced to brave a British naval blockade and risk a dangerous stretch of water at the beginning of hurricane season. In any case, there was no question of staying on this side of the Atlantic. Alexander von Humboldt, now the world's most famous scientific explorer—after a five-year expedition up the Orinoco River and along the slopes of Andean volcanoes—could not finally unpack his brand-new maps, his bottles of curare, his pressed acacias and lianas, anywhere but in Paris, the scientific capital of the Western world. Indeed, he had already sent many of his collections to France, and, as he admitted to a colleague, "I think only of preserving and publishing my manuscripts. How I long to be in Paris!"[1] Yet he also longed to experience Philadelphia and Washington. "For moral reasons," he wrote to President Thomas Jefferson, "I could not resist seeing the United States and enjoying the consoling aspects of a people who understand the precious gift of Liberty."[2] So Humboldt made for the Bahama Straits.

Unsurprisingly, on the night of May 6, 1804, the wind and water started to rise. Each wave, Humboldt noted, felt like a boulder, and he saw the most seasoned sailors fall to the deck repeatedly when the ship convulsed. Water flowed down the ship's stairways and under its doors. The skies were so dark that candles had to be lit all day on the poop deck. For about a week, cooking was impossible, so it was salted cod at almost every meal; much of the crew, meanwhile, subsisted exclusively on brandy, explaining to the passengers that it was best to drown one's sorrows before actually drowning. The passengers were not amused. Most of them were drenched, sleepless, and somewhat terrified. Humboldt's thirty-five remaining specimen cases, and his life, were clearly in danger, but the determined scientist at least knew how to distract himself: desperately trying to maintain his mental as well as his physical balance, he adhered as closely as possible to his routine of measuring air and water temperatures and examining the marine creatures swept aboard ship by the pounding waves. The wry entry in his scientific notebook said simply "that the Gulf Stream does indeed exist."[3]

In his private journal, to which he returned only after the storm abated on May 14, the thirty-four-year-old Humboldt was more expressive. "I have never been so concerned about my death as on the morning of May 9th," he scribbled. "I was thoroughly distraught. To see myself perish on the eve of so many joys, to see all the fruits of my travels perish with me, to cause the death of the two people accompanying me, to perish on a voyage to Philadelphia that was not entirely necessary . . ."[4]

As far as the twenty-first-century memory of Alexander von Humboldt is concerned, he may as well have gone down with his ship: many people have never even heard of him. But we're lucky he survived. In fact, he lived fifty-five more years, and his subsequent career, almost all of which he spent analyzing those precious specimens he had collected on his expedition, saw him become the nineteenth century's most influential scientist—especially in America—until at least 1859, the year he died and Charles Darwin published *On the Origin of Species*.[5] Darwin's explanation of the "struggle for existence" eventually seemed to eclipse Humboldt's more Romantic vision of the world, but the elder scientist's radical approach to nature and humanity makes him an astonishingly relevant figure for the twenty-first century. As Humboldt explained in 1799, at the very beginning of his expedition, his goal was to "recognize the general connections that link organic beings" and to "study the great harmonies of Nature." He seems to have been the first ecologist.[6]

Humboldt arrived at the port of Philadelphia just over two hundred years ago, on May 24, 1804.[7] Despite an aristocratic background, he had "simple unaffected manners," according to President Jefferson's personal secretary, William Burwell; Humboldt also came across as "remarkably sprightly" and "vehement in conversation."[8] Albert Gallatin, Jefferson's secretary of the treasury, reported to his wife that Humboldt spoke "more than Lucas, Finley, and myself put together, and twice as fast as any body I know, German, French, Spanish and English all together. But I was really delighted and swallowed more information of various kinds in less than two hours than I had for two years past in all that I had read or heard."[9] Throughout his six-week sojourn in the United States—he stayed only through the first week of July—Humboldt was hosted by members of the American Philosophical Society, the nation's most prestigious conglomeration of artists and scientists, founded by Benjamin Franklin. The painter and natural historian Charles Willson Peale, in particular, took a shine to the visiting explorer, and accompanied him as he shuttled back and forth in rattling stagecoaches between Philadelphia and Washington, to meet various luminaries in government and science. Clearly, Peale noted, this man could empathize with the American spirit of restlessness: "He has been travelling ever since he was 11 years of age," wrote the founder of Philadelphia's world-famous natural history museum, "and never lived in any one place more than 6 months together, as he informed us." Humboldt knew how to keep still for at least a few hours at a time, though, when in the company of friends: Peale got him to sit for a portrait, which today hangs in Philadelphia's College of Physicians.[10]

If any particular person in the young Republic could have quieted Humboldt's wanderlust for a more prolonged period, it would probably have been Thomas Jefferson himself. President not only of the nation but also of its foremost Philosophical Society, Jefferson seems to have been the real object of Humboldt's visit to the United States. In the aftermath of the post-revolutionary Terror in France, the best hope for democratic republicanism clearly resided with the American leader, and Humboldt yearned to connect with the man who had written the Declaration of Independence. Moreover, Humboldt loved the idea that the president of the United States was someone with whom he could talk shop. He ended his letter of introduction by referring to Jefferson's scientific writings: "As a friend of science," Humboldt fawned, "you will excuse the indulgence of my admiration. I would love to talk to you about a subject that you have treated so ingeniously in your work on Virginia, the teeth of mammoth which we too discovered in the Andes."[11] Here was someone obviously interested in Humboldt's brand of comparative, cosmopolitan research and, even better, in applying that research toward the more effective harmonizing of society and nature.

Jefferson was a much more settled man than the young Baron von Humboldt, with a deep attachment to his home.[12] Yet, as Humboldt sensed, his host at the presidential mansion had a wandering spirit. Since the 1780s Jefferson had been collecting books on the geography and exploration of western territories, dreaming of the seemingly limitless possibilities for American society. At his most idealistic, he saw the Garden of the West dotted with agrarian smallholders, envisioned the entire American hinterland as a cross-cultural, pastoral paradise where white migrants and Indians would intermarry and sustain an ethic of democratic participation in local communities. At other times, moments of pragmatism or even cynicism, he conceived of the West simply as a gigantic reservation for the Native Americans being forced out of the eastern states by continually expanding white settlement. In any case, more than any other political leader in the early Republic, Jefferson nurtured America's western obsession.[13] When the president purchased Louisiana from Napoleon in 1803 for fifteen million dollars, he created an American empire, doubling the territory over which he presided and focusing Americans' attention on the Pacific. A few months later, Lewis and Clark started up the Missouri River—just a couple of days before Humboldt's ship dropped anchor at Philadelphia.

Humboldt missed Lewis and Clark, but virtually every other American expedition of the nineteenth century would bear the mark of his influence—in part because the explorer Zebulon Pike (of Pike's Peak fame) copied Humboldt's main map on the sly, while it lay open on Secretary Gallatin's desk in Washington,[14] but mostly because the Prussian's writings became so widely popular in the United States. Editions of his books sold out repeatedly. Ralph Waldo Emerson called Humboldt "one of those wonders of the world . . . who appear from time to time, as if to show us the possibilities of the human mind"; Henry David Thoreau classified New England's climate zones according to Humboldt's model of plant ecology.[15] The peripatetic painters George Catlin and Frederic Church both traveled to South America specifically to seek out the native peoples and Andean landscapes described in Humboldt's books.[16] Humboldt's profound respect for indigenous Americans inspired Secretary Gallatin to write his foundational work of American ethnology, *A Synopsis of the Indian Tribes within the United States* (1836).[17] A couple of decades later, Walt Whitman would start to suffuse his poetry with the concept of "Cosmos," a term that suggested the world's overarching but mysterious harmony and that Whitman stole directly from Humboldt, who had plucked the word from ancient Greek to use as the title of his final work, the daunting subtitle of which was "A Sketch of a Physical Description of the Universe."[18] And during the Civil War, some abolitionists even published an antislavery letter from Humboldt to be distributed in New York and New England "for the Benefit of the Sick and Wounded Soldiers in the U.S. Armies."[19] It is quite possible that no other European had so great an impact on the intellectual culture of nineteenth-century America.[20]

In Washington, Humboldt and Jefferson compared their mammoths' teeth, discussed the sophistication of Indian languages, and debated the exploration of the American continent. The United States had just acquired a new border with New Spain; suddenly, there appeared at the president's dinner table a man fresh from the borderlands, a scientific traveler who had spent the past year digging in Mexican mines and archives. Jefferson naturally pumped his guest for information about the territory surrounding the Mississippi River, and the Red, and the Mexicana, and the Sabine—"Can the Baron inform me what population may be between those lines of white, red, or black people? and whether any or what mines are within them?" And Humboldt was ecstatic to oblige, to unroll his maps and lay out his notebooks—for a few days, anyway.[21]

Certainly, Humboldt would have been favorably impressed by Jefferson's description of the Lewis and Clark expedition. Perhaps Jefferson even showed his visitor the long list of instructions he had issued to Captain Lewis (a copy of which he had kept for himself). Humboldt undoubtedly approved of the president's insistence on a methodical topographical survey, complete with "celestial observations" (to determine latitude and longitude), temperature readings, and notes on flora and fauna. The most striking thing about Jefferson's attitude toward the West, though, was his respect for its human inhabitants. The president entreated Lewis to meet all native groups on equal footing and negotiate with them "in the most friendly and conciliatory manner," and then, if fortune permitted, to come back with information about their "language, traditions, monuments . . . , their laws, customs and dispositions," even "their relations with other tribes of nations." It was not Jefferson who inaugurated the myth of the American West as a blank, unpeopled space, ready for the taking.[22] Without doubt, the Lewis and Clark expedition served to consolidate the president's empire building (ethnographic information can be used in negotiations); yet, in sponsoring the trip, Jefferson seemed genuinely to be acting as both president of the United States and president of the American Philosophical Society. The zoological specimens Lewis and Clark collected and most of the materials they received from Native Americans wound up in Peale's museum in Philadelphia.[23]

The one thing about Jefferson to which Humboldt objected, and which probably sent him sailing back to Europe that much faster, was that the president was a slaveholder. Just a few months earlier, in an otherwise beautiful Cuban valley—sugarcane and coffee can look quite lovely—Humboldt had noted his horror upon realizing that "these plains are watered with the sweat of the African slave! Rural life loses its appeal when it is inseparable from the misery of our species."[24] Americans, he now learned, were perhaps more similar to the Spanish colonists than he had realized: clearly, they did not *fully* understand the "gift of Liberty." Humboldt, like virtually all his contemporaries, took the "improvement" of wild lands for granted as a symbol of human progress, so the sharing of scientific data that would facilitate that effort seemed to him a professional responsibility—though he did always argue, with a proto-environmentalist logic, that development ought to be undertaken with the greatest care, so valuable lands were not wasted. What was truly radical about his views, though, was his insistence that the benefits of such development be shared equally among all social groups—natives and blacks, too. The extension of national boundaries should never mean the extension of any kind of slavery, which he called "the greatest evil that afflicts human nature." Just before leaving the United States, in June 1804, he sent a letter to William Thornton, one of his new acquaintances from the Philosophical Society, in which he expressed his scorn for the "abominable law" that allowed the importation of slaves into the American South. If Americans pursued "the only humane course" and ended this practice, then the only negative consequence would be that "at the beginning you may well export less cotton. But alas! How I detest this Politics that measures and evaluates the public good simply according to the value of Exports! A Nation's wealth is just like an individual's—only the accessory to our happiness. Before being free, we must be just, and without justice there can be no lasting prosperity."[25]

For the rest of his life, Humboldt would hold up the United States as the world's model republic, but always with the caveat that it would have to abolish slavery in order to live up to its own ideals. Speaking with a *New York Times* correspondent shortly before his death in 1859, Humboldt said, "I am half American; that is, my aspirations are all with you; but I don't like the present position of your politics. The influence of Slavery is increasing, I fear. So too is the mistaken view of negro inferiority."[26] He could endorse American expansion to the West in the early nineteenth century because development under Jefferson was clearly preferable to development under Napoleon or the Spanish Crown: Humboldt knew that Napoleon had reinstated slavery in Haiti in 1802, and he had seen the brutality of Spain's plantation system with his own eyes. Indeed, he realized that Jefferson was morally opposed to the extension of slavery, and his respect for the president led him to write a letter in 1809— the two scientists would correspond regularly until Jefferson's death in 1826—apologizing if his friend had taken personal offense at the general attack against slaveholders that he had just published.[27] But, in 1804, Humboldt must have winced if Jefferson tried to portray himself as a kind and enlightened master. Back in Paris, writing up the narrative of his expedition, Humboldt would return several times to the subject of slavery: "I have observed the condition of the blacks in countries where the laws, the religion, and the national habits tend to mitigate their fate; yet I retained, on quitting America, the same horror of slavery which I had felt in Europe."[28]

America would expand in every direction in the nineteenth century, usually with overtly imperialistic goals. "East by sunrise," wrote a Philadelphia newspaper editor in 1853, "West by sunset, North by the Arctic Expedition, and South as far as we darn please."[29] But the actual explorers dispatched to America's various frontiers tended to have more complicated motivations.

Often, they focused on understanding the interrelationships of the peoples and landscapes they encountered in the wild, and they wound up questioning the values of their home civilization—in part, at least, because they were trying to follow in Humboldt's footsteps. Two hundred years after Humboldt braved the Bahama Straits, it is worth remembering his determination to make connections.

CHAPTER 1

"The Chain of Connection"

It is a surprising and memorable, as well as valuable experience, to be lost in the woods. . . . And not till we are completely lost, or turned round,— for a man need only to be turned round once with his eyes shut in this world to be lost,—do we appreciate the vastness and strangeness of nature. . . . Not till we are lost, in other words not till we have lost the world, do we begin to find ourselves, and realize where we are and the infinite extent of our relations.
—Henry David Thoreau, *Walden* (1854)

"When someone goes on a trip, he has something to tell about," goes the German saying, and people imagine the storyteller as someone who has come from afar.
—Walter Benjamin, "The Storyteller" (1936)

I couldn't find Humboldt. Admittedly, my ability to search was somewhat limited: I was on a highway in the middle of a desert. All I had to go on was a standard road map, which, like most road maps, never quite folded up properly, and a sign, about thirty miles back, suggesting I would come to Humboldt in about thirty miles. It was late September 1996, but it felt like midsummer. Bands of heat rippled just beyond the windshield. My ten-year-old station wagon, lugging all of my worldly possessions to California, was struggling with Nevada. The desert deceives: you see some mountains, a river, hints of green, and you think you're through it. I couldn't tell if my car was about to blow a gasket or not, but it needed a rest, and I was curious about the name of the town. Now if only Humboldt would appear by the side of the road.

It was *the* Humboldt, that I knew—cosmic theorist, climatologist, botanist, examiner of schist. Nevada also has Humboldt Mountains and the Humboldt River, named by the admiring American explorer John C. Frémont, who had traversed this basin in 1845, back when it was known as the Great American

Desert.[1] It seemed unlikely to me, though, that Frémont had managed to name any kind of settlement here. He was just passing through when he christened the river, during his third topographical survey, just before he got embroiled in the struggle over California at the tail end of the Mexican-American War. The town must have come later—though not late enough for the surveyor, or mapmaker, or local settler to have forgotten Humboldt's hallowed name. Perhaps I could find some sort of historical society, or at least a gas station selling local-pride postcards with instructive captions. But where was the town? There was a clear mark on the map; in this century, you should be able to trust the maps. We supposedly understand our place in the world.

A century and a half ago, when Frémont set out on his infamous Third Expedition, no reliable maps of this desert existed. "Contents almost unknown," read the chart published after Frémont's previous trip through the West.[2] Local trappers and traders had produced rough, annotated sketches, but Frémont, inspired by Humboldt, wanted to come to a broader understanding of this landscape, wanted to determine just how desertlike it actually was and compare it to other regions of the world where humanity had attempted to cope with aridity. Writing up his *Geographical Memoir* of the Third Expedition in 1848, he emphasized the surprising complexity of the land: "such is the Great Basin, heretofore characterized as a desert, and in some respects meriting that appellation; but already demanding the qualification of great exceptions."[3] Frémont himself sustained a kind of spiritual heatstroke in Nevada, and his reputation sagged as word spread of the violence he had allegedly countenanced on this expedition (trigger-happy Kit Carson was one of his most trusted men). But his new map and his *Geographical Memoir* helped to cement the reputation of his intellectual mentor, by making America's "desert" bloom with Humboldtian topographical features:[4]

> The most considerable river in the interior of the Great Basin is the one called on the map Humboldt river, as the mountains at its head are called Humboldt river mountains—so called as a small mark of respect to the "*Nestor of scientific travellers,*" who has done so much to illustrate North American geography, without leaving his name upon any one of its remarkable features. . . . It is a very peculiar stream, and has many characteristics of an Asian river—the Jordan, for example. . . . The stream is a narrow line, without affluents, losing by absorption and evaporation as it goes, and terminating in a marshy lake, with low shores, fringed with bulrushes, and whitened with saline encrustations. It has a moderate current.

The current of Alexander von Humboldt's geographical and environmental thought is still to be found in the crusty salt flats and spongy sinks of the

1 Alexander von Humboldt (in Venezuela, c. 1800), 1806. Friedrich Georg Weitsch.

Great American Desert—I am sure of it. But all the confluences of the nineteenth century, the palpable influence of Humboldt's theories on every explorer and scientist who wrote about nature, now constitute barely a rivulet in American intellectual culture.

When the exit for Humboldt, Nevada, finally appeared against the backdrop of crumpled, sage-dotted hills, I slid off Interstate 80 with a sense of relief and anticipation. The forest-green sign didn't *seem* like a mirage. My car, happily, was still functional, and, in addition to cooling down the radiator, I thought I might be able to slake my own thirst for Humboldtiana. But the exit ramp dead-ended at a dusty old trailer park. I couldn't even find a live human being, let alone a historical society.

As an explorer, I was astoundingly naive. But that's OK. Fieldwork is unpredictable; most explorations begin with dead ends.

———

Three years later. New York. You can find anything here.

In some obscure, no doubt outdated magazine article, I had read that at one point there was a statue of Humboldt in Central Park. So when my old friend Tom, visiting from out of town, suggested that we spend a blustery, late-December day exploring the big city, I proposed looking for Humboldt, and he laughingly agreed. In this day and age, you have to find new ways of exploring.

Fortunately, Central Park is a big place. Within just a few minutes, we were pleasantly disoriented. Perhaps it would have been most satisfying simply to stumble upon the statue in the winter twilight, but we eventually got tired of rambling aimlessly, and I felt guilty for making my friend spend his vacation in obsessive-compulsive pursuit of a hunk of metal or stone. So we proceeded to use logic. Shouldn't a memorial to a naturalist be located somewhere near the Museum of Natural History? We went to the museum's gift shop. We looked at a pristine, perfectly folded map. We found Explorer's Gate, at the corner of Seventy-seventh Street and Central Park West.

It was a bronze bust, dated 1869, made in Berlin. The inscription on the front said: "Humboldt." There was clearly a story here. Some of it I already knew, from previous research, but I saw gaps in my knowledge opening all around me, like cracks in the ground. I thanked Tom for the expedition. We had some dinner and discussed possibilities for further adventures, and then he took off for the West Coast. I headed to the archives.

Go down to the microfilm room. Go to 1869. There are only a few major papers to choose from: the *New York Times*, the *New York Herald*, the *Sun*, the *New-York Tribune*. Try the *Times*, for the sake of familiarity, though in 1869 it had a daily edition only eight pages long. Stop on September 15, an interesting news day. There is a one-word headline: HUMBOLDT. He got the entire front page—every single column.[5]

September 14 was the centennial of Humboldt's birth, though the great scientist had passed away more than ten years before, at the age of eighty-nine. We are talking about a dead, foreign, aristocratic intellectual, who visited the United States exactly once, for about a month, sixty-five years earlier. And yet his hundredth birthday was proclaimed across the whole North American continent: "Celebration Generally Throughout the Country," read the *Times*'s subtitle.

From San Francisco to Chicago to Boston, people read speeches, unrolled banners, and unveiled statues in Humboldt's honor. At two o'clock, on the edge of Central Park, an orchestra played Carl Maria von Weber's *Oberon* overture, and then C. E. Detmold, chairman of the Humboldt Monument Association, presented the bronze bust, just imported from Germany, to the park's commissioners.[6] In Boston, Professor Louis Agassiz of Harvard gave a two-hour oration explaining Humboldt's significance in the world of science and particularly in America. While Agassiz was best qualified to discuss

Humboldt's contributions in the fields of mineralogy, botany, and climatology, he instead made a point of emphasizing his mentor's commitment to three fundamentally American concepts: freedom (Agassiz noted Humboldt's abolitionism—despite Agassiz's own undeniable racism); independence ("he has bravely fought the battle for independence of thought against the tyranny of authority"); and union (Humboldt's cosmic theories connected all the world's "mutually dependent features," including human beings in all their diversity and strife, in one ecological web).[7]

This last scientific principle is what originally attracted me to Humboldt's writings and inspired my dusty Nevada detour. As an environmental writer in the mid-1990s, I had grown tired of forecasting ecological doom and instead started mining the past for a sense of intellectual continuity and for potentially useful ideas that might have been consigned to history's desert wastelands. According to most textbooks and scholarly studies, the word "ecology" was coined, and the science of ecology invented, by Ernst Haeckel, a German Darwinian, in the 1860s.[8] Yet Humboldt had clearly started thinking ecologically in 1799, at the start of his five-year scientific exploration of the Americas. As his empirical observations of nature mounted, he grew frustrated with "mere encyclopedic aggregation."[9] Indeed, in the introduction to his *Personal Narrative of Travels to the Equinoctial Regions of America, 1799–1804*, the first volume of which was published in Paris in 1814, he promised not to bore the reader with Linnaeus's Latin nomenclature, because his science addressed deeper questions: "The discovery of an unknown genus seemed to me far less interesting than an observation on . . . the eternal ties which link the phenomena of life, and those of inanimate nature."[10] Humboldt wanted to know why similar soils produced vastly different vegetation, why the vicious insects swarming in Venezuelan rain forests were not the same as those in the marshlands along the coast of Colombia, why the snow line of mountains at precisely the same latitude could differ by thousands of feet. The very impulse to ask these kinds of questions signaled his particular genius, as his older brother Wilhelm recognized early on: "He is made to connect ideas," Wilhelm wrote in 1793, "to see chains of things that would have remained undiscovered for generations without him." Almost half a century later, presenting *Cosmos* to his tens of thousands of readers, Alexander himself explained: "In considering the study of physical phenomena . . . , we find its noblest and most important result to be a knowledge of the chain of connection, by which all natural forces are linked together, and made mutually dependent on each other."[11] Tug on one strand in the web of life, and the whole structure quivers.

Unfortunately, the few scholars who have written about Humboldt in recent decades have tended to dismiss these scientific formulations as expressions of a mystical romanticism, or, worse, as attempts to conquer the globe through "intellectual annexation." By claiming to practice a "universal science," the argument goes, Humboldt was imposing Western, rationalist, colonialist concepts on peoples and places that in reality could not fit into any unified pattern.[12]

It is true that colonial science did often facilitate conquest. Maps can certainly be useful in planning an attack—whether an actual military invasion or something like an assimilation campaign. But what if Humboldt was just trying—as an environmentalist like myself might say—to "think globally"? In reading through his books, I was struck not by any sense of hubris or desire to obliterate diversity but, on the contrary, by his deep feeling of awe and appreciation for the great variety of landscapes and cultures that his obsessive traveling enabled him to experience. Nor did his image of cosmic connectedness seek to cover over realities of division or destruction. Noting the impact of Europeans on both the environments and the peoples of the Americas, Humboldt often condemned the colonial powers. In New Spain (Mexico), the scarred hills and struggling Indians spurred in him the constant "recollection of the crimes produced by the fanaticism and insatiable avarice of the first conquerors."[13] By tracing common patterns throughout the colonial world, Humboldt's "universal science" actually revealed the social and ecological damage wrought by colonialism. Seeking a "chain of connection," he did not ignore chains of bondage: his *Personal Narrative* contains one of the nineteenth century's most powerful attacks against the economic system that linked forced labor and plantation agriculture.[14] His was a complicated, dirty, difficult unity—one not too far removed from Abraham Lincoln's.

This social edge kept me coming back to Humboldt's ecology. I respected not just the earliness of his theorizing but its very nature, its inclusiveness, its analysis of social power dynamics as inherently connected to a society's interactions with its environment. This was not the simple, preservationist environmentalism I grew up with or even the ecological science I subsequently studied, but a powerful alternative, an intellectual torrent that swept through the Western world—the United States in particular—for about a century, before it evaporated in the desert heat of social Darwinism, which endorsed both human and environmental exploitation.

———

Humboldt's name does not appear only in the desert. The index of a good atlas might list it more than twenty-five times, referring to mountains in China and Venezuela, a glacier in Greenland, and naturally the Humboldt Current itself, off the coast of Peru.[15] Colleges, cafés, streets, public parks, and ships were all given Humboldt's name. Publishers fought over his manuscripts, and his books always sold quickly in their various editions (he wrote mostly in French—the language of his scientific colleagues—though a couple of his more important works were in his native German). Humboldt's name conjured up immediate images of exploration and natural science and could be dropped into almost any conversation or text without further explanation, as in Lord Byron's *Don Juan*, where Byron the Poet gently ribs Humboldt the Scientist for being more interested in measuring the blueness of the sky with his cyanometer than in chasing blue-eyed women.[16] A poem by Oliver Wendell Holmes suggested that Humboldt's influence equaled that of Napoleon, and many contemporaries made the same pairing.[17] Darwin himself sent data and manuscripts to the older scientist and claimed that he owed his entire career to his constant rereading of Humboldt's *Personal Narrative*.[18] Today, it's hard to get through high school without learning about Darwin's theories, yet such is the negative power of history that even professional scholars tend to know only vague details about Humboldt. ("Isn't there an ocean current named after him?")

A part of Humboldt would not be entirely displeased at his current obscurity. He was an intensely private person, despite his public charm. The renowned American historian George Bancroft, on meeting Humboldt in Paris in 1821, wrote that the scientist "does understand the art of talking to perfection,"[19] but one thing Bancroft didn't know was that Humboldt obsessively destroyed most of his personal correspondence. The letters that have survived—there are more than ten thousand of them—are almost all of a professional nature, requesting data or praising the work of his colleagues, and all seem to have been written hurriedly, in Humboldt's characteristic slant from the bottom left of the page to the top right. A twentieth-century graphologist analyzed his handwriting as revealing "a mixture of sobriety and fantasy, an unusual openness and sensitivity"; Humboldt himself tended to apologize for its being "microscopic" and "illegible," qualities he ascribed to the rheumatism he'd picked up "from having so long slept in the marshes of the Orinoco upon heaps of withered and damp leaves."[20] Whether in German, French, English, or Spanish—by the end of his life, he was also reasonably proficient in Latin, Greek, Russian, Italian, Danish, Hebrew, and Sanskrit—his letters almost always have copious grammatical mistakes, and the sentences tend to be unbearably long. Every now and then, one of his epistles reveals a spark of the intense emotion that lies just beneath the surface of so much of his published writing. Especially toward the end of his life, he was willing to express fury and disgust at the Prussian government's stultifying conservatism. Yet even his closest confidants felt that they never knew his innermost passions, his feelings for other people. His brother Wilhelm admitted that "there is between us, when it comes to his intimate life, a veil that neither of us would dare to lift."[21]

Humboldt himself acknowledged that even as a child, in the 1770s and '80s, he'd had "an anxious spirit," that his mother had seemed overbearing to him, that Wilhelm had developed more quickly and with more confidence.[22] The boys' warm and generous father died when Alexander was only nine, and after that, though both brothers received a broad education from the best tutors, Wilhelm was groomed to be a political leader, while Frau Humboldt insisted that the more timid Alexander prepare himself to go into finance. The brothers gave each other much comfort in both childhood and adulthood, but Alexander did not really start thriving until he was on his own. In college at Göttingen, finally able to wander off into the hills, to work deep into the night, to perform experiments on his own flesh, he started to recognize and accept his restlessness and use it to his advantage. "Uneasy, agitated, and never enjoying what I have finished, I am happy only when undertaking something new and when doing three things at the same time."[23] Humboldt traveled constantly, collecting plants and analyzing geologic strata, and in 1790, at the age of twenty, he published his first monograph, on the basaltic rocks of the Rhine Valley. Still under the influence of his mother, he did complete one year of training at the Academy of Commerce, but meanwhile he was teaching himself mineralogy, so successfully that he managed to secure a position with the government's Department of Mines. His stable, respectable job satisfied his mother, and it also provided him with opportunities to do research throughout Germany, as well as in Poland, Italy, and even the Swiss Alps. And then his mother died. "My heart," Humboldt wrote to a friend, "could not have been much pained by this event, for we were always strangers to each other."[24]

Suddenly freed from the expectation of pursuing a career and getting married—something he seems never to have considered, anyway—Humboldt, at the energetic age of twenty-seven, and endowed with a seemingly limitless inheritance, realized that he would never have to settle down. He could be a professional traveler. Yet he refused to be a mere tourist: he would not only see new places but also teach scientists to see in new ways. So there followed a burst of new studies and a series of new plans. Humboldt left the Department of Mines and dove into chemistry, physiology, meteorology, astronomy, and geography, while preparing to embark for the West Indies—or perhaps Egypt—or perhaps India and Tibet—or perhaps the South Pole. Each of these expeditions was very nearly launched—ships, captains, berths had all been secured—only to be blocked at the last instant by political developments.

Meanwhile, the search for precision instruments had led Humboldt to Paris, where he met a man who attracted him not only with his scientific perspicacity but also with his sense of adventure—the French botanist Aimé Bonpland, three years Humboldt's junior, but well trained, broad-minded, and, perhaps most important, without permanent ties to Europe. Thwarted in their attempt to board a ship bound from Marseille to Africa in the late fall of 1798, the two wanderers walked west along the Mediterranean, eventually crossed the Pyrenees, and made various measurements as they wended their way to Madrid, reserving the hope that they might be able to catch a packet boat from Cartagena to Tunis. But then, in one of the most remarkable coups of a charmed career, Humboldt proceeded to convince King Carlos IV to grant him unprecedented access to the Spanish colonies, those "most beautiful and most vast lands" of the New World. Bizarrely, Humboldt did not even have to refer to the potential material benefits of his proposed expedition—he just had to foot the bill. In memoranda that can still be read today at the dusty state archive of Madrid, he did not promise to locate new sources of mineral wealth, but merely "to study the Construction of the Globe" and "the influence of the atmosphere and its chemical composition on organic life" and "the similarity of the earth's layers in countries far apart from each other." By June 1799, Humboldt and Bonpland were aboard a ship rounding Spain's westernmost point, Cape Finisterre—the end of the earth.[25]

The two scientists, soon to become the most loyal of companions, would spend the next five years together, exploring what Humboldt usually called "the torrid zone" of the Americas.[26] Afterward, Humboldt based his operations in Paris, though he remained restless and rarely stayed for more than a few months at a time before setting out for surveying trips through various parts of Europe. In 1829, at the age of fifty-nine, he led a nine-month expedition into Siberia and Central Asia, during which he covered a distance nearly equivalent to half the circumference of the earth. The last three decades of his life Humboldt spent mostly back in Prussia. Even during these years, though, he haunted the academies of Paris and continued to climb European volcanoes. He was loyal to his place of birth, but he never truly settled there. In his perpetual wandering, Humboldt seems to have been living up to the twelfth-century ideal of Hugh of St. Victor: "The man who finds his country sweet is only a raw beginner; the man for whom each country is as his own is already strong; but only the man for whom the whole world is as a foreign country is perfect."[27]

The one country that consistently held Humboldt's affection, despite the flaws he saw in it, was the United States. In 1853, he wrote to the Arctic

2 Detail of a map of South America, 1897, showing the northwestern quadrant of the continent, where Humboldt concentrated his efforts.

explorer Elisha Kent Kane that he was "so devoted to America in heart and mind as to think of it as a second homeland."[28] The farthest west he got in the United States was Philadelphia, yet his loyalty to this country was clearly reciprocated by frontier Americans. The most remarkable concentration of Humboldt toponyms occurs in the United States. Here, Humboldt was a household name not just among elite geographers like Frémont but among bands of common settlers, who honored the so-called Rediscoverer of America with counties in Iowa and California and towns in Nebraska, Kansas, South Dakota, and Tennessee—as well as Nevada. In August 1859, a westbound migrant named Thomas Ambrose Cramer learned of Humboldt's death while he was following the Humboldt River through the desert. In his opinion, the waterway that bore the great man's name "does him little credit here. He was filled with wisdom and goodness, it only with mineral and vegetable poisons."[29] Somehow, this Prussian baron, born seven years before the American Revolution, became an unofficial American hero.

Scholars could spend years mapping the intricate web of Humboldtian influences in American science, literature, art, and politics. But Humboldt was a particularly inspiring figure for American explorers. In general, one doesn't think of explorers as reading books and debating ecological principles. They are men of action. In the mainstream media—the recent flurry of movies and books about Ernest Shackleton, for example—explorers tend to come across as having an indomitable will, unthinkable physical endurance, highly developed survival skills, and often a gift for charismatic leadership. Scholars, meanwhile, tend to portray them in much less glowing, though similarly non-intellectual, terms: especially in the nineteenth century, they were unthinking agents of empire, roguish flag-planters sent out to tame the frontier and thus facilitate various kinds of exploitation. As William H. Goetzmann, our foremost historian of exploration, suggests, the nineteenth-century explorer has become "a symbol of . . . a kind of superhuman Odyssean vision, and ultimately mankind's biblical urge to 'dominate the earth.' "[30] Yet these depictions fail to reflect the complexity of the Humboldtian explorer's engagement with the frontier.[31] A man like Frémont clearly saw landscapes through several different lenses, as he compared an actual desert to others he had seen or read about; as he attempted to determine how and why certain plants, animals, and even human beings could survive in certain microclimates; as he pondered the cosmos.

There is no doubt that explorers' expeditions sometimes paved the way for white settlement, for the removal or killing of native peoples, and for devastating transformations of natural resources and landscapes. What scholars of exploration have not fully recognized is that, just as often, the explorers themselves forcefully resisted the course of empire. Starting in the 1960s, Goetzmann shaped this field of study by making the useful point that exploration is always "programmed," that its purposes "are to a great extent set by . . . the current objectives of the civilized centers from which the explorer sets out on his quest." Goetzmann's framework, then, is the perfect analytical tool for understanding how Frémont came to be known as The Pathfinder, and why his route through the Great Basin was followed by thousands of migrants, like Thomas Ambrose Cramer, who referred to it as the Humboldt Trail.[32] But it does not account for the extent to which the "current objectives" of nineteenth-century American society might have been many-sided and intensely contested, rather than monolithic and preordained. Nor does it suggest the possible ways in which experiences on the periphery of American society might have undermined some of the goals that were hatched at its core.[33]

Not all Americans were converts to the idea of unlimited white expansion, for instance. The first Indian Removal Bill, hotly debated by Congress in the spring of 1830, almost failed to win a majority in the House of Representatives: the vote was 103 to 97. Opponents referred to removal as an attack against the unity of the American nation; to some radicals and even some moderates, Indians were not savage "Others" but potential members of a new, diverse society, crucial parts of a Humboldtian whole. Antiremoval agitators also used firsthand accounts of contact with natives to assert that groups like the Cherokee were "absolute, erect, indomitable," and deserving of sovereignty over their homelands.[34]

Direct contact with different cultures and new environments could have an unpredictable effect on people. No matter how explicitly "programmed" an expedition might be, exploration was always surprising. Indians could be ruthless enemies, sending arrows whirring past the doorway of your tent—or they could be trusted guides, showing you ways to make your tent blend in with the surrounding landscape. You might encounter swirling dust storms, malarial bogs, swarms of insects; you might find sweet-water springs, edible roots, natural shelters. Amid these conditions, nineteenth-century explorers, like many other cosmopolitans before them, had to develop new ways of living for months or years at a time and were often moved to reconsider old assumptions. They saw clear-cut forests and burned-out prairies and recognized the environmental impact of unconstrained enterprise. They breathed fresh air and learned about the Great Spirit and wondered if their compatriots back home had begun to worship smoke-belching factories. The Army's Corps of Topographical Engineers, which sponsored Frémont's expeditions, may have concerned itself primarily with facilitating settlement, but many of its topographers were convinced that settlement had gone horribly wrong. While they were not about to set aside the entire notion of developing the continent, they did engage in debates over how development should proceed. Lieutenant James W. Abert, for instance, whose views were shaped both by an education at the prestigious College of New Jersey (now Princeton University) and by his multifarious experiences of the frontier, wound up embracing a perspective even more Humboldtian than Frémont's. Crossing the plains in the winter of 1847, he laced his travel diary with questions about the conduct of his fellow (non-Native) Americans—and with perspectives and turns of phrase picked up directly from the natives. "It seems so shameful," he wrote on Thursday, February 4, "[what is happening to] these wide pasture grounds that the Great Being has planted for the wild beasts of the prairies. Well may the Indians look with hatred on those who go about spreading desolation. This winter, the buffalo have almost deserted the river, there is no grass for them; and the poor Indian, forced by the cold season to take refuge in the timber that alone grows on the streams, must now travel far away from the village to get meat enough for his subsistence. There should be a law to protect these noble pasture grounds."[35]

Following Humboldt, Abert—and many other American explorers—applied an ecological sensibility to the problem of frontier expansion, and wound up decrying abuses against comparatively powerless groups like Indians and the collateral devastation of environmental resources like forests and soils. The kind of outright radicalism expressed in Abert's field journal was relatively rare in the nineteenth century, but a general questioning of Manifest Destiny by men who studied natural science was not. As literary critic Raymond Williams has suggested, the writings of people who immersed themselves in nature 150 years ago cannot be "dismissed" as constituting a "merely romantic critique of industrialism." Scanning those writings today, no serious reader could fail to see "startling connections with the new ecological and radical-ecological movements."[36]

The more extreme arguments of Humboldtian explorers did not rule the day. While the ideas of some explorers deserve to be celebrated, these men were not pure environmental heroes.[37] All nineteenth-century science, including Humboldt's, was tinged with imperialism. And Humboldtian influence took many different forms: Frémont, for example, showed much less sympathy toward Indians than did his colleague Abert. What remains interesting and valuable about the intellectual struggles of these men is the very fact that they struggled, that they chafed against the society whose influence on them they could not deny. Many of them were so conflicted—and, at times, unlucky—that they wound up being perceived as utter failures in their day. All the more reason, then, to delve into their stories, to explore the Humboldtian current as an alternative to the mighty rivers of empire—to chart its radicalism, to measure its usefulness, and to discover which individuals, in particular, did the most to nurture it through the massive changes of the nineteenth century.

In April 1818, U.S. Secretary of State John Quincy Adams received a bizarre handbill in the mail from a man called John Cleves Symmes Jr. of Ohio. Symmes, a retired captain of infantry who served in the War of 1812, was just beginning to gain a national reputation, but he was already well known in his home state—as an utter quack. The handbill was addressed "TO ALL THE WORLD!" In it, Symmes declared that "the earth is hollow, and habitable within; . . . and that it is open at the poles 12 or 16 degrees." "I pledge my life in support of this truth," he wrote, "and am ready to explore the hollow, if the world will support and aid me in the undertaking. . . . I ask one hundred brave companions, well equipped, to start from Siberia in the fall season, with Reindeer and slays, on the ice of the frozen sea; I engage we find warm and rich land, stocked with thrifty vegetables and animals if not men, on reaching one degree northward of latitude 82; we will return in the succeeding spring." The proposed expedition, Symmes explained, would be dedicated "to my Wife and her ten children." And, finally, he named as his most prominent "protector" the distinguished "Baron Alex. de Humboldt."[38]

With government sponsorship not immediately forthcoming, Symmes embarked on a lecture tour to promote his Humboldtian expedition. During the mid-1820s he visited cities and towns in the old Northwest and on the eastern seaboard, leaning increasingly on the oratorical powers of his young, enthusiastic companion, called variously Jeremiah, James, or John Reynolds.[39] With Reynolds's first publication, an 1827 pamphlet titled *Remarks on a Review of Symmes' Theory*, and Symmes's death two years later, the apprentice became master, and over the next decade, signing his name simply "J.N. Reynolds," this somewhat mysterious figure started being identified as the foremost advocate of American exploration.[40] Reynolds toned down the hollow-earth rhetoric and dropped Symmes's obsession with the Arctic and instead focused on the many commercial, political, scientific, and moral reasons for launching an expedition to the South Pole. In 1828, Reynolds actually convinced John Quincy Adams, now serving as president, to authorize such a mission, but it was blocked the following year by the new administration of Andrew Jackson. Drained of patience and fired by a Humboldtian zeal— "What an extensive and interesting field is here opened for the contemplation of the philosopher and the naturalist!"—Reynolds wound up outfitting his own expedition, which sailed southward in late October 1829.[41] When he finally returned to the United States in May 1834, he deposited his collections at Boston's Society of Natural History and proceeded to captivate the American public with stories of his exploits and with further exhortations to explore the Antarctic regions.[42]

Reynolds was never enough of a scientist to add new dimensions to Humboldt's ecology, but he did earn immediate recognition as a philosopher of Humboldtian exploration: perhaps more than any other writer in the first half of the nineteenth century, he taught America the cosmopolitan value of seeing the world. In 1835, for having furthered what he himself called a "cause of magnitude and public utility, at home as well as in scenes of wild adventure," he was awarded an honorary degree by Columbia College in New York.[43] That same year, he published his *Voyage of the United States Frigate Potomac*, an account of the second half of his journey, to great fanfare, critical acclaim, and urgent sales: the book went through one edition per month for the first four months it was in print.[44] Meanwhile, Reynolds was busy taking his message to public auditoriums, universities, local governments, and the halls of Congress. On the evening of Saturday, April 2, 1836, he stood for about two hours before the assembled representatives of the United States and attempted to convince them to send out a fully funded scientific expedition to the South Seas. He noted that such a voyage would have significant commercial benefits, but, in the end, he admitted that his rationale was essentially Humboldtian: he wished "to collect, preserve, and arrange" specimens from all over the globe, "from the minute madrapore to the huge spermaceti," and then figure out how they connected to each other and to "man in his physical and mental powers, in his manners, habits, disposition, and social and political relations." Scientists should feel duty-bound "to examine vegetation, from the hundred mosses of the rocks, throughout all the classes of shrub, flower, and tree, up to the monarch of the forest," all in relation to "the phenomena of winds and tides, of heat and cold, of light and darkness."[45]

Reynolds's address was quickly published, together with supporting testimonials he had gathered from prominent citizens, as a three-hundred-page book. Edgar Allan Poe, already in correspondence with Reynolds, reviewed the publication, resoundingly endorsed its message, and promptly stole several passages from it for insertion into the novel he was then trying to finish, *The Narrative of Arthur Gordon Pym*.[46] A couple of years later, after the U.S. Exploring Expedition had in fact sailed, under Charles Wilkes, Reynolds paused for a while to consider his success in spurring the mission and his failure to be named its leader. Going over the notes from his previous adventures in the South Seas, he recalled hearing of a gigantic white whale known to attack large ships by barreling into them head-on, at full speed. The resulting short story, "Mocha-Dick; or, The White Whale of the Pacific," was published in the spring of 1839, when Herman Melville was nineteen.[47]

The figure sits atop Clover Peak, in the East Humboldt Mountains, Nevada, in 1868. He is a little bit fuzzy, bathed in sunlight, highlighted against the black granite cliffs that hover, seemingly inaccessible, in the background. What is in harsh, clear focus is the precarious pile of cracked boulders—stark, damaged rock—leading our eyes up toward the figure. This man is enjoying a peaceful, silent solitude, a communion with the largeness of nature that is hard to experience east of the Mississippi. Yet he is also overwhelmed by the western landscape, and he has clearly struggled to attain his position; the stable world falls off beneath him in a dark, dubious gravel slope, a steep declivity being slowly eclipsed as the sun sinks below the horizon.

The photographer, Timothy H. O'Sullivan, was fresh from the Civil War, where he had taken innumerable pictures of corpses. Today, we think of Ansel Adams as the prototypical American landscape photographer, but the darker-minded O'Sullivan actually founded this school of nature viewing, with help

3 "Clover Peak, East Humboldt Mountains, Nevada," 1868. Timothy H. O'Sullivan.

from colleagues like Carleton Watkins and Andrew J. Russell.[48] While Adams, photographing in the mid-twentieth century, generally erased humanity from the land in order to recapture a sense of pristine wilderness, O'Sullivan tended to include a tiny figure in his scenes, a ragged, gray man caught in the act of looking, a stand-in for the viewer of the photograph, a human intruder in some of the most inhuman, strife-ridden, harshly beautiful landscapes that Americans had ever witnessed. More often than not, the figure was Clarence King.

In 1879, at the peak of his career, King would become the first director of the U.S. Geological Survey, but he had been scrambling up boulder-strewn slopes in the Great Basin since 1863, when he lit out for the West at age twenty-one.[49] Armed with a degree from Yale's brand-new Sheffield Scientific School, King managed to catch on with the California Geological Survey as soon as he reached the Sacramento River. But it was the natural features of Nevada that really captivated him: strangely eroded tufa domes; scarred, twisted trees; blinding white cataracts; sulfurous hot springs emitting steam from jagged fissures in the earth; drifting sand dunes; accordion-like ridges of columnar trachyte; lonely mountain tarns far above tree line. Crossing the Humboldt Sink—where the Humboldt River sank below the ground—King

4 "Tufa domes, Pyramid Lake, Nevada," 1867. Timothy H. O'Sullivan.

5 "Trinity Mountains: Rhyolite Ridge," 1867. Timothy H. O'Sullivan.

thought back to his studies in New Haven and the year he spent in New York after graduation, when he read Humboldt's *Cosmos* and joined an avant-garde aesthetics society whose motto was TRUTH TO NATURE.[50] Humboldt himself celebrated the drawings and paintings produced by exploring expeditions as contributing "to a more active study of nature," and he suggested that the new technology of photography might go even further in enhancing people's appreciation for both the unity and the diversity of the natural world.[51] Inspired, King vowed to spend his life studying nature's "intricate mazes of fact" and the "harmonies of structure" to which those mazes would lead; he also endeavored to share his understanding of the land with as many of his fellow Americans as possible.[52] When he returned to Nevada in the late 1860s as head of the Geological Exploration of the 40th Parallel, he collaborated closely with Timothy O'Sullivan, and their artistic and scientific partnership lasted through 1876, when King organized an exhibit of O'Sullivan's frontier photographs for the United States' centennial celebration in Philadelphia.

Just as O'Sullivan combined a strict documentary skill with a distinct sensibility; just as Humboldt combined Enlightenment rationality with Romantic passion—so did Clarence King attempt to balance science and art. Learning to be a field geologist in the 1860s, he took time off to pen *Mountaineering in the Sierra Nevada*, a collection of highly imaginative stories about his experiences of frontier landscapes. Then, hailed by William Dean Howells and Henry Adams as America's next great nature writer, he retreated to the lab to work on an academic tome called *Systematic Geology*. In every piece of work he produced, King yearned to reconcile the analytic and the intuitive, the Apollonian and Dionysian worldviews—just as he yearned to express the paradox of serenity and catastrophe that he experienced in the wilderness. He never fully appreciated the social edge to Humboldt's thought, but he did further the Humboldtian project of trying to see nature truthfully—though, left behind in an era of intensifying specialization, he felt at the time of his death in 1901 that he had failed as both scientist and artist. By the 1880s, when King found himself trapped in the bureaucratic tangles of a government administrator's job, the complicated, global, Humboldtian perspective seemed to be losing its appeal in America. Yet King retained an awareness of what he was trying to accomplish. "The greatest realist," he wrote, in an 1886 essay about President Lincoln, "is he who can keep his feet always on the solid bottom while wading deepest into the foaming river of life."[53]

———

Almost all American scientists in the mid-to-late nineteenth century, no matter what subfields they waded into, considered themselves disciples of

Humboldt. One such author, Lorin Blodget, inserted a quote from the master himself on the title page of his own magnum opus, *Climatology of the United States*, to make a kind of textual frontispiece: "When the varied inflections of the Isothermal lines shall be traced from accurate observations in European Russia and Siberia, and prolonged to the western coast of North America, the Science of Distribution of Heat on the Surface of the Globe will rest on solid foundations."[54] Humboldt had urged travelers and explorers to take temperature readings in the air and water wherever they went, to aid in the development of his globe-spanning comparative ecology.[55] Indeed, his own measurements in the frigid waters off the coast of Peru had led to a much broader understanding of the ocean current that would later bear his name: it seemed to sweep surface waters northward toward the Galápagos Islands, allowing warmer, nutrient-rich pockets of ocean to rise up and feed Peru's famous anchovy fishery—whose disruption, today, is one of the first signals of an El Niño effect. By the mid-nineteenth century, scientists like Blodget had a good grasp of how the Humboldt Current, in the South Pacific, and the Gulf Stream, in the North Atlantic, interacted with other climatological factors to affect weather patterns, the distribution of plants and animals, and agricultural conditions. But, as Blodget suggested by highlighting Humboldt's reference to Siberia, there was still a lack of data about what he called, on page 219 of his book, "the Kuro-Siwa or Japan stream," which supposedly pushed warm water all the way from the Philippines into the Bering Strait.

The German geographer August Petermann called it the Black Current of Japan, and he believed that it had melted an entrance to an ice-free section of the Arctic Ocean.[56] Petermann's reputation was not as good as Humboldt's, but it was not as bad as that of Captain Symmes, and many explorers had in fact reported that there was sometimes open water beyond the initial ice floes.[57] More important, people were still desperate to believe that there was good agricultural land in the Arctic (there wasn't), and that the Northwest Passage existed (it did), and that maybe a ship could even sail to the North Pole (well, if the ice caps continue to melt . . .). In 1879, the explorer George Washington De Long and his engineer, George Wallace Melville, expressed a willingness to stake their lives on Petermann's theory. Only Melville, and one-third of the crew, survived.[58]

Melville was a navy technician rather than a naturalist in the Reynolds/King mold, but he had read Humboldt,[59] and during his disastrous Arctic expedition aboard the *Jeannette* he developed a distinctly Humboldtian attitude—not so much about nature's interconnectedness but about its power over humanity. Chastened by his experiences amid the ice, he taught rabid American expansionists to see northern landscapes not as potential commodities, nor as desolate wastelands, but as sublime signifiers of the harsh limits imposed by ecological forces. The boreal world spurred in him an almost spiritual reverence: "they filled me with an awe, those snowy summits bathed in the silver radiance of an Arctic moon, such as I had never known before." While his engineering projects attempted to harness the power of nature to serve human ends, his writings betrayed his suspicion that nature would never really be tamed. "The ice grinds, swirls, and piles upward," Melville wrote, as he watched the Lena Delta, in Siberia, freeze before his eyes, "while the river rises in its might and drives the ice before it like so much brushwood rolled before the wind."[60] Some rivers were simply unnavigable, some goals unreachable, some experiences unfathomable. As Humboldt had explained, the most important lesson of "communion with nature" was an awareness of "the narrow limits of our own existence."[61]

Eventually promoted to rear admiral, Melville served as engineer-in-chief of the navy for sixteen years. As the century wound down, and the younger scientist-explorers stopped reading Humboldt, Melville's environmental thinking tended to align itself with Theodore Roosevelt's conservation program. Engineers, in general, were well disposed toward Roosevelt's canonization of efficiency, the better-regulated, scientifically managed, "wise use" of natural resources.[62] Meanwhile, though, Melville remained a critic of American hubris—of imperialism in the Spanish-American War, of capitalism run amok, of Americans' lack of respect for science, nature, and native peoples. Throughout his narrative of the *Jeannette* expedition, *In the Lena Delta*, he emphasized that none of his crew would have survived the Arctic without "the friendly care and guidance of the natives."[63] On the surface, Melville may have appeared to be a classic, manly explorer-hero, with a sunny view of human progress and American ascendancy—but a black current ran through his career. The central message in his life's work was one of Humboldtian humility.

"How intensely," John Muir wrote in an 1865 letter, "I desire to be a Humboldt!"[64]

About two years later, Muir quit his factory job in Indianapolis and struck out southward, on foot, hoping eventually to reach the Amazon. He yearned "to see tropical vegetation in all its palmy glory," to experience the profusion of life-forms and diversity of landscapes described so viscerally in Humboldt's writings.[65] Even more important, though, was Muir's deeper Humboldtian goal, of exploring nature and culture in a new way, of reassessing humanity's ambitions, of broadening his sympathies enough to recognize the interconnectedness of the entire natural world and "the rights of all the rest of creation!"[66] Muir would not make it to South America until 1911, but the journal he kept during his journey through the post–Civil War American South established him firmly as an explorer-scientist in the Humboldtian mold. That journal, eventually published as *A Thousand-Mile Walk to the Gulf*, details the twenty-nine-year-old's meetings with wealthy planters and former slaves; it records his musings not only on the classic magnolia trees and honeysuckle vines he encountered for the first time but also on "Ebony Spleenwort," and alligators, and poisonous weeds, and mosquitoes. Muir had begun to see nature as "one great unit," and had begun to ask: "what creature . . . is not essential to the completeness of that unit—the cosmos?"[67]

Today, Muir is best known as the founder of the Sierra Club (in 1892) and the foremost defender of California wilderness. His legacy is invoked constantly by environmentalists, and his legend grows with each new glossy volume of photographs showing the falls of the Yosemite or one of the last remaining stands of giant sequoias. What is little recognized, though, is that Muir's embrace of wilderness preservation represented a significant retreat from his earlier, more Humboldtian approach to nature. Before Muir became an environmental lobbyist, he was an explorer, and among the most complete Humboldtian thinkers of the nineteenth century. Like his idol, he wandered through countless landscapes—the South, the Sierras, and, in particular, Alaska, Siberia, and the Arctic—not simply to get what he would later call "whole-souled exercise," but to stretch his conception of how human beings interacted with the natural world.[68]

In the harsh, confusing Alaskan environment, Muir noted that the "Eskimos . . . are keen questioners and alive to everything that goes on before them."[69] This active engagement with the land, in turn, helped the natives adapt to the most extreme conditions. The new bands of white settlers along the Bering Strait, meanwhile, were already disrupting the whale and seal populations, which in turn made life more difficult for the natives. During the second half of Muir's writing career, when he focused on producing a series of preservationist propaganda tracts, Indians disappeared from his landscapes. Living in a pragmatic age, he gave up trying to change the deeper ways in which people understood the world; instead, he focused exclusively on setting aside wilderness areas as museum pieces, which tourists could visit every now and then for a redemptive, restorative escape from the reality of industrial life. But in his exploring days, Muir treated the natural world as a constant context rather than an occasional refuge. Nature was humanity's habitat, something to which we were all connected and which all of us—white settlers, newly liberated African Americans, newly constrained Indians—would have to learn to share.

By 1911, when Muir finally followed in Humboldt's footsteps to the Amazon basin, he had committed all of his intellectual resources to preservationism, but he probably had second thoughts, because his stance caused him immense anguish during the entire last decade of his life—despite his status as a defender of nature and his awareness that his legacy would surely be carried on by the Sierra Club. His trip to South America provided only a brief respite from the most bitter, alienating struggle of his career, the fight over Hetch Hetchy Valley, in Yosemite National Park.[70] It was a fight Muir couldn't win. The people of San Francisco, already devastated by the earthquake of 1906, needed more fresh water and more electricity, and if that meant damming the Tuolumne River and flooding Hetch Hetchy, municipal leaders would not lose any sleep. Though Muir never admitted the moral precariousness of his position—that Hetch Hetchy deserved to be saved simply by virtue of its stunning beauty, its status, in his mind, as an earthly temple—he must have realized that in abandoning the social element of his environmentalism he was leaving himself open to charges of selfishness and elitism. At the end of 1913, after an eight-year battle, when both the House and the Senate passed a bill that would allow the city of San Francisco to encroach on national park land and block the current of the Tuolumne, the rhetoric of the day was "the greatest good for the greatest number."[71] Congressmen generally saw the conflict as pitting all San Franciscans against a few upper-class hiking enthusiasts from the Sierra Club, and Mayor James D. Phelan wondered out loud whether John Muir "would sacrifice his own family for the preservation of beauty."[72] A year later, Muir was dead.

———

Explorers are a lonely lot. When a person refuses to stay in one spot for very long, he is hard-pressed to develop lasting relationships. Muir portrayed himself as the gentle, happy spirit of California—but he was forever taking off for places like the Arctic and the Amazon, and if he did not in fact sacrifice his family for the sake of wilderness, he certainly sacrificed his family life. Of course, one reason many explorers leave in the first place is that standard social relationships don't work very well for them. Some begin to feel a sense of exile even as young adults, so, eventually, they banish themselves to the frontier. And then they return, transformed by their experience, to a community that cannot understand what they have been through. Humboldt, despite unimaginable fame and near-universal admiration, always stuck to the margins of society. In his youth, he spent much of his time outside the Prussian culture into which he had been born, consorting mostly with Jews like Moses Mendelssohn and Marcus Herz, who had no civil rights in Berlin but whose social circles represented the city's scientific and artistic avant-garde.[73] His brother, Wilhelm, confided to his wife the suspicion that Alexander would "hardly ever be happy, and never at peace, because I can't believe that he will ever form a real relationship."[74] Small wonder, then, that Humboldt so doggedly searched the lakes and jungles, the steppes and deserts, the rocks and rivers, for signs of connection.

In the righteous rush to condemn explorers for expanding the apparatus of imperialism, we have forgotten that they also expanded our sympathies. Often unable to find a comfortable niche in their own society, explorers through the ages, possessed by what Anatole France called "a long desire,"[75] sought links with cultures and environments radically different from those that shaped them. Many of them committed themselves to understanding— and portraying—each foreign landscape or society on its own terms. "I wished to acquire the simplicity, native feelings, and virtues of savage life; to divest myself of the factitious habits, prejudices, and imperfections of civilization," wrote Estwick Evans of his four-thousand-mile "pedestrious tour" through the American West in 1818.[76] Explorers tended to see themselves as interpreters, guides, mediators between worlds; the whole point was to fashion new bonds on the frontier and then share those bonds back home.[77] You return from a trip with stories, and stories have the capacity to take us beyond ourselves, to reshape the way we interact with the world. As Paul Zweig noted in a study of adventure myths, stories "beckon us out of the visible, providing alternative lives, modes of possibility. Merely listening to a story—'losing oneself' in it—creates a vision of other spaces and times." And stories can be binding forces. Because the explorer comes back to tell his tale, "his escape from society is a profoundly socializing act."[78] Good storytellers, like good ecologists, weave webs, enrapturing their audience with the delicate, sticky power of organic connectedness.

In Victorian America, connections were in short supply. Reynolds, King, Melville, and Muir all felt buffeted by rapid development, by the pressures of a society just entering a recognizably "modern" era. Their different versions of a connective, Humboldtian ecology arose from various strands of scientific thought, but they also grew out of a concern with social divisions: the deepening gulf between classes; the new division of labor; the overspecialization of the professions and, in the intellectual arena, the increasing separation between science and art; the constant tension surrounding race and ethnicity; the hardening of separate gender roles; explosive regional differences; and, especially, the apparent divide between civilization and nature. Industrialization put nature in exile, and it took fellow exiles to reach back out to wild environments and reincorporate them in a holistic, humanistic vision that respected them for what they were.

The current of Humboldtian thought branched out through America in the nineteenth century, trickling into expeditions bound for every part of the globe. Dr. James Eights, who accompanied J. N. Reynolds on his private expedition, discovered and analyzed the first fossil in Antarctica, and several other midcentury scientists tried to fit the South Seas and the newly recognized Antarctic continent into Humboldt's global ecological framework: Matthew Maury founded the science of oceanography; James Dwight Dana, a mineralogist with the Wilkes Expedition, proved that corals were in fact living organisms rather than rocks; and Charles Pickering, impressed with the incredible adaptations developed by each human group he encountered in the Pacific, insisted that "races" could not be arranged in any meaningful hierarchy. Out West, J. N. Nicollet pioneered the Humboldtian topographical survey in America; John Wesley Powell, who took over from Clarence King as director of the USGS, insisted that his studies of the geology, ethnology, and agriculture of the Arid Regions were inextricably connected. Northern expeditions attracted the Humboldtians Elisha Kent Kane, his surgeon/naturalist Dr. Isaac Israel Hayes, and Hayes's later rival Charles Francis Hall. Each was initially drawn to the Arctic by the search for Sir John Franklin, the infamous British explorer who vanished in the effort to find a Northwest Passage. But they sustained their interest in the northern regions by studying global climatology (the possibility of an open polar sea, in particular) and by making strong bonds with native peoples. Louis Agassiz would also have to be included on any list of America-based scientific explorers in the Humboldtian mode, as would the early ethnologist Henry Rowe Schoolcraft, author of *Travels in the Central Portions of the Mississippi Valley* (1825) and *Personal Memoirs of a Residence of Thirty Years with the Indian Tribes* (1851). And there's also John Lloyd Stephens, whose archaeological travels in Central America (and discovery of lost cities like Palenque) confirmed Humboldt's theory that this continent did in fact have an ancient history of highly developed civilizations.

None of the American explorers was quite as radical or broad-minded as Humboldt himself, but Reynolds, King, Melville, and Muir were certainly among the most compelling of the group—in part, because of their internal conflicts. Alienated Romantics, they were looking for both innocence and experience, both isolation and communion. They were curious. They were ambitious, but they recognized certain limits. Sometimes they celebrated the differentness of the native peoples they encountered, though at other times that differentness triggered a knee-jerk racism. Not one of them fit comfortably into an obvious social niche. They may have perceived exploration as a rugged, masculine alternative to their effete Victorian culture; they may also have been escaping the male competition of capitalism in favor of the fraternal cooperation that was crucial to any expedition. Each had good reason to resent the increasingly powerful national government—despite the fact that they all owed their careers, at least indirectly, to Washington connections. They all railed against capitalist greed and utilitarianism, defending exploration as science for science's sake. But none failed to pursue the main chance when it came along: Reynolds was a land speculator in Texas, King a mining consultant, Melville a commercial inventor, Muir a fruit rancher. They sought a more direct kind of experience—what Thoreau called "contact"[79]— by going to the ends of the earth; they then proceeded to capitalize on that experience by selling books about their adventures. What is perhaps most important, though, is that they continually went out into the cosmos, searching—and that, when they got home, they told stories of an interconnected, interdependent world, in which men were often dominating exploiters, but in which they could potentially be artful balancers and stewards.

I went back to Nevada, searching for connections, in October 2001. Connections can be hard to see. Humboldt sometimes referred to ecological links as "occult forces" because they work in such mysterious ways.[80] But it often helps, I find, to revisit old haunts.

This time, I wasn't just passing through: I had a few days and some research dollars. It was still a challenge, though, to figure out exactly what I was looking for. What could the desert teach? What else could I absorb from this already desiccated landscape? Was it perhaps more ethical just to leave it alone? Staring through my windshield at the cracked hills, the drifting dunes, the abandoned military installations, the fields of alfalfa in special irrigation zones, the trickle of the Humboldt River, I realized that you always notice different things when you go back to places, and then you are forced to determine how much you have changed versus how much the land has changed, and whatever your determination, you become more aware of history. Just as explorers sought connections with people and places distant in space, I was seeking people and places buried in the dunes of time. The past *is* a foreign country, but I also believe in certain continuities—qualities or ideas or struggles or visions that connect the past and present.[81] In the crucible of the desert, I wanted to explore that connection, that potential sympathy.

It was a bleak, lonely trip. I was in Nevada just a month after September 11, a week after the United States started bombing Afghanistan. In any case, with all the test sites around, it's hard not to think of bombs when you're driving through the Nevada desert.[82] This was a time of terror and war; even plague had been unleashed into the atmosphere, in the form of minuscule anthrax spores. How much relevance did my work have in such a climate?

Humboldt, Reynolds, King, Melville, and Muir imagined that their society could be different. Such rebels should always be welcome in our political and intellectual culture—but perhaps especially at times of stress, when our leaders, desperate for the semblance of national unity, tend to discourage dissent. At the heart of a hopeful democracy is the determination to remember debates from the past.[83]

What I had yet to do in Nevada, though, was truly connect with people like Frémont and King, who had tramped through these deserts so many years before. The great nineteenth-century historian Francis Parkman once asserted that at least some passages in historical treatises ought to depend "less on books than on such personal experience as should, in some sense, identify [the historian] with his theme"—and I couldn't agree more.[84] So I got out of my car. I wandered around with my camera, taking black-and-white snapshots at some of the same places visited by Timothy O'Sullivan. I consulted maps, guidebooks, local histories. I went to the Humboldt Museum and the Humboldt County Library. I even decided to attempt some mountaineering without the benefit of a marked path: I wanted to be forced, like Clarence King, to ask advice of local people, to follow animal trails, to read the landscape. As my goal I chose the summit of Job's Peak, which, according to an obscure archival document, had been the site of a bizarre and terrifying event. On September 22, 1867, as King adjusted his metal theodolite to determine how far up he had climbed, he was struck by a bolt of lightning. "I was staggered and my brain and nerves severely shocked," he later explained, and the entire right side of his body turned a light chocolate color.[85] Now *this* would be an interesting experience to recapitulate.

My problem was that I could not find Job's Peak. The same map I used in 1996, which had led me to the "town" of Humboldt, listed a "Job Peak" in the Stillwater Range, just east of the Carson Sink. More detailed maps showed a "Job's Peak" in the Carson Range, just east of Lake Tahoe. King's lightning letter did not specify a mountain range, but the broader paper trail suggested that he spent time in both the Stillwaters and the Carsons during the fall of 1867. The history books were split.[86] I wound up basing my final decision on the eleven or twelve guidebooks that I consulted, not one of which mentioned the Stillwater Range as being of any historical interest. So I drove down Route 206, skirting the Carsons, until I came to the development office of Job's Peak Ranch. An affable retired firefighter named Fred McBryde met me at the door, asked me if I might be interested in a luxury home, and handed me a glossy brochure. I opened the cover, and on the first page, in large green lettering, was a quote from John Muir.[87] Fred told me stories: he'd always intended to climb Job's Peak, he said, but now he was probably too old; a few youngsters made the ramble each year; it was named for Moses Job, who climbed it in about 1855 and planted an American flag at the top; its twin summit was called Job's Sister—in honor of Moses Job's sister—but no one actually knew which was which anymore; anyway, all you had to do was bushwhack up the canyon.

It was one of the hardest climbs I've ever done, and I didn't even reach the top. I had only a few hours to go from about five thousand feet to almost eleven thousand feet and then get back down again. Within a few minutes, my hands, forearms, and lower legs were bleeding from the stiff, thorny underbrush. Starting at about eight thousand feet, I had to rest every couple of minutes just to catch my breath and shake the dizziness from my head. And, of course, there turned out to be about seven false canyons that dead-ended at walls of granite. It was a clear day, though, and I relished the view back down into the Carson Valley. I was disappointed that I had to turn back before seeing the exact spot where King had set up his theodolite, which I of course envisioned as still blackened and scarred—but ultimately that wasn't the point. All I wanted was experience, and though it turned out to be one of the least pleasant hikes I've ever taken, it was perhaps the most intense. And I'm still grateful for that taste of nineteenth-century mountaineering—despite my subsequent research suggesting that I probably climbed the wrong mountain. (King almost certainly had his accident at the southern end of the Stillwater Range.)[88]

On my last day in Nevada, I went back to the Humboldt exit off I-80. This time, I had access to much more information. I had even learned that the state of Nevada itself had almost been called Humboldt.[89] In any case, many current maps still indicated Humboldt as a town, though some had no marking at all. I wanted to see if anything had changed.

This time, I noticed a sign—was it new?—for Humboldt Canyon. A minute later, I was on a dirt road, pointed up into the hills. Kicking up clouds of dust, trying to ignore the sound of rock against metal, I drove onward, and hoped that time would begin to move in reverse. At the end of the dirt road, I found what you find at the end of most dirt roads in America in the twenty-first century: beer cans and bullet cartridges. But I also found the remains of Humboldt, Nevada. There were no plaques, but I did see some beautiful mortarless stone foundations alongside a tree-lined stream trickling down the canyon. It was a calm, sunny day, and the location seemed almost idyllic. From my experience looking at photographs and digging in the archives, though, I knew that this town, like many others, had in fact been pretty industrial, even in the nineteenth century: there were probably some large smokestacks here, at the Atlantic and Pacific Mining Co., accompanying the stone houses, the blacksmith shop, the post office, the hotels, the saloons. I also knew that silver prospectors had arrived here in 1860, just after the discovery of the Comstock Lode. The earliest fortune-seekers negotiated a peace treaty with the Paiutes, and a year later the two hundred inhabitants of the canyon named their settlement Humboldt City, after the river that many of them had followed on their way out west. By 1863, Humboldt City had five hundred residents; by 1869, it was abandoned.⁹⁰ In silver we trusted, in Humboldt we busted.

Back down by the exit ramp, I found State Historic Marker No. 23, which described Humboldt House, a famous eating and resting stop on local stage lines and later on the Central Pacific Railroad. "It was truly an oasis in the great Nevada desert," the sign read, "with good water, fruit, vegetables, etc." But the history lesson ended on a somewhat ominous note: "Between 1841 and 1857, 165,000 Americans traveled the California emigrant trail past here. In 1850, in the dreaded 40-mile desert southwest of present Lovelock, over 9,700 dead animals and 3,000 abandoned vehicles were counted."

There were still no services anywhere, but I did notice one major development—or at least I thought it was a development: there was a huge open-pit gold mine in operation, just to the west of Humboldt Canyon. Was it possible I just hadn't seen it in 1996? "Well, we've been here since 1982," said the gruff man at the entry booth. Yes, operations had expanded considerably since 1996. No, I couldn't go in. No, there was no one around who could answer my other questions. The sign said "Florida Canyon Mine—a subsidiary of Apollo Gold." Later, I learned that Apollo had taken over from Pegasus Gold, which filed for bankruptcy in 1998 after getting cited repeatedly by the federal government for health, safety, and environmental violations.

On the north side of the highway, I looked for the trailer park, but it seemed to have disappeared. This time, I found a couple of ramshackle houses, with rusted trucks parked outside and livestock wandering through the yards. A middle-aged man stared me down through his open window. I pretended to consult one of my maps. If I had asked the man, would he have said that he lived in Humboldt? Even the Florida Canyon Mine, a stone's throw away, is listed under an address in Imlay, which, on the map, is several miles down the highway.

Eventually, I went down to the Humboldt River. Since it was already autumn, the water level was strikingly low. Steeped as I was in Nevada literature, I thought of a passage from Mark Twain's *Roughing It* and later looked it up: "People accustomed to the monster mile-wide Mississippi, grow accustomed to associating the term 'river' with a high degree of watery grandeur. Consequently, such people feel rather disappointed when they stand on the shores of the Humboldt. . . . One of the pleasantest and most invigorating exercises one can contrive is to run and jump across the Humboldt River till he is over-heated, and then drink it dry."[91]

I was thirsty and dusty; I felt like jumping in, immersing myself in history and watery grandeur, such as it was. The barest trickle, the slightest current, would have been enough to keep my interest alive. It had been a difficult trip—but still satisfying. My research had given me new eyes with which to view this landscape. The town was in ruins, and, a few miles westward, the river was dammed—everything had changed since 1869—but I still felt as though I could see what it had been like: I could begin to imagine the inter-play of culture and nature in the nineteenth century. Perhaps most important, I could now understand, in a deep, empathetic way, the explorer's daily trial—the experience of being lost and confused.[92] One of the great historical con-stants must be a sense of disorientation. This time, I'd done a thorough investigation, and there is definitely no present-day town called Humboldt, Nevada. We still can't get the maps right; there is still a need for revision. We may never completely understand our relationship to the world, but that rela-tionship compels us to keep exploring.

Exile: Napoleon's France

Born, 1769:
In Berlin: Friedrich Wilhelm Heinrich Alexander von Humboldt.
In the mountains of Corsica: Napoleone Buonaparte.

They were not destined to get along. Napoleon's 1797 campaign in Italy had spoiled Humboldt's plans to climb some volcanoes there, and, a year later, France's invasion of Egypt had forced the cancellation of Humboldt's proposed Nile expedition. By 1804, despite his desperation to see his colleagues in Paris again, Humboldt had begun to fear that Napoleon's megalomania might actually get in the way of France's intellectual progress. "Europe," he wrote to President Jefferson shortly before leaving the United States, "presents an immoral and melancholy spectacle."[1] He knew that the scientific, cultural, and republican capital he had left behind in 1799 was significantly different now: it had become, perhaps primarily, the Paris of the First Consul. And Napoleon was just in the early stages of his power mongering.

In 1806, Napoleon's army would invade and conquer Prussia, and subsequently Humboldt would be considered a spy in Paris. He lived in the French capital until 1827, but he had to be cautious the entire time. Napoleon's officers often opened Humboldt's mail and ransacked his apartment when he was gone. In 1810, Humboldt even received an order from Napoleon's police chief to leave the city within twenty-four hours or face prosecution. Fortunately, the intervention of Humboldt's friend Jean Antoine Chaptal, a chemist who also happened to have a high position in Napoleon's government, kept the order from being enforced.

It was in the fall of 1804, though, that Humboldt and Napoleon actually met face-to-face, for the first and only time. The tenor of this encounter would define their relationship over the next several years. Humboldt, having just made his triumphant return to Europe, was attracting the attention of the entire Western world, and was mobbed by intellectuals every time he made a public presentation of his adventures and his scientific findings. "It has rarely fallen to the lot of any private individual," wrote his sister-in-law, who happened to be in Paris at the time, "to create so much excitement by his presence or give rise to such widespread interest." Meanwhile, Napoleon was preparing to become emperor. In October, just after Humboldt first exhibited his drawings and specimen collections at the Jardin des Plantes, Napoleon invited the scientist to the manicured gardens of the Tuileries for a celebration of his imminent ascension to the imperial crown. When the exalted traveler arrived, the First Consul greeted him with a question: "So, monsieur, you collect plants?" Humboldt smiled in confirmation. "Ah," said Napoleon; "so does my wife." And he turned and walked away.[2] ■

PART ONE

EAST

Humboldt and the Influence of Europe

CHAPTER 2

Personal Narrative of a Journey

Radical Romanticism

In the distant regions, wherever the thirst for wealth has introduced the abuse of power, the nations of Europe . . . have displayed the same character. The illustrious era of Leo X was signalized in the New World by acts of cruelty that seemed to belong to the most barbarous ages. We are less surprised, however, at the horrible picture presented by the conquest of America when we think of the acts that are still perpetrated on the western coast of Africa.
—Alexander von Humboldt, *Personal Narrative of a Journey to the Equinoctial Regions of the New Continent*, vol. 1 (1814)

At the same time that we are earnest to explore and learn all things, we require that all things be mysterious and unexplorable, that land and sea be infinitely wild, unsurveyed and unfathomed by us because unfathomable. . . . We need to witness our own limits transgressed.
—Henry David Thoreau, *Walden* (1854)

In addition to plants, Humboldt liked to look at rocks. When engaged in geological or mineralogical studies, of course, he did much more than just look: he grabbed handfuls of stones and pebbles, or cooled chunks of lava, and sifted them, checking their density and weight; he sniffed at them; he sometimes licked them, tasting for salinity. But his initial preference was always to absorb a specimen visually, whether a huge outcropping of granite or the slightest flake of gold—to evaluate it both scientifically and emotionally, to give it a chance to enchant him. If he happened to find samples that were particularly novel or striking, like the sparkling sulfur crystals atop the volcano of Tenerife, in the Canary Islands, he would wrap them up and stow them in his satchel. In the Canaries, though, where Humboldt stopped in 1799, en route from Spain to the Americas, he was still something of a novice. What he at

41

first failed to notice about his beautiful, semitransparent crystals, perfect octa-hedrons, was that after a few minutes they became "bedewed with sulphurous acid": checking his satchel, he "soon discovered that the acid had consumed not only the paper which contained them, but a part also of my mineralogical journal." Another scientist might have concluded that his enchantment had blinded him, but Humboldt, even after losing data, still seemed rather enam-ored of his "bedewed" rocks. He grew more careful, over the years, but he never ceased gazing.[1]

Humboldt's eyes were drawn to volcanoes, in particular. To imagine the archetypal Humboldtian pose is to see the explorer on his stomach, balanced on the lip of a crater, bracing himself against periodic tremors, inhaling sulfurous fumes, staring into the rocky abyss. Everywhere he went, he climbed volcanoes, and their evocative names pop up in all of his writings: Stromboli, Etna, and Vesuvius in Europe; Pichincha, Chimborazo, Cotopaxi, and Tunguragua in the Andes; Orizaba and Popocatépetl in Mexico; Pe-schan and Ho-tscheu in Asia. "Volcanic phenomena . . . ," he wrote, "consid-ered in the totality of their relations, are among the most important topics in earth Physics. Burning volcanoes appear to be the effect of a permanent communication between the molten interior of the earth and the atmo-sphere that envelops the hardened, oxidized crust of our planet." By examin-ing a volcano's "superimposed layers" of rock, Humboldt could determine how geologic time had unfolded. And by seeking out "that intimate con-nection between so many diverse phenomena"—the frequency of eruptions and earthquakes; the angle of volcanic slopes; the latitude and elevation of different volcanoes; the location of volcanic hot springs; the different gases, liquids, and solids (including fish, in one instance) that had been erupted—he could explore the relationship between earth, air, fire, and water, between the internal and external. Then, after staring at enough volcanic rocks, he might ultimately be able "to recognize unity in the vast diversity of phenomena."[2]

The appeal of volcanoes, though, went beyond their ecological signifi-cance: Humboldt counted on them to provide some of his most intense expe-riences of the natural world. Having scratched his way to the top of Pichincha, the peak that towers majestically over Quito, Ecuador, Humboldt cast his eyes down into the void of the caldera and gripped the edge of his ledge. "No view," he reported, "could be as dismal, lugubrious, and frightening as what we saw at that moment."[3] The extreme elevation, the bursts of flame and puffs of ash, the black rock, the instability of the ground, the rotten-egg reek of sul-fur, all combined to break the spell of the mundane, the assumed, forcing

Humboldt to see in new ways. On the physical frontier between earth and sky, he also reached frontiers of feeling and thought, discovered what it meant to be avant-garde.

In science, Humboldt pushed the borders of several fields by attempting to combine them, on a global scale. Traveling as much as possible himself, and soliciting observations from other travelers—throughout the American expedition, he collaborated closely with his constant companion, the botanist Bonpland—he sought to determine how ocean currents affected mean temperatures all around the world, how volcanic formations affected vegetation patterns. Moreover, he applied a Romantic critique to the core beliefs of Enlightenment science, tempering the dream of impersonal objectivity with a conscientiously subjective gaze. He always "considered Nature in a two-fold point of view," he explained. "In the first place, I have endeavored to present her in the pure objectiveness of external phenomena; and, secondly, as the reflection of the image impressed by the senses upon the inner man, that is, upon his ideas and feelings."4

In some ways, his literary strategy of portraying humanity as overwhelmed by nature was simply an extreme expression of the Romantic Sublime: trepidation leads to inspiration.5 Atop Pichincha, though, the inspiration seemed in doubt. It is striking that Humboldt was still trying to record the frequency of the volcano's tremors—"we counted 18 in less than 30 minutes"—but the image he paints is certainly not of the heroic, confident scientist conquering nature through measurement. Rather, he is holding on for dear life, even as he clutches at any detail that might eventually lead to some enlightenment. As soon as he started setting up his instruments, he explained, "the sulfur fumes almost suffocated us . . . ; here and there, we even saw bluish flames dancing around."6 Could the Scientist be hallucinating? Humboldt's invocations of nature's sublimity are often not uplifting. Ultimately, they force his readers to experience a radical reversal of human dominance over the natural environment. Trepidation is not romanticized into an epiphany of pure human feeling but rather treated as confirmation of humanity's insignificance. Introducing his *Personal Narrative*, Humboldt warned his fellow Europeans that "in the New World, man and his productions almost disappear amidst the stupendous display of wild and gigantic nature."7 For Humboldt, the laws of nature trumped any human laws. His narratives of scientific exploration obscured Europe's great cities behind a tangle of tropical vines, buried them in volcanic ash. Studying rocks, Humboldt found humility.

———

If the New World brought Humboldt down to earth, Europe catapulted him to the heights of fame. Immediately after his five-year expedition to the Americas, and for the rest of his long life, the world's most renowned explorer grappled with the problem of how best to express humility and awe. Philosophers from Saint Augustine to Rousseau had addressed this question before, and Humboldt invoked both of those thinkers in 1806 when he wrote a short autobiographical letter entitled "My Confessions."[8] Yet he was never comfortable with such intensely personal forms: the focus on his own inner responses to the world, however self-effacing those responses might be, seemed inherently egoistical. Moreover, he disliked the religious overtones of the confessional mode. Though his emphasis on the unity of nature could easily be—and often still is—misinterpreted as reflecting a kind of Deism, Humboldt was an ardent agnostic.[9] He wished to articulate his smallness in the face of natural creation, not any Creator. So, more often than not, he produced relatively impersonal works that focused almost exclusively on the scientific findings of his expedition, starting with the groundbreaking *Essay on the Geography of Plants*, printed in Paris in 1805.[10] By repudiating the very idea of first-person narration, he thought, by explicitly disappointing the expectations of readers who had enjoyed the adventure-hero writings of explorers like Captain Cook and Mungo Park, he would not only advance the sciences more effectively but also make his subjection to nature more apparent.

Humboldt quickly discovered, though, that his monographs—whether on botany, zoology, astronomy, geology, geography, or meteorology—had their own drawbacks. He worried that his obsessive analysis might suck the life out of nature: upholding humility, he was neglecting inspiration. Moreover, monographic works were by definition specialized, narrow, limiting. He had to remind himself that his goal was to depict the different aspects of nature in all the glory of their mutual connectedness and dependence. So, in 1807, attending to some business in Berlin and suddenly finding himself trapped there by Napoleon's invasion of Prussia, he set out to write *Views of Nature* (also translated as *Aspects of Nature*), an utterly original book in which he juxtaposed aesthetic descriptions and scientific explanations of steppes, deserts, tropical river basins, primeval forests, and lofty mountain chains. There was no narrator and no narrative—these were generalized landscapes based on both Humboldt's experiences and his reading—but the tone and style were fundamentally human, full of warmth and sympathy. Humboldt offered his fellow Berliners an escape from "the relentless bludgeonings of our time," an opportunity to experience freedom by traveling to "a distant, richly endowed land" possessed "of a free and vigorous vegetation." Nature offered not only deep insights but also solace and sanctuary; the very image of a wild and over-grown landscape could move people spiritually, was perhaps even more valu-able in times of need than religion. *Views of Nature* would become Humboldt's favorite among his own books, in part just because it was immensely popular, but also because it so seamlessly blended science and art, so easily celebrated nature's grandeur without erasing humanity. Yet he was still dissatisfied. In places he found *Views* to be overly "poetical," self-indulgent, perhaps irre-sponsible in its escapism. It was not sufficiently humble, and, moreover, it was devoid of social and political issues. Its human element did not have enough of an edge.[11]

Shortly after returning to Paris in January 1808, Humboldt took up his notes on Mexico and began composing his *Political Essay on the Kingdom of New Spain*, another of his most ambitious works. This was a broad study in human geography, and here Humboldt developed an interdisciplinary form to which he could turn whenever he wanted to examine humanity's relation-ship to nature as an explicitly social problem. Tracing the demographic, politi-cal, economic, and agricultural trends affecting Mexico's development, he insisted that nature was the key to civilization, and that its fate not be left to in-dividual sentiment or abstract market forces. What was missing in the *Politi-cal Essay*, though, was any depiction of nature that might inspire respect: now Humboldt had completely abandoned his poetical landscapes in favor of the scientific management of natural resources.[12]

Always seeming to want to compensate for his latest perceived failure, Humboldt next revisited the power of American scenery, though this time he added a social element by venturing into the fields of anthropology and ar-chaeology. The resulting publication, *Views of the Cordilleras and Monuments of the Indigenous Peoples of the Americas* (1810), was a visually stunning celebra-tion of American mountains and of the accomplishments of civilizations that had explicitly adapted their culture to the mountainous American environ-ment. While many of his contemporaries dismissed indigenous Americans as worthless savages, Humboldt upheld their ability to build cities and roads in the sky. Inevitably, though, this book also failed to fulfill Humboldt's dream of an all-encompassing yet down-to-earth study, for it contained virtually no natural science.[13]

Even as he poured himself into all three of these avant-garde works of ex-perimental geography—*Views of Nature*, the *Political Essay*, and *Views of the Cordilleras*—Humboldt was also falling back on simpler, more humble scien-tific projects. Over the next few years, he published monographs detailing his zoological findings in the Americas, explaining the significance of his topographical data, and speculating on the implications of his new botanical specimens and their distribution at different latitudes and altitudes. It was not until 1814 that he finally published a book with a flexible enough form to be on the cutting edge of both science and literature—to depict the full range of humanity's complex relations with nature; to reveal the natural world as both inspiring and terrifying; to position the author in just the right posture of engaged awe.

Predictably, Humboldt himself was never satisfied with his *Personal Narrative*. He continually struggled with the balance between plot and analysis as he worked on the first three volumes, between 1812 and 1825, and he never published volume 4, which would have given us the story of his most impressive mountaineering feats in the Andes. He seemed worried that too much adventure might negate his scientific findings or cast doubt on his humility. In retrospect, though, nearly two hundred years into a tradition of nature-oriented travel writing supposedly invented by Darwin or Thoreau, the *Personal Narrative* seems a near-perfect solution to the problem of how to tell a story linking all the different branches of human and natural science, in an effort simultaneously to know nature and subject oneself to it. It took Humboldt thirty years and more than thirty publications to process his American expedition. But it was the *Personal Narrative* that allowed him most effectively to convey the meaning of the opportunity he'd had in the Americas—to explore the natural world by exposing himself to the full measure of its forces.[14]

———

Imagine him nearing the summit of Tenerife's volcano. It is almost sunrise, the climbing party having left its encampment at three in the morning, "by the sombrous light of a few fir torches." The day before, the five mountaineers, lugging sextants, barometers, chronometers, thermometers, specimen jars, and logbooks, had ascended from sea level to about ten thousand feet, through a chestnut forest, over tangles of brambles, past a region of "arborescent heaths," then ferns, then junipers and the highly flammable firs, across "an immense sea of sand" made of pumice dust, through narrow basaltic ravines, and finally into a barren region filled with huge black masses of obsidian that had been spewed out by the volcano itself. The cold of the night, in their high desert camp, compelled the men to light a fire, against whose smoke they then erected a cloth screen, most of which was consumed by stray flames just minutes after they tried to get to sleep. Now, hiking by torchlight, they toil up steep lava trails, the pebbles giving way beneath their feet: with every step forward, they slide a half step back. They pass through the strangely eroded region of the Malpays, or Badlands; they explore a glistening ice cavern; they stare at the sky for eight full minutes observing "a very curious optical phenomenon" that none of them can explain, a seemingly private light show, a horizontal oscillation of certain stars all shooting like rockets toward the same point in the sky and then flickering back to their original positions. Humboldt might pass it off as a hallucination—the altitude is beginning to make him nauseous and dizzy—if all five of them had not seen the same thing. Soon they enter the coldest part of the night, yet they are also approaching the fiery crater: "our hands and faces were nearly frozen, while our boots were burnt by the soil on which we walked." Humboldt pauses to absorb the scene, realizing that dawn is almost at hand, and sees far below them a perfectly uniform cloud formation, reminiscent of "a vast plain covered with snow." The other volcanic summits of the Canaries are "like rocks amidst this vast sea of vapours . . . , their black tints . . . in fine contrast with the whiteness of the clouds." Instruments at the ready, Humboldt hunkers down to watch the sunrise. He braces himself against the sharp, shiny stones, steadies his breathing as he peers eastward through the scope. He wants the exact time, the distance, the angle, the altitude—and the view. These are the moments—of vision and confusion, of familiarity and strangeness, of ecstasy and nausea—that he yearns for us to understand.[15]

His first-person narration allows him to depict himself in the act of seeing, and, in the end, this creative act is the subject of all his books. The richness of your life, Humboldt asserts, depends on what you're able to see. Can you detect the transparent strands linking the canaries of these islands to the tall, tufted *retama* shrub, whose scented blossoms attract wild goats and often decorate the hats of local goat hunters? Can you see why your crops will not grow without a healthy forest nearby? Have you ever watched the sunrise from atop a volcano in the middle of the ocean and observed the clarity of the light in the thin, pure air? "This transparency," Humboldt tells us, "may be regarded as one of the chief causes of the beauty of landscape scenery in the torrid zone; it heightens the splendour of the vegetable colouring, and contributes to the magical effect of its harmonies and contrasts." Moreover, on a good day, the quality of the air and light might leave you with a "lucid clearness in the conceptions, and a serenity of mind." This is the power of empiricism: watching Humboldt at work, we train our own eyes to see the truth—especially about our own smallness, our humble place in nature. "The earnest and solemn thoughts awakened by a communion with Nature intuitively arise from a presentiment of the order and harmony pervading the whole universe, and from the contrast we draw between the narrow limits of our own existence and the image of infinity revealed on every side, whether we look upward to the starry vault of heaven, scan the far-stretching plain before us, or seek to trace the dim horizon across the vast expanse of ocean."[16]

At the same time, though, Humboldt's obsession with viewing leads him to a trenchant critique of strict empiricism. The truth is not simply "out there"; not even the most careful scientists have access to some self-evident objective reality. Humboldt never figured out why those stars were dancing in the sky, but he relished the mystery as much as any scientific revelation. "Nature is an inexhaustible source of investigation," he asserts, almost giddily: you can always find something new and bizarre, "never yet examined." As common as celebrations of transparency in his writings are images of confusing mists, which cause the observer to become "lost in vain conjectures." The point is that being lost can be fun; just keep the conjectures coming. Skimming a Caribbean shoreline, he suggests that "the coasts, seen at a distance, are like clouds, in which each observer meets the form of the objects that occupy his imagination." This is as it should be—or at least as it must be. To see is to create a truth—to become conscious of the way your vision is always filtered through your imagination, and then to do justice both to the observed landscape and to your private sensibilities.[17]

Humboldt wrote his most powerful book by embracing his subjectivity: this is a *personal* narrative, in which his own story provides the unifying thread, as well as the political commentary. But he shows us a subjectivity fully immersed in and exposed to the world, not imposing a vision on it but engaged with it in a kind of visual dance. His favorite animal on Tenerife, the great empiricist confides to his readers, is one he barely glimpsed: "Of all the birds of the Canary Islands, that which has the most heart-soothing song is unknown in Europe. It is the capirote, which no effort has succeeded in taming, so sacred to his soul is liberty. I have stood listening in admiration of his soft and melodious warbling, in a garden at Orotava; but I have never seen him sufficiently near to ascertain to what family he belongs."[18] A master of precise language and of several different scientific disciplines, Humboldt never believed that increased knowledge would ruin a person's experience of nature—"I cannot, therefore, agree with [Edmund] Burke," he explained, "when he says that it is our ignorance of natural things that causes all our admiration and chiefly excites our passions."[19] Yet he was enough of a humanist to realize that people cannot live without metaphors, that the capirote is worth most to us as a symbol of freedom, that one of our deepest and most significant emotions is a kind of childlike wonder in the face of the world's mystery. "Imagination," Humboldt wrote on page one of his *Narrative*, "wanders with delight over that which is vague and undefined." Ultimately, at the top of the volcano on Tenerife, he found the view, and his response to it, indescribable. He wound up simply wallowing in his "sensations, which are the more forcible, inasmuch as they have something undefined, produced by the immensity of the space as well as by the vastness, the novelty, and the multitude of the objects, amidst which we find ourselves transported."[20]

What is perhaps most striking, throughout the *Personal Narrative*, is Humboldt's effort to see new things deeply, in context, and in connection to everything else he's seen and learned. The juxtaposition of his literal acts of observation and his *philosophical* observations, his adventures and his scientific "tangents," leaves us with a sense of swirling intellectual currents. Before we even arrive at the Canary Islands, we understand how the wind and water will carry us there, how the Equinoctial Current interacts with the Gulf Stream and the trade winds, how a given water molecule might make the circuit of the Atlantic, how trunks of tropical trees, seeds of the Jamaican mimosa, even Antillean tortoises, all drift ashore on the western coasts of Scotland and Norway. "Every motion is the cause of another motion," Humboldt explains, and every leg of his journey spurs a new strand of scientific theorizing. If Humboldt's travels provide the unity, his science provides the diversity: "To give greater variety to my work, I have often interrupted the historical narrative by descriptions. I first represent phenomena in the order in which they appeared; and I afterwards consider them in the whole of their individual relations."[21]

The key word here is "relations," for it is probably the most common term in Humboldt's book. When Humboldt gives us his analytical asides, which can last for twenty or thirty pages, they are not just random riffs but expansive solos meant to harmonize with all his other themes. In the original French, he refers to his book itself as a "*Relation*," a story related by one person to another: "only connect!" he seems to be saying. The scientific tangents add variety, then, but ultimately they reinforce the book's unity, because they are connected to each other and to Humboldt's travels in the same way that plants are connected to rocks, and wind to water, and the Old World to the New. By following these common threads, we begin to understand how Humboldt learns to feel at home anywhere on the planet. We watch him climb the volcano of Tenerife, and then, suddenly, we are visiting other volcanic regions, comparing "black, basaltic, and earthlike lava" to "vitreous and feldsparry lava," exploring obsidian and jade as alternatives to copper and iron, speculating as to what chemical reactions keep volcanoes burning under vastly different conditions. The story might get bogged down—when are we going to reach the "Equinoctial Regions of America"?—but the book's tangential structure has significant benefits:[22]

> I have endeavoured to render these researches interesting, by comparing the phenomena of the volcano of Teneriffe with those that are observed in other regions, the soil of which is equally undermined by subterranean fires. This mode of viewing Nature in the universality of her relations is no doubt adverse to the rapidity desirable in an itinerary; but . . . by *isolating* facts, travelers, whose labours are in every other respect valuable, have given currency to many false ideas of the pretended contrasts which Nature offers in Africa, in New Holland, and on the ridge of the Cordilleras. The great geological phenomena are subject to regular laws, as [are] the forms of plants and animals. The ties which unite these phenomena, the relations which exist between the varied forms of organized beings, are discovered only when we have acquired the habit of viewing the globe as a great whole.

Imagine learning not just how to think globally but how to *see* globally. Ralph Waldo Emerson said of Humboldt that "you could not put him on any sea or shore, but his instant recollection of every other sea or shore illuminated this."[23] The comparisons, the connections, the free associations that pop up everywhere in the *Personal Narrative* represent the development of a truly ecological vision. Humboldt recognized difference, even exulted in it sometimes, but he also saw "relations" where other explorers and scientists saw only bizarre, isolated phenomena, or inhuman wildernesses, or incomprehensible savagery.

Such a vision is deeply humane, and a worthy goal even in the twenty-first century. But it is virtually impossible to achieve. Of course, Humboldt's vast array of instruments helped tremendously: he could even compare the blueness of the sky in different climates by taking readings on his cyanometer, that newly invented apparatus which Lord Byron found so comical. In my own lifetime, I've had the benefit of comprehensive textbooks in geography and ecology, written in my native language; topographical maps and satellite photos; computerized Geographic Information Systems; and air travel. And I have tried conscientiously to think like Humboldt as I followed in his footsteps around the globe. Still, I struggle to see globally. Sometimes, with a certain amount of jealous bafflement, I conjure up the image of John C. Frémont, steeped in Humboldt's writings, looking out at the Pacific for the first time from what are now the Oakland hills, and naming the "Golden Gate"—a few years before the discovery of gold in California—simply because the terrain reminded him of the Golden Horn, where an inlet of the Bosporus spawned the prosperous port of Constantinople.[24] Humboldt came to Spanish America and noticed immediately that there was what he called a *"parallelism of direction"* among layers of certain granites, shales, and schists—that these layers were always inclined to the northwest, just as they were in Europe, making an angle with the globe's axis of about forty-five degrees![25] I have scanned the horizon from the Sierras, the Canadian Rockies, the Andes, the Alps, the Pyrenees, the Himalayas, even a volcano in China, looking for similarities in rock formations, patterns of vegetation, anything. Usually, the fundamental message I get from my taxed intellect and quivering senses is that the world is disorienting. Atop many mountains, though, I have also felt a kind of Humboldtian awe and peacefulness. And the arts and sciences, even theories of universal chaos, teach us that certain global patterns, beautiful in their connectedness, do exist.[26] Humboldt was one of the first people in the modern era to begin to sketch those patterns.

Of course, seeing globally has its drawbacks. Humboldt's efforts to find places on the planet that shared the same mean temperatures, for instance—he coined the terms "isotherm" and "isoline"—can be seen, in Anthony Pagden's words, as revealing a desire for "cognitive possession" of the world. Encircling the globe with his rationalistic, homologizing lines, packaging unruly realities in neat, easily transportable tables and graphs, Humboldt seemed to bend American extremes to fit European norms. To critic Mary Louise Pratt, Humboldt's global science comes across as pure hubris, the distorted dream of an arrogant colossus: Humboldt "assumes a godlike, omniscient stance," she asserts, "over both the planet and his reader." Indeed, claiming to know anything is a dangerous business, and trying to make a foreign landscape or culture comprehensible inevitably involves distortion, since the object of one's gaze must always be ripped from its original context and then re-posed in a book or painting or museum case. Moreover, the traveling scientist's attempts to make colonial worlds knowable, whatever his own intent, could easily be used by European powers to further their conquests: charts, surveys, censuses, even histories, often went hand in hand with repression and violent resource extraction.[27]

In the end, though, it is hard to imagine anyone navigating these intellectual shoals as carefully and conscientiously as Humboldt did. He was not a man of political action; he did not take direct steps to block colonialism. But he did register protests and show respect for the separateness of the colonies. He sought always to balance similarities and differences, trying to make the particularly challenging elements of American nature understandable without denying them their strangeness. His interest in isotherms never eclipsed his obsession with the extreme foreignness of the "torrid zone, where nature appears at once so rich, so various, and so majestic," where "everything . . . presents an exotic character."[28] Humboldt never posited a whitewashing unity *of* nature; rather, he deduced a "unity *in diversity,*" a unity that explicitly depended on the full richness of life he had seen in the tropics. He cared about each element of nature because every weed, stinging insect, and poisonous snake played a crucial role in what he came to think of as particular ecosystems, all of which seemed to thrive on biodiversity. "Ecosystem" and "biodiversity" are anachronistic terms—Humboldt did not use them—but they accurately reflect what he meant when he emphasized the unifying interconnectedness of nature.[29]

The downside of global thinking, in other words, is balanced by Humboldt's meticulously constructed ecological web, which warns us whenever a strand gets frayed or stretched too thin. Suddenly, the seeming arrogance of his desire to see globally leads to yet another lesson in humility, for a planetary ecology reminds us of our ultimate dependence on foreign lands, on the rest of nature. According to Humboldt, it even teaches us the limits of our own understanding. "The attempt perfectly to represent unity in diversity must . . . necessarily prove unsuccessful," he wrote, in *Cosmos*. "If . . . nature (understanding by the term all natural objects and phenomena) be illimitable in extent and contents, it likewise presents itself to the human intellect as a problem which cannot be grasped, and whose solution is impossible." Only when we actually try to see everything in nature, when we strain our vision by gazing toward "a horizon that endlessly retreats," do we come to recognize our own blindness.[30]

Humboldt took his first steps toward this insight while climbing the volcano of Tenerife. As he explains in the *Personal Narrative*, it was here, amid the vines, laurels, pines, *retama* shrubs, and grasses, that he started to become obsessed with global climatology. Passing through what he recognized as different zones of vegetation—an experience he would have again in South America, finding the bright orchids and tall hardwoods of the rain forest at the base of peaks like Chimborazo and Cotopaxi, and only the hardiest mosses and lichens on their snow-clad summits—he realized that climbing certain mountains was like walking from the torrid equator to one of the frigid poles: vertical geography mirrored horizontal geography. Developing his "habit of viewing the globe as a great whole," he identified climate as a unifying global force, proving, in a sense, that we can never escape the weather. Wherever you might wander, the distribution of plants will be tied to the effects of climate, which justifies the boundaries Humboldt proposed for "temperate," "tropical," and "boreal" vegetation zones all around the world. As early as 1805, in the *Essay on the Geography of Plants*, he was fostering the planetary consciousness he so cherished by forcing his readers to see the links between climate and every other component of the environment. His beautifully colored plate illustrating Chimborazo's vegetation zones, published in conjunction with the essay, remains an eloquent introduction to the science of plant ecology. (For the rest of his career, he would produce innovative graphic representations of his scientific insights—mountain profiles showing the relative height of the Alps, Andes, and Himalayas, for instance— designed to foster a global vision.) And his innovations in comparative climatology underlie current attempts to understand the threat of global warming. Even more important, though, are his conclusions regarding what we can *never* understand about such "occult forces" as the planet's climate.[31] These insights have become the basis for our wisest brand of environmentalism— the kind that recommends, quite simply, a modicum of respect for natural forces like heat and cold, which are ultimately out of our control. This was the grand lesson of the Canary Islands, burned onto the soles of Humboldt's shoes.

———

Humboldt launched perhaps his most formative journey in 1790, at the age of twenty, in the company of Georg Forster, whose experiences on Captain Cook's second expedition (1772–75) had led him to the belief that open-minded travel was the key to science. In Humboldt, Forster found an open-minded disciple—one who as a child had devoured the elder scientist's *Voyage Round the World* (1777). Humboldt, fresh from his schooling at the University of Göttingen, eagerly accepted Forster's invitation to explore the Rhine Valley, in part as a way to postpone the career in finance planned for him by his mother. During the course of three and a half months, the two men collected rocks in Holland and Belgium; traveled to Dunkirk, where Humboldt saw the ocean for the first time; crossed over to England to meet Sir Joseph Banks and visit his gigantic tropical herbarium; and then passed through Paris to soak up the fervor of the French Revolution, whose participants were about to celebrate the first anniversary of the storming of the Bastille and whose zeal converted Humboldt into a lifelong democratic republican. And, years later, Humboldt would attribute "the passion that seized me for everything connected with the sea and for visiting tropical lands" to "the companionship I enjoyed on this journey." He even claimed that it was Forster "to whom, for the most part, I owe what little knowledge I possess."[32]

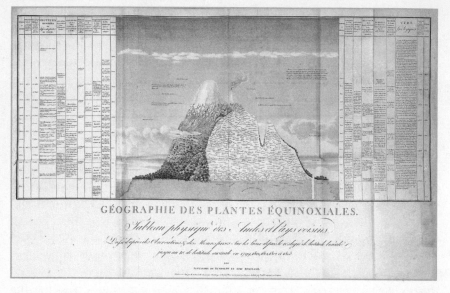

6 Chart showing the climate zones of Chimborazo, Ecuador, 1805. Alexander von Humboldt.

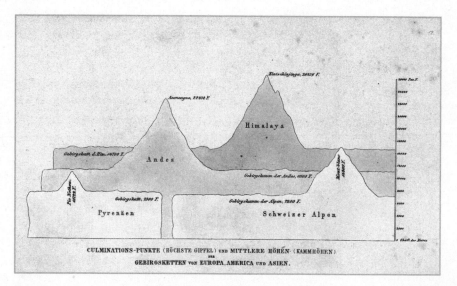

7 Chart comparing the height of mountain ranges, 1825. Alexander von Humboldt.

Forster not only elaborated on the experiences that lay behind his *Voyage* but also trained his student in interdisciplinary fieldwork. Examining geological strata along the lower Rhine, the companions considered how soil conditions and climate might offer clues as to whether the local basalt had originally formed as underwater deposits in a primordial ocean or as magma produced by the intense pressure of volcanic fires. They interviewed politicians and peasants, descended into mines, collected revolutionary pamphlets, witnessed parliamentary debates, visited museums, libraries, factories, farms, parks, concert halls, cathedrals, castles, prisons, observatories. At times they simply opened themselves up to the wonders of nature and culture; usually, they analyzed and discussed their every observation, straining to see connections between microclimates, wool prices, soil conditions, crop rotations, even the quality of local beer. Their trip was an exercise in blending perception and reflection. When Forster turned the journey into a book, he titled it *Ansichten vom Niederrhein* (Views from the Lower Rhine), and Humboldt was explicitly referring back to this work in 1807 when he wrote his *Views of Nature*—*Ansichten der Natur* in the original German—in occupied Berlin. Among other things, Forster had taught Humboldt to use language carefully and suggestively: the word *"ansichten"* is a near-perfect evocation of his approach to travel literature, for it refers both to visual depiction and to "views" in the sense of personal opinions. It was in Forster's company, then, that Humboldt first developed his subjective, Romantic empiricism as well as his broad ecological vision.[33]

In *Cosmos*, Humboldt would credit Forster with founding "a new era of scientific travelling, having for its object the comparative knowledge of nations and of nature in different parts of the earth's surface."[34] Here was travel not as adventure or conquest but as a multilayered intellectual challenge that forced an individual to understand human experience in all its facets—and this was the goal Humboldt set for himself in the 1790s. His circumstances did not immediately permit him to go too far afield after his trip with Forster, but he conscientiously began preparing himself to be the most perspicacious traveler he could be in case the right opportunity should ever arise. Even as he studied commerce and then took his position in the Department of Mines, he published tracts on botany, geology, and galvanism, and he read widely— Kant, Rousseau, Diderot, Herder, Goethe.[35] He even perfected his sketching technique. In his midtwenties, he repeatedly applied electrodes to his own body, raising huge, purulent welts, in an attempt to determine the effect of electricity on the blood, nerves, and lymphatic system. He hiked all over Europe, camping out on mountain flanks and practicing with his sextant to double-check latitudes, taking regular measurements, sometimes all night long, of oxygen levels, the atmosphere's electrical charge, temperature shifts, barometric pressure, humidity. As a mining inspector, he studied underground mosses and wrote a short essay on the significance of light to plant growth and pigmentation. His research in the mines also led him to propose a pension plan for sick and injured miners and to invent a new kind of respirator and safety lamp. Throughout these years, certain of Humboldt's friends and teachers repeatedly warned him that he was working too hard and in too many different fields. But Humboldt saw his adventures in various disciplines simply as natural steps toward the fulfillment of his vision. "How can you stop a man," he asked, "wanting to find things out and understand the world around him? In any case, a wide variety of knowledge is vital to a traveler."[36]

In addition to knowledge, though, Humboldt felt he had to acquire and nurture a certain sense of openness to the world, and, again, it was Forster's example that guided him toward this approach to scientific travel. Forster never accepted pat answers; he thought the role of the intellectual, and especially the traveling intellectual, was to upend the status quo, to pose new questions. In troubled times, Forster, a classic Enlightenment rationalist, nevertheless turned to the mystical dark arts, joining a local Rosicrucian order and actively practicing alchemy, desperate to believe "in the existence of a spiritual world and the possibility of communicating with it."[37] He also became a pariah in his homeland, because he moved to Paris in the early 1790s and embraced the most radical phase of the Revolution, during which the French army attempted to annex the part of Germany where he had formerly lived. But Forster, like Humboldt, had *always* felt like an outsider: his father had moved their family to England when Georg was just eleven, and he had sailed with Cook—on an utterly disorienting voyage, which lasted four years—at the age of seventeen. Almost instinctively, Humboldt learned from Forster to cultivate his sense of exile and his need to seek out new perspectives. Maybe the Rosicrucians had some answers; maybe the Tahitians did. Reading up on the history of exploration, Humboldt found that while men like Cook generally portrayed themselves as godlike figures bringing culture and enlightenment to the simple, hedonistic natives of the tropics, Forster emphasized the diversity of the native groups he encountered, explicitly trying to dispel stereotypes of noble and ignoble savages through detailed descriptions of different tropical civilizations. Writing of a time when one of his shipmates executed a Tahitian for an act of petty theft, Forster commented that "the first discoverers and conquerors of America have often, and very deservedly, been stigmatised with cruelty, because they treated the wretched nations of that continent, not as their brethren, but as irrational beasts, whom it was lawful to shoot for diversion; and yet, [even] in our enlightened age, prejudice and rashness have often proved fatal to the inhabitants of the South Sea." What a shame it was, Forster wrote, that his fellow Europeans had "humanity so often on their lips, and so seldom in their hearts!"[38] Perhaps the most important lesson Humboldt learned from his mentor, then, was that travel provided crucial opportunities to question the assumptions of one's home culture and come to fresh understandings of the world—and that such opportunities were available only to those travelers who embraced disorientation.

———

As Humboldt's ship sailed deeper into the tropics, leaving the Canaries behind, the sky began to change. Familiar stars blinked out, and the scientist's old routine of nightly observations took on a brand-new meaning: Humboldt was suddenly forced to rethink celestial navigation, to see completely different patterns in the heavens. Perhaps for the first time, he came to realize how little he would be able to count on during this trip, how little he actually knew. "Nothing awakens in the traveller a livelier remembrance of the immense distance by which he is separated from his country," he noted, "than the aspect of an unknown firmament." As the winds grew calmer and the heat more oppressive, Humboldt grew more restless, desperate to understand a surreal, confusing world. One day the strange phosphorescence of the ocean broke up into little pieces and suddenly became flying fish, which leapt twelve, fifteen, eighteen feet up in the air and landed in piles on the deck. At about the same time, a few sailors and passengers fell into a delirium, seemingly the result of a malarial fever. Humboldt was dismayed to find that no one had brought along any quinine bark and that the "ignorant and phlegmatic" Gallician doctor believed in bleeding his patients until they revived or passed away. Finally approaching the shores of Spanish America, Humboldt tried his best to see parallels with familiar terrain, only to be surprised later at how often his perception was clouded by his expectations. He *did* make striking connections; he also made mistakes. "The aspect of the mountains of Paria, their colour, and especially their generally rounded forms, made us suspect that the coast was granitic; but we afterwards recognized how delusive, even to those who have passed their lives in scaling mountains, are impressions respecting the nature of rocks seen at a distance." They had apparently reached their goal, but not even the captain could figure out where exactly they needed to go to find a port, because the land "was so differently marked in the French, Spanish, and English charts."[39]

Humboldt tried to embrace the foreignness of his surroundings, the languorous rhythm of the tropics. "Nothing can equal the beauty and mildness of the climate of the equinoctial regions on the ocean . . . ," he wrote. "What a contrast between the tempestuous seas of the northern latitudes and the regions where the tranquility of nature is never disturbed!"[40] But he knew he was overstating the case. Despite his tendency toward environmental determinism, he realized that inner tempests could rage no matter what weather conditions prevailed on the outside, and a ship ravaged by tropical diseases is enough to make anyone yearn for the choppy waters of more temperate climes. When one of the sick passengers actually died, perhaps from typhus, Humboldt felt the loss deeply and wondered whether the poor nineteen-year-old boy should ever have left Spain. At this particular moment, nature offered no solace. "A profound calm reigned over these solitary regions, but this calm of nature was in discordance with the painful feelings by which we were oppressed." Fortunately, Humboldt was able to take some comfort in a short shipboard funeral, "an affecting ceremony, which brought to our remembrance those times when the primitive christians all considered themselves as members of the same family." Facing death far from home, churning inside but gazing out at placid, glassy waters, the like of which he had never seen before, Humboldt fell back on a prayer for unity.[41]

The moral, or social, aspect of Humboldt's unifying vision was spurred explicitly by his travels to the torrid zone. Europeans celebrated the wonders of their civilization, and Humboldt acknowledged those wonders, but he also asked his readers to leave Europe for a moment and recognize the horror that Europeans tended to bring with them around the globe. In Ecuador, Humboldt noted in his journal his developing suspicion that perhaps "the very idea of a Colony is immoral," because colonies were so frequently sites of exploitation. The main "success" of "European governments" in the colonies had been "to spread hate and disunity." *Liberté, égalité,* and *fraternité* had not held sway for very long even in France; such ideals had never governed policy in the colonies.[42] As Humboldt hinted with regard to the young Spaniard's last rites, Western traditions bore the seeds of a connective unity, but they had become corrupted during Europe's long period of expansion. "The Christian religion, which in its origin was highly favourable to the liberty of mankind, served afterwards as a pretext to the cupidity of Europeans," he explained, noting the proximity of missions and slave markets in every part of America to which he traveled. His countrymen congratulated themselves on having abolished certain "feudal institutions," yet such institutions "still press heavily on the people" of the colonies. In fact, the violence of European colonists had "swept off whole tribes from the face of the earth"—in Humboldt's opinion, simply because Europeans could not accept differentness. Humboldt understood on the deepest level how strange surroundings and unfamiliar people could spur anxiety and insecurity, but he urged his readers to seek common ground.[43]

On July 15, 1799, just off the coast of what is now Venezuela, two canoes, each hollowed out of a single tree trunk, approached Humboldt's ship, and the disoriented scientist had his first real opportunity to test his beliefs about unity. Fortunately, he was already beginning to welcome the surprises of travel and to second-guess most European assertions about the world. Gazing down at the thirty-six Guayqueria Indians, Humboldt noted that "they had the appearance of great muscular strength, and . . . projected on the horizon, they might have been taken for statues of bronze. We were the more struck with their appearance, as it did not correspond with the accounts given by some travellers respecting the characteristic features and extreme feebleness of the natives." Eventually, the crew welcomed the Indians on board, and, communicating in a blend of Spanish and an improvised sign language, explained that they were lost. In response, the Guayquerias offered gifts of fresh coconuts and "very beautifully colored fish"; "the master of one of the canoes" said he would be happy to pilot the ship to the port of Cumaná, less than a day's sail away. Humboldt was impressed by this man's obvious warmth and competence, and he immediately began questioning him about the other "riches" displayed in the canoes—bunches of plantains, a calabash drinking cup, "the scaly cuirass of an armadillo." Many travel narratives of the early nineteenth century describe mute, nameless Indians, eyes wide with disbelief at tales celebrating European technology in capital cities. But in Humboldt's story the Europeans stay up all night listening to the Indian describe *his* home culture. "I feel a pleasure in recording in this itinerary," Humboldt wrote, "the name of Carlos del Pino," a man "sagacious in his observations," who wound up serving as a guide and an interpreter for Humboldt's expedition over the following sixteen months. That very night, Carlos taught them the local words for several common plants; for the verdant coast and the cool plateau and the desertlike plains; for jaguars and boa constrictors and electric eels; for the different saltwater fish caught by the Guayquerias for food. Then, the next morning, having guided the ship safely to port, Carlos pointed out the vivid plumage of the egrets, alcatras, and flamingos sunning themselves amid the cacti and coconut trees. Forty-one days after setting out from Spain, Humboldt stared up in awe at "the mountains of New Andalusia," familiar of form but bathed in the foreign glow of a tropical sunrise—and "half-veiled by mists."[44]

Humboldt loved those luminous mists, and the tangles of the tropical rain forest, and the potent Andean volcanoes. For the rest of his life he told of "the feelings of admiration and delight which penetrated us when we first touched this animated South American soil." It is not the dark, mysterious wilderness that gets penetrated by enlightened, manly heroes, in this powerful reversal of classic exploration rhetoric; rather, the scientists themselves get penetrated, by classically Romantic emotions.[45] And Humboldt continued to expose himself to all the baffling landscapes and unfamiliar peoples of Spanish America, over the next five years, on the Venezuelan steppes, in the mines of Mexico, on the Peruvian coastline. He did occasionally adopt the mantle of the scientist and do his part to dispel mists and myths—by showing, for instance, how the legend of El Dorado had been propagated by highly imaginative explorers (Sir Walter Raleigh being the most infamous) who wished to exaggerate the significance of their discoveries, and by clever Indians who wished to get rid of said explorers: yes, yes, the Gilded King and his City of Gold are just up this river, beyond the waterfall, on the other side of the mountain. More often, though, Humboldt just gave himself over to "the extent and power of organic life." He wallowed in "the mysteries which natural philosophy cannot solve," and never more so than during his three-month-long expedition up the Orinoco basin, the story of which takes up the entire second volume of his Narrative.[46]

On March 30, 1800, at four in the afternoon, after several months of botanizing on the coasts and meeting with native groups on the arid llanos, Humboldt and Bonpland finally squeezed into their lancha, a small riverboat that they would recall as luxurious once they were forced to switch to a tiny canoe at the Great Cataracts of the upper Orinoco. Their Indian pilot and four rowers had even nailed up some ox hides to make a small cabin, under which the Europeans stowed their instruments and their initial provisions—eggs, plantains, cassavas, fresh cacao, tamarinds, oranges, and some sherry. They would also catch fish, shoot birds, and collect turtles' eggs, which Humboldt called "an aliment more nutritious than agreeable to the taste." Yet he relished those eggs, because he was starting to learn how to enjoy novel sensations. Traveling on six different rivers, upstream and down, searching for the alleged link between the Orinoco and Amazon systems, Humboldt was finally immersed in a universe of wonders equal to his imagination. "You find yourself in a new world, in the midst of untamed and savage nature," he wrote, referring to the first day of this journey. He would stare, entranced, at jaguars, tapirs, and peccaries, which peered back at him from the thick, wild brush alongside the river. Crocodiles lay in wait in occasional muddy clearings;

manatees and freshwater dolphins swam alongside the boat; often there were piranhas and vampire bats.[47]

On otherwise "calm and serene" nights, the "deep silence of those solitudes" was routinely broken by "the cries of so many wild beasts howling at once"—moaning apes, squawking parakeets, screeching sloths—that Humboldt felt fully confirmed in his assertion that every aspect of nature was linked together: after all, the entire jungle seemed to be conspiring to keep him awake. Yet he did not want to overemphasize the chain of connection here, because part of what he loved about the torrid zone was its ability to produce utterly unique circumstances, deliciously disorienting in their inexplicability. "In examining the physical properties of animal and vegetable products, science displays them as closely linked together; but it strips them of what is marvelous, and perhaps, therefore, of a part of their charms." Let the beasts howl, for whatever mysterious reason, while Humboldt turns over in his hammock to face the fire, opens his journal, and writes of trees that produce a sap as nutritious and mildly flavored as milk; of a man whose nipples flowed with enough *actual* milk to feed his motherless child for five months; of the Ottomac Indians, who "swallow a prodigious quantity of earth" to stave off hunger when no fish are available, without losing any weight or falling sick; of a small island in the middle of the Orinoco whose sandy beaches are completely covered with tortoises, all laying their eggs at the same time. "That which speaks to the soul, which causes such profound and varied emotions, escapes our measurements as it does the forms of language."[48]

There were times on this river trip, though, when Humboldt was ready to renounce his sense of wonder and his respect for the exotic. "Untamed and savage nature" is not always flowing with milk. At various moments, Humboldt and Bonpland were caught in flash floods; abandoned by their guides; stalked by jaguars; stung by electric eels; prostrated by fever, fatigue, diarrhea, and dehydration; bereft of all food except dried, unsweetened cacao; and simply, hopelessly, lost. Humboldt found the Orinoco beautiful and inspiring, but also difficult and terrifying. The inevitable truth, he decided, was that "man is ever wrestling with nature"—and with his own kind as well. It may have been easy to accept the foreignness of the statuesque Guayqueria Indians who helped Humboldt land at Cumaná, but the cannibals he met along the Orinoco truly seemed like a different order of humanity. How could he trust such people? He did refer to some of them as surprisingly "mild and humane," but they still made him fear for his life.[49]

Indeed, Humboldt recorded life-threatening incidents on almost every day he spent on the Orinoco and its tributaries—as when he bathed in waters that he later found to be infested with crocodiles. Once he took a lesson from a tribe's poison master in the chemical preparation of curare, and then bottled up a few specimens to take back home, learning that the final product could be eaten—in fact, it made a good stomachache remedy—but that if it entered the bloodstream directly, it was immediately fatal. A few mornings later, Humboldt stopped short as he rolled out of his hammock and groped in the dark for a sock. It felt sticky: some curare had spilled. Humboldt's feet, at the time, were infested by burrowing chiggers, and bleeding profusely.[50]

Fairly often, Humboldt felt that the insects might kill him. "It is impossible," he explained, "not to be constantly disturbed by the mosquitos, zancudos, jejens, and tempraneros, that cover the face and hands, pierce the clothes with their long needle-formed suckers, and getting into the mouth and nostrils, occasion coughing and sneezing whenever any attempt is made to speak in the open air."[51] Trying to avoid this plague of flying creatures, Humboldt buried himself in dirt, slept in trees, rubbed his body with crocodile fat and turtle-egg oil, slapped himself silly, and even enclosed himself in a smoking clay oven. From the Romantic Sublime we have sunk to the ridiculous, nature at its most overwhelmingly exasperating. Humboldt was seemingly just as aware as Darwin of "the struggle for existence."

Even while acknowledging the need to wrestle with nature, though, Humboldt never ceased wrestling with the problem of simultaneously recognizing humanity's connections to it. He had what John Keats called "Negative Capability," the presence of mind to accept uncertainty, to express two contradictory opinions at the same time in the search for a more complex truth.[52] Looking for patterns, Humboldt asserted that "Nature in every zone follows immutable laws." Then again, "it cannot be too often repeated that nature, in every zone, whether wild or cultivated, smiling or majestic, has an individual character."[53] In the equinoctial region, the natural world shows its darkest side, scares us, disorients us—"yet amid these strange sounds, these wild forms of plants, and these prodigies of a new world, nature everywhere speaks to man in a voice familiar to him." Follow Humboldt up the river, watch as that jaguar pokes its head out of the forest to drink: you'll stare into its eyes, and suddenly you'll sense both its viciousness and its innocence. Better than any modern writer I've come across, Humboldt captures the miraculousness and bafflement that have characterized my own wildlife sightings and my immersions in unfamiliar natural worlds. He understands how experience can undermine any classic understandings of natural beauty or peacefulness, how a landscape can be "at once wild and tranquil, gloomy and attractive."[54] He grants nature its own separate reality, acknowledges that its workings are vastly complicated and utterly different from our own, yet reminds us that this is where we come from.

Humboldt's trip up the Orinoco was a success geographically: the Orinoco and Amazon basins were in fact connected, just as Humboldt had thought, and he had the satisfaction of both paddling and charting the river that linked them (the Cassiquiare).[55] But the real significance of the expedition for Humboldt, as well as for modern environmentalism, was that it had shown him nature's most off-putting and threatening secrets, and he had managed to pull through without needing to demonize those parts of the natural world.[56] He attempted to appreciate tropical nature on equal terms, across very real boundaries, and then acted as a go-between, a translator, a link. His goal was to foster what we think of today as cross-cultural understanding. He offered his readers an opening in the forest wall: he gave them a chance to see.

If you're an experienced traveler, you know how to turn bad trips into good stories. You're also accustomed to the gradual process by which traumatic incidents become fond memories, for the simple reason that you survived them. In Humboldt's case, despite the pains of his Orinoco expedition—"the inconveniences endured at sea in small vessels are trivial in comparison with those that are suffered under a burning sky, surrounded by swarms of mosquitos, and lying stretched in a canoe"[57]—the journey would be recast in his mind as something of a pleasure cruise, during which the explorer felt truly at home for the first time in his life. From 1804 on, he would often keep the temperature in his private rooms at eighty degrees Fahrenheit.[58] But perhaps it should not be surprising that Humboldt identified so well with nature's dark secrets, put such a positive spin on all that was distant, foreign, different, abnormal. He had his own secrets; he knew what it was like to be exiled, on the margins of society.

Always on the move, always searching for answers, Humboldt seemed to adopt an outsider status, even while being celebrated throughout Europe. Travel opened up new worlds to him, and gave him the opportunity to broaden the horizons of his readers and the people who listened to his lectures. Yet his transience, his restless discomfort, was also a source of tension. He often complained of his "moral and mental isolation" and his "disjointed life."[59]

Humboldt tried to sublimate his alienation by throwing himself into his work. "No strong passion will ever sway me with an overwhelming power," he wrote in an early letter. "Serious occupation and the *calm induced by an*

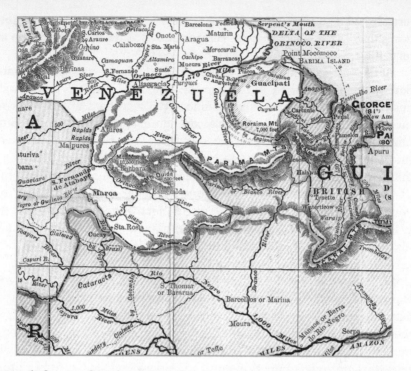

8 Detail of a map of South America, 1897, showing the region of Humboldt's Orinoco journey. Note the Cassiquiare River, near Maroa, connecting the Orinoco to the Rio Negro and thus to the Amazon.

absorbing study of nature will preserve me from the temptations of life. . . . You can judge for yourself whether I am strong enough to walk alone on this world's slippery path." This strategy often worked. From Cumaná, Humboldt wrote to Wilhelm that he was relieved to be alone in nature and free of social obligations: "The distractions in civilized countries that result from the commerce of men never bother me here; nature, meanwhile, offers me new and interesting things without cease." At the same time, though, he was lonely, missing in particular the stimulation of interaction with intellectuals who were just as curious about the world as he was. "The only thing one might regret in this solitude," he explained, "is . . . being deprived of the advantages which result from the exchange of ideas."[60] Humboldt had a broad, gregarious, compassionate personality, and he lived for the experience of connectedness, yet he could never shake his sense of himself as somehow not fitting in. Something drove him constantly to escape, to create distance from his friends and family and associates and culture. The idea of holding a stable job in an

established institution, of becoming enmeshed in a network of formal relationships, was anathema to him. Perhaps even more frightening was the idea
of getting married.

At the center of Humboldt's feeling of marginalization was his fairly well-
known preference for intense bonds with other men. (This was likely the subtext of Lord Byron's jibe about Humboldt's lack of interest in blue-eyed
maidens.) Certainly, part of the pleasure Humboldt took in his expedition
through the Americas was his close collaboration with his fellow traveler
Aimé Bonpland: during five years of constant companionship, often under
difficult circumstances, their relationship never faltered, and Humboldt filled
his letters home with praise for the work and devotion of the French botanist.
Meanwhile, in Ecuador, Humboldt met a young man who seemed to captivate
more than his mind. Though Don Carlos Montúfar had little to contribute to
the scientific expedition, Humboldt insisted that he come along and ended up
paying his expenses all the way back to Paris. Even before setting out for the
colonies, though, Humboldt had established a pattern of close association
with male friends that sometimes spilled over into romantic passion. Of the
letters written during his twenties that still survive, not a few betray the
pathos of his impossible love affairs. "Other people may have no understanding of this," he wrote to Reinhard Haeften in 1794. "I know that I only live
through you, my good precious Reinhard, and that I can only be happy in your
presence." Not even Haeften's marriage the next year could stifle Humboldt's
feelings: "Even if you must refuse me, treat me coldly with disdain, I should
still want to be with you." To another object of his young desire, he confessed
that "when I measure the longing with which I wait for news from you, I am
certain that no friend could love another more than I love you."[61] Such expressions of emotional intimacy and dependence were not entirely uncommon
among young men of a certain class in Humboldt's day, and it is impossible
to know exactly how intimate Humboldt ever got with his various partners.
What is undeniable, though, is that he never settled into the type of domestic
sexual relationship that was expected of him, and his awareness that his private life was anomalous left him, at times, pained and alienated.[62]

Once Humboldt had established his scientific reputation in Paris, he was
able to lead a relatively stable social and romantic life. For several years he
lived, worked, and traveled through Europe with the physicist Joseph-Louis
Gay-Lussac, cultivating an intense intellectual bond with him seemingly modeled on his earlier collaboration with Bonpland. His long relationship with
the astronomer François Arago was just as close, though not as overt, since
Arago was married. And he maintained relatively passionate connections
with numerous young artists and writers whom he met at François Gérard's

studios and Chateaubriand's salons.[63] Yet, despite his fame and the ease with which he circulated among Europe's best and brightest, he still had to live and love underground to a certain extent, and that took its toll. Arago often had to reassure him: "Can it be that you doubt my inalienable affection?"[64] Humboldt spent a good part of his boundless energy simply covering up the truth about his personal affairs, worrying that Napoleon was not the only one who questioned his masculinity. In the end, he was clearly most at ease when far from home, in a virtually incomprehensible world, free of all the expectations of European culture.

———

"We plunge into a vast solitude," Humboldt wrote, on embarking for the interior of Spanish America, and when he met people living in those regions of exile, he could generally empathize with them. "They are considered as savages because they choose to remain independent"—and Humboldt was tempted by the same choice.[65] It was mainstream Europeans who had it wrong, who were in the process of losing their ability to connect with the cosmos. European spirituality had never meant anything to Humboldt; he inclined more toward the natural religion of the Indians: " 'Your God,' said they to me, 'keeps himself shut up in a house, as if he were old and infirm; ours is in the forests, in the fields, and on the mountains of Sipapu, whence the rains come.' "[66] In urban capitals, the masses of people living indoors were now cut off from crucial lessons in humility. Humboldt preferred to be "in those distant regions, where men yet feel the full value of the gifts of nature." Moreover, those appreciative men and women had the kind of local scientific knowledge that cosmopolitan transients like himself could only taste. They could cure fevers with roots and leaves; they could distinguish the water—by flavor alone—of several different rivers; and they could navigate rapids, thick jungles, dusty plains seemingly devoid of any landmarks, with startling accuracy: "The Indians, I repeat, are excellent geographers."[67]

Humboldt's respect for the natives he encountered did not stem from a true cultural relativism. He was disgusted by cannibalism and other practices he considered vicious or even inhuman, such as infanticide based on "false notions of propriety and family honour" that forbade the raising of twins. Indeed, he occasionally mocked Rousseau's idealization of "man in the state of nature," because he was such an environmental determinist that he inevitably saw the dark side of the jungle reflected in its inhabitants. To his credit, though, he did remark that the natives' "acts of cruelty" were "less frequent than they are believed to be" in Europe, and when he complained of their "indolence" he explicitly denied that this was a racial characteristic, assuming rather that it was induced by the sweltering, stagnant atmosphere of the tropics. And he never mentioned what he considered to be the Indians' blameworthy traits without also applauding their admirable ones, without noting "the contrast between the virtue of a savage and the barbarism of civilized man!"[68]

Humboldt certainly took for granted the superiority of certain aspects of European "civilization": he espoused a kind of Jeffersonian agrarianism, believing in the inherently uplifting power of working the soil, and he consistently celebrated Western arts and sciences as humanity's greatest achievements. What he desperately wanted to communicate to his European readers, however, was that every single human group, no matter how superficially bizarre or savage, was equally capable of contributing to those arts and sciences, and that the whole point of getting to know far-off peoples was to understand their special contributions, which could not have been developed in any other place by any other group. Humboldt wrote in support of "the diversity of nations; that unfailing source of life and motion in the intellectual world." The corollary to this conviction, of course, was a belief in "the ties of consanguinity, by which [each man] is linked to beings to whose language and manners he is a stranger." Again, there was unity in diversity. It was his readers' duty, by absorbing the ethnological observations in his books, to appreciate and then see beyond those strange languages and manners, finally "to discover some of those family features by which the ancient unity of our species is manifested."[69]

Humboldt encountered native groups in the New World who impressed him as "free" and "self-governing," not to mention "wise" and "industrious," and he was eager to let his readers know that many Indians practiced a sophisticated form of diversified farming, encompassing the plantain tree, cassava, and cotton. "It is a common error in Europe, to look on all natives not reduced to a state of subjection, as wanderers and hunters. Agriculture was practiced on the American continent long before the arrival of Europeans." Indeed, he argued, most of the native peoples whom "we designate under the name of savages, are probably the descendants of nations highly advanced in cultivation." Many of Humboldt's books—and letters—have long passages testifying to the grandeur and advancement of pre-Columbian societies, describing Mexican hieroglyphics, calendars, and pyramids, as well as Incan gardens and fortresses. Humboldt also took great pains to plead for the preservation of all native languages, often citing the resonance and appropriateness of both their common and their proper nouns. Predictably, he found it especially pleasing when groups "named places after the quality of the soil, the shape of the rocks, the caverns that gave them shelter, and the nature of the tree that overshadowed the springs."[70] He also sought explicitly to counter

European assumptions about the underdevelopment of languages in the New World. "I have been quite occupied with the study of American languages," he wrote to his brother, a noted linguist, "and have seen the utter falsity of what La Condamine said regarding their poverty. The language of the Caribs [for example] is at the same time rich, beautiful, energetic, and polite. It is not at all lacking in expressions for abstract ideas; one can speak of posterity, eternity, existence, etc." Later, more publicly, he broadened this line of argument: "What some learned writers have asserted from abstract theories, respecting the pretended poverty of all the American languages, and the extreme imperfection of their numerical system, is as doubtful as the assertions which have been made respecting the weakness and stupidity of the human race throughout the New Continent." And for good measure, Humboldt asserted in his *Personal Narrative* that "there is scarcely any work of modern literature that might not be translated into the Peruvian."[71]

Embedded in this defense of native Americans, of course, was an attack against the savagery that characterized their treatment at the hands of conquistadores and colonists—and against the general brutality and hypocrisy of Europeans. The Spaniards, Humboldt explained, were undoubtedly "one of the most polished nations of Europe," yet their policy toward the Indians added up to "compulsory misery": "If the Spaniards visited [the New Continent's] shores, it was only to procure, either by violence or exchange, slaves, pearls, grains of gold, and dye-woods; and . . . the trade in the copper-coloured Indians was accompanied by the same acts of inhumanity as that which characterized the traffic in African Negroes."[72] While his readers might observe, then, that some Indian groups had "lost that vigour of character and that natural vivacity which in every state of society are the noble fruits of independence," the explanation for this regression should be clear. They had been "humiliated but not degraded," Humboldt wrote, "by long oppression."[73] Again, these careful explanations are notable for their lack of racism: Humboldt did not naturalize some kind of "vivacity" as the essential characteristic of "savages." Rather, he tied both the Indians' ability to thrive and their subsequent difficulties to social conditions. For Humboldt, it always came down to power, to *relations*.

———

The theme of Europe's oppression of other nations is a constant in Humboldt's travel writings. But his sojourns in Cuba and Mexico, in the years following his trip up the Orinoco, were particularly significant in the formation of his views on colonialism. These two countries were much more "developed" than the Orinoco region, and, in Humboldt's experience, wherever there was more development in the colonies, there was likely to be more human suffering—and more devastation of nature. The two types of exploitation, Humboldt found, were always joined. Volume 3 of his *Personal Narrative*, which covers Cuba, is a perfect companion piece to his *Political Essay on the Kingdom of New Spain*—in fact, a large chunk of it was republished later as the *Political Essay on the Island of Cuba*—because it takes a similar interdisciplinary approach in analyzing the horrors of colonialism's infrastructure, which focused on the extraction of resources and depended directly on slavery and plantation agriculture. "A colony," Humboldt argued, "has for ages been considered useful to the parent state only in so far as it supplied a great number of raw materials."[74] Here was the initial insight leading to his social ecology.

Humboldt's own complicity in colonial power structures cannot be denied. The first European to bring guano to the Old World and demonstrate its effectiveness as a fertilizer, Humboldt almost single-handedly spurred the infamous "guano boom" of the early nineteenth century, which enriched countless European investors but had a devastating impact on the economies of Chile and Peru. Similarly, his careful survey of mining regions sparked a rash of investments, especially by the British, in Mexican gold and silver mines.[75]

Humboldt was scornful of the British investors, though, and several long passages in his *Political Essay* on Mexico describe the mining industry as crippling the entire country. His actual recommendation was to cut back on the mining of precious metals and to invest more in agricultural crops to feed the Mexican population, or at least in the mining of iron and lead, which might benefit all of Mexican society rather than a few members of the elite classes. "If the labor of man in America has been almost exclusively directed to the extraction of gold and silver," Humboldt lamented, in 1811, "it is because the members of a society act from very different considerations from those which ought to influence the whole society."[76]

Humboldt's withering attack on colonialism at times reads as though it were written in the twenty-first century by a left-leaning expert on international environment and development issues. He anticipated current critiques of unjust land distribution, cash cropping, the tragedy of the commons, violence against isolated indigenous groups who refuse to submit to governmental settlement plans, and even oppressive work environments.[77] "All the vices of the feudal government," he explained, "have passed from the one hemisphere to the other. . . . The property of New Spain, like that of Old Spain, is in a great measure in the hands of a few powerful families who have gradually absorbed the smaller estates. In America as well as Europe, large commons are condemned to the pasturage of cattle and to perpetual sterility." Emphasizing economic factors that imperialistic organizations like the World Bank and the World Trade Organization still refuse to acknowledge, Humboldt questioned Mexico's reliance on foreign commerce. Although profits from mining and cash cropping usually allowed Mexico to purchase certain necessities from other countries, the vulnerable members of Mexican society tended to go hungry whenever trade faltered for diplomatic reasons or when the international market for Mexico's main commodities happened to crash. Again, Humboldt attacked the colonial elites for pursuing personal profits rather than attending to the welfare of the country: "Whenever the soil can produce both indigo and maize, the former prevails over the latter, although the general interest requires that a preference be given to those vegetables that supply nourishment to man over those which are merely objects of exchange with strangers." His definition of national wealth was based on self-sufficiency. "The only capital of which the value increases with time consists in the produce of agriculture," he insisted; "nominal wealth becomes illusory whenever a nation does not possess those raw materials which serve for the subsistence of man or as employment for his industry."[78] All colonial difficulties, in short, could be ascribed to the "restless and suspicious policy of the nations of Europe. . . . The scarcity of necessary articles of subsistence" was a problem limited to the poor, and it arose only "where the imprudent activity of Europeans had inverted the order of nature."[79]

Europe's worst inversion of the natural order, though, was slavery. Even in 1858, at the age of eighty-eight, Humboldt remembered the brutalities he had witnessed in Cuba and wrote about how he abhorred "the sufferings of our colored fellow-men, who, according to my political views, are entitled to the enjoyment of the same freedom with ourselves." His science, he believed, demonstrated that all races were equally human. Harnessing the evidence he had gathered through ethnography, linguistics, geography, and natural history, Humboldt would close the first volume of *Cosmos* with a bit of moral philosophy. "Whilst we maintain the unity of the human species," he asserted, "we at the same time repel the depressing assumption of superior and inferior races of men. There are nations more susceptible of cultivation, more highly civilized, more ennobled by mental cultivation than others—*but none in themselves nobler than others*. All are in like degree designed for freedom."[80] In the *Personal Narrative*, meanwhile, Humboldt focused on painting a vivid picture of slavery, forcing his readers to see and understand exactly what their society rested on. He gave us the slave market at Cumaná, where the sellers rubbed coconut oil on their merchandise and the purchasers examined the slaves' teeth, "forcing open their mouths as we do those of horses." Later, Humboldt focused on one particular slave, name unknown, who had been "torn from his family in his native country, and thrown into the hold of a slave-ship," where he was whipped and forced to dance and sing. And finally Humboldt showed us the *"sugar or slave colonies"* of Cuba, where "the most fruitful soil which nature can furnish for the nourishment of man" was wasted on a luxury crop; where men, women, and children toiled through each day, planting in the rainy season, harvesting in the scorching sun, always stooped low to the earth, guarded by barking dogs and armed men; where the least lucky were forced indoors to tend the grinders, boilers, and distillers; where about one in five died from the heat and filth and labor each year.[81]

Approaching the end of volume 3, Humboldt explained the moral responsibility he felt to expose as frauds those people who tried to "veil barbarous institutions by ingenious turns of language"—the Jeffersonian planters, for instance, who claimed that they provided "patriarchal protection," that slavery could be good for slaves. To Humboldt, a scientific expedition was no mere tour of the colonies, observation no mere positivist exercise. When he took it upon himself to describe horrific social conditions, he did so with the goal of abolishing them. "It is for the traveller who has been an eyewitness of the suffering and the degradation of human nature," he explained, "to make the complaints of the unfortunate reach the ear of those by whom they can be relieved."[82] Humboldt wanted to change his readers, wanted them to launch their own voyage of discovery, wanted them to see themselves in a new way—to see how they were linked to the soil of Cuba, and to the slaves who worked it, and to the tribes who used to call it their home, before Europeans had forever extirpated them from the fabric of the cosmos.

Perhaps the best you can hope for at the end of a journey—or a book—is that you are inspired to move on. After years of staring at rocks, this is what Humboldt found: "notwithstanding the care with which we interrogate nature, and the number of partial observations which present themselves at every step, we return from the summit of a burning volcano less satisfied than when we were preparing to visit it. It is after we have studied them on the spot, that the volcanic phenomena appear still more isolated, more variable, more obscure, than we imagine when consulting the narratives of travellers."[83] Perhaps Humboldt worried that to publish volume 4 of the *Personal Narrative* would be to give his readers an artificial sense of closure and certainty. His life's defining journey had in fact ended in 1804, but he didn't want the larger journey to be over. Letters to friends suggest that he may have worked up a long manuscript version of volume 4, but, if so, it has not survived, and the only reasonable explanation seems to be that Humboldt destroyed it (or never wrote it).[84] In his readers' minds, then, he managed to remain perpetually in midexpedition. Moreover, as his scientific ambitions continued to expand, perhaps he wished to remind himself that this project would never be neatly wrapped up, that the web was too large and intricate for anyone ever to understand. Or perhaps he abandoned the *Personal Narrative* for a more personal reason: perhaps he just needed a change. By the time volume 3 appeared, in 1825, Humboldt had spent almost his entire inheritance, not to mention two full decades, on publications relating to the American expedition. The next few years would see him close up shop in Paris, move back to Berlin, deliver the public lectures in which he first propounded his theory of the Cosmos, and launch an expedition to Siberia. He turned sixty in September of 1829, somewhere on the dusty, scorpion-infested Kazakh steppes.

CHAPTER 3

Cosmos

Unification Ecology

The imagination is . . . solemnly moved by the impression of boundlessness and immeasurability, which are presented to the mind by every sea voyage. All who possess an ordinary degree of mental activity, and delight to create to themselves an inner world of thought, must be penetrated with the sublime image of the infinite when gazing around them on the vast and boundless sea, when involuntarily the glance is attracted to the distant horizon, where air and water blend together, and the stars continually rise and set before the eyes of the mariner. This contemplation of the eternal play of the elements is clouded, like every human joy, by a touch of sadness and longing.

—Alexander von Humboldt, *Cosmos*, vol. 1 (1845)

Shall I not have intelligence with the earth? Am I not partly leaves and vegetable mould myself?

—Henry David Thoreau, *Walden* (1854)

For much of his life, Ralph Waldo Emerson kept a print of Mount Vesuvius hanging in the front entryway of his house. He had actually climbed the great volcano once and made a point of noting the sensation of warmth in his soles and the experience of peering down at the caldera, "the hollow of salt and sulphur smoking furiously beneath us."[1] On that early spring day in southern Italy, Vesuvius just puffed and rumbled, hinting at but ultimately muffling the destructive force of its internal heat. Emerson's print, though, showed a full eruption. Perhaps it was meant to inspire him to live with more fire and intensity, to embrace the present—to imagine that at any instant, like the villagers of Pompeii and Herculaneum, he too might be swallowed by ash and frozen in place.

Emerson started living for the moment when he started losing his faith.

73

He started losing his faith when his wife started to die, in the fall of 1830, when he was twenty-seven. Before Ellen, the church had been his world; then Ellen became his church; now his buttresses were fracturing. Deprived of his imagined future, he tried to savor every present moment with her. When she slept, he read, to distract himself. In December, Emerson began reading Humboldt's *Personal Narrative*.[2] Facing utter aloneness, he turned not to God but to the physical world. Through much of Western society, science was beginning to challenge religion, and, from now on, immersion in nature would anchor Emerson's philosophy.

Within two months, Ellen was dead. Emerson felt a void that could not be healed by any eventual salvation. He couldn't live in the future tense. Within two years, he had left his pulpit. A couple of years after that, he would start writing his first book: *Nature*.

During the months between Ellen's death and his decision to abandon the church, Emerson reached a watershed in his intellectual and spiritual explorations. In March 1832, on one of his daily visits to Ellen's tomb, he suddenly ripped the lid off her coffin and stared at her remains. Then, pondering ashes and dust, he urged his congregation, in consecutive sermons, to focus on the "virtues," even the "pleasures," that were *"near at hand."*[3] In May, the month he turned twenty-nine—Humboldt's age when he left Europe for South America—Emerson fell back on volcanic imagery to suggest the necessity of living for the present: "We walk on molten lava on which the claw of a fly or the fall of a hair makes its impression, which being received, the mass hardens to flint and retains every impression forevermore." Physical science, not Christianity, had provided him with a means of reconciling the immediate and the eternal, the external and the internal. It was time to withdraw from organized religion. He exclaimed: "Is it not better to intimate our astonishment as we pass through this world if it be only for a moment ere we are swallowed up in the yeast of the abyss? I will lift up my hands and say Kosmos."[4]

———

Humboldt gave the lectures that would eventually become *Cosmos* in Berlin, between November 1827 and April 1828. Subjects to be covered ranged from volcanoes and marine algae to sunspots and shooting stars, from the origins of language to the development of the telescope. In unexpected numbers, Berliners came out to see if the old man, newly returned to his hometown, could actually weave all these topics together. Some wondered whether he could even compose sentences in his native language anymore. Whatever the spur to their curiosity, they overflowed the lecture hall at the University of Berlin—founded by Humboldt's brother Wilhelm in 1810 and today known as Humboldt University—to such an extent that Humboldt felt compelled to repeat his series of talks at the city's Music Academy. The diversity of his crowds provided inspiration: speaking "before a mixed audience (King and bricklayers)" allowed him to further one of his highest aims, the "popularization of science. With knowledge comes thought, and thought imbues people with earnestness and power." Humboldt had come home to a climate of political repression, in which censorship was not uncommon, so he sought to warm the air with the buzz of ideas. Form fit function: speaking to the assembled masses, he endorsed freedom of speech and assembly, developed an accessible, cosmopolitan science that could become "the common property of all classes of society."5 Guiding his listeners over oceans and down rivers, through jungles and up mountains, pausing to admire Indian civilizations and track the orbit of comets, Humboldt used his lectures to unite human beings with the earth and sky—and with each other. Ultimately, he hoped his listeners would rebel against the authoritarianism and Christian orthodoxy of the Prussian state and create a society in which every individual could freely experience the kind of "astonishment" Emerson found in nature.

Emerson drew the word "Kosmos" directly from Greek—"cosmos" would not become established in English until volume 1 of Humboldt's book was translated, in 1846[6]—to suggest a sense of overarching beauty, harmony, and orderliness. The ancient Greeks distinguished between the "ouranos," the innermost region of the universe, a place of constant change located within and just beyond the earth's atmosphere, and the "cosmos," which was composed of celestial bodies that seemed to float through space in regular, predictable patterns. From the perspective of human beings on earth, the stars and planets obeyed the pure laws of physics, while in our sublunary lives we had to cope with the messiness of biology and chemistry. Humboldt's innovation was, without denying this messiness, to recognize that there was also a certain order in the terrestrial realm. And this discovery, in turn, led to the possibility that human societies could draw the deepest kind of inspiration not just from the eternal heavens but also from the ground beneath their feet, the here and now. Humboldt's cosmology brought spirituality down to earth.[7]

In his lectures, Humboldt made sure to acknowledge that his location of the spiritual in nature was not entirely new, that in fact many cultures had already developed a "sense of the all-powerful unity of natural forces, and of the existence of an invisible, spiritual essence manifested in these forces, whether in unfolding the flower and maturing the fruit of the nutrient tree, in upheaving the soil of the forest, or in rending the clouds with the might of the storm."[8] Yet he never referred to this "essence" as God—a conscientiously radical omission, for which he was duly attacked. In the end, his science seemed to be offered as a thinly veiled means of supplanting organized religion.9

What Humboldt wanted his listeners and readers to worship was not whatever force might have been responsible for the cosmos, but the cosmos itself, the beautiful whole that could not exist without each of its parts, the overall community of which human beings were members. The most-repeated words and phrases in the introduction to *Cosmos* all serve to develop the thematic obsessions Humboldt had laid out in the *Personal Narrative:* this is a science of "mutual dependence and connection," "the mysterious relations existing among all types of organization," "the connecting links of the forces which pervade the universe."10 It was an awareness of this mutuality, this interdependence, that Humboldt counted on to spur a kind of religious awe in his audience. He attempted to "paint" the physical world in such a way as to provide "that enjoyment which a sensitive mind receives from the immediate contemplation of nature . . . heightened by an insight into the connection of the occult forces." These forces would always *remain* occult, speaking to "the soul" with a kind of "mysterious inspiration" that calls to mind Humboldt's favorite bird of the Canary Islands, the unseen capirote. What mattered was simply a vague understanding that this bird would not be singing its haunting, elusive melody if not for its ability to eat a particular plant, which could not have thrived except in this volcanic soil, at this particular latitude and altitude, where the fiery earth rose out of the sea to meet the sky. Once you are engaged in "the earnest endeavor to comprehend the phenomena of physical objects in their general connection . . . , moved and animated by internal forces," then "communion with nature, mere contact with the free air, exercises a soothing yet strengthening influence on the wearied spirit, calms the storm of passion, and softens the heart when shaken by sorrow to its inmost depths."11 There is spiritual comfort in mutual relationships.

———

For those with a less-developed sense of Romanticism, meanwhile, Humboldt's cosmology also had a pragmatic side. After all, human societies were dependent on the world's web of interconnections for their material well-being. Though some critics have dismissed Humboldt's emphasis on interrelatedness as unscientific and misleading, a kind of pantheistic pipe dream—"occult forces indeed!" exclaimed one exasperated scholar—it is nevertheless at the center of most environmental debates in the twenty-first century.12 Our understanding of nature's "connecting links" has led to the regulation of pesticides (which always seep into the groundwater and travel through the food web); the protection of obscure species (without which the "higher" species could not exist); and campaigns against deforestation (which exacerbates climate change). At least one scholar has contended that Humboldt's science "corresponds to none of our modern scientific disciplines or specialties," and certainly Humboldt took a much broader, more general approach to the world than most of today's scientists do. Indeed, in Humboldt's cosmology lies a trenchant and valuable critique of scientific overspecialization. At the same time, though, many geographers still cite him as one of their forebears—David Harvey, author of *Justice, Nature, and the Geography of Difference*, recently wrote an article called "The Humboldt Connection"—and the final formulation of his philosophy clearly has a modern descendant in current ecological studies. When I taught basic ecology courses at a community college in the late 1990s, I used a textbook called *Environmental Science: A Study of Interrelationships*.[13]

The ecological thinking Humboldt expressed so overtly in *Cosmos* had long before led him to a nascent conservation ethic. Back in 1801, walking through Venezuela, Humboldt came upon the renowned "lake of Valencia, called Tacarigua by the Indians," high in the mountains and at the center of "a basin closed on all sides." Lake Tacarigua was the beneficiary "of a little system of rivers, none of which have any communication with the ocean." With watersheds all around, this mini-ecosystem seemed made to "preserve verdure and promote fertility." Yet the lake's water level had been sinking for years, and the entire basin was threatened with what Humboldt called simply "desiccation": "We find vast tracts of land which were formerly inundated, now dry." Local inhabitants and previous visitors assumed that the valley's water was secretly seeping down to the sea through some underground outlet. But Humboldt, leaning on his model of interdependence, immediately blamed "the destruction of forests, the clearing of plains, and the cultivation of indigo."[14]

Diversion of water from nearby streams in order to irrigate indigo crops was one obvious proximate cause; the key issue, though, seemed to be deforestation. Humboldt explained that the surrounding soils, once deprived of the trees' root systems, had a greatly diminished capacity for water retention, so they could no longer recharge the springs that fed the lake. In addition, he noted, flooding and soil erosion had increased dramatically: "As the sward and moss disappear with the brushwood from the sides of the mountains, the waters falling in rain are no longer impeded in their course . . . ; they furrow, during heavy showers, the sides of the hills, bearing down the loosened soil, and forming sudden and destructive inundations. Hence it results, that the clearing of the forests, the want of permanent springs, and the existence of torrents, are three phenomena closely connected together." These interconnections spurred Humboldt to recognize nature's limits and insist on caution in the use of natural resources, for the sake of posterity. "By felling the trees which cover the tops and sides of mountains," he asserted, "men in every climate prepare at once two calamities for future generations; want of fuel and scarcity of water."[15]

Part of Humboldt's concern over the use of resources came directly from his reading of Thomas Malthus's 1798 *Essay on Population*, a forecast of doom that Humboldt called "one of the most profound works of political economy ever written." Echoing Malthus's fear that population growth would eventually outstrip food supply, Humboldt noted that not all soil and climates are appropriate for growing crops, that "natural riches . . . are but sparingly scattered over the earth." Humboldt's analysis of environmental scarcity, though, was more sophisticated than Malthus's, because he emphasized over-consumption and mismanagement of resources rather than inevitable demographic pressures. People went hungry in the early nineteenth century—as in the early twenty-first century—not because the earth could not provide enough food, but generally because social inequalities prevented certain people from growing or purchasing it. Either the Latin American colonists had deliberately deprived the poorer classes, or, alternatively, they had initiated careless practices because they simply didn't understand the local ecology, the "occult forces" that linked the forests and the rivers and the crops. "Man cannot act upon nature," Humboldt warned, in the introduction to *Cosmos*, "or appropriate her forces to his own use, without comprehending their full extent, and having an intimate acquaintance with the laws of the physical world."[16]

Use of natural resources, in other words, should be guided by the same kind of unifying ethic that gives meaning to the mere contemplation of nature. Just as our vision is enriched when we see the plants and the soil behind the bird, so must we consider our impact on entire ecosystems, no matter how seemingly insignificant a slice of them we mean to employ. Humboldt never went so far as twentieth-century ecologists like Aldo Leopold, who insisted, metaphorically, that "to keep every cog and wheel is the first precaution of intelligent tinkering."[17] But he did take care to remind readers of *Cosmos* that no plant or animal was "merely an isolated species," and his cosmopolitan inclusiveness helped spur the development of environmental ethics. Throughout his writings, he emphasized the importance of even the most frail and finespun filaments in the web of life, those "phenomena which naturalists have hitherto singularly neglected." "Our imagination," he explained, "is struck only by what is great; but the lover of natural philosophy should reflect equally on little things." After his experience on the Orinoco, Humboldt was not exactly imbued with sympathy for mosquitoes, yet he felt compelled to remind prejudiced Europeans that even "these noxious insects . . . , in spite of their minute size, act an important point in the economy of nature."[18] "Economy" comes from the same Greek root as "ecology": "oikos," meaning household. Again, Humboldt forced his readers to remember that we must all live under the same roof.

His overcoat closed against the Siberian night, Humboldt looked out at the so-called rooftop of the world. He faced south, perched on a small, barren promontory, and took his customary astronomical observations. Directly ahead of him, below a clear, star-filled sky, was the seemingly endless steppe that terminated in the Tibetan Plateau. To his left, in the southeast, lay Mongolia and the Gobi Desert. On his right towered the Tien Shan range of western China, reaching elevations of about twenty-five thousand feet, as imposing as the volcanoes he'd seen in South America. This was the greatest altiplano in the world, a high desert microcosmos—bare, twisted, desiccated badlands surrounded by ice-capped granitic peaks, hot and dusty during the day but cooled by glacial winds at night.[19]

It was late August 1829, about a month before his sixtieth birthday, and never again would he be so far East. Humboldt had left Berlin in April, a year after completing his lecture series at the Music Academy, traveling by coach with his two scientific companions, Gustav Rose and Christian Gottfried Ehrenberg. The journey was far from easy: there were blizzards, river crossings in makeshift sailboat ferries, roads that were little more than herding trails, and Cossack escorts who tended to be imperious and drunk. Yet, as always, Humboldt managed to find an almost religious comfort in his observations of the landscape, and in this case he tried immediately to share that comfort with his brother, whose wife had died in March: "Nature can be so soothing to the tormented mind," he wrote to Wilhelm; "a blue sky, the glittering surface of lake water, the green foliage of trees may be your solace. In such company, it is even possible to forget the reality of one's personal existence. It lends wings to our feelings and thoughts." Of course, with characteristic two-mindedness, Humboldt also admitted that nature could *cause* torment. By the beginning of August, he was telling Wilhelm of his efforts to "flee" across Siberia's "monotonous grasslands," where the party "suffered greatly from heat, dust, and yellow mosquitoes. . . . The insects on the Orinoco can hardly be worse."[20]

The scientists knew they had come to the end of their road at about one in the afternoon on August 19, when they were met at a post along the Irtysh River by a young Chinese official, whom Gustav Rose described as "rather tall and gaunt, wearing a blue silken overcoat which reached to his ankles and one of those famous wool caps that come to a sharp point . . . , with peacock feathers stuck into it horizontally to proclaim his rank." Goats, sheep, and camels kicked up sand as they wandered out beyond the flapping yurts of some nomadic Kirghiz herders. The mandarin, whose name turned out to be Tschin-fu, greeted them warmly and made a sign that they should follow him into his personal tent, where they took seats on crates upholstered with thick rugs and were soon sipping tea, "offered in the Chinese fashion, without milk and sugar." Thanks to their Russian escorts and a Mongolian interpreter whom they had brought from Buchtarminsk, the three Prussians were able to exchange information with their host: it took a while for the questions and answers to go from German to Russian to Mongolian, which Tschin-fu could understand, and then from Chinese all the way back to German, but eventually Humboldt explained that he and his friends were geologists especially interested in mining, and the mandarin told them that he had just reached this border post, which he was to patrol for three years, after having traveled four months on horseback from Peking. To ensure amicable relations all around, the entourage then went to see the Mongolian officials on the far bank of the river and the Cossack fishing wardens on an island in midstream, and finally paid a quick visit to the local temple, a modest wooden structure, perfectly square, with two altars and a picture on the wall of golden Buddhist idols.[21]

Back on the Chinese side, Humboldt set up camp and opened his tent to the mandarin and his attendants for a visit and the presentation of a gift. Tschin-fu brought some tobacco and insisted on sharing his pipe. There was a moment of embarrassment as Humboldt tried to explain that he and his friends didn't smoke, but luckily one of the Russian escorts came forward and eagerly offered to light up as the visitors' representative. More awkwardness followed, though, as Tschin-fu repeatedly refused, with ritualistic politeness, to accept Humboldt's gift of fine blue cloth. Fortunately, once the Prussians finally prevailed upon him, the mandarin seemed genuinely pleased, and he immediately announced that, given his guests' obvious erudition, he had hit upon the perfect present to give them in return. With a quick command, he sent one of his men back to his own yurt to procure four beautifully bound volumes, which he then offered to Humboldt, who could not have been more delighted. They constituted a classic of Chinese literature, entitled *Sankuetschi*, an eighteenth-century historical novel about the three kingdoms into which China was divided after the fall of the Han Dynasty. Humboldt thanked his host and explained that he would bring the books back to his brother, whose linguistic studies had earned him a certain amount of expertise in Chinese—a fact which, in turn, caused Tschin-fu great joy. At Humboldt's suggestion, the mandarin wrote his name in the first volume as a greeting to Wilhelm, using a pencil that Humboldt offered as another small gift, since the mandarin had never seen such an implement before and found it quite charming. Tschin-fu was less enamored of the Madeira wine, sticks of sugar, and zwiebacks laid out by the Prussians, but he politely sampled each bit of refreshment before standing up and taking his leave.

Soon afterward, the scientists packed up and moved on. Despite the cordial meeting with Tschin-fu, they were worried about "arousing suspicion among the Chinese."²² Humboldt's genuine belief in cultural exchange was tempered by a pragmatic respect for boundaries. Having traveled among cannibals, Humboldt had learned to cultivate an acute sensitivity, an ability to determine his hosts' level of openness to the blunt interference that exploration always entails. Pushing too far was not an option. Perhaps, at this tiny outpost on the edge of so many different worlds, he also remembered that there had been Portuguese officials combing the jungle for him near the Rio Negro, with orders to arrest him if he stepped or floated beyond Spanish territory.²³ In any case, he was wary of penetrating areas where his investigative gaze might not be welcome, so he led the expedition back to the northwest, in search of at least a trickle of water where the travelers could pitch their tents and make a meal. No appropriate campsite appeared until about midnight. Sensibly, once the party had settled down, Rose and Ehrenberg went straight to sleep. Humboldt spent the rest of the night stargazing.²⁴

———

At twenty-nine, Humboldt had scampered up volcanoes and studied electric eels with an almost reckless verve, but at fifty-nine he explored cautiously. His energy remained boundless and his health fine; as one of his Russian companions explained, the old scientist could still outstride his younger colleagues: "When the carriage was unable to proceed he would set off on foot, climbing high mountains without any signs of fatigue, clambering over rough terrain like one accustomed to fieldwork."²⁵ But during the entire Siberian expedition, and for the final three decades of his life, Humboldt seemed to revert to the subdued, conflicted melancholy of his first thirty years. In moving to Berlin, he had sacrificed his most important friendships and professional relationships, and he had committed himself to a new position as privy chancellor to the king of Prussia. Independently wealthy in Paris and defended by his circle of scientific innovators, Humboldt had always retained his intellectual freedom and his radical bravado. Now, though, he was truly beholden. Pronouncements about social injustice would have to be toned down. He despised the Prussian regime and aristocratic pretension in general, declaring his honorary titles—"Excellency, Privy Chancellor, Baron"—to be "pestilential . . . German misdemeanors."[26] Yet he could not afford to lose his nicely remunerative position at court, for twenty-five years of publishing folio volumes devoted to his American expedition had left him utterly bankrupt. He *had* to be careful about what he did and said.

The tendency toward compromise that marked his mature years has led scholars to suggest that he was a "liberal conservative,"[27] but I think of him more as a pragmatic radical. In his younger days, after all, he had supported Latin American revolutions and coached the young Simón Bolívar, perhaps even training him to combine the rhetoric of earth science and social change: "A great volcano lies at our feet," Bolívar wrote, in seeming tribute to Humboldt (the two friends had in fact climbed Mount Vesuvius together). "Who shall restrain the oppressed classes? The yoke of slavery will break, each shade of complexion will seek mastery."[28] Humboldt always lamented the bloodshed involved in revolutionary change, and expressed his personal preference for the gradualist approach—a rising tide or shifting current rather than an eruption—but he also endorsed revolution over oppression. Even as an old man in Friedrich Wilhelm IV's court, he successfully defended the civil rights of Jews, and was almost single-handedly responsible for a piece of legislation automatically freeing any slaves who might ever cross into Prussia. In 1848, along with a minister, a priest, and a rabbi, he helped lead a massive funeral procession for 183 revolutionaries who had recently been killed by soldiers on the barricaded streets of Berlin, during one of the republican uprisings that were sweeping Europe that year. Certainly, most people at court considered Humboldt "a Jacobin, who carried the tri-colored standard in his breeches pocket"; he often referred back to the French Revolution—and his experience of it, in the company of Georg Forster—with nostalgia. And to defend himself against radical critics who in 1849 insisted that he ought to resign his position at court, he published an article in a leftist journal explaining his attempts at reform from within and expressing his hope that his successes would ultimately prove "that in the center of Europe is the France of 1789."[29]

Still, Humboldt's later years were deeply marked by a sense of constraint. In the 1830s and '40s he often went back to Paris and stayed for months at a time, but he could never fully escape his duties to the Prussian king. And the writing he did about his Siberian expedition is perhaps the most impersonal, and least impassioned, of his career. The very terms of the trip had made him wary from the beginning, especially since the entire enterprise was bankrolled by Czar Nicholas I, who primarily wanted Humboldt to provide information about the mining potential of the Ural Mountains. Moreover,

Humboldt's reputation for social criticism had spurred the czar and his minister of finance, Georg Cancrin, to request explicitly that the explorer stick to natural science and refrain from making any controversial statements about Russian feudalism. Enticed by the opportunity to escape his chancellor's duties for a few months and see Asia for the first time, Humboldt reluctantly agreed "to report more on products and institutions than on people."³⁰ In the end, he ceded the responsibility of writing up a narrative of the expedition to his friend Gustav Rose, and his own two books about Russia, *Fragments of Asiatic Geology and Climatology* (1831) and *Central Asia* (1843), are bone-dry scientific monographs bursting with temperature tables and isothermic charts and data on terrestrial magnetism.

Even in these intimidating tomes, though, capable of stymieing anyone's radicalism, Humboldt managed to discuss the negative effects of "changes man has produced on the surface of continents by felling forests and modifying the distribution of water sources"; to mention the situation of oppressed minority groups like the Uyghurs; and even to insist that the best scientific method was always to try to understand geographical regions "according to the tradition of indigenous peoples."³¹ Humboldt was reined in—he did not write as freely here as he had in his *Personal Narrative*, where he repeatedly denounced slavery and colonialism and even mentioned serfdom, assailing "the state of oppression to which some classes are still subjected in the north and east of Europe"³²—but he did not fundamentally change any of his political positions. And, while generally fulfilling his promise to the czar with regard to the official expedition reports, he nevertheless snuck some more direct commentary on Russian social injustices into a couple of other publications in the 1830s and '40s. His *Critical Examination of the History of the Geography of the New Continent* (1836–39), for example, though focused on "the crimes which in the conquest of America have forever sullied the history of humanity," also touched on the duty of Europeans to see atrocities within their own borders and to continue the "long struggles . . . for the enfranchisement of the serfs and the general amelioration of the condition of laborers." Indeed, Humboldt went so far as to note that "in Siberia, the exploitation of the celebrated Kolivan mines, in the southwest of the Altai Mountains, is still in part based on the system of the *mita*," the Spanish colonial practice of forcing natives to travel hundreds of miles to serve as mining labor, slaves to "the savage cupidity of whites."³³

It is true that, in order to fulfill his missions to the governments of Spain and Russia, Humboldt had to point out "exploitable resources."³⁴ He did ultimately help the czar find diamonds in the Siberian outback—in part because he simply accepted the inevitability of development, the human need for minerals, fuel, building materials. At times, he felt, people were even capable of using resources in ways that added to the embellishment of nature: the "rich cultivation" of pastoral landscapes could provide comfort and cheer under the right circumstances.[35] Once, in the tropics, when his Indian guides pointed out some particularly beautiful trees, with red and golden wood, Humboldt wondered aloud if such specimens would "one day be sought for by our turners and cabinet-makers." Yet his vision of resource use depended on free institutions, and on an ethic of prudence, care, respect, and egalitarianism. When colonists violated these principles, whether in their direct abuse of nature or in their abuse of less powerful people, Humboldt usually condemned them. His dream of European craftsmen turning tropical hardwoods into sophisticated furniture sounds exploitative in the twenty-first century, but it should be placed in the context of his generalized contention that "forests are destroyed . . . everywhere in America by the European planters, with imprudent precipitancy." Indeed, significant differences exist between selective cutting for furniture production and clear-cutting for coffee plantations or cattle ranching. Humboldt's approach was to seek ways of extracting nature's bounty that were socially just and that maintained the environment's inherent balance—goals which, today, are too often superseded among environmentalists by a simple desire to keep certain (beautiful) places from being used at all.[36] To the pragmatic Humboldt, who also had an unconditional love for the natural world as a whole, the point was to achieve *sustainable* use—"a more enlightened employment," as he explained in *Cosmos*, "of the products and forces of nature."[37]

Humboldt's compromises with dictatorial rulers, his practice of dedicating books to them, his direct employment in the Prussian court, all cast doubt on his ultimate integrity and radicalism.[38] At times, the scientific opportunities offered by such monarchs seem to have distracted Humboldt from his concomitant concerns with social justice. In responding to Czar Nicholas's hope that he might discover diamonds in the Ural Mountains, for instance, Humboldt excitedly explained that the geological stratification of the Ural region greatly resembled that of some famous Brazilian diamond deposits, so he could virtually guarantee a positive outcome. Of course, the very success of his powerful scientific technique ultimately increased the substantial gap between rich and poor in Russia and gave wealthy landowners further excuses to exploit vulnerable minorities as miners. On the whole, though, it seems quite likely that Humboldt was using his sponsoring dictators—Carlos IV, Nicholas I, Friedrich Wilhelm III (who ruled Prussia from 1797 to 1840), and Friedrich Wilhelm IV (1840–58)—to virtually the same extent that they were using him. In the case of his expedition to the New World, Humboldt truly was grateful to Carlos IV for giving him access to all the American colonies, but his dedication of the *Political Essay on the Kingdom of New Spain* to the Spanish monarch rings awfully false once you actually start reading the book, which is filled with scathing criticisms of Carlos's particular form of colonialism. Indeed, writing to Thomas Jefferson, Humboldt explained quite forthrightly that "my book was dedicated to King Carlos IV so as to pacify the attitude of the Madrid government toward certain individuals in Mexico who furnished me with more information than the court would have regarded proper."[39] And he thought of Nicholas I as yet another despicable tyrant; his dedication of *Central Asia* to the czar was simply "an unavoidable step."[40] Humboldt apparently made a pragmatic calculation about science and political power at a very young age: as a scientist, he might have to serve governments he despised, but in making himself indispensable to them, he could also attain the freedom to say almost anything. He would be a pawn, but he would never get knocked out of the game, and even pawns can go on the attack. His science could locate diamonds, but it could also light volcanic fires— or at least cause a shift in certain currents.

On his way out of Russia, Humboldt took advantage of the czar's satisfaction with his services by asking him for a favor of his own—a scientific favor, which he hoped would ultimately illuminate the relationship between earth and sky. First, though, he had to endure several seemingly endless speeches at the Moscow Academy of Science—mostly in Russian, of course—including one poetical oration of which all he could understand was that he was (inexplicably) being celebrated as a modern-day Prometheus, bringer of fire to humankind. His Siberian expedition, he explained to Wilhelm, had "so greatly impressed the chimerical notion of my usefulness to everyone that I feel quite overcome by the amount of drudgery imposed upon me by this position." When Humboldt finally got his turn at the podium, he presented data on terrestrial magnetism collected in the remote hinterlands and astronomical observations like those he'd made on the night after his expedition had reached the Chinese border. He also requested that Russian scientists establish permanent observation stations to measure geomagnetic and meteorological trends throughout the country. A few days later, he repeated this request to Czar Nicholas in St. Petersburg. And within six years, a chain of scientific outposts under Russian control stretched out not just along the route of Humboldt's expedition but all the way to Alaska, where government technicians regularly measured wind currents, snow and rainfall, barometric pressure, temperature levels, and the local magnetic declination.[41] Thus Humboldt laid the groundwork for the kind of internationally coordinated global monitoring that we all take for granted in the twenty-first century. Such cosmopolitan collaboration had never existed before, but, now that it did, Humboldt was finally ready to work on *Cosmos*.

Back in Berlin, in the early 1830s, he wrote a series of letters firming up his ties to scientists and explorers in Europe and the United States and pleading with British officials to send observers to as many of their overseas possessions as possible: Canada and Jamaica, Australia and New Zealand, Ceylon and St. Helena. By the end of the decade, Britain had responded enthusiastically, equipping their monitors with instructions written by Humboldt himself. Soon the seventy-year-old scientist was receiving data from all over the world. (An American reviewer of *Cosmos* noted that the "crusade" to measure magnetism globally, "undertaken at the solicitation of Humboldt by many of the governments of the Old and New World . . . , will alone furnish . . . 1,958,000 observations for every three years of its operation.")[42] And, little by little, Humboldt began locating and analyzing trends, testing them against cutting-edge theories in all the new scientific disciplines that were sprouting up around him. In general, his informants maintained a reliable correspondence with him: communications had improved drastically since the turn of the century, when Humboldt had gone about two years in the Americas without receiving any mail from Europe. Now, he claimed to write some two thousand letters a year and to receive three thousand, and though he often felt overwhelmed by the mails, he also noticed lapses on the part of his far-flung observing teams. He was appalled, for instance (as he noted in a *Cosmos* footnote), when the Tasmanian crew, making piety their first priority, ceased taking readings on a wild magnetic storm at precisely midnight on a Saturday, in honor of the Sabbath. Saints of the empire be damned; Humboldt would have preferred Tasmanian devils.[43]

Meanwhile, though, the Royal Society did him the favor of launching an Antarctic expedition in 1839 under Sir James Ross for the explicit purpose of studying terrestrial magnetism, and Ross succeeded in fixing (for the time being) the location of the ever-wandering magnetic pole. Eventually, Sir Edward Sabine, who oversaw the British observation efforts with a distinctly Humboldtian flair, combining astronomical interests with a study of geophysics, managed to trace magnetic variations noted on earth to changing patterns in sunspots. The dancing, multicolored sheets of light known as the auroras, making life that much more strikingly beautiful at the earth's poles, were caused by storms on the sun, millions of miles away. As Humboldt explained, in the first volume of *Cosmos*, we must never forget to look up, to appreciate the celestial: "the terrestrial must be treated only as a part, subject to the whole." In his final work, then, he would "not proceed from the subjective point of view of human interests" but rather "[begin] with the depths of space and the regions of remotest nebulae," and only then "gradually descend through the starry zone . . . to our own terrestrial spheroid."44

Volume 1 of *Cosmos* was finally published in 1845, in German, and the first official English edition appeared the following year—translated by Elizabeth Juliana Sabine, Sir Edward's wife. Humboldt had been preparing to write this book since his 1827 lectures in Berlin, and the first volume's long introduction derived directly from his lecture notes, explaining that "the study of a science that promises to lead us through the vast range of creation may be compared to a journey in a far-distant land." And a sacred, universal journey at that—to read this book was to go on a pilgrimage. Yet he did not want his readers to founder in the deserts and vast, starless expanses of abstract science. He would strive to be "scientifically true, without losing myself in the dry regions of knowledge."45 Back in 1834, when he originally chose the title "Cosmos," he wrote to his friend Varnhagen von Ense of the shift in his thinking that led him away from a simple focus on "Gaea" toward a project that would be more comprehensive and, perhaps, more appealing to the spiritually inclined, something that would encompass both "Heaven and Earth": "I have the extravagant idea of describing in one and the same work the whole material world—all that we know today of celestial bodies and of life upon the earth—from the nebular stars to the mosses on the granite rocks—and to make this work instructive to the mind, and at the same time attractive, by its vivid language."46

In its final, published form, volume 1 of *Cosmos* did in fact take its readers through the universe, with an encyclopedic inclusiveness and with dazzling attempts to unite not only astronomy and geophysics but also celestial dynamics and human economics, geography and meteorology, climatology and hygrometry. An overview of the multivolume work to come, the book touched on the entire "mingled web of free and restricted natural forces," starting with Andromeda and ending with the zoophytes floating "amid the eternal night of the depths of ocean." Throughout, Humboldt wove in the data sent to him by his colleagues around the world, as he tried to cover the bits of blank canvas in his portrait of the universe, to shed light on the blind spots in his global vision. And, toward the back of the book, he included a section on culture and politics, asserting once and for all "the unity of the human race." Of course, all his travels, both geographical and intellectual, had taught him that in practice humanity was anything but unified: "So the wanderer is followed through the entire wide world, sea and land, like the historian through all the centuries, by the uniform, hopeless picture of divided mankind." But Humboldt clung to hope, which perhaps lay in the human gift for communication. The final pages of *Cosmos* are devoted to language, that common marker of human-ness, in direct tribute to Humboldt's brother. Perhaps thinking back to his surreal tea party on the Siberian frontier and to Tschin-fu's delight upon hear-ing that Wilhelm, thousands of miles away in a European study, could actually read Chinese, Humboldt ended the first installment of his magnum opus by quoting his brother directly, to celebrate Wilhelm's noble goal "of establishing our common humanity—of striving to remove the barriers which prejudice and limited views of every kind have erected among men, and to treat all mankind, without reference to religion, nation, or color, as one fraternity, one great community."[47]

Volume I of *Cosmos* was dedicated, prudently, to King Friedrich Wilhelm IV, but the guiding spirit behind the book was Wilhelm von Humboldt. Alexan-der had been working up the manuscript and showing it to friends in 1833 and '34, when he originally chose his title. Then, in April 1835, after a brief, de-bilitating illness, Wilhelm died—and *Cosmos* was delayed for a decade. His brother's linguistic studies in many ways represented and confirmed Alexan-der's faith in the unity of humanity; Wilhelm was also his best critic and most trusted friend. When his brother passed away, Alexander wrote to François Arago the simple, stunned line: "I am very much alone." At least, as he told the philosopher Friedrich von Schlegel, Wilhelm had died "with the compo-sure and cheerful acceptance of one blessed by great mental talents," like a seasoned explorer, perhaps, maintaining a calm optimism even when hope-lessly lost or stranded. But Alexander's period of mourning stretched longer, as he battled for his own composure: "All around me is like a desert—so com-pletely desert that there is no one . . . to understand why I grieve."[48]

Over the next ten years, Humboldt seemed systematically to avoid working on the book that, he knew, would have meant more to his brother than any of his previous publications. Only in the spring of 1843, after coming back to Prussia from another trip to France, did he finally revisit his manuscript. He was seventy-five, "in the late evening of an active life," when he wrote the pref-ace to *Cosmos*, in November 1844, at the king's Sans Souci palace in Potsdam. Humboldt explained that he was presenting the work "with a diffidence in-spired by a just mistrust of my own powers," that he had tried, in fact, to for-get about this project. But, like his brother's memory, the book haunted him. "I have frequently looked upon its completion as impracticable, but as often as I have been disposed to relinquish the undertaking, I have again—although perhaps imprudently—resumed the task."[49]

When the printers at last completed their job, in early 1845, the first volume of *Cosmos* erupted onto the scientific scene in Europe and triggered a massive shift in cultural and intellectual currents. Its widespread success caused one reviewer to insist that, "were the republic of letters to alter its constitution, and choose a sovereign, the intellectual sceptre would be offered to Alexander von Humboldt."[50] Popular anticipation had led the publisher to issue twenty-two thousand copies of the first edition alone (which sold out in two months); three translations into English followed immediately, and editions in ten other languages appeared within the next decade.[51] No other such compendium of up-to-date scientific knowledge could be found in the Western world, and, because of Humboldt's fame and popularity, this one was bought not only by scholars but by the general reading public. Everyone could relate to his emphasis on "unity and harmony," the religious yet scientific sense he conveyed of "the grandeur and vast expanse of nature, revealing to the soul, by a mysterious inspiration, the existence of laws that regulate the forces of the universe."[52]

In Germany, a group of savants, their imaginations sparked once again by Humboldt's intellectual travels, immediately started compiling a companion volume to *Cosmos*, with explications, elaborations, and further references. Each time Humboldt put out a new volume, so too did his fans, until the *Letters on Alexander von Humboldt's* Cosmos became a set of five tomes exactly mirroring the original. Indeed, it was with the publication of *Cosmos* that the Humboldt industry really took off in both Germany and France. The Poulain chocolate company, founded in 1848, hawked its authentic South American cacao by stamping advertisements with a miniature portrait of Humboldt. And German travel agents made references to Humboldt that resonated for decades after his death. Not long ago, on a research trip in Berlin, I sat down for dinner in a small neighborhood restaurant, only to notice a vintage turn-of-the-century poster on the wall that promoted a vacation cruise through the tropical regions of Latin America—the "Cosmos" tour.[53]

In Britain, reviewers of *Cosmos* often found Humboldt the generalist to be inferior to their country's more specialized scientists, notwithstanding the respect granted to him by Darwin, the geologist Charles Lyell, and the botanist Joseph Hooker. James David Forbes, professor of natural philosophy at Edinburgh, was particularly peeved that Humboldt had "attempted to cast ridicule upon the English government and English men of science" by publishing that infamous Tasmania footnote about a certain Sabbath eve in September 1841.⁵⁴ Other British philosophers, meanwhile, were seizing upon Humboldt's seeming scorn for religion not as a sign of his underappreciation of the empire's scientists but as sufficient evidence to convict him of heresy. Even as science gained ground in Western society, many scholars, including some of Humboldt's own disciples, like Louis Agassiz of Harvard, continued to bend over backward to ensure that their theories appeared consistent with Christian doctrine. And those authors who neglected to pacify their religious readers risked the wrath of a still-powerful clergy. Responding to Humboldt in the *Westminster Review*, John Crosse suggested that "a sketch of the universe in which the *word* 'God' appears nowhere, but the *spirit of God* is supposed everywhere, will perhaps be regarded as dangerously Atheistical by the stickler for *the word*."⁵⁵

———

Several British reviews of *Cosmos* were reprinted in American periodicals, but it was mostly the raves that reached these shores. And American readers tended to be less bothered by Humboldt's agnosticism. Indeed, while Humboldt was careful not to mention any Creator, he also made sure not to say anything definitively sacrilegious, so that moderate Christians, of which there were many in the newly industrializing United States, could focus on the "spirit of God" in his cosmology and not worry so much about the "word." A writer for the *Methodist Quarterly Review* defended Humboldt by noting that, in *Cosmos*, "there is no attempt, open or secret, to establish or to refute religious truths."⁵⁶ Many Americans also found elements of his theories that seemed to resonate peculiarly well with the political culture of the United States and with the wild American landscape. In some ways, it was in Boston, New Haven, New York, Philadelphia, and Washington, rather than in Paris, Berlin, and London, that Humboldt's cosmopolitanism made the most sense. In the 1840s, America saw the rise of immigration, the penny press, the abolition movement and the women's movement, the rapid expansion of railroad and telegraph lines, and the Hudson River School of painting: issues of unity, social justice, scientific progress, and natural beauty were foremost in many people's minds.⁵⁷

The most dramatically positive response to *Cosmos*, though, and also the most perceptive, was penned by James Davenport Welpley for the *American Review*, a self-proclaimed "Whig Journal of Politics, Literature, Art, and Science." As historian Daniel Walker Howe has argued, one of the defining characteristics of Whiggery, the party-based political movement that opposed Jacksonian Democrats between 1834 and 1854, was its "determined assertion of the *interdependence* of different classes, regions, and interest groups."⁵⁸ So someone like Welpley was particularly well positioned to appreciate *Cosmos*, in Welpley's own words, as "an exposition of the very spirit of liberal culture." At one basic level, Humboldt defended the pragmatic value of science by virtue of "the utility of new discoveries, rising every day to the notice of the world." Even more important to Welpley, however, was the fact that the ultimate goal of Humboldt's investigations seemed to match the social goal of the United States: to construct "a model of the universe" by "grasping and blending the infinite variety," by learning how to appreciate "the unity of this wonderful diversity" with which we are blessed.[59]

For Welpley, the first sign of this convergence was Humboldt's stated desire to be a popularizer, to let "science be made accessible to the people." Not a single sentence of *Cosmos*, according to Welpley, is "contaminated with pedantry or technicality. The author seems to feel a sympathy with man, as well as with savans; and finds a greater satisfaction in giving, than in hiding knowledge." To a twenty-first-century ear, Humboldt's prose seems plagued by pauses and flourishes. But in his own day, even writing in High German, whose grammar does not permit anything *but* a tortuous prose style (there has never been a Prussian Hemingway), Humboldt was undoubtedly a paragon of clarity—as his sales figures attest. And, to Welpley, the politics of clarity spoke directly to the increasing divisions in American society, transcended class, race, ethnicity, and religion to "make common cause against ignorance and prejudice." Scholars of Humboldt's ilk were "knit among themselves, and with the multitude, in a bond of humanity more powerful even than a community of belief. . . . If the world is ever to be harmonized, it must be through a community of knowledge, for there is no other universal or non-exclusive principle in the nature of man." Like many Whigs, Welpley could sound both paternalistic and naive, but in this case his idealism is something Americans can be proud of. Embedded in his celebration of Humboldt is his belief in one of the key issues on the Whig political platform, a principle we now take for granted: free public education.[60]

The liberating and unifying effect of Humboldt's theories was not limited to the social sphere, however. Welpley also understood *Cosmos* as a means for average Americans to reconnect to the landscape, to experience "the pleasures of observation, the effect of natural scenery," to put themselves "in harmony with the world." He delights in Humboldt's "combination of educated taste and free fancy," his assertion that "from the contemplation of nature arise two kinds of enjoyment": the *"sensuous,"* which taps the "Imagination," and the *"intelligent,"* which stimulates "Understanding." Here, Welpley felt, was the "universal benefit, extended equally to all," of art combined perfectly with science. "Not suffering the poets to blind, or the naturalists to overwhelm us," Humboldt adduced his facts in such a way as to enhance rather than erase his readers' sense of magic and wonder. Instead of pulling apart the structures of the universe strand by strand, in a mode of debunking exposition, *Cosmos* gradually built up a web of intricately beautiful interconnections. So, in the end, each individual reader "is made to feel that he is connected, by the very nature and substance of his body, with every part of the universe. . . . His eye associates him with the remotest stars; his muscular sense places him in union with the gravity of the world. . . . Knowing the nature of electricity and magnetism, he finds himself in a state of equilibrium, living by the antagonism of the great powers—the opposition of air, earth, and sea. Thus, by intellect, he is in a manner blended and reconciled with all existence. He is no longer agitated with the divorce of spirit and matter."[61]

It is no wonder that Emerson and the other writers of what we now call the American Renaissance responded so intensely to this spiritual ecology, as they sought to define a new culture grounded in the local landscape. Indeed, Welpley himself recognized that Humboldt's appreciation for the experience of wild scenery was enough to get Americans interested, that "no farther inducement need be added, to the cultivation of a nearer acquaintance with nature . . . —not, at least, to those who concern themselves with the hope of a national literature." Humboldt was Transcendental.[62]

Emerson's famous philosophy of the "transparent eyeball," of the self blending with nature, was only one of the more recent American reflections of Humboldtian science that Welpley probably had in mind when he wrote his review. In truth, American writers had been responding to Humboldt's work ever since the thirty-four-year-old explorer first sparked the interest of President Jefferson in 1804. For Jefferson, Humboldt represented the ideal Enlightenment polymath, but most Americans, even most intellectual Americans, were less broad-minded than Emerson and Jefferson, and tended to seek out specialized elements of Humboldt's approach for inspiration and further study. Senator Charles Sumner of Massachusetts looked to the explorer as an exemplary antislavery activist. The crackpot John Cleves Symmes assumed that Humboldt, the great physicist of the earth, would surely support his expedition to the globe's open poles. Washington Irving, known as much for his epics of the far western fur trade as for his colorful legends set in the vicinity of Sleepy Hollow, seems to have learned travel writing from Humboldt. To George Catlin, who traipsed through America's heartland in the 1830s desperately painting Indians, using his art as a kind of salvage anthropology, Humboldt represented the effort to appreciate and understand American natives as the bearers of complex and vibrant cultures. Indeed, virtually all

Americans who studied Indians in the first half of the nineteenth century, dating back to Albert Gallatin, used Humboldt's musings on the origins of native peoples and the antiquity of their civilizations as starting—or sticking—points for their research.[63]

It is hard to find a well-known American scientist in the first half of the nineteenth century who did not exchange letters with Humboldt. Benjamin Silliman and James Dwight Dana at Yale; Asa Gray and Louis Agassiz at Harvard—all are now considered to be founders of American science, and all were intellectually indebted to Humboldt as a long-distance mentor. The Swiss-born Agassiz even turned to Humboldt for direct financial assistance, to cover his initial studies in Paris, and, much later, to help with the establishment of Harvard's Museum of Comparative Zoology. Humboldt also sponsored his German colleague Prince Maximilian of Wied and the Swiss painter Karl Bodmer on their expedition through the American West in 1833 and 1834, which, together with George Catlin's various travels, produced the most important visual documentation of Plains Indian culture. In part through another German acquaintance, Balduin Möllhausen, a cartographer and naturalist who joined one of the U.S. surveys in the 1850s that were sent out to determine the best possible route for the transcontinental railroad, Humboldt also made contact with all the important topographers and explorers connected to the American government. Even before this intimate link, though— Möllhausen would eventually marry the daughter of Humboldt's live-in assistant—Humboldt had been in touch with John Charles Frémont and his boss, John J. Abert, head of the Corps of Topographical Engineers (and father of James Abert), as well as the leader of the Coast and Geodetic Survey, Alexander Dallas Bache (the distinguished great-grandson of Benjamin Franklin), and Bache's most important scientific rival, Matthew Fontaine Maury. A navy man, Maury followed Humboldt's directives regarding the study of temperature patterns, wind and water currents, and the global relationship between sea and sky, and thus became the first oceanographer.[64]

Given this depth of influence, it does not seem a stretch to think of almost all official American exploration, at the time of *Cosmos*'s publication, as essentially Humboldtian. Indeed, it should no longer seem surprising that James Abert had such intense sympathy for the plight of the Plains Indians and the prairie lands on which they depended. As the Democrats, led by President James K. Polk, drove the country into the Mexican-American War in 1846, Lieutenant Abert was busy making lists of local animals, insects, and plants, sometimes in English, sometimes in Spanish, sometimes in Cheyenne; taking temperature readings; writing up his geological speculations; inspecting ancient Indian burial mounds; and even consorting with the enemy. He found many of the Mexicans he met in the Southwest, during his survey of the lower plains, to be extremely congenial, and, as he learned Spanish, he started adopting it as the language framing his field notes. On "Miércoles Nov Quarto" (Wednesday, November 4, 1846), he explained that his Mexican companions had "pointed out a plant which they called 'oregón,' a species of wild marjoram? that doubtless gave the name to our northwest territory."

Abert was clearly proud of his country's territorial expansion, and he was not above viewing Indians in stereotypical terms at times—when tribes committed brutally violent acts against each other, for instance. But he was also convinced that Americans ought to be treating Indians very differently. After talking with Yellow Wolf about the chief's request to the U.S. Army to have a fort built for the struggling remainder of his band, Abert wrote: "It seems but just, since the people of the U.S. have driven off the buffalo by continually travelling through the country, by multiplying roads, and by killing great numbers, that the Government of the U.S. should aid these Cheyennes." And, like many another explorer in the field, Abert found himself in certain ways preferring the direct, honorable life of "savages" to his previous existence on the rapidly industrializing East Coast. "When I viewed this simple grave," he wrote, on the morning after Independence Day, 1846, upon seeing a small circle of rocks gathered around a leaning stick, "my mind turned to the proud monuments . . . in our great cities which are daily leveled with the ground to give place to some improvement. Here on the wild prairie the Indian and the rude hunters pass by this spot and not for the world would they remove one stone. Who now shall we call the rude man? the wild man?" In the depth of winter, given shelter by an Arapaho chief, Abert even experienced some of the niceties of "civilization": "It is surprising how comfortable their lodges are. They do not smoke and a very little fuel suffices to warm them." Comfort combined with environmental efficiency: now *this* qualified as an "improvement." While America boldly extended its empire in the mid-nineteenth century, its government explorers, agents of ambivalence, gradually extended Humboldt's social ecology.[65]

Welpley's review of the first volume of *Cosmos* appeared in June 1846, just as the twenty-five-year-old Abert entered the field. On the first page of the same issue of the *American Review*, meanwhile, was an article by the prominent Humboldtian Whig D. D. Barnard registering his protest against the Mexican-American War. "After a period of political tranquility," he sighed, "now of more than 30 years duration, the peace of our country is broken. We are involved in war. The gates of the Temple of Sacrifice are once more thrown open, and all who love to worship at a shrine of blood are invited to enter." Barnard saw Mexico as "a sister republic of the New World," home to "a people who are not only neighbor to us, but with whom we have been, and, let us not hesitate to say it, with whom we ought now to be, friends." You might say he was aware of the two countries' unavoidable interdependence. Humboldt, meanwhile, had been made an honorary citizen of the Mexican state in 1826, and he had an equal amount of affection for all the free republics of the American continent. About the United States' war against Mexico, he expressed the opinion that it was merely a maneuver by Southern Democrats to "expand their devilish slave system" into new territory.[66]

The mutual understanding between Humboldt and America's Whig intellectuals in the 1840s was remarkable, but it did not actually peak until the appearance of the English translation of the second volume of *Cosmos*, in 1848. Even many-sided men like Emerson had still grasped just individual strands in Humboldt's theoretical web. Emerson was more of a Romantic than Humboldt, and significantly less of a scientist. Thanks to Humboldt, he had understood nature's "unity in variety," but he also insisted in 1836, explaining the title of his first book, that "Nature always wears the colors of the spirit."[67] Humboldt, meanwhile, as if in response to this kind of misapprehension of his delicate balancing of fancy and reason, of the internal and external, clarified his approach in volume 2 of *Cosmos*, asserting his "belief in the power of the external world over the emotions of the mind." The person who travels openheartedly into the natural world, the explorer who truly attempts to see, can actually be transformed by the experience; he is not doomed simply to impose his expectations onto the landscape. The key is to train people to break out of their circumscribed lives, to immerse themselves in foreign ecosystems, to allow themselves to be overwhelmed. And then, inspired by the newness of their surroundings, they will be better able to describe what they see with enough precision to transfer the power of nature to someone else. Poets and artists must combine "the vivifying breath of imagination" with "sufficient sharpness and scientific accuracy." Emerson never quite understood this possibility, but Thoreau did. And so did the landscape painter Frederic Church. And so, too, did America's entire first generation of landscape photographers and explorer-scientists.[68]

Volume 2 of *Cosmos* traces through time the many different forms of "physical contemplation of the universe," which Humboldt depicts "as a means of exciting a pure love of nature." A tour de force of intellectual history, it transfixed many Americans in part just because they were already developing a passion for wilderness. As they read about how the idea of the cosmos developed through the ages—how the nomadic Arabs developed astronomy, how the Chinese constructed their elaborate gardens, how the telescope and pendulum changed the world—they took pride in the proliferation of natural history societies in the United States and in the natural wonders, Niagara Falls being the most celebrated, that America offered its inhabitants. Landscape painting took over as the dominant art form in the 1830s and '40s, as the Hudson River School, led by Thomas Cole, joined forces with the New Englanders of the American Renaissance to carve out an indigenous high culture distinct from the tired old productions of history-obsessed Europe. Humboldt broadcast an appeal, in volume 2, to botanical gardeners, to painters who would make sketches directly from nature, to the brand-new army of artists who understood how to take "photographic pictures." America answered his call.[69]

After volume 2 of *Cosmos* appeared in the United States, the enthusiastic response generated a rush of Humboldt reprints and new editions. By 1850, American readers had access to two new translations of *Ansichten der Natur*, both of which (*Aspects of Nature* and *Views of Nature*) Thoreau purchased. A new version of Humboldt's *Personal Narrative* appeared two years later, and the Humboldt cult had clearly established itself on these shores. In 1853, Frederic Church left for Colombia and Ecuador specifically to confirm Humboldt's sense of "what an inexhaustible treasure remains still unopened by the landscape painter between the tropics," to adopt the scientist's mission of effecting "an extension of the visible horizon" through "an acquaintance with the nobler and grander forms of nature, and with the luxurious fullness of life in the tropical world." And by 1860, a writer in the popular magazine *The New Englander* could speak of Humboldt's fame as a truism: "take down from the shelf any modern cyclopaedia, or dictionary of biography;—you will find the career of Humboldt pictured in the most brilliant colors. . . . Ask any schoolboy who Humboldt is, and the answer will be given." In America, Humboldt was the Albert Einstein of his day.[70]

_____

Humboldt's ecological awareness circulated much more widely in antebellum America than is commonly thought. His fears about the impact of deforestation, for instance, are echoed in the writings and paintings of George Catlin and Thomas Cole. In an "Essay on American Scenery" (1835), Cole noted that "the ravages of the axe are daily increasing—the most noble scenes are made desolate, and oftentimes with a wantonness and barbarism scarcely credible in a civilized nation."[71] It was in the 1850s, though, that Americans responded to Humboldt in some of the most enduring and sophisticated ways, as Thoreau scholar Laura Dassow Walls has suggested. Setting aside

Whitman, and Washington Irving, and John Lloyd Stephens (the rediscoverer of ruined Mayan cities), and Bayard Taylor (America's first professional travel writer)—Humboldt disciples, all—Walls helps us arrive at Thoreau as probably the most Cosmic of American authors.[72]

Just as the French poet Charles Baudelaire, author of *Flowers of Evil*, found a soul mate reflected in the horrific work of Edgar Allan Poe (himself an admiring reader of *Cosmos*), so did Thoreau dedicate himself to interpreting Humboldt. American Renaissance authors, even amid their efforts to create a unique and independent literature, were still in thrall to certain European influences. Many scholars of American literature have expressed confusion at Thoreau's anti-Romantic break with Emerson and dismay at his seeming descent into scientific list making in the 1850s, but Walls argues persuasively that Thoreau died in 1862 while still trying to work out in his journals and essays what Humboldt's social ecology meant for the United States.[73]

The Thoreau of this period, as revealed even in the material he added to *Walden* between 1852 and 1854, was immersed, body and soul, in natural science and in nature itself, trying to live according to Humboldt's model of interdependence—identifying specimens, measuring depths, adapting to ever-changing conditions of soil, climate, atmosphere. He wanted to swim with the currents swirling around him. And, like Humboldt, he swam either alone or with close male companions, like William Ellery Channing, his favored partner for hikes, whether in Concord or on Cape Cod, or once even in Quebec. Perhaps, in avoiding conventional sexual relationships and thus stereotypical gender roles, Humboldt and Thoreau—and so many others of their ilk—were also avoiding the classic dichotomy between the realm of science and reason and the realm of art and emotion. Caught up in "the power of the external world" as seen through both intellect and instinct, Thoreau was beginning to feel more comfortable with his social discomfort.[74]

In the rambling, seemingly random writings of his last decade, as Walls suggests, Thoreau tried to capture "the murmur of multiple voices and actions, not the ecstasy of transcendental disembodiment . . . , not the imperial eye and the power of the all-dissolving mind but the immense beauty of, and urgent need for, the vision of all and the wise action of the collective." He celebrated natural facts, local details—corn growing in the night, crickets chirping the different tones of the scale, huckleberries, frogs, water wheels—and sought to sympathize with them while still allowing for their difference from himself. He broke down his vision of the whole into all of its component parts; made sure he understood each of the particulars; and then built the whole back up into something even more complex, rich, overwhelming. To be slightly off balance oneself was the only way to make a cosmos of the chaos.

"In the true natural order," Thoreau wrote in his journal, "the order or system is not insisted on. Each is first, and each last."[75]

Thoreau played the hermit, toward the end of his life, but at the same time he was developing the role of ecological participant—a step up, surely, from mere observer. And despite his posture of withdrawal from official literary and scientific circles, he was still consulting with the Boston Society of Natural History, and thus self-consciously contributing to the social construction of knowledge. He also continued delivering public lectures in which he attempted to shake his fellow citizens out of *their* isolationism, urging them to share in his disorienting epiphany of full physical and spiritual contact with the minutiae of the environment. "I wish to speak a word for Nature," he told the men and women of Concord, "for absolute freedom and wildness, as contrasted with a freedom and culture merely civil,—to regard man as an inhabitant, or a part and parcel of Nature, rather than a member of society. . . . Give me a culture that imports much muck from the meadows."[76]

Even more noticeable than the shift from Emersonian Transcendentalism to Thoreauvian Humboldtianism, meanwhile, was the parallel passing of the artistic torch from Thomas Cole to Frederic Church, which took American art from the shores of the Hudson to the flanks of Ecuadoran volcanoes. In volume 2 of *Cosmos*, when Humboldt quotes one of his literary colleagues on the difference between older works of art, "interwoven with allegorical allusions," and more modern pieces, some of which make us feel "as if we could breathe the free air of nature, or feel the reality of the mountain breath and the valley's shade," he could easily be referring to the contrasting paintings of Cole and Church. Cole was to some extent a Humboldtian painter, as Emerson was a Humboldtian philosopher, but his depiction of nature was generally subservient to his social message: there was almost always a (Christian) moral, as is evident from titles like *The Expulsion from the Garden of Eden*, *The Voyage of Life*, and *The Course of Empire*. Dark skies hover over twisted, misshapen trees; angels appear on forested pathways; and crawling vines reclaim the ruins of a decadent cityscape. In stark contrast, as one contemporary art critic pointed out, Church made "of each tree, shrub, and flower an individuality which can only be the result of careful botanical observation." Church's vegetation, like Humboldt's, stood primarily for itself rather than for any divine blessing or reprimand. The 1860 *New Englander* article about Humboldt pointed out the affinity between scientist and painter by asserting that in a given chapter written by the Prussian explorer, a reader "may see the 'Heart of the Andes' almost as vividly as in that master-piece of American art, the landscape of Church."[77]

The Heart of the Andes, exhibited in 1859, proved to be more popular, more successful, than any painting previously unveiled in the United States—despite its foreign subject matter.[78] Why did this South American landscape hold so much appeal for North American audiences? Its very size is intimidating: five and a half feet high by ten feet wide, the canvas is nestled in a thirteen-by-fourteen-foot frame. And it is a disorienting picture. You peer into it as if balanced on a precipice. There is no traditional foreground, no way of entering the landscape. You're at the base of a watershed, and all the snowmelt from those towering peaks threatens to rush down the cataract that disappears into the bottom edge of the picture and wash you all the way out to the Pacific Ocean. Perhaps the cataract itself is a comfort, since its horseshoe shape perfectly recapitulates the view Church painted of Niagara Falls just two years earlier to memorialize the power and grace of America's wilderness—and thus of America itself. Yet the rest of the canvas dwarfs this central waterfall. There is no place for the eye to find "repose," complained one reviewer, because the artist has left "every square inch of the canvas . . . covered with nature's statistics."[79] You step back to take in the scope of the panorama but find that you can no longer appreciate the detail. You move forward and stare through your opera glasses at the minuscule birds and butterflies, but quickly become overwhelmed by the painting's vast scale. The scene covers every climatic zone on earth, from teeming, tropical lowlands to frozen summits. A cross at the edge of a cliff marks either a sacred shrine or the scene of someone's plunge down the falls. There is a path here, but it quickly disappears into unimaginably thick jungle. There is a small, peaceful church—the artist's signature detail—in the distance, but it is Catholic, in the Spanish-colonial style, and appears to be utterly inaccessible from below. Erosion and decay have exposed the gnarled roots of a prominent tree. Condors or vultures circle over rocky peaks, just below a layer of dark clouds. The vegetation is more tangled and lush, the birds and insects more colorful and diverse, the mountains higher and more imposing, than anything you might have seen in the United States.

At the same time, though, perhaps the painting's exoticism draws you in—and quickly becomes familiar. Reviewers lingered on "the tangling luxuriance of the foliage . . . , the infinite play of color upon the infinitely various lines of the mountain mass." And by casting their imperial gaze on this alien landscape they discovered in it "the characteristic magnificence of American scenery." The art critic at the New York Times emphasized the picture's "marvelous, much-embracing unity," its "harmony of design": all the "multitudinous details" came together in a pattern that perfectly balanced flowing water, fertile earth, inspiring air, and volcanic fire. "What seems, at first, a chaos of

9 *The Heart of the Andes*, 1859. Frederic Church.

chords or colors gradually rises upon the enchanted mind a rich and orderly creation, full of familiar objects, yet wholly new in its combinations and its significance."[80] *E pluribus unum.*

The current consensus about nineteenth-century American landscapes is that, like exploration narratives, they tended to be expressions of either the urge to conquer nature and expand empires—Manifest Destiny—or the urge to escape from a rapidly industrializing modernity into an edenic wilderness. Both of these themes are present in *The Heart of the Andes*, but the painting also attempts to establish a direct, respectful, long-term relationship with the natural world. The undertones acknowledge a potential heart of darkness— the wilderness is wild, and human beings are capable of violence—but the goal is peaceful coexistence. And human violations of that peace have significant consequences. Some art historians have pointed to the sunbeam illuminating the artist's signature carved into the tree at the lower left of the canvas as evidence that Church was celebrating his own act of creation, his ability to leave a mark, to bring nature under imperialistic human control. What few have noticed is that the tree itself, though surrounded by the fecundity of the jungle, is stunted, dying—desiccated.[81]

Church himself explicitly tried to make the canvas worthy of Humboldt's many theories and directives, from the climatic determination of vegetation zones—referenced especially by the artist's inclusion of Mount Chimborazo in this scene—to the perfect balancing, in Humboldt's words, of "the severer forms of science and the more delicate emanations of fancy," the blending of "the elements generated by the more limited field of . . . direct observation, and those which spring from the boundless depth of feeling." In Ecuador's Chillo Valley, near Mount Pichincha, Church felt somewhat overwhelmed by the orchids and philodendra, the morning glories and elephant ears. Mist rose from several waterfalls; long membranous birds' nests hung down from trees entwined by columnea and codonanthe vines; yellow butterflies flitted around his head; the sweet scent of passionflowers began to make him dizzy. Even Humboldt had grown confused here: "in the midst of this abundance of flowers and leaves, and this luxuriantly wild entanglement of climbing plants, it is often difficult for the naturalist to discover to which stem different flowers and leaves belong." Later, though, dipping nature's details into the varnish of his imagination, Church was able to develop a composite scene, replacing Pichincha with Chimborazo, shaping the flora and fauna and traces of civilization, exploring the ways in which our communities fit into the cosmos. The union of science and art—physics for poets and painters, chemistry for Concord and Cambridge—led to the union of nature and humanity.[82]

As Church's friend Theodore Winthrop put it, however, in his exhibit pamphlet, *A Companion to the "Heart of the Andes,"* this picture's unity was not of "the bridal-cottage picturesque" variety; Church did not indulge in the "pastoral insipidity" of depicting New England farms nestled amid "jocund plains and valleys." Unity in diversity is a complicated, challenging proposition. Sounding a lot like Thoreau, Winthrop insisted that Church's scene "has power to confound conceit into humility." Things are a bit different in South America, after all. "Here," Winthrop wrote, "there is simplicity in complexity—order in bewilderment." Here "we are at home and yet strangers."[83]

———

Humboldt seemed more and more a stranger in Berlin as he grew older. The burden of longevity lies, in part, in watching your brother, your friends, your familiar way of life, die off. "I live to bury all my kith and kin," he wrote in 1856, after attending the funeral of Wilhelm's eldest daughter.[84]

The last three volumes of *Cosmos*, published in 1850, 1858, and (posthumously) in 1862, were highly technical. It is hard to find any of Humboldt's characteristic vigor and enthusiasm in them. After the failure of the Revolutions of 1848 in both Prussia and France, Humboldt seemed to lose even more faith in his social and political programs and decided to stick with further elaborations of current scientific theories and data. (During the 1848 Berlin uprising, which Humboldt had joined, he found that he "continued to enjoy much affection in the lower classes of society," but he was perhaps disillusioned on the night of March 18, when, "placed between two barricades, I was attacked four times by armed men, who did not know me and had not read *Cosmos*.")[85] The last volumes he wrote sound, well, old—as if his poetical side had dried up. Or perhaps melancholy had overtaken him. There were signs of his despair even in volume 2. When going through the history of exploring expeditions that had "served as means of encouraging the scientific study of nature," for instance, he stopped for a while on his old friend and mentor Georg Forster, seemingly to celebrate his having launched "a new era of scientific voyages, the aim of which was to arrive at a knowledge of the comparative history and geography of different countries." The conclusion of Humboldt's long paragraph on Forster, though, seemed to cancel out the elder scientist's considerable accomplishments: "But alas! even to his noble, sensitive, and ever-hopeful spirit, life yielded no happiness."[86]

By 1859, the year of his death, Humboldt was ready to be forthright about his sense of oppression and exhaustion. With new scientific and technological developments exploding onto the scene every day, with shifting social currents threatening to drown him, he felt a need to be left in peace. On March 15, he sent out a plea to be published in the newspapers:

> Laboring under extreme depression of spirits, the result of a correspondence which daily increases, and which makes a yearly average of from sixteen hundred to two thousand letters and pamphlets on things entirely foreign to me—manuscripts on which my advice is demanded, schemes of emigration and colonization, invoices of models, machinery, and objects of natural history, inquiries on balloons, demands for autographs, offers to nurse and amuse me—I once more publicly invite all those who desire my welfare, to try and persuade the people of the two continents not to be so busy about me, and not to take my house for the office of a directory, in order that, with the decay of my physical and intellectual strength, I may enjoy some leisure, and have time to work. Let not this appeal, to which I only resorted with reluctance, be interpreted with malevolence.

In New York, Humboldt's old acquaintance Washington Irving, also in the last months of his life, came across this notice a short while later and was quick to express his sympathy. He had known Humboldt as "a very amiable man," with "a great deal of bonhommie"—someone who had always welcomed visitors and quickly answered letters. The great scientist's mind had clearly not deteriorated—who else could write such exquisitely long sentences?—but Irving understood Humboldt's need to narrow the scope of his thoughts.[87]

In his final decade, Humboldt drew at least a little solace from a series of amiable visits by some of Irving's most distinguished compatriots. "Among the multitude of all nations who visited Humboldt in his last years," wrote one of his first biographers, just weeks after his death, "none were so warmly received as those who came from America: to be an American was an almost certain passport to his presence. . . . He seemed to have a love for . . . their freshness and enthusiasm." At his modest apartment on Oranienburger Strasse, in downtown East Berlin—now a Thai-fusion bistro with an official plaque on the cement facade paying homage to the cosmopolitan Humboldt—the great scientist welcomed, among others, the celebrated orator Edward Everett and the literary historian George Ticknor from Boston; the painter George Catlin, whom Humboldt had called in volume 2 of *Cosmos* "one of the most admirable observers of manners who ever lived among the aborigines of America"; Bayard Taylor, the well-traveled man of letters renowned for having talked his way onto Commodore Perry's Japan expedition in 1853; and Yale's Professor Benjamin Silliman, "the most prominent and influential scientific man in America during the first half of the nineteenth century."[88]

Typically, visitors were received by Seifert, Humboldt's assistant; brought up to the "plain" second-floor apartment; and then escorted through the library to the study, where they might sit on a small green couch, surrounded by books and maps, to wait for the master. Professor Silliman noted a striking portrait of Wilhelm on the wall of the library. In the study, the walls were lined with bookcases stretching from floor to ceiling. Humboldt's desk and the long table opposite were piled with letters, folios, and rolled-up charts, tables, and manuscripts. Beneath the orderly mess, the desktop was perpetually covered with notes and calculations, which Seifert simply planed off when Humboldt required more blank space. A stuffed bird, meanwhile, looked down on the proceedings from a pedestal in the corner. The whole apartment was probably uncomfortably hot, given Humboldt's desire to be constantly reminded of the tropics.[89]

Silliman, traveling in 1851 with his son and his son's friend George Brush, who, a decade later, would be one of Clarence King's most influential professors at Yale's Sheffield Scientific School, reported that Humboldt "met us with great kindness, and perfect frankness . . . and he is as affable as if he had no claims to superiority." Moreover, the old scientist, "although associated intimately with kings, is evidently a friend to human liberty, and rejoices in the prosperity of our country." This joy was tempered by anxiety, though, as Silliman discovered upon receiving a letter from Humboldt just before his departure from Berlin. Fully aware of the Compromise of 1850, which had allowed California to enter the Union as a free state but had also established a harsher fugitive slave law and arranged for the remaining lands won in the Mexican-American War to be governed as territories with no restrictions on slavery,

Humboldt saw America as being at a crossroads. On one hand, he hoped that the country's many frontiers "shall come to resemble (in their institutions) New-York, Boston, and Philadelphia," and thus that the expansion of the United States would ultimately serve to "facilitate the movement of knowledge." Then again, he explained, "I have moral reasons to fear the immeasurable aggrandizement of your confederacy," given "the temptations to the abuse of power, dangerous to the Union," not to mention the terrible impositions "upon the indigenous races which are in a course of rapid extinction." It is tempting to wonder what Humboldt might have contributed to American politics had he moved to Washington and become an adviser to presidents instead of kings.

In the end, Silliman left Berlin "greatly gratified" by his visit and quite taken with Humboldt:

> Such are the modest and unassuming language and appearance of one who has, in person, explored a larger portion of our globe than any other living traveller; of a philosopher who has illustrated and enlarged almost every department of human knowledge: general physics and chemistry, geology, natural history, philology, civil antiquities, and ethnography. . . .
>
> He has endured the extreme vicissitudes of opposite climates, and seen men, animals, and plants, under every phase and aspect. His published works are a library. His faculties combine the enthusiasm of poetry with the severity of science. . . .
>
> Such is the philosopher, who of all living men belongs not so much to his country as to mankind, and who, when he departs will leave no one who can fill his place.

The Yale professor, though, a deeply religious man, was also a bit saddened by Humboldt's refusal to engage with Christian tradition. It was fine, even admirable, to find spiritual comfort in scientific interconnections. But the two of them were old men. Could science grant an afterlife? "We dismiss him," Silliman wrote, "with the hope that he may inherit blessings beyond the grave, and find in a higher state of being, that his large measure of human knowledge is infinitely surpassed by the spiritual illumination and revelations of that glorious world."[90]

When Bayard Taylor presented himself to Humboldt in 1856 and 1857, he, too, wound up thinking a great deal about the old man's imminent departure from this world. At first, Taylor focused on Humboldt's "clear blue eyes," his intellectual "restlessness," his knowledge of every "occurrence of the least interest in any part of the world": "he talked rapidly . . . , unconscious which language he was using, as he changed five or six times in the course of the conversation. He did not remain in his chair more than ten minutes at a time, frequently getting up and walking about the room, now and then pointing to a picture or opening a book to illustrate some remark." But the subject of death kept coming up. When Taylor offered greetings from Washington Irving, Humboldt was forced to remember that he had first met the American author some thirty-three years previously. "I have lived so long," he mused, "that I have almost lost the consciousness of time. I belong to the age of Jefferson and Gallatin."

Humboldt liked to talk about his old age, but he never seems to have been particularly concerned about death or any possible afterlife. He had suffered a bit of a stroke in between Taylor's two visits, but referred to it dismissively as "an electric storm on the nerves, perhaps only a flash of lightning." This was someone, after all, who had shot electricity through his own body, and who had measured electric storms on three continents: he should know. Some sort of death would come eventually; meanwhile, he would do his earthly labor. Indeed, on bad days, he thought of death as "the termination of the state of ennui we call life." In many ways, Humboldt suffered through his later years as a melancholy isolate, yet, at the same time, he maintained an admirable work ethic—he told Silliman that he slept only four hours a night—and a warm, dry sense of humor deemed notable by almost all of his acquaintances. He truly did take comfort in his science of mutuality. And he seemed to come to life whenever there was an opportunity to engage in conversation, to meet someone new from a far-off land—to connect.[91]

———

Alexander von Humboldt died on May 6, 1859, in his Berlin apartment. He had no children. One of Wilhelm's daughters, though, Gabriele von Bülow, sat by his bed and nursed him through his final illness.

Silliman had been right: no one could fill Humboldt's broad shoes. And, since he had never held a university position, he did not even have any official disciples. Despite his worldwide acclaim, his legacy was by no means assured. But, in the meantime, at least he had been understood—especially by Americans. As one Connecticut journalist explained, referring to Humboldt's Berlin funeral: "The company of American students who followed him to the grave, in advance of the high officers of the Prussian State, were fitly recognized as the representatives of their country." The memorial services held across the United States, meanwhile, the public eulogies, the written testaments of scholars and intellectuals, all helped to spread what was already an acute understanding of Humboldt's cosmology. Americans felt something for him: the connection was mutual.[92]

On May 24, at a meeting of the American Academy of Arts and Sciences in

Boston, Louis Agassiz suggested that Humboldt's highest goal had been "to present to others what nature presented to him—combinations interlocked in such a complicated way as hardly to be distinguishable." A few days later, there was a tribute at the New-York Historical Society, during which the "overflowing" audience heard speeches by the historian George Bancroft, Alexander Dallas Bache of the Coast Survey, and the geographer Francis Lieber, among others. The librarian of the sponsoring organization, the American Geographical and Statistical Society, also read letters from Benjamin Silliman, Agassiz, Matthew Fontaine Maury, and D. D. Barnard, the eloquent Whig who had protested so vehemently against the Mexican-American War. Doctor Joseph P. Thompson launched the proceedings by gesturing to the many paintings, photographs, and sculptures of Humboldt scattered throughout the lecture hall, explaining that he and his colleagues had come together "to commemorate the name of him whose statuette graces this table; [and] whose portrait [hangs] yonder, painted in his youth at Quito by an artist of that country, and now in the possession of Mr. Church, who has so recently illuminated for us the Heart of the Andes." Then Professor Lieber took the podium to establish the human implications of Humboldt's cosmology: "It has ever appeared to me that this great man, studying nature in her details, and becoming what Bacon calls her interpreting priest, elevates himself to those heights whence he can take a comprehensive view of her connection with man and the movements of society, with language, economy, and exchange, institutions and architecture, which is to man almost like the nidifying instinct of the bird. . . . He was with all his erudition and the grandeur of his knowledge, eminently a social man." Later, Arnold Guyot, respected author of *Earth and Man*, made Humboldt's connection to the United States even more overt:[93]

> And we in America, we pre-eminently, by Providence, the cosmopolitan people, and the people of the future, let us rear here among us a monument such as he would approve; let us honor the science that he cherished; let us faithfully and harmoniously continue the work of the exploration of this New World that he has so well begun . . . ; and let this be the homage and fit tribute of the Western Hemisphere to Alexander von Humboldt.

Already, in the 1850s, various American authors had been composing tributes to Humboldt and his Americanness, and they would continue doing so for decades to come. Walt Whitman had proclaimed himself "a kosmos" in 1855 in his epic poem *Song of Myself*, and in 1881 he added a poem to a new edition of *Leaves of Grass* whose title was "Kosmos." Now, butting up against the universe, he described himself this way: "Who includes diversity and is Nature, / Who is the amplitude of the earth, and the coarseness and sexuality of the earth, and the great charity of the earth, and the equilibrium also."[94]

On September 14, 1869, cities all over the world celebrated the Humboldt centennial,[95] and Robert Ingersoll, the attorney general of Illinois, paid tribute in Peoria. Eventually known far and wide as "the great agnostic" and America's foremost orator, Ingersoll spoke for science, reason, order, knowledge, law, fact. He also believed in politics and free debate: "GIVE ME THE STORM AND TEMPEST OF THOUGHT AND ACTION," he said. While someone like Guyot felt comfortable invoking American destiny and exceptionalism in relation to Humboldt, Ingersoll felt that the key to Humboldtian science was the attempt "to do away with that splendid delusion called special providence." Humankind is not a chosen species; Americans are not superior; there is no divine program; our job is to consider carefully and then to use public forums as we argue about the best ways in which to arrange our society and our relationship to the universe. To Ingersoll, Humboldt fit in well with the United States because he emphasized "infinite diversity in unity," because he proposed the building of a whole based on an honest, generous confrontation with difference. "We honor him," Ingersoll explained, "because he honored us—because he labored for others."

Remarkably, Ingersoll's use of Humboldtian science in his campaign for secular democracy also did justice to Humboldt's spiritual side. He railed against religion and "THE DEAD CALM OF IGNORANCE AND FAITH"— but not against the soul. Indeed, he claimed that "the glory of science is, that it is *freeing* the soul." Humboldt could help people achieve a kind of spiritual balance by illuminating their connectedness, their mutual dependence, in the web of life. Science, approached on Humboldt's terms, did not stand for technological progress, for the aloof, rational manipulation of nature, for the shattering of mystery. Humboldt himself "was an admirer, a lover, an adorer of Nature," and Ingersoll, too, meant to inspire love, and an almost-religious humility spawned by a vision of "the infinite shore of the universe."

Ingersoll saw that the great experiment of the United States could go terribly wrong if the country oriented itself toward religion instead of science and politics. He spoke at a time of great unrest, and on behalf of tens of thousands of newly freed slaves, and Jewish immigrants, and indignant women. But he would never have been so successful a speaker if he hadn't also known how to preach. People could listen to his attack on religion because he was clearly moved by something that could produce just as much zeal. He had rhythm, and style. And he could speak in sentences as long as Humboldt's:[96]

We associate the name of Humboldt with oceans, continents, mountains, and volcanoes—with the great palms—the wide deserts—the snow-lipped craters of the Andes—with primeval forests and European capitals—with wildernesses and universities—with savages and savans—with the lonely rivers of unpeopled wastes—with peaks and pampas, and steppes, and cliffs and crags—with the progress of the world—with every science known to man, and with every star glittering in the immensity of space.

Eureka: The Death of Edgar Allan Poe

Edgar Allan Poe, with uncanny appropriateness, took a long time to die. Like so many of his characters, he seemed to waver on the border between a nightmarish consciousness and an even worse dream state, still anchored in reality but experiencing it through the filter of wild hallucinations. In his last lucid moment, he told the doctor attending him at Washington Medical College, in Baltimore, that he would appreciate having his brains blown out at the doctor's earliest convenience. Poe then lapsed into delirious writhing, which lasted almost two full days. Finally, his body exhausted, he fell back onto his bed and commenced repeating a single two-syllable word—a name, seemingly—every few minutes, until his death several hours later.

No one figured out what Poe was saying. The doctor had a guess, but all he could agree on with the two nurses who had been holding Poe down for those last two days was that the dying writer, on the frontier of the universe, somewhere between the earth and the heavens, or between heaven and hell, seemed to be looking for someone.[1]

Poe had spent much of his life searching. Then, in 1848, the year before he died, he published *Eureka*—the implication being, of course, that he had made a discovery. This bizarre, often inscrutable, 130-page prose poem turned out to be his last major work. Its cosmological argument, he explained, was the following: *"In the Original Unity of the First Thing lies the Secondary Cause of All Things, with the Germ of their Inevitable Annihilation."* That's on page one. But immediately before that, even before the preface, there is a page that says only this: "With very profound respect, this work is dedicated to Alexander von Humboldt."[2]

It's hard to say exactly what Poe owed to his reading of *Cosmos*, though it, like *Eureka*, is certainly a book about mortality. Perhaps he discussed Humboldt with Frederic Church when he visited the artist's studio in New York during the winter of 1848–49 (one of Church's disciples described Poe as "a slender, nervous, vivacious, and extremely refined personage").[3] In any case, Poe seems to have interpreted *Cosmos*—at least its first two volumes, the only ones he lived to see—as a book about the paradoxes of exploration and science. Here was a treatise proposing to discover the entire universe's interconnections, to see holistically, whose author also happened to believe that nature is "illimitable" and "likewise presents itself to the human intellect as a problem which cannot be grasped, and whose solution is impossible."4 Humboldt's attitude toward science, in other words, matched Poe's attitude toward art, which he described, in "The Poetic Principle," as "the desire of the moth for the star." Just as Humboldt sought out extreme ecosystems, without expecting fully to understand them, so Poe drove himself and his characters toward final frontiers, geographic and psychic borderlands—precipices, whirlpools, polar seas, dreams, trances, opium hazes—in search of some sort of ultimate knowledge, hints of the unconscious, of what might lie on the other side. He wanted to capture an "ecstatic prescience of the glories beyond the grave." In one of his earliest published stories, "Manuscript Found in a Bottle," the narrator is an explorer guiding his ship ever southward, until it gets caught "in a current; if that appellation can properly be given to a tide which, howling and shrieking by the white ice, thunders on . . . with a velocity like the headlong dashing of a cataract." He is horrified, "but a curiosity to penetrate the mysteries of these awful regions, predominates even over my despair"; he cannot let go of his desire for the universe's "never-to-be-imparted secret, whose attainment is destruction."5 To live is to be propelled onward by the mysterious, all-powerful current of Death and its dark, subconscious impulses; to live as an artist or a scientist or an explorer is to surf the current. Poe seems to have become a writer so as to explore his death wish—and to express the ultimate unintelligibility of human experience.

The narrator of another of his stories, "A Descent into the Maelstrom," poses exactly the same dilemma: "I positively felt a *wish* to explore its depths, even at the sacrifice I was going to make; and my principal grief was that I should never be able to tell my old companions on shore about the mysteries I should see."6 Though Poe's debt to Humboldt may remain hazy, it should at least be possible, given the image of watery abysses in both of these stories, to notice his dependence on the theories of John Cleves Symmes. The imaginary force giving momentum to the voyages of these two narrators—not to mention that of the title character in Poe's only novel, *The Narrative of Arthur Gordon Pym*— might well be called the Symmes Current, for it is dragging the poor explorers into the earth's hollow core, propelling them through one of the globe's vast polar openings.7 Poe did not know Symmes, but he knew his Theory of Concentric Spheres quite intimately, through the lectures and writings of Symmes's disciple J. N. Reynolds. In the very same issue of the *Southern Literary Messenger* in which he gave the world the first installment of *Pym*—January 1837—Poe also published an article about the proposed "Surveying and Exploring Expedition to the Pacific Ocean and South Seas"—Reynolds's pet project.8

Poe's novel ends impossibly, in the first person: "We rushed into the embraces of the cataract, where a chasm threw itself open to receive us."9 Did Pym somehow survive to make fantastic discoveries—Eureka!—and contact the lost civilizations of Symmes's Hole? Regardless, the implication is that his ultimate union with nature contained the germ of his inevitable annihilation.

Poe definitely didn't survive. He did travel to the brink and try to send a message back for more than two days. He went to what he called—in his aptly titled *Marginalia*—the "border-ground," where "the confines of the waking world blend with those of dreams."10 He peered into the maelstrom. And then he acted out a scene that he could have written himself, in which he repeated that one incomprehensible name over and over again, that one nugget of truth hinting that some sort of order governed the universe, that everything was connected, that meaning lurked just below the surface. In the end, though, as in so many of Poe's tales, truth, order, connection, meaning, remain inaccessible. We'll never know what Poe was trying to say.

When pressed, though, Dr. J. J. Moran, the attending physician, insisted that Poe was calling out for a man named "Reynolds."11 ■

PART TWO

SOUTH

J. N. Reynolds and the "More Comprehensive Promise"
of the Antarctic

CHAPTER 4

"Rough Notes of Rough Adventures"

Exploration for Exploration's Sake

After toiling on for more than a league, we reached the base of the main ridge, when a scene was presented, on which a connoisseur in volcanoes would certainly have luxuriated.
—J. N. Reynolds, "A Leaf from an Unpublished Manuscript" (1839)

I fear chiefly lest my expression may not be *extra-vagant* enough, may not wander far enough beyond the narrow limits of my daily experience, so as to be adequate to the truth of which I have been convinced.
—Henry David Thoreau, *Walden* (1854)

On October 17, 1829, about a month after Humboldt turned sixty in Siberia, thirty-year-old J. N. Reynolds sailed from New York Harbor on the brig *Annawan*, bound for the Antarctic. A little more than a year later, he found himself staring into the abyss of a highly active South American volcano.

The *Annawan*, named for a famed Wampanoag chief killed in 1676 by British colonists, in the heart of the maritime region of New England that would become dominated by the whaling and sealing industries 150 years later, constituted one-half of what came to be known as the Palmer-Pendleton Expedition to the South Sea. The other ship was the *Seraph*, also a brig, and these two vessels were unofficially but consistently escorted by the schooner *Penguin*.[1]

Organized almost entirely by Reynolds, the venture was approved by the House of Representatives as an official voyage of scientific exploration in January 1829; blocked by the new president, Andrew Jackson, in March; and then frantically reconfigured over the summer to attract the backing of wealthy shipowners—such as Nathaniel Palmer and Benjamin Pendleton—in the sealing business. Finally, the expedition set off with the dual goal of searching for land in Antarctic waters and procuring enough seal skins to pay for itself. The compromise seemed sound, but in practice the scientific and commercial interests were never reconciled. As Reynolds later explained, the exploring mission almost immediately encountered both external and internal resistance: "After having cruised for months in . . . tiny barks amid 'thick ribbed ice' in the regions along the Antarctic circle," the scientists grew restless for something to study, and the sealers started to worry about earning their shares. "Storms and fogs," as well as an "astonishingly great" number of icebergs, made the high latitudes nearly impossible to navigate. Captain Pendleton called the voyage "a lengthy cruise of much anxiety and suffering." His men, aboard the *Seraph*, were starting to get mutinous, having been "much worn down by fatigue and from their being almost constantly wet in this region of rough sea and cold rugged weather, with at the same time alarming symptoms of that dread disease the scurvy making its appearance." Pendleton himself, according to Reynolds, was "not fit for any thing bold—he is a great *coward*. . . . I am, in truth, disgusted with [him]. And his total want of manly *daring*." Reynolds yearned to push farther southward, but he eventually succumbed to Pendleton's doubts and settled for the opportunity to do some inland exploring, along the same fiery mountain range that had so captivated Humboldt.[2]

The ships would find a "snug harbor" so that the men could focus on getting furs, something both captains were eager to do. (Reynolds thought that the other skipper, Palmer, would have been "a good fellow—if he did not love money a little too much.") Meanwhile, Reynolds and his associate John Frampton Watson, a doctor from Philadelphia, would get dropped off on the very southern tip of Chile.[3] On October 25, 1830, Reynolds wrote a letter "from the base of the Cordilleras" to his friend Samuel Southard, who had been secretary of the navy when the expedition was approved. The young explorer, emboldened by recent exploits and exhilarated at having escaped to such an exotic place, seemed to enjoy showing his brash side to the elder statesman. "I have left my vessel," he wrote,

for seven months, with the view of penetrating West Patagonia from this point towards the Straits of Magellan. This is an untrodden Field, and rich in every department of Natural history. I intend passing along near the base of the Cordilleras, and thro' that powerful and warlike race of natives, known under the general term of Araucanians—. These are the people, who forever held the Spaniards in Check, in despite of their insatiable thirst for gold and dominion. . . .

There are two active volcanoes within our range, both of which I hope to ascend, and leave the American Colours on the top of the burning craters.

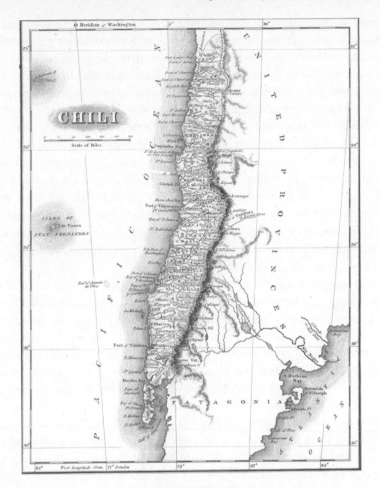

10 Map of Chile, 1823.

No foreigner or even Spaniard has ever traversed this route, so that whatever little I do, will at least have the merit of being new. . . .

The enterprise, here, is looked upon as extremely hazardous; but I have really become so accustomed to adventures of one kind or another, that I am, perhaps, culpably negligent of such considerations.—I have been lost from my vessel, in an open Boat, for several days and nights, amid the gloomy regions and eternal ice around the Shetland Isles—I have landed on icebergs while floating in the ocean, and ascended precipices when one misstep—could have ended my earthly career. . . .

I saw in one place, the skelletons of 50 whales perfectly petrified, 400 ft. above the level of the sea and 8 miles from it. I have probably been the first to visit an old Peruvian City, on the borders of the Desert,—the walls are still visible, and in the burying ground, contains, I might say fifty thousand human skelletons, lying partially covered with sand, and changing and drifting about by the force of the winds. The subjects are curiously envolved, by many folds of ingeniously wrought textiles. The hair is very perfect and the very pins by which the folds are fastened up are still remaining. . . .

Indeed the compass of a letter is quite inadequate even to touch on such subjects;—while the apparently egotistical enumeration of a few of my little excursions, you will readily perceive, is only intended for the indulgent eye of an old and very good friend—. . . .

Poor lonely and unprotected, as I am, in a foreign country, I am determined, never to return, with the character of *tameness* applied to all My operations—Nor do I fear it; the present excursion will not be without interest.[4]

There is an unabashed patriotism here, and a youthful, male anxiety, a desperate need to prove the worth of his backwoods American stock. Though Reynolds and Humboldt set off on their major expeditions at virtually the same age, Reynolds did not arrive in South America the way his role model had, as a confident, well-trained, committed scientist, and he would never achieve Humboldt's acute understanding of environmental interconnections and social injustices. But his letters and journals do reveal a genuinely Humboldtian zeal for precise observation and expansive views, for contact with the world, for connective experiences.[5]

By mid-November, Reynolds and Watson, with an interpreter, two guides, horses, and five days' provisions, were gazing up toward the volcano of Antuco, which was, "at this period, probably the most active on the globe." The journey turned out to be confusing, awe-inspiring—rife with yawning chasms and "gigantic fragments" of rock, with hissing gases and jumbles of human and geologic ruins, all suggesting the region's violent history.[6]

Reynolds published only relatively short excerpts from his journals, never managing to collect them all into the long narrative they would have constituted, and he constantly apologized for how "roughly sketched" they were, for his failure to "clothe [them] in more becoming attire."[7] Yet he had a talent for blending images of his immediate surroundings with evocations of the geological past. Walking the fault lines of Romanticism and praying for tremors, Reynolds also juxtaposed vivid scenes and imaginative leaps with repeated reminders that no art, literary or graphic, could translate these experiences, and no human imagination could "conceive the magnitude" of nature's changes: "the beds of lava were, in extent, more enormous than any we had before seen; and we felt how difficult it was to imagine, how impossible adequately to describe, the tremendous commotions that must have shaken and rent the solid walls of that gigantic furnace, when the fused volumes leapt in blazing cataracts from its summit. . . . To suppose one side of a mountain composed of solid rock, suddenly torn off by an explosion, would hardly account for the mass of heterogeneous material which lay around."[8]

The climbers' initial attempts at actually scaling the volcano were turned back by bad weather. On the first evening, the storm clouds "were heavier and darker around the summit . . . , forming a curious contrast to the stream of fire issuing from its crater." Then, "as night gathered, the wind began to blow powerfully from the north, and a scene occurred which can never be erased from our memory. A violent conflict of the elements, witnessed from an elevated position among the Andes, is terrific and even awful. Perhaps, in this instance, there was something in the loneliness of our situation, which added to the natural grandeur of the spectacle." They slept hardly at all: "lightning issued in one tireless flash," and "the peals of thunder were fierce and deafening, as they reverberated along those everlasting colonnades of rock—'the masonry of God'—whose spiral capitals were probably surrounded by the blue ether, far above the region of storms." Reynolds, who mixed conventional religion with a kind of Humboldtian nature worship, knew there was clear sky in the upper atmosphere, the dwelling of the Creator, but in his earthbound condition he couldn't see it: this was no time for an ascent.

The next afternoon, though, brought a calm atmosphere, and they were able to climb a little farther up. But, "having passed places where a single false step of our horses would have precipitated them and us hundreds of feet below," they were abandoned by their terrified guides in the early evening. Then it rained and snowed all night, and "the crater, as if it acknowledged a secret sympathy with the elements, was more active than it had been during the day, discharging, at short intervals, gushes of smoke and fire, with explosions like those of heavy artillery." In the morning, Reynolds and Watson were covered in snow, and decided that they had to go all the way back to the village where they had started.

A few days later, on November 22, the two Americans set out again with the sun in their faces, unaccompanied but resolute. There was quite a bit of new snow on the trail, and they "sunk to a considerable depth in plunging through it." But they made progress, setting up their crude evening camp at an impressive altitude on the west-southwest side of the peak. Then, at about 4:00 a.m., after a bit of rest, they resumed their

difficult and fatiguing journey. . . . Our thermometer stood at 53° at starting. It soon fell to 49; the snow became more compact at 44; at 39 the whole surface of the lava was covered with it, and sufficiently hard to support our weight; while at 30 . . . it became slippery. . . . The difficulties of the ascent multiplied at every step, and we were compelled frequently to cling with our hands to the projecting rocks, and thus draw ourselves up from point to point. The last three thousand feet of the acclivity must have formed an angle, varying from 40 to 60 degrees, with the horizon. The volcano contin-ued very active, "letting off its steam" at intervals of about five minutes, and discharging vast quantities of stones and ashes, which sometimes came rolling by us down the steep with great velocity. With cautious steps we at length climbed so near the summit as to be amid the suffocating vapors emitted from the fissures in the rock, and amongst the loose unstable frag-ments forming the apex and mouth of the principal crater. At each outburst of the volcano we were enveloped in smoke, and as we advanced still higher, the heat became insufferable.

For the last 1500 feet the surface was destitute of snow, and the stones, too hot to be touched by the hand, crisped the soles of our shoes as we pressed them. . . . Added to this, at the height we had attained, respiration was difficult from the rarity of the atmosphere, and this inconvenience was much increased by the heat and smoke; so that our cheeks grew pale and our lips blue, accompanied by faintness and sickness.

With the last of his energy, Reynolds planted his American flag, and then he and Watson quickly "made such observations as we deemed interesting and important." Here at the summit, "the thermometer, which had once fallen as low as 30 in our ascent, when held aloft in the hand rose to 115." They tried to measure the width of the crater but had great difficulty, "as the mouth of it was constantly sending forth dense columns of bituminous vapor." Finally, "we estimated the height of the volcano at 10,000 feet above the level of the river; and the distance we had climbed at least 3 leagues." And then they went down.

When I first read Reynolds's description of his descent, I was startled at how similar it sounded to my own experience of descending the 14,400-foot Mount Rainier. I have never been on an actively smoking volcano, but this dormant one in Washington State certainly provided its share of sublimity. For me, it was a descent into pure exhaustion and a kind of blind faith that gravity would bring me home. Even amid the exhilaration of reaching the summit, I had barely been able to muster enough energy to snap a couple of photographs. Wracked with altitude sickness, I spent my victorious half hour mostly curled in the fetal position. Setting up a telescope or barometer would have been unthinkable. Then, dragging myself down, I stumbled and flailed every two or three steps, as did Reynolds and Watson once the snow started to

melt. "The sun shone with unshadowed splendor, and the day being comparatively calm, the snow wasted rapidly. Ere we had descended a mile, it became
too soft to bear us, and we sank to our knees, often deeper, at every step.
Owing to the steepness of the declivity, we sometimes fell, and were for a moment submerged in the drifts." They reached the town of Antuco that evening, but for the next three days they "were confined to our apartments. The
dazzling whiteness of the snow, rendered still more intensely brilliant by the
reflected sunbeams, had made us partially blind. Our eyes, much swollen, became excruciatingly painful. Our lips blistered, and the skin peeled from our
faces, as though from the effect of scalding."9

The perseverance, the self-sacrifice, the permeable border between earth
and sky, the overwhelming power of nature, the desperate measurements—
many of the details of Reynolds's narrative call to mind Humboldt on Pichincha or the volcano of Tenerife. Reynolds clearly knew how to use the
Humboldt field manual. And now he had the scars to prove it.

———

It is hardly surprising to find that Reynolds spent so much of his expedition
climbing volcanoes and contemplating craters, peering down fissures and
cracks to see what might be happening at the planet's core. He had long been
fascinated by the heat of internal forces. In the late 1820s, after all, he had delivered hundreds of lectures about the possibility that the earth was hollow—
an idea which, if not generally accepted by America's scientists, was certainly
of great interest to them. If you flip through the massive and eclectic collection of Benjamin Silliman's pamphlets, which are bound together in decaying
brown tomes at a Yale University archive, you'll find that pamphlet number 16
in the eighteenth volume is *Remarks on a Review of Symmes' Theory*, written in
1827 by a "Citizen of the United States"—who turns out to be J. N. Reynolds.10

Captain Symmes, the theoretician, was a man obsessed. A decorated soldier from the War of 1812, he gave up his position as a popular Indian trader
on the Mississippi River and essentially abandoned his ten children—who
nevertheless seemed to adore him, and who defended his legacy decades
later—for the last eleven years of his life, from 1818 to 1829, in order to wage a
constant publicity campaign on behalf of the "Symmes Theory of Concentric
Spheres and Polar Voids." In the backwoods schoolhouses and town halls of
Ohio, Kentucky, Indiana, and Missouri, Symmes spun his custom-carved hollow globes, performed tricks with magnets and iron filings, and expounded
on tree rings and lava tubes and ocean currents, insisting that all the universe
was founded on concentricity and centrifugal force. He pictured it with perfect clarity: just as our very bones and "the minutest hairs of our heads" and
"the stones called Aerolites" are all hollow cylinders, so is our planet a hollow

sphere, with other spheres nested inside it, each maintaining highly specialized life-forms, and each, because of the way it turned on its central axis,
accepting a tremendous flow of water at its poles, which then cycled back
through desalinating tunnels to burst forth again as equatorial springs and
sources for mighty rivers like the Nile and Amazon.[11]

People paid attention. As early as 1820, a still-unidentified author who
called himself Captain Adam Seaborn published a clunky first-person narrative called *Symzonia*, either in tribute to Symmes or in mockery of him. The
book traces the very first voyage to the world of the inner spheres, which turns
out to be populated by a higher race of beings who scorn their huge supplies
of gold and walk around in loose white garments that resemble togas.[12]

Part of Symmes's conviction arose, quite simply, from an overactive imagination. The great Humboldt acknowledged that "fancy" played a key role in
scientific progress, but only someone a little daft would conclude, after watching martins build their nests near houses in downtown Cincinnati, that, since
the birds seemed to "delight in the society of man" and no one knew "from
whence they come and whither they go," they must migrate to other human
habitations in the bowels of the earth.[13] Indeed, in the first volume of *Cosmos*,
the normally generous Humboldt saw fit to poke fun at the deceased captain:
the infamous "hollow sphere," he wrote, "has by degrees been peopled with
plants and animals, and . . . it was further imagined that an ever-uniform temperature reigned in these internal regions. . . . Near the north pole, at 82° latitude, whence the polar light emanates, was an enormous opening, through
which a descent might be made into the hollow sphere, and Sir Humphrey
Davy and myself were even publicly and frequently invited by Captain
Symmes to enter upon this subterranean expedition: so powerful is the morbid inclination of men to fill unknown spaces with shapes of wonder."[14] Yet
Humboldt himself had fed Symmes's fire through his early publications by
providing isothermal evidence that the equatorial regions were not always the
hottest on the planet. One of Symmes's most learned boosters even cited
Humboldt's description of fish being erupted from a volcano as possible evidence for Symmes's idea that another world existed inside the earth's core.
Certainly, the captain's contention that there were open polar seas rather than
ice caps was based not on his imagination but on his compulsive reading of
exploration narratives, many of which did mention that there were whales
and seals and birds and even signs of vegetable life at the highest latitudes attained, and that the ocean was in fact more navigable once you passed
through the thickest ice fields between about 60° and 65° north or south.[15]
And it was this possibility of the earth's internal warmth bubbling up toward
the poles that most intrigued J. N. Reynolds.

Symmes and Reynolds met in Ohio, probably in 1824, when Symmes was busy lecturing and his new admirer was editing a small-town newspaper. The whole state was abuzz with concentric spheres and exploring expeditions, and in Cincinnati—the city that, overlooking the Ohio River trade route, served as capital of the old Northwestern frontier—the renowned Western Museum boasted a portrait of Symmes by none other than John James Audubon. Members of the Ohio delegation had even started asking Congress, as early as 1822, that Symmes be put in charge of a mission to the North Pole. This was a time of radical experiments, after all, the era of Joseph Smith's revelations about Mormonism, and of Robert Owen's utopian community at New Harmony, Indiana. As P. T. Barnum would discover just a decade later, citizens of the young Republic were ready to look and listen, as long as you had an entertaining story. Symmes had tapped people's interest enough to become a celebrity. Some Ohioans chuckled as they explained, of some misplaced item, that it had disappeared "down Symmes's Hole," but ultimately they didn't care if the captain was wrong or right: they simply busied themselves holding benefits for him and composing poems in his honor ("Has not Columbia one aspiring son, / By whom the unfading laurel may be won? / Yes! history's pen may yet inscribe the name / of SYMMES to grace her future scroll of fame"). So Reynolds, in his midtwenties—he was born in 1799, the year Humboldt launched his American expedition—approached Symmes with respect and a certain amount of awe. By September of 1825, though, fired by ambition and imagination, Reynolds had assumed most of the old veteran's lecturing duties and was insisting that the two of them take their show on the road to the major cities of the East. Symmes prickled at his disciple's tendency to emphasize open polar seas over habitable inner spheres, but the captain was growing more and more enfeebled, and Reynolds did have a way with words.[16]

Reynolds also had a way of collecting father figures. Later on, he would cling to Samuel Southard, and just before Symmes there was Francis Glass, an obscure scholar of Latin and Greek who took on students in "a good neighborhood of thrifty farmers" in western Ohio, at a log-cabin schoolhouse "resembling more a den for druidical rites, than a temple of learning." Reynolds, having already attended the University of Ohio, went to Glass for private tutoring when he was living nearby, and wound up spending every evening with his new mentor, during whose "excursive flights" he "sat a delighted listener for hours." Since childhood, he had sought intellectual inspiration wherever it was available on the Ohio frontier. His strict stepfather, whose last name (Jeffries) Reynolds never adopted, had wanted to keep him at home to earn money for the family, so the boy dug ditches in his spare time to gain the privilege of attending school, and he openly complained of having no true parent "to guide and protect me through life." When Symmes came along, Reynolds saw him not only as an imaginative scientist but as a conduit to a larger, more cosmopolitan world, and so he embraced the theory of the hollow earth, even though it meant abandoning his beloved Glass. The old Northwest was manifestly a frontier of settlement; Reynolds was looking for a slightly different kind of borderland.[17]

––––––

"There are still within the mighty bosom of the universe," Reynolds wrote in his pamphlet on Symmes's Theory, "many unexplained phenomena." And once he added his eager voice to the captain's road show, more and more Americans expressed a desire to accompany the two lecturers on their search for the explanations, their scientific voyage to the planet's core. (The French-man Jules Verne, incidentally, intrigued by Charles Baudelaire's translations of Poe, based several of his novels—including *Journey to the Center of the Earth*—on the ideas of Symmes and Reynolds.) In the fall of 1825 and winter of 1826, the pair were a huge hit in Pittsburgh, Philadelphia, and Harrisburg, where fifty legislators were inspired to take up their case: the excited Pennsyl-vanians wrote a joint letter to Congress asserting that the impetus behind Reynolds's proposed expedition was "quite as reasonable as that of the great Columbus," not to mention "better supported by facts." Reverend Andrew Wylie, president of Washington College and a highly esteemed Pennsylvania intellectual, gave Reynolds a particularly strong boost by confirming that his lectures were "in the strictest accordance with the economy of Nature."[18]

Meanwhile, though, Reynolds was outgrowing his relationship with his latest stepfather. Contending publicly for a seafaring expedition before he had even come in contact with the sea, Reynolds became increasingly impatient with the theoretical aspects of Symmes's argument and insisted that he be given free rein to focus on the practical matter of securing funds for a voyage, which he might then be able to accompany. But Captain Symmes, staunchly committed to the purity of his theory in all its hydrological ramifications, would have none of this patricide: it would be concentric spheres or nothing.

By the time the two men arrived in New York that spring, they were lectur-ing separately. The old veteran plodded along for another few months, but soon he was so sick that he had to retire to a friend's home in New Jersey, where it took him two years of recuperating just to gain enough strength to travel back to his family in Ohio. In any case, it was Reynolds who was attract-ing all the audiences. His candlelit lectures at New York's Tammany Hall in May and June were the sensation of the season. One newspaper referred to him as "a man of education, science, and talents," noting also that "he has a fine voice, good diction, considerable eloquence, and . . . an honest, fervent zeal." While pressing for an exploring expedition, Reynolds even managed to sneak in all the most intriguing elements of the hollow-earth theory, thereby attracting New York's self-styled intellectual sophisticates, such as the critic from the *Mirror* who wrote about himself in the third person:[19]

> A gentleman of this city, who, never having heard the theory of the concentric spheres properly explained, had always viewed it as the wild chimera of a half-disordered imagination, lately attended one of Reynolds' lectures. He went, as he himself confessed, in hopes of hearing something sufficiently absurd to give good exercise to his risibles; but soon felt more inclined to listen than to laugh, and by the time the discourse was finished, became a thorough believer in what he had lately derided. Such sudden conversions, perhaps, are not the most permanent; but they are sufficient to prove that the above theory is more worthy of investigation than of ridicule.

By August, Reynolds had moved on to the nation's capital, addressed himself to Secretary of the Navy Southard, and refined his philosophy of scientific exploration. Though Symmes's first circular had proposed a trip to the Arctic, Reynolds now decided that he would leave the possible Northwest Passage to the British and push exclusively for an expedition to the wide-open South Sea. Symmes, meanwhile, had become so incensed at Reynolds's encroachment on his intellectual territory that he now wished to challenge his disciple to a duel. One newspaper editor, who found the whole situation more worthy of ridicule than of investigation, suggested that Congress "grant two appropriations, one to Reynolds for the antarctic regions, another to Symmes for the arctic regions, furnish them with cannons, and let them fight their duel" through the earth's hollow core, using "snow-balls as ammunition."[20]

Reynolds simply ignored such abuse. "An immense tract of the southern hemisphere remains unexplored," he remarked, "and to use the words of the intrepid Parry: 'Who can tell what there is where man has never been?'" His note to Samuel Southard, dated "Washington City, August 3rd, 1826," simply invited the secretary to his lecture that evening and stated his support of an Antarctic voyage "for the sake of discovery on the broad and liberal principles of science." These new core principles turned out to be based squarely on Humboldt's cosmopolitan empiricism: we don't know what's out there and how it all fits together, Reynolds argued, so we'd better join the universal effort to collect facts, make contact with other nations, and start figuring out the best ways of living in harmony with nature. "Inquiries concerning the figure of the earth we inhabit, are among the noblest speculations of the human mind. They enlarge our views, and frequently bring remote parts of the earth into a knowledge and interchange with each other." In the end, from a scientific perspective, the proposed expedition was "a plain practical common sense undertaking," by means of which we could acquire "much useful information in the hydrography and geography of the Antarctic regions; as well as many important and interesting observations on the atmospherical, magnetical, and electrical phenomena." Four days after first being contacted by Reynolds, Secretary Southard, for one, was convinced. On August 7, he wrote that "the voyage, if well conducted, cannot fail to be profitable to science—and we all, individually and nationally, owe a debt which it is time for us to set about discharging in the best way we can."[21]

Reynolds was on a roll. The arguments he used to convince Southard would also win over the nation's leading scientists and would eventually create a huge swell of public enthusiasm for a fully equipped Antarctic expedition. It had been twenty years since Lewis and Clark returned from the Pacific, and most Americans had forgotten their exploits, because, practically speaking, they had accomplished so little (neither one was a particularly astute scientist or compelling writer).[22] But Reynolds consistently reminded his audiences of Jefferson's complex rationale for sending them out, and added his own appeals to Americans' pride in bold, revolutionary undertakings, to their zeal and initiative and curiosity and perseverance and courage. More than any other person, Reynolds was responsible for spurring America's embrace of exploration for exploration's sake in the nineteenth century. The expeditions sent out west during the previous two decades—most famously, Zebulon Pike's to the Southwest (1806–7) and Stephen Long's to the Red River of the South (1819–21) and the Red River of the North (1823–25, at the end of which Long ran into Symmes and Reynolds in Ohio)—had served the fundamental purposes of establishing borders with Mexico and Canada and guiding future settlement. But the United States had no existing boundaries in the South Sea, and no need of new territories there. Reynolds managed to capture public sympathy with a much more abstract program, with a challenge to know the universe, "to step beyond the *beaten track*."[23]

Resistance came largely from strict-constructionist Southern Democrats who did not see an exploring expedition as addressing national interests. To South Carolina Senator Robert Hayne, chair of the Senate Committee of Naval Affairs, such a venture would signal the "abandonment of the fundamental [Constitutional] principles" that so far had allowed America to avoid "unnecessary connections abroad." Hayne thought it seemed "altogether superfluous to attempt the discovery of unknown lands" when the already-owned West was still largely unmapped.[24]

Reynolds, meanwhile, added a commercial twist to his argument, pitting the liberal Northern maritime regions against the conservative South: the United

States, he said, owed this expedition to the whalers and sealers of New England, who were having trouble finding new concentrations of furs. Perhaps a voyage to the South Seas would provide valuable information "concerning the Winter retreats of those sea animals, which are peculiarly interesting as sources of commercial prosperity." Science, in other words, was perfectly compatible with the expansion of the American economy—though privately Reynolds worried that the interests of commerce might interfere with his scientific goals and admitted that "'tis most barbarous to pull the fine skins off these poor animals—I don't think I should like the business." Like Humboldt, he said what he had to say to gain support for his enterprises, and, even more than Humboldt, simply tried to ignore any unsavory consequences.[25]

When it came to patriotic reasons for exploration, meanwhile, Reynolds did not have to cover up his true feelings. His flag planting on Chilean volcanoes, though not an endorsement of American expansionism, revealed his genuine desire that Americans attain the glory of scientific discovery that had always been the exclusive province of the European powers. It meant: we were the first ones here, the first to take measurements, the first to add this volcano to the international register of Known Quantities, the first to speculate about its ultimate significance. Reynolds felt his expedition could define a national culture in the spirit of Washington Irving and James Fenimore Cooper, by following European models but going a step further and mixing in a little American grit and wildness. Elite Americans of the 1820s were desperate to answer scornful European claims that there was no proper civilization in the New World—thus the fascination with wilderness, which could potentially surpass Europe's ruined castles in sublimity. Humboldt himself had commented that, "whilst contemplating these grand scenes, we feel little regret that the solitudes of the New World are not embellished with the monuments of antiquity." Cooper's novels were blatant imitations of Sir Walter Scott's romances, but Cooper added to the mix the rugged character of Natty Bumppo, Leatherstocking, the romantic half-Indian frontiersman. (Cooper also depended on certain American publications for inspiration: much of *The Prairie*, written in Paris, is lifted directly from one of Stephen Long's official expedition reports.)[26] Reynolds's voyage would be modeled on similar British undertakings, but would be accomplished by "Plain Republicans," without pomp and circumstance, without "the splendor and cost" of fancy European-style expeditions, which supplied the crew "not only with the necessary conveniences, but even with . . . proper articles to keep up with the London fashions in masquerades." Regardless of whether you can understand the science, Reynolds exhorted his fellow Americans, you can still support exploration out of "national feeling and pride." Here was a real opportunity for cultural Independence.[27]

For those who did have a scientific bent, though, Reynolds kept pushing his various theoretical constructs—especially the open polar sea, whose existence no one had been far enough south or north to disprove. His pragmatic approach to exploration might have led him simply to drop Symmes's Theory altogether,[28] but he found that the Theory spurred curiosity, and, besides, it gave him access to a whole set of arguments—mostly about the necessity of open-mindedness—that he didn't want to abandon. Need he remind his listeners of all the seeming lunatics who were eventually proven to be geniuses? "The most sublime conceptions of Newton, of which we are all so very tenacious, when first promulgated to the world, were pronounced the splendid visions of a madman! Galileo, that incomparable philosopher, died in prison for teaching that the earth is round! . . . Franklin was laughed at when he declared, that he could elicit lightning from the clouds." The key for America, Reynolds thought, was to cultivate "that inquisitive feeling, which characterizes alike the savage and the sage." Of course, his expedition would set sail "not definitely to test the truth of any theory, but for discovery on the most extended principles of science"—but could he be blamed for warming to the idea of floating freely through the upper latitudes? "Let the cold and phlegmatic call this enthusiasm if they please. Those who feel deeply on a subject of this magnitude, should not be censured for the ardor of their *feelings*." And if his conviction were to "excite the 'lash of criticism,'" let it come. Perhaps it may awaken a spirit of inquiry" in the country. At the very end of his *Remarks on a Review of Symmes' Theory*, addressing the reviewer who had pilloried his former mentor, Reynolds made it clear that what he desired most in America was the cascading current of free and open debate: "I have no objections to your writing as many reviews and criticisms as you please about the New Theory; . . . *but don't block up the Polar regions with ice!*" Then he tacked on one final, providential plea: "In the full flow of good feeling and good nature, I wish you a prosperous and pleasant Editorial career; and, if it should so happen, two or three years from this time, that you sit down to write a review of a Journal to the Southern Polar regions, it is hoped no remembrances of the present controversy will cause other than the milk of human kindness to flow from your pen."[29]

———

Reynolds's success flowed directly from his enthusiasm as an agitator. Just like Robert Ingersoll about forty years later, Reynolds sought to stir things up into a whirlpool of conversation and speculation wherever he went. He believed in participatory democracy. After gaining Southard's endorsement, he went to work on President John Quincy Adams, who surely remembered receiving Symmes's first circular back in 1818, when he was secretary of state.

Adams was convinced by November—he approved of Reynolds's lectures "as exhibitions of genius and of science"—but he feared that the expedition would "have no support in Congress." So Reynolds started traveling up and down the eastern seaboard on a kind of barnstorming campaign, "to concentrate the force of public opinion, and to *attack the people's men, by the will of their masters the sovereign people.*"³⁰

For the next three years, Reynolds floated from Annapolis to Albany, from New York to New Bedford, and even down South into enemy territory—to Richmond, Raleigh, Norfolk. He gave lectures, circulated petitions, met with mayors and city councils and state legislatures, solicited letters of support, knocked on doors, and generally made a nuisance of himself. Yet he transmitted his convictions with a highly effective combination of passion and humility. In New York, for instance, he portrayed himself as "a youth in years and a child in science" (he was in his late twenties) who was indebted to Captain Symmes for much of his knowledge and who hoped merely to inspire "some mind better qualified to do the subject justice."³¹

Public opinion quickly began to rise in his favor. "The current is now turning," he wrote to Southard from Charleston, South Carolina—which, in turn, buoyed up his energy, determination, and confidence. From Raleigh, he scrawled, simply, "Veni, Vidi, Vici—" and once, upon leaving Washington just after meeting with Southard, he scribbled a note to his new father figure saying, "I *can* and *will* do all you require, and more than you expect." Once he'd found a surrogate parent who really was capable of guiding and protecting him, Reynolds seemed to discover all kinds of talents in himself, and soon he was learning more about polar exploration than Symmes had ever known. He told Southard that when he wasn't talking to the experts, he was reading their logbooks and scientific treatises, that he had committed himself to the serious study of "all that belongs to the Pacific. . . . The field is immense before me, affording a wide range for the merchant, the philosopher, and statesman." And the range of Reynolds's knowledge and understanding was reflected in the broad-minded testimonials that he collected in his travels and submitted to publishers and members of Congress. Various citizens and elected officials wrote about their sympathy for "our hardy fishermen," who "frequently suffer shipwreck, with many privations and loss of property"; about the need for both "a more perfect knowledge of the Geography of the South" and "new channels for commercial enterprise"; about Americans' debt to the "foreign nations who have long turned their attentions" to the charting of islands and shoals and currents in the Pacific; and about the duty of a federal "Government whose political existence is in a great measure dependent on the general intelligence of the People" to respond to the "great number" of citizens from "different sections of our country" who believe that "an expedition should be fitted out and without delay."[32]

Especially impressive was Reynolds's success in gaining the respect and favor of the country's leading scientists. The solid classical education he had received in Ohio clearly placed him in the higher ranks of a society whose school system was still just developing. It was second nature for him to para-phrase Shakespeare in his letters: *"A ship, A ship, my kingdom for a ship!"* But he was entirely self-taught in science, so there was no obvious reason why he should have received the support of people like the famed entomologist Thomas Say and the artist Titian Peale, both of whom had accompanied Stephen Long out West, or leading professors like Samuel Mitchill and James DeKay of New York's Lyceum of Natural History (whose collections eventually led to the founding of the museum that sits just north of the Humboldt bust at Explorer's Gate, on the edge of Central Park). Reynolds was especially proud to tell Southard that "Dr. DeKay, who ranks among the ablest and most practiced *Natural Historians* in the United States, is about to offer his services to the Expedition—It shows much zeal in him, as he is a married man, and in easy circumstances—It shows, also, the strength of public feeling!"[33] By May of 1828, a House resolution recommended actually sending out an expedition: it seems that another Washington speech given by Reynolds in February, followed by his testimony before the House Naval Affairs Committee, had served to convince everyone both of his scientific bona fides and of his legitimacy in claiming to know the will of The People.[34]

Meanwhile, Reynolds was also conducting extensive research into the commercial implications of his proposed expedition by interviewing captains and owners of whaling and sealing ships—each of whom enthusiastically endorsed the voyage's potential for rendering their efforts safer and more profitable. And in late June 1828, Southard hired Reynolds as a special agent of the Navy Department to go back "to New London, Stonington, Newport, New Bedford, Edgartown, Nantucket, and other places where information might be found of the Pacific Ocean and South Seas," to fish out any advice from the whalers and sealers that might help the expedition navigate the final approval process in Congress or, eventually, navigate the higher latitudes. Reynolds's resulting report documents his relentless research techniques and his commitment to a broad, comparative Humboldtian framework: he did with islands, reefs, rocks, winds, currents, and marine mammals what Humboldt had done with mean temperatures and vegetation patterns—not as elegantly, to be sure, but just as thoroughly. His efforts to collate data even led him to one of the earliest statements of American concern about the limits of natural-resource extraction: "The scarcity of the whale on the common whaling ground may be easily accounted for, when it is understood that . . . our own whalemen take about eight thousand a year . . . and more than two thousand are mortally wounded that cannot be taken, making ten thousand a year destroyed by us."[35]

Reynolds was nevertheless amazed at the accomplishments of the rugged New Englanders. "I am now in the region of romance," he wrote to Southard; "I have now in my possession more knowledge of the South Seas and Pacific Ocean, by ten times, than the British Expeditions indicated when they returned. I have met with Captains just off a whaling voyage, who are more than 70 years of age—doubled Cape Horn 20 times, taken 30,000 pounds of oil, and travelled more than one million miles by sea!" Indeed, the unofficial but precise notes of the whalers and sealers revealed that "the English charts, and those of other countries, are as yet very imperfect." But Reynolds remained aware of how much more the British, French, and Russians had done than the Americans in the realm of official maritime exploration. "Perhaps it does not become us," he muttered, "to be hypercritical upon other nations, as we have as yet no maps or charts of our own to compare with them." In any case, Reynolds was indebted to the old sea dogs of southern New England, and he pledged to do all he could to recognize and add to their efforts in the charting of distant waters. The sealers, in particular, who worked in smaller vessels and higher latitudes, captured his imagination by telling him of "voyages and adventures, too, . . . of the most daring kind." Indeed, some of them had "been beyond 70° S. latitude in a few instances, . . . [where] they experienced moderate weather, a clear sea, and no land or ice to the south." Here was his best evidence yet of the open polar sea. So, regardless of the great discoveries already made by these worthy mariners, here was also an open invitation for further exploration: "the field for discoveries is still prolific." The most salient feature of Reynolds's report, in some ways, is its sense of mystery and contingency. His own discovery in New England was that currents can carry you in unexpected directions, that, especially once you hit uncharted oceans, you can never be sure where you'll find an island with fresh water and where you'll find a sunken reef—can never be sure whether the natives will be hostile, whether the pack ice will have melted, whether the most straightforward mission will return successful.[36]

Reynolds returned from the region of romance in March 1829, when Andrew Jackson became president and canceled the expedition, which Jackson's cronies jokingly referred to as the (now-burst) "South Sea bubble." (Scientists like James DeKay, meanwhile, referred to Jackson's government as "an administration which is almost beneath the contempt of the community.")[37] A

Southern Democrat with little interest in science or sealing and a particular dislike of Samuel Southard, the new president took no notice of Reynolds's popularity and felt no qualms about squashing the project to which the young lobbyist had devoted the prime of his life. Predictably, Reynolds was devastated. But, in the tradition of explorers who somehow maintained their determination despite the debilitating effects of high altitude, or sulfur fumes, or swarms of mosquitoes, he quickly absorbed his public humiliation and turned back to his connections in the private sector.

Reynolds was not enamored of Captains Palmer and Pendleton, but they had made some important discoveries in the South Sea (Palmer's Land, for one), and Pendleton certainly attracted his attention by telling stories of Deception Island, which "abounds with volcanoes" and thus with "several places where a man may stand on ice and snow, and cook his dinner in water that boils a few feet below him." Reynolds probably needed little further enticement to sign on with Palmer and Pendleton, but they both also came recommended by the man Reynolds described as "the Father of all Sealers," Captain Edmund Fanning. It was Fanning who ultimately served as the godfather of the private expedition: in the summer of 1829, he provided the initial funding and organizational impetus for the founding of a new South Sea Fur Company that included exploration as one of its goals. Later, in a very popular compilation of his maritime narratives, Fanning would remind the public that he had been lobbying Congress to sponsor an Antarctic voyage since before the War of 1812, but meanwhile he did everything he could to make Reynolds's dream come true as soon as possible. So Reynolds gave more speeches, collected subscriptions from supportive citizens and scientific organizations (the New York Lyceum donated five hundred dollars), and hired Dr. James Eights, a highly trained geologist from Albany, as well as the Philadelphian John Watson, and a couple of assistants. In September, having procured the ships and at last glimpsed the possibility of completing his frantic preparations, he wrote to Southard: "I am in glee for work, have stomach for a task, and am quite eager to be doing—I shall never be worth anything in the graver concerns of real life, until this South Sea scrape is over." The New York papers, meanwhile, urged their readers to contribute instruments and books to the expedition and eventually reported that, owing to the enlightened liberality of New Yorkers, the *Annawan* would in fact sail with an impressive scientific library.[38]

A writer for the *New-York Mirror* referred to Reynolds as "the indefatigable originator of the scheme" and acknowledged "his laudable efforts to extend our nautical knowledge." Reynolds felt that his government had betrayed him, but it was public opinion that mattered most, and the *sovereign people* were still on his side. "We had hoped," the *Mirror* column concluded, "that the national government would have illustrated its history by patronizing this plan of discovery, and we have to regret their refusal as a national loss. On Mr. Reynolds, and the daring spirits who will accompany him on his perilous but praise-worthy exploit, we cordially invoke favoring gales, continued health, and the most unbounded success."[39]

As of October 1829, the record for Farthest South had been held by the British sealer James Weddell for more than six years: he reached 74°15′ on February 20, 1823, aboard the brig *Jane*. William Edward Parry's 1827 Farthest North—82°45′—seemed significantly more impressive, but the renowned British admiral had accomplished his feat on a sledge. His two failed attempts to seek out the Northwest Passage under sail had left him bitterly frustrated and kept the world largely in the dark about the northernmost tip of the globe. Still, human beings knew even less about the Antarctic—though not for lack of trying. The European obsession with finding a navigable northern route between the Atlantic and the Pacific Oceans had been matched by unbounded curiosity about *Terra Australis Incognita*, whose mythical status had been at least partly responsible not only for Captain Cook's voyages in the 1760s and '70s but also for late-eighteenth- and early-nineteenth-century Russian expeditions under Krusenstern, Bellingshausen, and Kotzebue, and French expeditions under Bouvet, Kerguelen, Freycenet, La Perouse, and D'Urville (who was looking for the lost La Perouse). The unofficial sealing expeditions, meanwhile, with their smaller, more maneuverable vessels, had probably navigated the southern ice most successfully—but sealers were reluctant to reveal their trade secrets. And, certainly, no sealing voyage had ever been accompanied by scientists. Though a couple of captains, including Nathaniel Palmer, had reported sighting landmasses that by the 1840s would be confirmed as belonging to a great, ice-covered continent, no one was really sure at this point whether Antarctica actually existed. There were good reasons, in other words, for Reynolds to believe that a scientific sealing expedition might be able to reap a significant harvest after plowing the deep for a couple of years.[40]

It took only a few months for him to abandon his liberal idealism about the compatibility of science and commerce. Once confronted with an actual crew of sealers, aboard an actual ship, actually rounding Cape Horn, Reynolds found it impossible "to inspire these men, with the feeling, that there is something worth living for, besides money."[41] If he had lived into the twentieth century, he probably would have *wanted* to agree with the assertion of pragmatist John Dewey that "commerce itself, let us dare to say it, is a noble thing. It is intercourse, exchange, communication, distribution, sharing of what is

otherwise secluded and private."[42] Humboldt himself, writing of international trade, noted "the favorable influence which the proximity of the ocean
has incontrovertibly exercised on the cultivation of the intellect and character
of many nations, by the multiplication of those bands which ought to encircle
the whole of humanity." But Reynolds would never be able to forget the way
his exploratory goals were sabotaged by what he thought of as simple greed—
though his crew can hardly be blamed for wanting to earn their living. Only a
government-sponsored mission would be free enough to do the necessary
work, he decided—"I see something, almost every day," Reynolds wrote to
Southard, "that makes me regret the failure of our National Expedition"—
because, in the end, science and private business concerns could *not* be reconciled. Even Captain Pendleton agreed: "I am now convinced from experience
in this enterprise, that an exploring expedition, under private means, never
can produce any great or important national benefits; the same must be
under authority from the government."[43]

The Palmer-Pendleton expedition, in the end, would contribute virtually
nothing to the annals of exploration or to the international effort to expand
the scope of science. As the famous American botanist John Torrey explained,
"it turned out just as several of us suspected, that the Expedition was destined,
not for discovery, & for scientific purposes—but to catch seals!" Reynolds
would reuse the lure of commercial advantages to argue for another Antarctic
expedition in the 1830s, but he would never again actually believe that commerce was noble, even if it *could* knit the world together in a kind of Humboldtian web. In the immediate term, the prioritization of commerce meant
that he had been deprived of what would turn out to be his only opportunity to
search for the open polar sea, to explore the Antarctic's strange borderland of
internal fire and external ice, to pursue his own private Humboldtian goal of
metaphysical discovery. "We have scarcely entered the *Vestibule* of Nature's
great Temple," he had written in 1827, and now, though the altar was in sight,
he was frozen in at the pews.[44]

It must have been an especially bitter pill to swallow, twelve years later,
when his nemesis Charles Wilkes, whom Reynolds called, with telling redun-
dancy, "exceedingly vain and conceited," returned from the United States Exploring Expedition to announce that there *was* in fact an Antarctica: one of his
ships had coasted it for hundreds of miles. But Reynolds could take some
small solace in the knowledge that he had at least predicted such a landmass,
which coexisted in his mind with the warm, ice-free waters above 70° south.
In 1828, he had rightly suggested that Palmer's Land, which "extends farther
[south] than any one has yet penetrated," might be part of a big enough continent to "afford spacious harbors . . . like Hudson's or Baffin's Bay."[45]

Meanwhile, James Eights, the naturalist Reynolds had hired, published several impressive studies, dealing not only with the existence of Antarctica but also with rocky conglomerates, the first Antarctic fossil ever analyzed, the "clearness of the atmosphere," and a new ten-legged species of "sea spider." Drawing on his geologic expertise, Eights, upon finding "a few rounded pieces of granite" on the shores of the New South Shetlands, immediately asserted that they had been "brought unquestionably by the icebergs from their parent hills on some more southern land, as we saw no rocks of this nature, *in situ*, on these islands." Indeed, anticipating by several years an observation made famous by Charles Darwin, Eights insisted that these so-called erratic boulders likely indicated the composition of the Antarctic substrata. Eights was not perfectly certain that a massive continent would be discovered to the south, but he believed that, at the very least, "extensive groups, or chains of islands" lay just beyond the region of thickest ice: "The drifting fuci we daily saw, grow only in the vicinity of rocky shores, and the penguins and terns that were almost at all times about us, from my observation of their habits, I am satisfied never leave the land at any great distance." Comfortable in his professional specialization as a natural historian, Eights competently went about his business and left Reynolds to puzzle out the deeper meaning of exploration and its relation to free trade.[46]

_____

"But what place is exempt, what creature safe, from the intrusion of man! Boast as he may of his humanity, he is in a state of perpetual warfare with every living thing which can satisfy his wants or pamper his appetite, for luxuries; and his path, almost the world over, may be tracked by blood." Such were Reynolds's fiery thoughts while stranded on a small, icebound Antarctic island in January 1830. Yet there is absolutely no question but that Reynolds's own efforts to spur exploration exacerbated the slaughter of animals and facilitated humanity's general pillaging of the natural world. He noted the factory-style destruction of whales by Americans—and then showed whalers where to find more blubber. And even his jagged barbs at complacent American liberalism, at insipid New England pastoralism, at the rise of an oppressive, stultifying industrial society, at humanity's general barbarity, are usually couched within comforting patriotic narratives. His tale of being abandoned on the "savage and cheerless" Antarctic coast, "this domain of solitude and desolation," because a sudden gale prevented him and a few others from rowing their boat back to the *Annawan*, eventually closes with an entirely predictable rescue. Reynolds tended to challenge his readers with uncomfortable situations but then to render those challenges merely exotic, in the end, by reassuring everyone that all is well and America remains blessed and

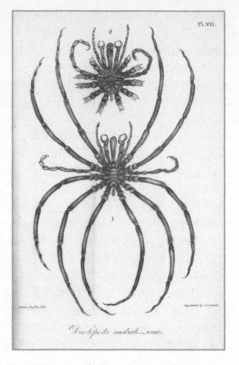

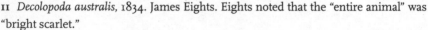

11 *Decolopoda australis*, 1834. James Eights. Eights noted that the "entire animal" was "bright scarlet."

triumphant. And yet—it is the challenges that ring out in his prose. Reynolds didn't publish this story of abandonment, these particular "Leaves from an Unpublished Journal," until 1838, the year Wilkes's expedition departed without him, so it should come as no surprise that he included some social criticisms.

Here he is, marooned with a bunch of sealers, pausing to observe his compatriots' prey: "The sea-elephant rarely runs, or turns to battle. . . . When the club is lifted above his head, or the spear pointed at his heart, he merely raises his weeping eyes with a look of supplication to his murderer, and awaits the deadly blow with the resignation of a martyr."[47] The scientist Eights confirmed that "a singular character in the habit of these animals is the faculty they possess of shedding tears when in any way molested," so Reynolds was just reporting the facts[48]—but his protest still registers. He writes in a style utterly foreign to someone like Eights, a style that calls to mind his love of emotional debate. Reynolds was much less radical than Humboldt, but, in the trenches of national politics and in the icy mists of the Antarctic, which curled up around him "like the smoke from the crater of a volcano," he learned some distinctly Humboldtian lessons about the limitations of human science and about the paradoxical universe, which is both welcoming and terrifying, resplendent and prepotent, ratifying and repudiating. We are part of it all; but which part do we see reflected in the ice? It may be natural to hunt, to intrude, but if we declare war and attempt conquest, we are setting ourselves up for a fall. Even the humble goal of understanding nature is a delicate and perhaps dubious proposition.

Despite his crew's ultimate triumph over nature's perils, Reynolds's primary experience of this Antarctic island seems to be befuddlement: he and his companions are engulfed, enshrouded, beclouded, and even the light of science is virtually powerless to burn off an Antarctic fog. Reynolds frantically takes temperature readings of the water and air and "maintained an unceasing watch," but "the faint radiation of the sun seemed hardly enough to penetrate the veil of mist that rested on the sea." It is as if all the men are as snowblind as Reynolds and Watson will be several months later while descending the volcano of Antuco. Whatever they try to examine remains "as impervious to the eye as the darkness of midnight."49

Occasionally, even without the comfort of what Humboldt called "a free and vigorous vegetation,"50 Antarctica reveals certain charms, and its explorers are granted "a flood of dazzling splendour, which was reflected and multiplied from a thousand gleaming pinnacles of ice and snow." Under such conditions Reynolds is tempted to succumb to the brilliancy, to trust the universe, go with the flow: "we leaned for a season on our oars, gazing in mute admiration on the wonders around us." But conditions can change quickly in the high latitudes. Suddenly, Reynolds and his companions had "drifted, by the force of the current, several leagues, and were not a little surprised to find ourselves, as the fog scattered . . . , completely hemmed in by icebergs, a portion of them in motion, others aground, looming like mountains around us." You can make your camp on the graceful curve of an inlet, but beware the "danger of being whelmed by the avalanches of snow and ice, which were constantly liable to be detached from the overshadowing peaks on which they reposed." And as soon as you start to feel sympathy for the whales, you can be sure they'll correct the impulse: "An old *fimbacker* rose directly under the bows of the brig, and heaved her for some distance in the water, as if determined to test the weight and prowess of the new intruder into regions where he had probably reigned and ranged undisputed for more than half a century." Even here, Reynolds retains a sense that he and his men are the ones in the wrong—yet it is hard to bond with a creature that seems capable of tipping your ship. How exactly should humanity relate to the challenges posed by nature?

Though it was something of a cliché even in the nineteenth century, I believe

Reynolds when he winds up insisting, again, that the Antarctic really does present an "unspeakable grandeur," that in the end "the description we have attempted" will fail to convey any kind of truth to "those who have had no experience of the Titan scale upon which nature has operated, and is continuing to operate in these regions." I had the same feeling in the Himalayas, on the Tibetan Plateau, gazing up at the rooftop of the world. In Washington State, if you rent crampons at the "base" of Mount Rainier, you'll already be well above sea level, so you're separated from the summit by only about nine thousand feet. But in Nepal it's common to stare up at summits more than twenty thousand feet above the trail on which you're trekking. The scale is utterly inhuman. Extreme environments remind us that our grasp of nature, as a whole, will inevitably remain rough. Ultimately, Reynolds's experience of the Antarctic led him to the Humboldtian conclusion that nature demanded both hubris and humility, both an unrelenting commitment to understand and an acceptance of our inability to achieve understanding—an immersion in "the magnitude and surpassing splendour of the *realities*" and a dependence on "all the wild legends of fairy magnificence" that our "imagination" and "fancy" can conjure.[51] How can you adequately articulate your ultimate comeuppance? It is something, as Poe might say, that we can only hope to grasp in death.

From January to April 1830, the Palmer-Pendleton expedition skirted the 60th Parallel, searching for uncharted islands and breaks in the ice pack. Assaulted by wind, water, and disease, the men got restless, and the ships made no progress southward, nor any new discoveries. So the captains consulted, and then sailed north to Valparaiso, Chile. From here, Reynolds made a few inland excursions through northern Chile and southern Peru, but he never described them, except in the brief letter to Samuel Southard that mentioned the bizarre Peruvian graveyard. Then, in July, when the ships set out to get their load of sealskins, Reynolds and Watson were put ashore in the land of the Araucanians. After several months of wandering, they met up with the expedition again the following April, and Watson shipped home. But Reynolds, completely enamored of South American volcanoes, decided to stay and explore the hinterlands more thoroughly.[52] He remained in Chile until October 1832, when he signed on as private secretary to Commodore John Downes, whose ship, the frigate *Potomac*, stopped briefly in Valparaiso when Reynolds happened to be there. Two years later, Reynolds was in New York, writing up a narrative of the voyage of the *Potomac*, and promising his readers that "the particulars" of the Palmer-Pendleton expedition and of his "travels by land through the Republic of Chili, and the Araucanian and Indian Territories to the south, will be given to the public in another volume." In the end, of

course, he never actually offered the reviewer of Symmes's Theory an opportunity to criticize his South Seas and South America travel journals. And the
journals themselves seem to have been lost, so all we have are the little pieces
of them that Reynolds published as "Leaves from an Unpublished Journal" (a
title he used several times, in slightly varying forms) or "Rough Notes of
Rough Adventures"—a few intriguing reflections of his personality and intellectual development.[53]

We know, for one thing, that in January 1831, two months after Antuco,
Reynolds and Watson "planted our staff, with a flag attached, rude in construction, but still bearing the stripes and stars," atop Araucania's other active
volcano, near Villarica. This scientific and aesthetic conquest was similar to
the previous one, though it presented the added challenge of a "region of
abysses," where the climbers had to make constant detours around "immense channels, or ravines in the snow, some of them 100 ft in depth, and
varying from 10 to 30 ft in breadth." Overall, the experience seems to have left
Reynolds with his usual sense of confusion, ambivalence, and awe. On the
one hand, he thrills to the visual opportunities presented by exploration:
"What a view did the pinnacle . . . command! Surely the traveller receives an
ample reward for his labors in the magnificent spectacles they enable him to
contemplate." But on the other hand, once he takes the time to explain his
surroundings in detail, it becomes clear that he can't actually see very well at
all. "The benumbing effect of the keen piercing wind," the abashed scientist
explains, "prevented us from making our observations as minutely as we
could have desired." Reynolds eventually gets to the panorama, which is vertiginous and filled with a "silent and awful grandeur." Meanwhile, though, he
is entranced by his view of the crater itself: "From this orifice smoke issued at
intervals, with much impetuosity; but in rushing out did not fill the entire
opening. Sometimes the wind caused it to roll back, when the dark column
would divide into several branches, which curled upward as if rising from
various apertures, but were afterward reunited in one volume as at first."
If this is the kind of "magnificent spectacle" Reynolds expects travelers to
want to see, then he must be imagining a very specific kind of Humboldtian
traveler—one attracted to the most challenging sublimity, one capable of recognizing unity in diversity and of finding a sense of connection in even the
darkest borderlands.[54]

Perhaps the most significant discovery Reynolds made on his expedition,
though, arose not from his interactions with icebergs and volcanoes but from
his confrontation with the fearsome Araucanian Indians—his one exploration experience that seems to have upended his expectations more than any
other. In 1843, introducing "Rough Notes of Rough Adventures" to the readers

of the *Southern Literary Messenger* (formerly edited by Edgar Allan Poe), Reynolds once again described the Araucanians as "unconquerable." The "armies of Spain," he explained, "while in the zenith of her power, after over-running Mexico, Peru and Central America, were vanquished and driven back by the invincible Araucanian, who, with naked valor, triumphed over the steel clad warrior." In other words: prepare yourselves for a tale of brave, noble savages, who may be violent but who are at least full of worthy manliness, thanks to their closeness to the wildness of the natural world. By the end of this ten-page narrative, though, Reynolds's readers would have realized that, as he put it, this was not just another in "the torrent of reprints, poured out in diluting streams by the mighty agency of steam." Ever-obsessed with public debate, Reynolds wanted to remind his fellow citizens, with their steam-powered presses and their mass-circulation newspapers and their rapidly expanding factories, that they still had to shape a distinctly American culture for themselves—one that ought to be based on the American scene and which therefore ought to include its native peoples, who were not simply warlike brutes nor even simply noble savages. In 1830, when Reynolds first entered Araucania, President Jackson had signed the fateful Indian Removal Act, which would determine American policy for the next several decades, and in 1838, as the Wilkes Expedition was sailing, the Cherokees walked the Trail of Tears. Now, in 1843, Reynolds decided to offer the public an image of the infamous Araucanian Indians as an independent, peaceful, sovereign nation.

Reynolds never completely escaped the role of the predictably patronizing Victorian observer: he described one of the first Araucanian chiefs, or "Caciques," he met, as a "warrior of some celebrity," having "a countenance full of those contradictory traits of character, which often render the aspect of the savage as admirable as it is appalling." But it is remarkable to watch as Reynolds gradually questions his own assumptions. This warlike, "splenetic" cacique, "named *Uaiquimilla*," paces and mutters when he hears that Reynolds wants to explore in his area—but then Reynolds suddenly realizes, with the help of his interpreter, that Uaiquimilla's rage is "related to the former wars of his people with the Spaniards, and his belief that we, like them, were come to search for gold and silver." So the violence of these Indians is not some sort of natural characteristic, but a moral response to a political situation, and it is now incumbent on the invading white man to prove that he is not like the Spanish gold seekers—or even, say, like American sealers, intent upon extracting valuable natural resources from distant lands. Indeed, when Reynolds moved deeper into the heart of Araucania and found people who "had been in less communication with the Spaniards than any we had yet seen," people who, in fact, had completely avoided "the disastrous wars of the early conquerors," they turned out to be "more friendly," and they did not "evince that jealousy and distrust which had occasioned us so much difficulty among the northern tribes."

The rest of Reynolds's experience among the fierce Araucanians was essentially a pastoral gambol. You thought they were savage hunters? No—their "farms displayed a perfection of agriculture that would have done credit to a civilized people." If he had been walking past Christian settlements in New England, he might have expressed scorn for the "beautiful meadows and shaded lawns," but here he saw only rustic charm, enhanced by a slightly wilder and larger landscape than New England could offer: "abundant crops [wheat, corn, potatoes, vegetables], interspersed with groves of wild apple trees, diversified the country through which we were passing; while the adjacent inland hills were crowned by forest trees of gigantic growth." Native communities fed him dinners of lamb and green peas, and in the morning the entire village came to see if he had slept well in the humble bed they provided. And besides being hospitable, they seemed amazingly savvy in the political realm: "These people . . . not only kept aloof from the colonial contests, but, what is better and more wonderful, they have continued almost from time immemorial at peace with all their neighbors. They lived comfortably on the fruits of their industry." The Araucanians were even "well clothed: indeed we had not seen an Indian poorly clad since crossing the river Imperial." It was only the touch of empire, in other words, that turned Indians into warring savages.

In the end, it could be said that Reynolds's encounter with the Araucanians was not really a legitimate confrontation with difference or "otherness": they turned out to be nice, gentle Victorians. But his framing of the story reminds us that he could never count on anything in his travels: what if the next cacique turned out to be the one who kills all intruders? When he stumbled upon another settlement and was told that he'd be brought before "Legen Pangi, which in the Indian tongue signifies The White Lion," he was forced to wonder how this man got his name. It turned out to be a reference to the seventy-year-old's shaggy mane of hair, though, and Legen Pangi was perhaps the most peace-loving of all the caciques, having secured a kind of retirement for himself: "His chief delight was to pass day after day on the banks of his favorite lake, and occasionally to ride out and look at his fine herds." Thus, Reynolds discovered, "'The White Lion' appeared to be a true philosopher of Nature's school." In addition, "we were informed that he was the first Indian of note who had expressed a wish that the colonies might succeed in their struggle for independence; though he never sanctioned any interference on the part of his people. . . . For more than 50 years his nation had not found it necessary to fight a single battle." And such diplomacy, even on the part of a nation that *did* have people of Spanish descent living just beyond the edge of its territory! "There are many Spanish families," Reynolds explained, perhaps presenting his ideal model of coexistence between white settlers and native peoples, "who live in harmonious intercourse with the natives, and to whom the soil affords the means of subsistence with little toil. They raise a few cattle, make and drink no inconsiderable quantity of cider, and in short revel in all the easily acquired luxuries of a lazy border life."

Of course, frontier life in the United States in 1843 would never have been described as "lazy." There were certainly places—middle grounds—where peaceful coexistence lasted a few years, but the American government was quickly turning the frontier into a war zone. And though George Catlin was valiantly trying to persuade Americans to see the value of Indian culture in his paintings, most popular depictions of Indians rendered them as utterly primitive. It must have been all the more striking, then, for readers to arrive at this strange conclusion to Reynolds's "Rough Notes":

> We had now been near seven months within and on the borders of the Araucanian country; had succeeded in penetrating all parts of the territory, from the ocean to the Andes; had ascended its principal rivers, and visited its lakes and the wild recesses of its mountains, so long deemed inaccessible to the foot of civilized man. We had seen the haughty tenants of the soil, beyond the pale of civilized influence, exhibiting their true character in the common routine of their simple, we will not say savage, life; and we felt that while the jealousy which had sometimes annoyed and retarded us was attributable to the remembrance of ancient outrage, the hospitality we had so freely and frequently experienced was the natural offspring of a noble generous nature.

Here was the triumphant explorer, who had gone where no outsider had gone before, thus proving the superiority of his race and his nation—except that his race and his nation were distinctly inferior to the kind, peaceable people he found in the "wild recesses" of the American continent.[55]

CHAPTER 5

"Mocha-Dick"

The Value of Mental Expansion

To give a history of discovery is to sketch a living picture of the universe. . . . On every side the barriers of prejudice were trodden down. The temperate zones were no longer deemed the only habitable portion of the globe. The torrid zone, instead of enclosing sandy deserts, scorched up by the intolerable heat of a vertical sun, was found to teem with organic life, and to possess a population even more dense than that of the temperate zones, together with a soil equally well adapted to the support of animal and vegetable life. The frigid zones were no longer begirt with perpetual snows, where nature, as if to amuse herself in the loneliness of her solitude, exhibited the wildest and most fantastic forms. Navigators advanced toward the north, and found that during the partial summer, plants grew, flowers bloomed, and that human beings made it their permanent residence and home throughout the year.

—J. N. Reynolds, *Address on the Subject of a Surveying and Exploring Expedition to the Pacific Ocean and South Seas* (1836)

We must learn to reawaken and keep ourselves awake, not by mechanical aids, but by an infinite expectation of the dawn.

—Henry David Thoreau, *Walden* (1854)

In May 1839, during the fortieth year of his life, J. N. Reynolds published "Mocha-Dick; or, The White Whale of the Pacific," which he described as yet another "Leaf from a Manuscript Journal." It is now almost completely forgotten, except by devoted scholars who want to know exactly what Herman Melville read at bedtime. Like its much more famous descendant, though, this little story of a recalcitrant whale and the people who hunt it is ultimately a microcosmic assessment of American industry, culture, and politics, as well as a cosmic meditation on humanity's relationship to a huge, inscrutable nature. And "Mocha-Dick" also contains, on its very first page, one of the most striking images in early American literature—of a whaling ship, "seen from a distance of three or four miles, on a pleasant evening, in the midst of the great Pacific. As she moves gracefully over the water, rising and falling on the gentle undulations peculiar to this sea; her sails glowing in the quivering light of the fires that flash from below, and a thick volume of smoke ascending from the midst, and curling away in dark masses upon the wind; it requires little effort of the fancy, to imagine one's self gazing upon a floating volcano."[1]

Off the Chilean coast, near the isle of Mocha, on the border between water and land, Reynolds saw another opportunity to explore the border between people and nature, and he found his countrymen embroiled in a potentially eruptive situation. Primal forces were at play. The tough, reliable New England whalers had dipped their boats into the sea, stirred up the Pacific, bloodied their harpoons, and made off with their prey. Now, with their ship borne along by the Humboldt Current, they were boiling blubber down into the oil that would light their homes and drive the ongoing modernization of their country. They burned fuel to make more fuel, to harness the universe's energy in a perpetual cycle of fire.[2] But, as Reynolds perceived, bubbling cauldrons of whale oil, like live whales whose home waters have been invaded, behave in unpredictable ways. The extraction of natural resources is never an easy business: mines can cave in, mud can slide, oil can spill and burn.

For a man so intimately acquainted with the power of volcanoes, such imagery must have been fraught with meaning—though Reynolds leaves us to decipher it for ourselves. Are the whalers fighting the current, playing with fire? Or have they themselves, through their manliness and ingenuity, attained the force of nature? Perhaps American industry is just about to explode onto the international scene. Or, then again, perhaps the ship of state is about to go down in flames. Are Reynolds's industry-minded countrymen more or less savage than the Araucanians?

Reynolds himself, having adopted a thoroughly unsettled—some might say uncivilized—lifestyle, had never held a real job. It is unclear how he even supported himself in the late 1820s and 1830s. He had edited a newspaper, given lectures on the earth's core, lobbied for the exploration of the South Seas, sailed to the Antarctic, written a book, given more speeches, and finally, in the year before he published "Mocha-Dick," he had been deemed unworthy of a position on the official expedition he had at last succeeded in launching. Now Reynolds would have to find a way of continuing to be an itinerant intellectual in a society with very little room for nonindustrial activity. In the Northeast, factories threatened to rule the day. Men of any ambition at all

12 Detail of a map of Chile, 1823, showing the isle of Mocha and the southern Andes, including the two volcanoes Reynolds climbed and wrote about, Villarica and Antuco. (Antuco is called Tucapel here—not to be confused with the Antojo volcano.)

were specializing in one particular area of business and then attempting to reign over it by investing in mass production. But Reynolds was a natural amateur, a renaissance man with what seemed to be illimitable interests. Moreover, having dabbled in newspapers early in his life, he was keenly aware that, with the advent of the penny press, writers were now a dime a dozen: "the torrent of reprints" that he so detested signaled the commercialization of literature, the new need for writers to compete in a tangled, overgrown marketplace of ideas.3

As America entered an age of mechanical reproduction and gave birth to a culture saturated with print, only a few professional writers managed to eke out a living—Edgar Allan Poe, Washington Irving, James Fenimore Cooper—and they had to produce stories, poems, sketches, novels, as if they themselves were running under steam power. Someone like Poe, who was consistently impoverished, wound up experiencing crises of both status and originality. When you go back to your desk in the dark library and attempt to turn your mental crank, how can you be sure you'll produce something worthy of a society that cares only for newness? Perhaps you'll merely sit there being tortured by a raven, which repeats the same ominous word over and over again. Or perhaps you'll end up as another Rip Van Winkle, who likes to tell "endless sleepy stories about nothing" and thus becomes a favorite among the village children but never quite makes ends meet; whose wife constantly upbraids him for "his idleness, his carelessness, and the ruin he was bringing on his family"; who finally falls asleep for twenty years and then awakes, bizarrely, to behold "a precise counterpart of himself" and to "doubt his own identity" and finally to realize that he has slept through a "revolutionary" era and that his society has now passed him by. American literature, filling the lives of the reading public in the 1820s, '30s, and '40s, was itself filled with repeated words and doubled authorial characters, expressing a basic anxiety about what happens to writers and ideas when a nation is defined by industrial production and reproduction.[4]

Reynolds was not even sure that he was a writer, but he knew he wasn't an industrialist. And, as of 1839, he suddenly realized that he wasn't going to get to be an explorer again. After several years of being attacked in the press by important Jacksonian politicians, he had finally been stripped of his reputation and his most important opportunities. He responded by having a giant white whale attack a floating symbol of his explosive society. But he had still not given up on America or on the core ideas that had hurled him into the public eye. Indeed, ever since his return to the United States in 1834, by imaginatively revisiting his South Seas experiences, by reinvestigating the borderlands of whaling grounds and volcanic ruins and coastal currents, Reynolds had been developing the debates he first stirred up in the 1820s—about exploration and natural resources and about the borders between science and commerce, literature and industry. It was not until the mid-1840s, when America turned seriously to the problem of settling the West, that the country's foremost Humboldtian explorer faded from view. Meanwhile, even into the early 1850s, Reynolds's influence rumbled to the surface again and again in the realm of culture—and especially in the quest-driven tales penned by Poe, Irving, and Melville.

———

Reynolds's own literary fame came immediately upon the publication of his 1835 book about the circumnavigation of the *Potomac*, the ship whose crew he had joined in Valparaiso in the fall of 1832 after completing his explorations of South American volcanoes. Yet the book for which he was best known was also the one he seemed to care about the least. It was a commissioned volume, a project meant to document the completion of the *Potomac*'s mission in Asia, the daring feats accomplished by Commodore Downes long before he picked up Reynolds in Chile.[5] Once the *Potomac* arrived in Boston in May 1834, Reynolds did set about his task enthusiastically, but not before firing off a letter to Samuel Southard, now a senator, suggesting that what he really wanted to do was prepare once again to sail southward:[6]

> We arrived here on the 23rd [he wrote, on May 25]. . . . To me it seems more like some delusive dream, than reality, after an absence of more than four years. Four years, what a period in one's life, and yet a mere speck in the annals of time. Four years, what have I not suffered and gone through during that time? My rambles by sea and by land have been more extensive, than you can form any idea of. . . . I was delighted to hear of [your] election to the Senate, and am still more pleased to see how you are paying off and settling old scores with your political opponents—I wonder if they would allow the . . . South Seas [Expedition] now? . . . Do you know I am more anxious than ever to go on that enterprise, and much as I am delighted by my return, I would most cheerfully embark tomorrow;—it must yet be done though three years hence!!!

Over the next several months, besides writing the *Voyage of the* Potomac, which entailed a careful review of the ship's log kept by the previous secretary to the commodore (who had died aboard ship), Reynolds read over his notes on his own expedition, sorted his natural history specimens, and traveled throughout New England in a renewed effort to convince scientists and lawmakers to join him in lobbying for an official (noncommercial) expedition to the South Seas. He met with politicians in Rhode Island, saw Benjamin Silliman again in New Haven, and visited the Lyceum in New York, surprising old friends with fossil fragments and tales of smoking volcanoes. By August, Reynolds was telling Southard that if the senator would merely raise the possibility of an expedition in Washington, then "I will bring a few States to back it, at the next Session." And by mid-December, Reynolds's ambitious schemes were once more spurring debate in the halls of Congress.[7]

Meanwhile, ensconced at Tremont House in Boston during the colorful fall months, Reynolds continued his writing and became friendly with the scientists of the Boston Society of Natural History. Founded just four years previously by doctors and naturalists who, like Reynolds, tended to be neither mere amateurs nor pure professionals, the society was one of the earliest of many such organizations in the United States designed "to promote a taste, and afford facilities for the pursuit of Natural History, by mutual cooperation, and the collection of a Cabinet and Library." There were lecture series, seemingly modeled on Humboldt's talks in Berlin in the late 1820s; different periodicals designed for experts, students, and the general public; and frequent exhibitions, with the main "cabinet" of specimens open "every Wednesday from 12 to 2 o'clock, for the free admission of any persons, whether young or old, who might wish to examine it." Reynolds seems to have made frequent visits to the cabinet and library, which were housed just down the street from his hotel, in "a large room in Tremont Street, over the Savings Bank," where specimens were arranged for optimal entertainment and instructional value.

The entire institution must have enchanted Reynolds: just months after he had departed what seemed like a hopelessly backward nation, whose government would not heed the people's cry for exploration, here had arisen a group of men explicitly devoted to science for science's sake, to the cooperative gathering of facts and manufacturing of theories, to the free and wide diffusion of ideas for the sake of spurring public enthusiasm and debate. The society could hardly be described as radical—its members were largely Harvard-educated Boston Brahmins—but it did carve out a friendly, fraternal space beyond the border of competitive commerce. In 1834, it was just the place for Reynolds, as it would be for Thoreau fifteen years later, when the Concord eccentric entered his own intense Humboldtian phase.

As early as July, according to the society's first accession book, Reynolds began to hand over his collections. His first gift was "an Herbarium, containing 51 species of Plants from Chili and 15 from the Gallipagos [sic] Islands," and a page later the society's recording secretary, D. Humphreys Storer, M.D., noted in his perfect cursive that Reynolds had also donated "a sheet of Colored Drawings of rare Fishes." By November, the society had elected Reynolds a corresponding member and offered him its official thanks, and in later years his collection would be considered one of the cabinet's three most important foundational gifts. Reynolds turned over his most significant stash to the society in October: accompanying a mysterious skull, there were "441 specimens of Birds from Chili, Peru & S. Shetland Islands; Botanical Specimens from Chili, Peru, Araucania & the Gallipagos Islands; Several boxes of Minerals and Organic Remains from the Southern Andes; A large and valuable collection of shells, among which were many rare species; . . . Nests & Eggs of various S. American Birds; . . . [a] Box of Insects from Chili; [and] Colored Drawings of numerous insects, fishes, fruits, etc. collected in the Pacific Oc. and S. America—." Remembering the attacks of his political opponents, who had questioned his qualifications for organizing any kind of scientific expedition—and they would do so again in coming years—Reynolds thus deposited the evidence of his Humboldtian commitments with people he could trust not only to uphold his values but also to help solidify his reputation.[8]

Their help became almost unnecessary, however, once *Voyage of the Po-tomac* appeared in the spring of 1835. The first edition, which sold out in a month, attained its popularity thanks to its patriotic subject matter rather than its authorship: almost the entire literate public knew that the *Potomac* had been sent out in the summer of 1831 to punish an act of piracy perpe-trated against an American trade ship on the coast of Sumatra, and now read-ers wanted to hear the whole story in all its gory detail, complete with its triumphant, happy ending. In February 1831, some Malays in the port of Quallah-Battoo had stolen aboard the *Friendship*, out of Salem, Massachu-setts, and killed three crewmen in an attempt to seize the ship and take back a cargo of pepper; exactly a year later, the *Potomac* arrived in Quallah-Battoo with Commodore Downes in command, demanding indemnification for all losses and immediate imprisonment of all guilty parties. But, as the majority of the local rajahs had no interest in parley—a sentiment they made clear by almost immediately firing upon the approaching Americans—a battle en-sued. There were fatalities on both sides, but many more among the Malays, with the marines of the *Potomac* taking several forts and chasing large groups of natives into the jungle. Reynolds's *Voyage*, in other words—on the surface, at least—was more tale of conquest than exploration narrative.[9]

The book's ongoing popularity, though—edition after edition flew out of the hands of booksellers over the next several months—was matched by criti-cal acclaim, and the nation's intelligentsia embraced Reynolds not only for producing a "graphically sketched" story but also for self-consciously "travel-ling beyond the record of the voyage"—for adding details of ethnography, natural history, and politics—and thus making "a valuable addition to our geo-graphical libraries." The reviewer for the *Southern Literary Messenger*, though he somewhat scornfully recalled Reynolds as the principal defender of Symmes's "indefensible" theory of the earth, nevertheless insisted that "his book will gain him reputation as a man of science and accurate observation." Reynolds had been handed a project designed simply to impart glory, and he turned it into a broad-minded Humboldtian narrative. While often celebrat-ing certain American values and accomplishments, the book, especially in its tangents and asides, also becomes a deeply ambivalent examination of Ameri-can power. The ultimate message of Reynolds's dedication, for instance, is one of humility: "With all the enterprise of our countrymen, their navy and commercial marine, still we can say,—'Of this huge globe, how small a part we know.'"[10]

———

J. N. Reynolds was obsessed with the unknown. Just like Humboldt. "A little mystery . . . ," Reynolds had once written to Samuel Southard, "is very important in all great undertakings." Admittedly, when Reynolds explicitly mentioned the name of "the learned Humboldt" or referred to "the observations and analysis made by Baron de Humboldt," he was invoking not inscrutability but rather his mentor's reputation for knowledge, accuracy, and fanatical empiricism. Humboldt had discovered that the water of the bay at Callao, Peru, for instance, was "two degrees colder than any other on the coast [of South America]," presumably by taking temperature readings at every port he encountered; at Lima, he explained temperature fluctuations by linking them to fluctuations in atmospheric pressure, taking barometer readings all through the day and night to prove his point. So, Reynolds said to his readers, here, let me give you these latitude and longitude markers; this data on winds and currents and tides; these comparative altitudes; these demographic tables—it is what Humboldt would have done. There is a massive meteorological chart in Reynolds's appendix giving daily weather data for the *Potomac*'s voyage—though only for the last year or so of the trip, while Reynolds himself was on board to take the measurements. Yet the real lesson Reynolds drew from his reading of Humboldt had to do with the *limits* of measurement. *Voyage of the* Potomac is a book about "contrary currents," about "fickle and treacherous elements," about the experience of watching as "a towering iceberg, shrouded in a cold mist and fog, [moved] slowly on, by the power of deep currents, from the gloomy and cheerless regions of the south." Impenetrable fogs, unfathomable currents, ever-changing winds, unpredictable eruptions: these are the constants of Humboldtian travelogues. The Antarctic had taught Reynolds that much scientific theorizing represents "a futile attempt to investigate secret things." So, modeled squarely on Humboldt's *Personal Narrative* and *Political Essay*, Reynolds's most famous work drifted along the border between empiricism and Romanticism, fact and mystery, science and religion—not to mention the border between pride in the idea of American independence and shame over his country's actual reliance on the exploitation of blacks and Indians, between a blind belief in the superiority of white European civilization and an intense respect and sympathy for every culture and society in the world.[11]

"What a field is here presented for philosophical speculation!" shouted Reynolds, echoing an earlier exclamation from his *Remarks on a Review of Symmes' Theory*. Now he was investigating all the interlinked earthquakes and volcanic eruptions that had ravaged South America's Pacific coastline, the rugged strip of land washed by the gentle Humboldt Current. We are back to the question of internal fire. "Is the whole range of South America, west of the Andes, resting on and slumbering over unfathomable caverns of combustible materials? And are not these connected beneath?" The idea, as usual, was to shake people up. Do you really know what is happening beneath your feet? As Humboldt had explained, "the history of the globe informs us, that volcanoes destroy what they have been a long series of ages in creating. . . . Happy the country, where man has no distrust of the soil on which he lives!" Reynolds, then, was still following in his master's footsteps: "Those who dwell in a tranquil country, seldom visited by the slightest terrestrial vibrations, can with difficulty form an adequate idea of those terrible convulsions of the earth which ravage and lay waste the largest and most splendid cities, and overturn the very mountains, in countries less favoured by nature in this respect. We are accustomed to look upon the earth with a conviction that it is solid and fixed beneath our feet, and few of us can realize that it has been, and is still, in some parts of the world, subject to undulations more terrific than the mightiest surges of the rolling ocean." An active citizen, though, should always be interested in overturning his convictions; an avant-garde artist should go looking for surges and undulations.[12]

For Reynolds, it is motion that defines life. There is nothing more difficult for him to bear than a becalmed ship, "the tedious monotony" of a "lingering passage." And he lives in fear of tropical cities where the air and water are still. "Look at its position!" he wrote of Batavia (now Jakarta, the capital of Indonesia): "So near the equator,—surrounded on all sides by stagnant waters, fens, bogs, and oozy ditches—every street intersected by canals," which had to "find their tardy way to the sea, by channels which had scarcely any current!" Give him rolling rivers and roaring gusts. "How indispensable to the salubrity of our earthly dwelling-place, are the ever-moving and changing winds. They may be called the exercise of our atmosphere, by which it preserves its healthful principles, and shakes off the terrible evils of stagnation"— the very evils you may be all too familiar with if you spend every day working in a factory or bank in the northeastern United States, if your life is as repetitive as that of Herman Melville's character Bartleby the Scrivener.[13] Reynolds would have you see the world, stand into its refreshing gales, ride its currents, feel its temblors. His long tangents on phenomena like the Gulf Stream and the trade winds, as in Humboldt's writings, are meant to spice up the narrative by simultaneously deepening both his readers' understanding and their interest in learning more. The artistic images and scientific treatises in the book always beautifully reinforce Reynolds's ultimate goal—the goal of getting his readers moving, and thinking, in new ways.

Stranded at home in 1830s New England, you may have a hard time appreciating the worth and importance of different peoples and societies. Much of *Voyage of the* Potomac, reflecting Reynolds's powerful experiences in Araucania, deals with Americans' preconceived notions about so-called savages.

Comparing white dealings with indigenous peoples in South Africa and the United States, for instance, Reynolds insists that "the true character of the natives in both countries has been but little understood, and much misrepresented." All too often, "the kind reception and hospitality of the natives have been requited by acts of rapine, cruelty, and oppression." Indeed, Reynolds hoped that by pointing out the admirable qualities of native peoples around the world, he might elicit more sympathy among his fellow citizens for "the much wronged and oppressed aborigines of our own country." Maybe his urbane readers simply didn't remember the massacres of Indians in New England, the execution of noble Annawan at the end of King Philip's War; maybe they weren't fully aware of frontier realities, of the daily brutalities and injustices endured by Indians in the West, not to mention slaves in the South.

Most of his readers did know that the *Potomac* had been sent out to put the Malays of Sumatra in their place, and Reynolds offered a triumphal narrative, complete with certain damning stereotypes about the Sumatran natives—but he also suggested that the natives were not entirely to blame: "The Malay— treacherous, cruel, and vindictive as he is—fierce and unrelenting as the tiger of his own mountains, by which he is often destroyed—is still a being entitled to the sympathy and compassion of the civilized world; and we cannot but pity his condition, even when his vices demand a measure of punishment at our hands. . . . For three centuries, what has been the history of Europeans trading on his coast, under the direction of heartless, grasping monopolies, but a record of oppressions, cruel exactions, and abominable injustice!" If the Malays now seemed more animalistic than human, it was only in response to the "insatiable avarice" that had defined how the "civilized" Europeans had treated them.

Your ignorance of the fires burning inside the earth, then, may well be matched by your ignorance of what lies inside the earth's people. Reynolds did not do as much scientific work as Humboldt to prove the ultimate equality of all human beings, and he was too politic to launch an outright attack on the slaveholders of the American South, but he did condemn the "civilized" countries' "horrible traffic in human flesh" as the ultimate "disgrace of humanity." And he thought it ridiculous that some white people deemed other races to be "too ignorant or unfit for self-government." It was never a question of fitness, to Reynolds, but of oppression. You could form an accurate judgment of native peoples only if you traveled far enough to see how they arranged their society beyond the influence of any invaders. As to the invaders themselves, they could not be judged by their arts and industries alone: you also had to consider their moral abuses. At times, Reynolds could not avoid thinking in terms of social hierarchies and making generalizations about the "heathen mind" and "civilized improvement." But, like Humboldt, he also pointed out that all "civilization" rested upon savage exploitation of both people and natural resources. Take the case of Spain: "From whence issued those immense streams of wealth which flowed from the colonies into the lap of the mother country, during the three hundred years of her tyranny and dominion, but from the poor and subjugated Indian? Who can reflect, without horror, on the destruction of eight millions of these wretched beings, who, in Peru alone, perished under the cruel and unjust exactions of the *Mita*?" Reynolds never attained Humboldt's ecological sophistication, but he understood *social* ecology, and he understood how exploration—and exploration narratives— could make people more curious about the wondrous capacities of both nature and humanity.[14]

As usual, in *Voyage of the* Potomac, Reynolds's fundamental message was: we desperately need to keep exploring. The United States must not stagnate; his compatriots must keep pushing for "discovery and advancement in every department of human science and knowledge."[15] With Samuel Southard chairing the Senate's Committee on Naval Affairs, and with the country enjoying an economic climate marked by rapidly expanding investment, it seemed a propitious time for Reynolds to redouble his efforts to procure a South Seas exploring expedition.

In the early evening of Saturday, April 2, 1836, as the sunlight disappeared from Washington's flowering trees, a sense of mystery filled the main hall of the Capitol. Reynolds would have it no other way. Hundreds of candles flickered as the congressmen and other interested citizens bustled around the chamber and whispered in anticipation of the upcoming lecture. Did this agitator, this ceaseless wanderer, still believe in Symmes's Hole? Did he have any interesting scars from falling ice or boiling lava? Wouldn't this whole enterprise be too risky, too uncertain, too expensive? Finally, at about 7:30, Reynolds stood up—"a firmly built man, of medium stature, with a short nose, and a somewhat broad face"—and took the podium. His opening quotation was from George Washington's Message to Congress on December 7, 1796: "The assembly to which I address myself, is too enlightened not to be fully sensible, how much a flourishing state of the arts and sciences contributes to national prosperity and reputation."[16]

For the next two hours, Reynolds tried to uphold exploration as the ultimate expression of art and science. "As long as there is mind to act upon matter," he insisted, "the realms of science must be enlarged; and nature and her laws be better understood, and more understandingly applied to the great purpose of life."[17]

The lecture was a tour de force of stamina and persuasive eloquence. It would be published a few months later, and Poe would steal passages from it word for word to include in *The Narrative of Arthur Gordon Pym*. Can you imagine any twenty-first-century speech shown on C-Span being lively enough to make it into a popular work of fiction? Reynolds knew how to set scenes and tell stories, how to spark imaginations. And there was something poetic about him. For all his electric charisma—and this is another way in which he seems eerily similar to Humboldt—Reynolds came across as a rather melancholy person. A listener at another of his lectures recalled his de-meanor as "in a high degree respectful and earnest and withal very sad, as though some great sorrow lay upon his heart, which won our sympathy."[18]

The congressmen were sharp-minded, pragmatic people focused pri-marily on the virtue of economy; but they had also gotten caught up in the nation's recent fascination with the economy of nature. It would be commer-cially worthless to attain the South Pole, and for that reason alone some repre-sentatives would vote against Reynolds's expedition. Others, though, might be intrigued by the bizarre prospect of seeing the world from the angle of 90° south—"yea, to cast anchor on that point where all the meridians terminate," Reynolds cried—"where, amid the novelty, grandeur, and sublimity of the scene, the vessels, instead of sweeping a vast circuit by the diurnal move-ments of the earth, would simply turn round once in twenty-four hours!" Surely this would be a worthwhile and original experience—to be carried, to be moved and yet not moved, not by mere wind and tide and current, but by the very spinning of the earth?[19]

Reynolds recapitulated many of the arguments he had made in the late 1820s: regardless of whether or not the expedition ever reached the South Pole, it offered the "more comprehensive promise" of expanding trade; pro-tecting the country's whalers and sealers; making a scientific contribution to the international community; and, quite simply, enhancing America's reputa-tion. In particular, Reynolds elaborated on the country's debt to its marine in-dustries and urged sympathy on his listeners for "families deprived of parents and children" by oceanic "accidents" that could easily have been avoided. "I put the question to every liberal-minded, intelligent individual, within these walls: is it honorable, is it politic or wise—waiving the consideration of hu-manity and duty—to look supinely on, while our citizens are exposed to ship-wreck in seas, on coasts, and among islands, of which they possess no charts capable of guiding them aright?" Reynolds gave his audience vivid images of a rapidly changing world, in need of constant monitoring: "the dark volcano, bursting from the depths of the sea, pours its broken fragments and molten lava above the level of the waters, and, by the decomposition of its surface, is rapidly converted into an island."

Reynolds also found it expedient to manipulate his audience's fear of island-dwelling cannibals. Overall, his rhetorical treatment of natives remained judicious: he explained that many islanders were in fact "affable, hospitable, and honest" and that, when violence did occur, "these untutored beings have not always been the first offenders." Indeed, he hoped that the expedition he proposed would also serve diplomatic ends: "it will be a matter of national pride that our country should be the first to . . . retrieve the character of civilized man, and in some measure atone for the accumulated injuries which centuries have seen of daily increasing enormity." Acknowledging that whalers and sealers often did not behave responsibly, he figured that increased contact between official American explorers and natives might actually help "protect the latter against the wanton cruelty of men claiming the appellation of *civilized*." But Reynolds also suggested that the expedition should bring a frigate along, as well munitioned as the *Potomac* had been, to use military intimidation when diplomacy didn't work, and to gain revenge when evidence arose that American sailors had been captured, tortured, or even "massacred by savages." Some of the most visceral passages in the *Address* tell stories of the deceptions and brutalities suffered by whalers at the hands of natives inhabiting paradisal islands. This is what sometimes happens on frontiers, though: Reynolds's cautionary tales about vicious cannibals smiling and waving and "bringing bread-fruit and yams," for all their obvious racism, were also realistic warnings to Americans that they had to pay closer attention to what was happening at the edges of their country, where societies clashed. Which groups were warlike and which friendly? Which had developed intriguing alternatives to "civilized" life? As in Herman Melville's first novel, *Typee* (1846), whose narrator admits his disgust at being held captive by savages but also delights in his distance from "the vices, cruelties, and enormities of every kind that spring up in the tainted atmosphere of a feverish civilization," the stereotypes of exotic savagery in Reynolds's writings ultimately force his readers to realize how narrow their horizons are.[20]

For those Americans unwilling to hear even the slightest criticism of their civilization, meanwhile, and for those congressmen especially concerned with concrete results, Reynolds simply insisted that a South Seas exploring expedition would bring home priceless benefits, in the form of both information and material resources. In the conventional language of the day, he touted American empire and enterprise, celebrating the national character of a people compelled "to pierce unexplored forests and tame the howling

wilderness," painting a picture of ever-expanding markets for American manufactures and constant supplies of oil and furs and all the "great diversity of
the products of the earth's surface," which could "be turned to good account
in the trade between inhabitants of its more distant portions."

With the pragmatic goals firmly established, though, Reynolds felt comfortable reinforcing his commitment to more abstract scientific aims—even,
perhaps, to a mode of life that questioned the very values of commercial capitalism, that repudiated his "merchant" society, "engrossed in the pursuit of
wealth." That simple participle, "engrossed," suggests Reynolds's nightmare
of a citizenry that has become fat and lazy, confining itself to narrow, stagnant
pursuits. "Let us not, then, forget that wealth to nations, as well as individuals,
is a means, and not an end; and that the most awful reverses have befallen
those who have disregarded this unchangeable law, and forgotten that the accumulated harvest of riches arising from past exertions, was intended as the
seed of future enterprise. No! we cannot remain stationary. If we cease to
move onward, that instant we retrograde, and our prosperity, like the stone of
Sisyphus, will bear us along with it down the precipitous descent, into the
depths of national effeminacy." It was a sign of cultural weakness, in other
words, to use money simply to make more money. "Shall the reproach for
ever rest upon our character, that we can do nothing, think of nothing, talk of
nothing, that is not connected with dollars and cents?"[21]

Reynolds thus saw his expedition as an alternative to the rat race, an escape
to a realm of "high and daring adventure" populated only by rugged men. No
matter how good the charts ever got, in the "unfrequented bays" of the South
Seas, "everything depends upon a kind of instinctive, intuitive sagacity and
foreknowledge of approaching danger, which nothing but a constant exposure to appalling hazards can ever give." Rugged men also ran the capitalist
system on shore, but they were men without ideals, narrow-minded slaves
to profit—men whose ruggedness would lead to "effeminacy." At the very end
of his talk, Reynolds actually appealed to *women* as the true visionaries
of human society, as the people most closely "identified with this subject" of
exploration—for "it was from the sagacity and generosity of one of your sex,—
the high-minded Isabella, queen of Spain,—that this continent was discovered at the time it was," while other "monarchs hesitated, and ministers
looked on with cold and calculating indifference." In this wonderful confusion of gender roles (which sex is more sagacious?), what emerges is an endorsement of a vigorous Humboldtian humanism, an instinctive embrace of
the universe, a celebration of curiosity. Reynolds read out Jefferson's instructions to Lewis and Clark in full during his talk, and insisted that the document
ranked "second only to the Declaration of our Independence" in displaying

the president's genius, because it revealed his engagement with absolutely everything, with *"animate and inanimate nature, the heavens above, and all on the earth beneath!"* What is most fully human, what connects men and women, is the pursuit of goals for the sake of the pursuit itself. Often, Reynolds explained, "we have been asked the question, 'What advantage has Great Britain derived from her endeavours to find a northwest passage, and what does she still promise herself in the prosecution of a design which, even if accomplished, can never lead to any practical benefit in carrying on the commerce of the world?' We answer that the question, *cui bono?* [for whose benefit?] should never be put in affairs of this kind. Scientific research ought not to be thus weighed." Scientific research ought to be pursued because we have always felt the urge to explore mysterious questions, and because such explorations lead not only to knowledge but to an endless proliferation of even more delicious mysteries.

And with a mysterious smile to the ladies—none of whom ever seems to have charmed him as much as his adopted father figures or his expeditionary companions—Reynolds bowed and walked into the standing, cheering crowd.[22]

———

Members of the House and Senate argued furiously over the value of exploration. In the maritime regions, of course, Reynolds enjoyed celebrity status, and the New England representatives embraced his cause, making use of distinctly Humboldtian rhetoric to insist that what was good for their local sailors was good for the entire country: "Our vision is limited to a narrow circle, but if we would impartially consult our understandings, they would teach us that there often exists a mutual and close connexion between interests apparently disconnected and remote." Reynolds's friends from Ohio, Congressmen Thomas Corwin and Thomas Hamer, also gave the expedition a huge boost, lining up the landlocked western states in support of their seafaring son. Hamer not only told his colleagues that he had "known Mr. Reynolds from his boyhood" and that they came from the same "neighborhood" but also made the point that this man of "pure principles and fair character" had designed the exploring expedition to benefit their home region as much as any other in the country, by potentially expanding the "market for the surplus productions of our fertile soil." Meanwhile, as usual, a harsh, mocking opposition arose from Jacksonian Democrats in the South. Representative Albert Hawes of Kentucky asked if the House "should next have a proposition for a voyage of discovery to the moon," expressing his disdain for Reynolds and his "chimerical and harebrained notion." What could possibly be the purpose of this scheme? "To take the vessels and seamen of the United States, and send them to the South seas, exposing them to all the diseases, hurricanes, and mishaps of that climate; and for what? For nothing on the face of the earth." If Reynolds was interested in what went on *inside* the earth, he could go visit such places on his own dollar.[23]

Overall, though, the debate was going Reynolds's way: the current of public opinion was again sweeping him toward his dreamed-of destination. In March, Senator Southard had released a resoundingly positive Report of the Committee on Naval Affairs, which endorsed the expedition as furthering "the common cause of all civilized nations—the extension of useful knowledge of the globe which we inhabit." Reynolds himself, meanwhile, had acquired testimonials from commercial organizations like the East India Marine Society; from whaling and sealing captains; from newspaper editors; from various naval officers, including Commodore Downes of the *Potomac*; from several state legislatures; and from his most famous scientific friends, his fellow Humboldtians Benjamin Silliman, Asa Gray, Charles Pickering, and James DeKay. Silliman wrote a particularly detailed letter, urging the expedition to take along experts in meteorology, astronomy, zoology, geology, mineralogy, topography, osteology, entomology, ornithology, botany, and ethnography, not to mention people capable of making accurate drawings of plants, animals, and people, and of studying magnetism, volcanic activity, and ocean currents.[24]

As evidenced by Reynolds's published correspondence with his scholarly colleagues, American science had blossomed in the 1830s—and even President Jackson, proud of his nation's growing ability to compete with countries of the Old World, had been won over to the cause. In 1832, just three years after having squashed Reynolds's initial expedition, Jackson facilitated the resuscitation of the U.S. Coast Survey—a program launched by Jefferson back in 1807 but dismantled in 1818 as a matter of economy. Now, in 1836, as Edgar Allan Poe noted in a review of Southard's widely circulating report, "we have astronomers, mathematicians, geologists, botanists, eminent professors in every branch of physical science. . . . We possess, as a people, the mental elasticity which liberal institutions inspire, and a treasury which can afford to remunerate scientific research. Ought we not, therefore, to be foremost in the race of philanthropic discovery?" This decade marked a radical turning point in the country's scientific development, because suddenly, with Jackson the unlikely sponsor, the government became interested in collaborations with universities and nonmilitary academies. In his *Address*, Reynolds had expressed his anxiety about how unappreciated men of science and literature were in his new, commercial society: "we have been a by-word and a reproach among nations for pitiful remuneration of intellectual labours." But he had spoken in the kind of language people like Jackson could understand, and within a decade the goal of putting American science on a par with that of France, England, Russia, and Prussia would lead to the establishment of the Smithsonian Institution and the general rise of Humboldtian science in Washington, through people like the Aberts, John Charles Frémont, Matthew Maury, and Alexander Dallas Bache. The immediate effect of Reynolds's rhetoric, though, was to help his own personal cause. On May 14, 1836, just six weeks after his speech to Congress, President Jackson signed the bill authorizing the United States Exploring Expedition.[25]

If Reynolds rejoiced on that day, the new law inspired only despair for the man assigned to bring it into effect: Mahlon Dickerson of New Jersey, the staunchly Democratic secretary of the navy. Dickerson's first response was to discuss the matter with Jackson in a cabinet meeting and remind him that this whole mess had been brought about by Southard, who had not only caused Dickerson fits in New Jersey politics but always managed to raise the president's ire as well. Jackson remained firmly committed to the project, though, and even insisted that the secretary make room for civilian scientists. Well, Dickerson said, let me at least insert someone more qualified, a navy man, in place of the meddling Reynolds, who'd had the nerve over the winter to go to New Jersey and talk the state's House of Assembly into writing an official resolution in favor of his pet project. While the president took some time to consider this request, Dickerson noted in his journal on July 7 that he was "plagued with the Exploring Expedition." Four days later, he wrote: "Plagued with J. N. Reynolds." But President Jackson had already composed an official order securing Reynolds a spot in the scientific corps. "It will be proper that Mr. Reynolds go with the expedition—" he told Dickerson. "This the public expect."[26]

Nobody expected the Panic of 1837. The Ex Ex, as it came to be called, had not yet sailed, for Dickerson had done everything in his power to delay preparations. He repeatedly hired potential officers and then bullied them into resigning; offered scientists positions but refused to tell them when they would start getting paid; and just generally twiddled his thumbs and lived up to his surname, dickering about. Jackson had seemingly transferred power quite smoothly to Martin Van Buren in March (though the new president would soon be nicknamed Van Ruin), and Dickerson had held onto his cabinet position. But then, in May, the banks started failing, and by the end of the summer 90 percent of the country's factories had shuttered their windows. Thousands of unemployed workers marched in the streets of New York, demanding relief, or at least flour to make bread. The entire economy crashed. Many suspected that the main cause of the crisis had been speculation in western lands. American expansion—not to mention exploration—no longer seemed so certain.[27]

Even before the economic downturn, many Americans were skeptical of far-western prospects. The era of Manifest Destiny would arrive soon enough—within a decade—and booster literature would convince tens of thousands of people that their industriousness would be sufficient to transform the West into a well-manicured Garden. Meanwhile, though, it was still a Desert. In October 1836, Washington Irving had published *Astoria*, the second volume in his trilogy depicting life in the far West, a life marked by long periods of isolation, dangerous encounters with native peoples, and a rugged dependence on hunting, trapping, and trading—a life, in other words, not too far removed from that embraced by South Seas whalers and sealers. Irving included a long section on "The Great American Desert," which he characterized as "a region almost as vast and trackless as the ocean . . . , desolate sandy wastes wearisome to the eye from their extent and monotony." Astoria itself was a lonely fur-trading outpost established (in absentia) by the wealthy merchant John Jacob Astor at the mouth of the Columbia River in 1811—a daring venture that quickly became a colossal failure. Irving's book attempts to make Astor into a heroic, visionary entrepreneur, but the story is ultimately a tragedy, complete with a devastating Indian conflict provoked by the crew of Astor's supply ship, the *Tonquin*, which culminated in an onboard explosion that killed the last remaining sailor and more than a hundred natives.[28]

Astoria appeared at exactly the same time as the volume containing Reynolds's *Address* and all his supporting documents, and Edgar Allan Poe reviewed both books in the January 1837 issue of the *Southern Literary Messenger*—in which he also published the first installment of *The Narrative of Arthur Gordon Pym*. Suddenly, in the perverse imagination of one of America's greatest writers, there came together an obsession with psychic and spiritual frontiers and an obsession with spatial and geographic frontiers; a fascination with novelty, originality, discovery; and a compulsive need to explore the possibly horrific consequences of extreme exploration, of expansion to the far South and far West.

Poe believed, with Humboldt and Reynolds, that frontier territories brought out the savagery in civilized societies. Though considered by most scholars to be a conservative on racial issues, Poe made a point of expressing the dissident opinion that his government ought to change the name of the United States to the Indian-derived "Appalachia," as partial recompense to this country's "Aborigines, whom, hitherto, we have at all points unmercifully despoiled, assassinated, and dishonored." Pym's shipmates, at different points in their voyage, engage in mutiny, cannibalism, and, when they arrive at the bizarre island of Tsalal, the attempted exploitation of native labor to secure a cargo of local resources (the bêche-de-mer, a sea cucumber highly valued in the China trade). Commentators have tended to interpret the violent, jet-black Tsalalians as African American slaves rising up in a revolt against their masters, but their names, dwellings, relationship to the environment, and system of commerce all seem modeled on Native American culture. Pym also finds arrowhead flints on their island near some ancient engravings. Moreover, the climactic confrontation between Pym's shipmates and the Tsalalians, involving a shipboard explosion, is stolen almost verbatim from Washington Irving's description of the *Tonquin* disaster on America's far-western frontier. Look to your borderlands, Poe says to America, and you should expect acts of resistance and revenge perpetrated by rightfully indignant native peoples, as long as you continue invading their territories and trying to cash in on their resources. Your pioneers will soon receive their comeuppance.[29]

Seeking a new origin myth for America, Pym went as far south as he could and prepared to hurl himself down a polar abyss, as if he were exploring Abyssinia, the unmapped, southernmost province of the ancient Near Eastern world, the mysterious lower regions from which the Nile rises. Down here, in Wordsworth's famous image, was Imagination itself, "That awful Power . . . from the mind's abyss . . . like the mighty flood of Nile / Poured from his fount of Abyssinian clouds / To fertilise the whole Egyptian plain."[30] Indeed, if Captain Symmes turned out to be right, Captain Pym might actually be able to ride his canoe from the South Pole all the way to Egypt, via the center of the earth. Maybe America, or Appalachia, had to revisit the desert, the true cradle of civilization.

The South had led Poe to the West. Frontiers, he explained in a letter, offered the novel possibility of "a national as distinguished from a sectional literature." He was sick and tired of "being ridden to death by New-England," disgusted with its stay-at-home pastoralism and monotonous industry and with Emerson's false promise, in his 1836 book, *Nature*, of "an original relation to the universe."[31] Originality came from isolation and inner fire, from rock and dust.

How ironic, then, that *The Narrative of Arthur Gordon Pym* is so unoriginal a book, with entire chapters cribbed not only from Reynolds and Irving but from several other sources as well, including published accounts of the Lewis and Clark expedition.[32] Poe's Abyssinia was the vastly expanded print culture of the 1830s. But at least he knew it. Clearly annoyed, like Reynolds, at the "industrialization of literature" for the emerging mass market, Poe did sometimes satirize the hack work of (cannibalistic?) writers who simply spliced together passages from other people's books—yet he hailed them as geniuses if they did their splicing artfully.³³ In the modernizing world, he seemed to suggest, dishonesty was the only remaining means of being honest. Flooded with information, all writers inevitably recycled material; all experience was inevitably secondhand; all voyages of discovery were inevitably repeat trips.

By the mid-1840s, even Emerson had thrown up his hands: "an innavigable sea washes with silent waves between us and the things we aim at and converse with," he wrote in his essay on "Experience."³⁴ Perhaps American expansion would cause an implosion of the Soul; perhaps immersion in primal landscapes was simply impossible in an industrial society. Emerson could well have reached such a conclusion upon reading the end of Pym's narrative. Nothing makes sense at the South Pole. Pym is stranded on the edge of night and day: the last date he records is March 22, the South's autumnal equinox, which he would have experienced as an all-day sunrise, the half-dark, half-bright orb balanced precisely on the horizon for the full twenty-four hours. "The terms *morning* and *evening* . . . ," he notes, "must not, of course, be taken in their ordinary sense." But most of all, he is blinded and bewildered by the utter whiteness of the Antarctic, by the "milky depths of the ocean" and the "luminous glare" of the sky. Giant white birds fly by, shouting "Tekeli-li!"—whatever that might mean. Finally, the canoe is about to pass through the white curtain and plunge into the abyss, when a ghostlike human figure arises in the mist, with skin "of the perfect whiteness of the snow."³⁵ Blankness everywhere, an indecipherable hieroglyph, a systemic, atmospheric rebuke to humanity's knowledge and originality. Poe sees only the gaps on the geographer's map, the artist's naked canvas, the writer's empty page. Eureka: the cosmos is an overwhelming chaos; in this "original unity" is the "germ of inevitable annihilation."³⁶ Everything is too bare, stark, white—the water, the air, the birds, the people—even, presumably, the whales.

———

Reynolds was not as unhopeful as Poe: he had the optimism of an explorer with actual field experience. All through his battles with Mahlon Dickerson, all through the Panic of 1837, he still believed that his expedition would sail, and that he would sail with it.

Dickerson technically had full control over the recruitment of the scientific corps, but Reynolds stayed several steps ahead of him. All through 1836 he had been visiting and corresponding with the nation's most prominent scholars, asking them if they wanted to accompany the expedition and soliciting written recommendations for the corps's guiding principles. Intellectuals of all stripes had responded enthusiastically: Nathaniel Hawthorne arranged for his friend Franklin Pierce, the senator and future president, to tell Reynolds that he would be more than happy to accompany the expedition as a correspondent or chronicler of events. (Hawthorne, like most of his characters, wound up staying in New England.)[37]

When Dickerson finally got around to hiring the brain trust, then, many of the scientists expressed concern that the secretary seemed to have less enthusiasm for the venture than they did. Benjamin Silliman, for instance, was somewhat surprised that he was being consulted again and definitely annoyed that the preparations hadn't been taken care of yet. The Yale chemist had known and trusted Reynolds for ten years. In June 1826, he had attended one of the early Symmes Theory lectures and reported publicly that "Mr. Reynolds . . . handled his subject like an accomplished scholar."[38] But Dickerson seemed like just another inefficient bureaucrat.

In March 1837, both houses of Congress resoundingly approved the new Navy Appropriations Bill, giving Dickerson even more money to work with in launching the expedition, and Reynolds went to Washington to follow up with the just-inaugurated President Van Buren. By April, he'd secured a new official appointment from the new chief executive. And then, in May, the Panic struck, and Dickerson dickered—and Reynolds panicked.

The constant delays—the reports that the specially built expedition brigs were leaky and unstable—the loud grumblings from the navy about the intrusion of civilian scientists—the fact that the scientists had been appointed but were not getting paid—and now, a financial crisis—all led Reynolds to cease even pretending that he could cooperate with Dickerson. Now he would do what he did best: take his case to the public. For much of the next two years, he published a series of harsh, mocking letters in New York newspapers asking the government to explain why its appointed officials were destroying what the elected representatives of the people had voted to create. "The official acts of a public functionary may be fully canvassed by the humblest citizen," he explained in his first missive, on June 9, 1837, and then he proceeded not only with a full canvassing but with an all-out, minutely detailed attack on both Dickerson and his colleague Joel Poinsett, the secretary of war, who took over responsibility for the Ex Ex in January 1838.[39]

In a climate of constant political agitation, in a city where, for the first half of the nineteenth century, public spaces were frequently swarming with protests and speeches, Reynolds did not have a hard time attracting support.[40] By September, New York papers were noting that, while Dickerson delayed, Reynolds's *Address* had already spurred a new French expedition to the South Seas. And a writer for the *North American Review* reiterated Reynolds's contention that "at present there are none, to whom the Pacific is a source of greater interest than to the inhabitants of the United States."⁴¹ The first appointed commander of the expedition, meanwhile, handed in his resignation, explaining that he was tired of coping with "uncompromising opposition" and "procrastination the most extraordinary." In response, several members of Congress introduced resolutions complaining about "the present organization of the Navy Department, to say nothing of the incompetence at the head of it."⁴²

Dickerson fought back by writing his own public letters, arguing, in particular, that civilian scientists would only impede the smooth functioning of an efficient naval operation—letters that, in retrospect, sound eerily similar to speeches given by late-twentieth-century politicians who sought to ban women or gays from the military. The secretary also resisted the basic democratization of science, insisting that any decent gentleman amateur, possessed of the proper breeding and broad-mindedness, could use his general skills in natural history to replace any three or four of the young specialists recommended by Reynolds. Why should the government hire James Eights to study only fossils and Asa Gray to study only plants, when Dickerson could simply ask an interested banker to cover both fields and to come on board essentially as a volunteer, since he did not depend on science for his income?⁴³

Reynolds thus felt he had to defend not only his own honor—"how he became a *savant* remains yet to be discovered," noted Dickerson of his archrival—but also the very practice of modern natural science. He admitted that he was but a self-taught generalist, best suited to be the expedition's historiographer. But he had nevertheless received the endorsements of the nation's top scientists, and he and they could easily explain the significance of each of the specialized fields they had identified as being crucial to the mission of the Ex Ex. For one thing, Reynolds argued, the area to be surveyed "includes every climate, from the prolific torrid zone to the extreme limits of animal and vegetable life"—and surely it was ridiculous to expect one person to be an experienced, competent tropical botanist *and* glaciologist?⁴⁴ Even more important, though, was Reynolds's conviction that specialization did not mean isolation, that the group would have to work together as a "scientific association," with each man able to focus enough on his own field to see just how rich it actually was but also able to remain connected enough to his colleagues to see just how interdependent all of their fields had become. "The zoologists, therefore, will not be merely collectors sent out to grasp up animals and preserve specimens for home inspection and dissertation, but men of high acquirements sent out *to study the organic world alive*. . . . The same zoological laws influence ourselves and the meanest insect; and the muscle which moves a finger at our wish, we know not by what secret connexion between will and motion, acts from the same cause as that which controls the foot of a fly." As science developed and ramified, no individual could be a Humboldt by himself, but a highly organized team could adopt a Humboldtian combination of empirical and speculative approaches, addressing both minute and cosmic concerns.[45]

One particularly egregious mistake Dickerson made was to put Lieutenant Charles Wilkes in charge of acquiring all the necessary instruments and books, many of which were to be found only in Europe. Wilkes had antagonized civilian scientists ever since Reynolds's proposed expedition in the late 1820s, and now he happily abetted Dickerson in his effort to undermine the Ex Ex. The lieutenant came back from Europe with a few navigational instruments and some devices that would help him in his own astronomical studies, and little else. In private, Asa Gray expressed the frustration of the whole scientific corps: "that gentleman is not a naturalist, and appears not to have consulted any naturalist abroad." Reynolds, as usual, went public. In a letter to the *New-York Times*, he reminded Dickerson of Wilkes's documented claim that his handpicked instruments "comprise *all that can in any way be useful* for scientific purposes on *any expedition,* and *are all of them of the very best construction.*" The hubris, and the wrongheadedness, were extraordinary: "These assertions," Reynolds posited, "which *Humboldt* would not have ventured, are further confirmed in the next sentence, where it is said, '*I trust they will be found fully adequate to the wants of the expedition.*'" How was it, Reynolds wanted to know, that Wilkes had forgotten the microscopes? "Finding no such articles on the list, I suppose it was considered that everything earthly, aerial, or aquatic too small to be seen with the naked eye was too insignificant for the notice of *savans* on 'any expedition!'" And then there was the problem of the expedition's library. "I am not a little puzzled," Reynolds wrote, "with this heterogeneous *mélange* of scientific works which have been brought hither. . . . The naturalists will require *working books, manuals, and models;* and these, sir, have not been provided." At the very least, Wilkes should have included "the complete works of Humboldt. In a word, the catalogue is in itself sufficient evidence that no naturalist had any share in its adoption."[46]

Reynolds's public sarcasm did not serve him well in April 1838, when, after six other captains had either declined the position or resigned in disgust, Lieutenant Wilkes was presented with the command of the entire expedition. It was an unprecedented, and much resented, decision, for thirty-eight other navy lieutenants had more experience at sea than Wilkes. But he had made a not-so-secret deal with Dickerson and Poinsett to halve the scientific corps and arrange the whole affair according to their wishes. Imagining the nature of the backroom Washington politicking, Reynolds even went so far as to publish a mock mini-drama featuring Dickerson, Poinsett, and Wilkes as the main characters, at the conclusion of which Dickerson exclaims, "Wilkes, remember your promise to call an island after me!"⁴⁷ Many people were upset when Wilkes succeeded in eliminating several specialists from the scientific corps. Asa Gray, for one, simply resigned, and the fired scientists—including some of the most experienced, like James Eights—would spend years unsuccessfully petitioning the Navy Department for all the pay they had lost over the previous several months, during which time they had quit other jobs and devoted all their energy to preparing for the voyage.⁴⁸ There was also intense public pressure applied to the directors of the Ex Ex with regard to Reynolds's ultimate role. Magazine editors lionized the tireless agitator; Senator Southard and Representative Corwin insisted that the expedition could not succeed without him; and forty-three members of Congress signed a petition urging that Reynolds be chosen to oversee the scientists and write up the summary of their results. With the ships finally scheduled to leave harbor on August 15, though, Poinsett wrote to Reynolds on the first with official confirmation that he had been dropped from the corps: he had become too controversial a figure and would only cause unrest among the crewmen. "Your desire to accompany the expedition is natural," said Poinsett, "and, under ordinary circumstances, your having, in some measure, originated the design, would give you a strong claim to be indulged in your wishes; but all subordinate considerations must yield to the paramount one of conducting the expedition to a successful issue."⁴⁹

The Ex Ex eventually entered the history books as the Wilkes Expedition. Its legacies were many, and many of them problematic. Wilkes abused his men, promoted himself from lieutenant to captain, and at times indiscriminately attacked native islanders, burned villages, and destroyed crops. Midshipman William Reynolds (no relation) noted that "our path through the Pacific is . . . marked in blood," and he deplored not only his commander's cruelty but also Wilkes's sense of imperial superiority. "I could not help thinking," William Reynolds wrote of the natives, "how much better it would be to let them go their own old way, but No, No! We must have all the world like us." When the ships finally arrived back in New Jersey in 1842, Wilkes faced court-martial after court-martial and in the end received an official reprimand from the government. Many people read his long narrative of the expedition's accomplishments, but his writing was forgettable, and the Ex Ex itself quickly faded into obscurity in the public mind.⁵⁰

On the other hand, despite its less-than-ideally outfitted scientific crew, the

Ex Ex finally confirmed the existence of Antarctica, produced numerous new charts, and brought back more than 160,000 natural history specimens, spurring the continuing development of American science. One of Wilkes's most important contributions was the 1841 mission he sent to the Oregon coast, the site of the Astoria disaster, where a small but enthusiastic group of scientists, taking a much more peaceful approach than that of their commander, wandered through the volcanic Cascade Range down to the snow-covered Mount Shasta in Alta California, collected plants and minerals, and traded with the local Indians. Their findings would spur John Charles Frémont's second expedition to the far West just two years later, as activity on the Oregon Trail began to heat up. The Ex Ex's overall stash, meanwhile, justified the founding of the Smithsonian Institution and a National Herbarium in Washington. Even Asa Gray got involved in helping to sort through the botanical specimens, eventually authoring volume 15 of the expedition's official scientific report. On pages 723 through 726, he described an interesting new genus in the English ivy family, comprising two new species: one that the explorers found in a ravine on Oahu, in the Sandwich Islands, and one that had been pointed out by a native chief in the thick, interior forests of Savaii, "one of the Samoan or Navigators' Islands." Gray named the plants *Reynoldsia sandwicensis* and *Reynoldsia pleiosperma*, to honor his friend J. N. Reynolds "for the unflagging zeal with which he urged upon our Government the project of the South Sea Exploring Expedition, and also for having made, under trying circumstances, an interesting collection of dried plants in Southern Chili, many years ago."[51]

From a certain perspective, all Reynolds got for his efforts were a couple of species of English ivy. But names are important. Actually, from Reynolds's point of view, the ivies might just as well have borne the moniker of the chieftain of Interior Savaii; the truly significant decision made by Gray was to call the whole South Sea venture by its Reynoldsian name—and never to refer to it, under any circumstances, as the Wilkes Expedition.

———

LETTER FROM THE AUTHOR TO HIS ESTEEMED READERS, with regard to the honorable J. N. Reynolds and the decision to leave him off the scientific roster of the United States Exploring Expedition. Accompanied with testimonials from various writers and editors intimately familiar with the subject in question.

Esteemed Readers:
I feel it my bounden duty to present to you a few memorials, signed by citizens of the United States, attesting to the injustice done to J. N. Reynolds in 1838. Ultimately, the lapse of years makes it difficult, if not impossible, to pass judgment on

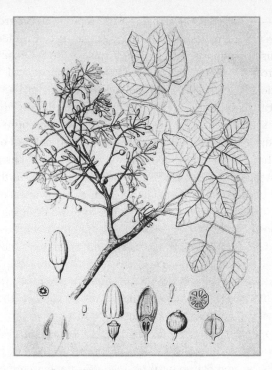

13   *Reynoldsia sandwicensis*, 1854. Asa Gray.

*the merits of a life so actively lived and so intricately layered. The historian can
hope merely to paint as full a portrait as possible. So, simply that you may know
more closely what his contemporaries thought of him, is the goal held in mind, in
the presentation of these testimonials to Mr. Reynolds's broad-minded passion, by*

> *Your obedient servant,*
> *A. J. Sachs*

"Nothing can deprive him of the merit that attaches to the originator, the defender, and the successful advocate of this important measure. The miserable imbecility, the dogged and asinine obstinacy of Mr. Secretary Dickerson,
and his inveterate hostility to the expedition, Mr. Reynolds has kept before
the public—adhering with the most laudable perseverance to its original
objects, and resolved that they should be duly consummated. *But the sailing
of the Expedition is the triumph of Mr. Reynolds.* To him the people will ascribe
all the credit."

—Editor, *New York Enquirer*, 1838[52]

"Here lives—amongst political pamphlets, reports of congressional debates, of executive departments, and of investigating committees— . . . the celebrated historiographer of the famous 'Cruise of the Potomac'—the originator and projector, though not allowed to be a participator, of 'The Exploring Expedition.' . . . It was an impolitic movement on the part of the government to deny to this gentleman the place in the expedition, which has recently sailed, so properly his due: for never had any administration a more formidable opponent than the present popular political orator of National Hall, has proved himself to be to that now in power. His political lectures are listened to by throngs, whom he addresses, not in the language and manner of a demagogue—not by allusions to his own real or imaginary wrongs;—far, very far from it. He invites debate and discussion from the side he opposes, and he urges home his arguments upon the multitudes who crowd to hear him, with soberness and discretion, and not in the tone of a mere party haranguer of the populace. In manner he is very graceful, impassioned, and impressive; in the choice of language, discreet, well-prepared and classical; and in argument close, subtle, fair, clear, and convincing."

—J. F. Otis, "Current Calamosities," *Southern Literary Messenger*, 1839[53]

"To him, we say—and to him in fact *solely*—does the high honor of this triumphant Expedition belong. Take from the enterprise the original impulse which *he* gave—the laborious preliminary investigation which *he* undertook—the unflinching courage and the great ability with which *he* defended it when attacked—the unwearying perseverance with which *he* urged its progress, and by which *he* finally ensured its consummation—let the Expedition have wanted all this, and what would the world have had of it but the shadow of a shade . . . ? One thing is certain—when men, hereafter, shall come to speak of this Expedition, they will speak of it not as the American Expedition—nor even as the Poinsett Expedition, nor as the Dickerson Expedition, nor, alas! as the Wilkes Expedition—they will speak of it—if they speak at all—as 'The Expedition of Mr. Reynolds.'"

—Edgar Allan Poe, *Graham's Magazine*, 1843[54]

"But to return to Mocha Dick—which, it may be observed, few were solicitous to do, who had once escaped from him."[55]

It was just after the sailing of the Wilkes Expedition that Mocha Dick, the mother of all white whales of the Pacific, appeared for the first time in American literature. He was a brooding, brutish creature—not the kind of whale you'd want to cross. "Numerous boats are known to have been shattered by his immense flukes, or ground to pieces in the crush of his powerful jaws." At the same time, however, his reputation was for peacefulness when he was left to his own devices: "though naturally fierce, it was not customary with Dick, while unmolested, to betray a malicious disposition." Just beware the wrath that would rise up from the leviathan "with the first prick of the harpoon." And when he "flung the water from his nose . . . , its expulsion producing a continuous roar, like that of vapor struggling from the safety-valve of a powerful steam engine," he became both an incarnation of utter wildness and the possessed demon of Yankee industry.[56]

Other whales had attacked ships: many Americans were familiar, for instance, with the destruction of the whaleship *Essex*, shattered by the brow of a huge, anonymous South Pacific sperm whale in 1820.[57] But Mocha Dick was different—was immediately recognizable—was *"white as wool!"* Reynolds may have based this specific whale on a real oral tradition, or he may have made him up by combining several different traditions, about various whales that were white, or aggressive, or called Dick by the whalers (in 1834 Emerson wrote of an aggressive white whale named Old Tom).[58] Certainly, "Mocha-Dick," more so than any of Reynolds's other published "Leaves" from his journal, is a self-conscious literary construction: it is a story within a story, told for the most part not by Reynolds himself but by a First Mate whom Reynolds encountered off the Chilean coast. Reynolds clearly crafted all the details and capitalized on their symbolic potential. Like Poe's South Pole, then, the character of Mocha Dick the white whale remains forever mysterious, hauntingly unreadable. Perhaps, Reynolds seemed to suggest, it could never be clear what Dick's fury said about humanity's relationship to nature.

Reynolds, like most writers of the American Renaissance, who both scorned mass-market literature and prayed that their books would sell well enough to support them, had a hard time deciding what to think about the expanding commerce and industry of his day—not to mention the expanding scope of American settlement. Frontiersmen, like whalers and sealers, lived a rugged, romantic life under the stars, in step with the animals they hunted. And yet they were also generally guilty of acts that Reynolds had to consider savage. Still, it was surely better to hurl the harpoon than to live the life of "some over-fed alderman." For the First Mate who tells the story of Mocha Dick, for "the untamed brightness of his flashing eye," for "his features, on which torrid sun and polar storm had left . . . a tint swarthy as that of the Indian," Reynolds clearly had an undying admiration.[59] He would never get over his basic ambivalence about the project of American expansion and industrial development, but he was sure that at least attempting to live more like the Araucanians would be a healthy tack for overcivilized Americans. In the end, what Reynolds cared about most was people's willingness to see the world, to break their routine, to challenge themselves.

It seems clear that Reynolds succeeded in stirring things up. One month after "Mocha-Dick" appeared, Horace Greeley asserted that "we hardly know of a more fluent and affluent writer than Mr. Reynolds. The mind of the reader rushes along with his, as if, to use a sailor's expression, he 'had it in tow.'"[60] In "Mocha-Dick," then, Reynolds presented the whalers and sealers he knew so well as challenges to the stay-at-home entrepreneurs of New England, New York, and Washington, as invitations for them to get caught up in the current of his free-style adventures. "The varied records of the commercial world can furnish no precedent, can present no comparison, to the intrepidity, skill, and fortitude, which seem the peculiar prerogatives of this branch of our marine. . . . They are the natural result of the ardor of a free people; of a spirit of fearless independence, generated by free institutions. Under such institutions alone, can the human mind attain its fullest expansion, in the various departments of science, and the multiform pursuits of busy life."[61] Ultimately, like Poe, Reynolds embraced *mental* expansion.

So, when Reynolds's story finally reaches its climax and the First Mate, having survived Dick's wrath, finally slays his nemesis, maybe he is a noble Araucanian killing the lethargic, hubristic white race. Or maybe he is killing the Old World, the colonialist influence of Europe, the tough, stodgy regime that for so long had prevented Americans from sailing freely over the oceans and from giving birth to their own arts and sciences. Or maybe he is simply killing the steam engine, asserting the power of a less mechanized, purer lifestyle, of a more direct, original relationship with the natural world.

There is something unnerving, though, about the whaleman's ultimate triumph. "To get the harness on Dick" was perhaps to go too far in taming nature, to destroy any future possibility of engaging with wildness.[62] Reynolds himself loved peering into volcanoes; would he have them all dormant? Indeed, it seems justifiable to say that Melville knew better how to end this story, with the great white whale still alive, still mysterious, still awe-inspiring to anyone daring to sail the vast reaches of the Pacific. Yet there was a sense in which Reynolds *had* to kill Mocha Dick, for the whale seems to have taken on yet another layer of symbolism as Reynolds examined his nine-year-old field notes and wrote up this tale in 1839. Maybe the First Mate, in destroying Dick, was simply wreaking Reynolds's revenge on the U.S. government. It seems an unlikely coincidence, after all, that the whale should share his initials with Reynolds's most despised political enemy. But it probably would have been too obvious for Reynolds to call him Mocha Dickerson.

———

In 1840, Edgar Allan Poe published the first six parts of a novel whose avant-garde explorer-hero had the same initials as J. N. Reynolds. "The Journal of Julius Rodman," which Poe never completed, purports to be "an account of the first passage across the Rocky Mountains ever achieved by civilized man." Not only does the fictional Rodman beat Lewis and Clark to the Pacific by a good fifteen years, but he also journeys through "an immense extent of territory [Western Canada and Alaska] which, *at this day* [1840], is looked upon as totally untravelled and unknown, and which, in every map of the country to which we can obtain access, is marked as '*an unexplored region,*' . . . the only unexplored region within the limits of the continent of North America." Thus, in Poe's fiction, did Reynolds receive a measure of redemption, in a sense finally getting credit for his avant-garde role in American exploration. And, once again, Poe was perceptively connecting the far South—Rodman was also the surname of a prominent New Bedford whaler—to the far West, suggesting that pioneering, like whaling and sealing, was exhilarating but exhausting, that frontier life by definition entailed a certain amount of antisocial behavior. In the pages Poe devoted to framing Rodman's journal, he described the explorer as "possessed with a burning love of Nature," and it seems that Rodman "worshipped her . . . more in her dreary and savage aspects, than in her manifestations of placidity and joy. . . . He was urged solely by a desire to seek, in the bosom of the wilderness, that peace which his peculiar disposition would not suffer him to enjoy among men. He fled to the desert as to a friend." Rodman was as restless as Reynolds; as enamored of the natural world and also as disgusted by his modernizing society; as attracted to the wild, dark side of the universe. Together, it seems, Poe and Reynolds, with help from Humboldt, had given birth to America's consummate Romantic hero, the wandering wilderness lover, the never-satisfied road-tripper, the social critic in search of a more direct, original connection with the primeval.[63]

In Poe, these Romantic tendencies push the individual explorer to the brink of a lonely death, but in Reynolds they somehow lead to a slightly more social vision, to the Humboldtian fantasy that, on the frontier, individual representatives of "civilized" nations might embrace a kind of ecological interdependence, might do justice to nature and to native peoples and eventually learn to live like the Araucanians. This possibility is latent in almost everything Reynolds ever wrote, even if the vision is never realized—in "the fertile soil of Mocha," in the community that develops aboard ships, in the way that whalers become "tethered" to their prey, in the "genuine, heart-stirring harmony" to be heard upon approaching "a rookery of fur seal."[64] As the literary critic Robert Pogue Harrison has commented, the 1800s in general were "a century of nostalgia, to be sure, but also of visions of future alternatives which history for some reason never fulfilled"; some members of Reynolds's society at least "dreamed of a truly radical and redeemed modernity." Indeed, the entire century at times takes on the appearance of "brooding storm clouds drifting over a drought-stricken land without discharging their moisture."[65]

Herman Melville, of course, is virtually dripping with moisture, and *Moby-Dick* (1851), the ultimate elaboration of Reynolds's concerns, is full of moral exploration and mental expansion.[66] It is a story of men who have "by the stillness and seclusion of many long night-watches in the remotest waters, and beneath constellations never seen here at the north, been led to think untraditionally and independently, receiving all Nature's sweet or savage impressions fresh from her own virgin voluntary and confiding breast." Captain Ahab, like Captain Symmes, fixates on an idea that drives him to the ends of the earth, to a goal as white and magnetic as either Pole—and, indeed, by the end of the book, the sailors' navigational compass seems to have reversed itself through some fluke of terrestrial magnetism, and the world, like the white whale, becomes unreadable. The captain's quest is in a sense an ideological rejection of capitalism: when Starbuck, the first mate, asks him how much Moby Dick will fetch back home "in our Nantucket market," Ahab's simple response is, "Nantucket market! Hoot!" His is an operation that exists outside of commerce—and yet his ship, the *Pequod*, named for "a celebrated tribe of Massachusetts Indians, now extinct," is a fully functioning, industrial-capacity whaler, and Ahab himself is the tyrannical manager on the factory floor, and his idealistic quest is actually a mission of vengeance and murder of the kind that could easily lead to the extinction of a particular tribe of whales.

If Ahab is the doomed Poe character, though, then Ishmael is the Reynolds character. He boards the *Pequod* not to kill but to avoid suicide: going to sea is his way "of driving off the spleen" and the "damp, drizzly November in my soul"; it is his "substitute for pistol and ball." Like Ahab, he is an individualist, an *"isolato,"* "not acknowledging the common continent of men, but . . . living on a separate continent of his own"—yet at the same time Ishmael recognizes that even *isolatoes* are necessarily "federated along one keel." Ishmael comes to see the whaleship as a potentially utopian version of the United States, or even the cosmos, a place of unity in diversity, where he becomes "wedded" to Queequeg, who is described as a "soothing," "sensible," and "sagacious savage." Indeed, this South Sea cannibal is joined aboard the ship by a Massachusetts Indian, a black African, an African American, several "Parsees" from the Near East, and members of almost every imaginable class of white Americans. And all of these workingmen, despite a definite factory-style hierarchy, are joined together in their labor: "all varieties were welded into oneness" as they read the ocean's waves, chased a whale, finally slew their prey, and then sliced it up and boiled it down and stood in circles squeezing the beast's crystallized sperm back into liquid, and squeezing each other's hands, drenched in communion and solidarity and what Ishmael calls "an abounding, affectionate, friendly, loving feeling" that was quite capable of effacing all "social acerbities." Approached ecologically, in the open air, on the open sea, work could bring redemption and fellowship rather than oppression. It is still an impossible vision of society, in part because it excludes women, but there is at least a deep Humboldtian hope in Ishmael's epiphany.

In the twenty-first century we are in a sense distracted by the environmentalist dream of ceasing to extract natural resources altogether, but the dream of such nineteenth-century radicals as Humboldt, Reynolds, and Melville was that we might get what we need from nature in a more original, direct, unalienated way, in a way that would yield a fuller sense of contact and connection. And though Ishmael is orphaned and alone at the end of *Moby-Dick*, he at least retains a memory of his "wild, mystical, sympathetical feeling" toward the universe. Thanks to his actual, physical contact with nature and with his fellow men, he has made the discovery that "nothing exists in itself," that "it's a mutual, joint-stock world, in all meridians." At the end of his far-ranging explorations, Ishmael's experiences, though often indecipherable, have converted him to a kind of rugged, cosmopolitan democracy, a politics of interdependence.[67]

———

J. N. Reynolds was awash in labor politics in 1839. In March, a public letter appeared in the *New York Evening Star* signed by hundreds of citizens (all of whom had heard Reynolds speak before) asking if he would offer a lecture series to the city's "mechanics and working men at large without distinction of party." The politics of work had become Reynolds's new cause, and, after gaining admission to the New York bar that summer, he embraced labor issues almost as fully as he had embraced exploration.[68]

Reynolds did not have South Seas exploration completely out of his system, though. During the actual years of the Dickerson debacle, he had described himself as "tired" and "disgusted": "I am sick—sick—shall close this letter and throw my pen in the fire," he wrote to his friend Thomas Hamer, the congressman from Ohio. By 1840, Reynolds was starting to get over it,[69] but he was still writing about Antarctica and natural history in his spare time, and in 1841 he came out with a volume called *Pacific and Indian Oceans; or The South Sea Surveying and Exploring Expedition: Its Inception, Progress, and Objects*.[70] Wilkes himself was due to return within a year, though, and most readers were more interested in waiting to hear what the Ex Ex had actually accomplished than in rehashing the scandals that had preceded its sailing. Reynolds's time was finally up. In 1843, he did go back to his field journal once more, to publish "Rough Notes of Rough Adventures," but after that he seems at last to have let the subject slide.

Within another year or so, Reynolds would essentially withdraw from public life. In 1844, he threw his whole spirit into Henry Clay's presidential campaign against James K. Polk. His speechifying on behalf of workers culminated in a rousing lecture at New York's Great Procession of the Whigs on October 30, just before the election, when Reynolds addressed a crowd of "FREEMEN OF THE UNITED STATES!" To vote for Clay was the only way to see "our People thrifty and happy; Labor actively employed and fairly rewarded; the Workman dreading no man's frown." And Reynolds also made a point of attacking Polk's expansionist ambitions in Mexico, Oregon, and especially Texas: "Are you opposed to the incorporation of a Foreign Nation into your own, in violation of our solemn Treaties, and at the cost of an expensive and protracted War? Stand, then, by the Great Statesman who was among the foremost to oppose, and the most earnest to condemn, that nefarious project!" Polk won; Reynolds retired.[71]

On the Whig lecture circuit he had met D. D. Barnard, the great Humboldtian activist, and Reynolds no doubt joined Barnard and "all good Whigs" in opposing Polk when the president led America into the Mexican-American War two years later.[72] But Reynolds had become deeply cynical by then, and he did little but practice law until his death in August 1858—several months before Humboldt himself passed away. Reynolds's one other big project, in the last decade of his life, was a real estate scheme—in Texas, of all places. Fed up with his imperial, commercial society, he decided he might as well make some money off its hubris. "After our political defeat in 44," he explained in 1850 to his friend Tom Corwin, who was now serving as secretary of the treasury, "I was disgusted with all political matters—dropped them for ever—and in very spite made investments in Texas. May not a Christian use what a robber throws in his way—provided he did not help steal it—I believe now as I believed during the Campaign of 44 that the whole mode of annexation was iniquitous. It was the poisoned fountain from whence the bitter streams of domestic strife have opened and God only knows when they will cease to run."[73]

Reynolds saw that his country's unchecked expansion, and the accompanying question of the expansion of slavery, would bring the kind of conflict that Henry Clay's Compromise of 1850 could not hope to prevent. Like Clay, he prayed for national unity, but now he began to suspect that the nation's diversity was becoming too overwhelming. With Frémont's expeditions, and the surge of pioneers taking to the Oregon Trail, and Polk's success in enlarging Oregon Territory and adding Alta California and New Mexico, and now the gold rush, the whole country was looking West in 1850 and wondering what would happen—to the settlers, to the fortune-seekers, to the slaves, to the Mexicans, to the still-wild Indian tribes. For Reynolds, in his resentful middle age, the current was moving too fast.

He picked up various partners in law and business during his last years, but, like Humboldt, he never married; his life seems to have been almost entirely devoid of women. Perhaps he simply preferred men. Or, when he wrote of "the bitter streams of domestic strife" flowing through his country, perhaps he was also suggesting the alkalinity of his own domestic life. As with Humboldt, though, if Reynolds ever penned any revealing letters about his personal affairs, they have been lost or destroyed. The closest he ever came to sharing his pain was in telling his friends of the long bouts of incapacitation he suffered, apparently as a result of various tropical diseases he had picked up—again, like Humboldt—in South America. At the end of his life, in 1858, when he fled to a spa in Canada on a doctor's recommendation yet found himself unable to recover his health, J. N. Reynolds must have felt as alone as Ishmael at the end of *Moby-Dick*, when the only man to survive the *Pequod*'s final voyage was carried away, clinging to Queequeg's coffin, on the gentle current of the Pacific Ocean.

Watersheds: 1859–1862

In July 1859, after closing up the New York exhibition of *The Heart of the Andes*, the painter Frederic Church found himself floating off the east coast of Labrador and just south of Greenland, at the base of an iceberg. The ice was constantly melting and shifting, roaring and heaving. New cracks appeared every few minutes, revealing sapphire-colored seams. Blackish green waves slapped against the sides of the towering berg, polishing its sharp corners, while white torrents of snow and ice came hissing over glassy promontories, and trickles from melting overhangs dripped with a metallic pinging into the sea. Amid this perpetual din, the explosion was unexpected. The whole iceberg seemed to convulse, and there was a sudden thunderclap—a "terrific crack," wrote Church's companion, the Reverend Louis Legrand Noble, "a sharp and silvery ringing blow upon the atmosphere," which left the two men in a daze. A second later, though, they were leaping to their feet in the front of the boat and shouting, "Row back! Row back!" It was an avalanche. "At once, the upper face of the berg burst out upon the air, as if it had been blasted, and swept down across the great cliff . . . with a wild, crashing roar, followed by the heavy, sullen thunder of the plunge into the ocean, and the rolling away of the high-crested seas, and the rocking of the mighty mass back and forth, in the effort to regain its equilibrium." The two men feared for their lives, but the crew responded quickly, and, as luck would have it, their skiff "breasted the lofty swells most gracefully."

As Noble explained in his travel narrative *After Icebergs with a Painter*, the rest of that particular day was much calmer—despite an ongoing struggle against seasickness. Mostly the two friends—Noble, like Theodore Winthrop, had written a companion pamphlet for Church's *Heart of the Andes* exhibit—just gazed at "the Alpine berg," as they called it: "It resembles a precipice of newly broken porcelain, wet and dripping, its vast face of dead white tinged with green, here and there, from the reflection of the green water at its base." Church had a hard time keeping steady enough to complete his oil sketches, but he persevered, marveling at the "supernatural splendors" of the ice, water, and sky, bathed in an ever-changing light. A couple of times they did row up close again, to admire the details of this giant mountain of water, newly shed by the churning ice fields of Greenland's coastal peaks. They even discovered a small fount, from which they took a long drink. "We content ourselves," Noble wrote, "with catching a panfull of water, fresh from the great Humboldt glacier, quite likely, and cold and pure it is."[1]

Church had originally planned to spend that summer visiting Humboldt himself while *The Heart of the Andes* was touring Europe. On May 9, he had written to his friend the travel writer Bayard Taylor apologizing for his plan to leave New York before Taylor returned from his own wanderings. Church's "principal motive" in heading overseas so quickly, he explained, was "to have the satisfaction of placing before Humboldt a transcript of the scenery which delighted his eyes sixty years ago—and which he had pronounced to be the finest in the world." On May 19, however, the *New York Times* reported that Humboldt had died in Berlin. Church wrote to Taylor: "A friend communicated the sad intelligence of Humboldt's death—I knew him only by his great works and noble character but the news touched me as if I had lost a friend."[2] The painter decided to go ahead and send his landscape to Europe at the end of the month as planned, but he no longer felt any desire to accompany it. A different kind of travel was called for. In 1853 and 1857 he had gone south to the Andes, following in Humboldt's footsteps. Now he would go north.

As it happened, on the trip from Boston to Nova Scotia, aboard the *Great Republic*, Church and Noble had encountered one of Humboldt's most distinguished disciples, Louis Agassiz of Harvard. Just days before, Agassiz had overseen a ceremony to mark the founding of his university's Museum of Comparative Zoology—which would eventually inherit many of the collections of the Boston Society of Natural History. Now Agassiz was headed back to his native Switzerland for the first time in thirteen years, for a vacation, after having exhausted himself with the struggle to establish the museum. One of the world's foremost experts on glaciation, Agassiz had spent years traipsing around the glaciers of the Alps but had never seen an iceberg, and he envied Church his expedition. In later years, while Church produced bold images of giant ice formations, Agassiz continued developing his bold theory of the Ice Age, in the face of constant scientific skepticism. And within just a few months of his European vacation, after receiving a brand-new copy of *On the Origin of Species* sent by Darwin himself, Agassiz became America's most prominent anti-Darwinist, in direct opposition to his Harvard colleague—another Humboldtian—Asa Gray, the botanist whom J. N. Reynolds had chosen for the Ex Ex. Now, with Humboldt dead and his students in disagreement,

14 *The Icebergs; or, The North,* 1861. Frederic Church.

and with Darwin suggesting that human beings were condemned to perpetual conflict, natural history had reached a crossroads.³

So had America. Church's northern masterpiece, eventually known as *The Icebergs,* was first exhibited in the United States in 1861, as the Civil War was just getting under way—at which time Church chose to call his landscape, simply, *The North.* All the proceeds from its New York exhibition went to the Patriotic Fund, which supported the families of soldiers serving the Union cause.

At its opening, the painting seemed devoid of humanity (the broken mast in the foreground was actually added later). The main feature was—and remains—the huge, blank, blinding iceberg, a kind of lifeless white whale. "We shall be surprised," wrote one reviewer, "if those of acute sensibilities do not look upon it at first with a positive feeling of pain, akin to that which we sometimes feel in the presence of terrible visions of sleep."⁴ And in its final form, including the dwarfed piece of wreckage, the image posed a direct challenge to survival, captured the tragedy of exploration and expansion. If this was meant to be nature, then nature was not something for which human beings were particularly fit.

Church's next major work was hardly more comforting: it was a monumental portrait of another South American volcano described by Humboldt, which Church depicted as a smoldering cone spewing huge black fumes into the sky. He called it *Cotopaxi,* but, exhibited in 1862, it may as well have been named "The South."

15 *Cotopaxi*, 1862. Frederic Church.

Meanwhile, the opening of the April 1861 showing of *The North* coincided with the publication of a review of the painting by Theodore Winthrop in *Harper's Weekly*. Even in his long essay about *The Heart of the Andes* two years before, Winthrop had made clear his preference for wild, challenging landscapes. A first-class Yankee, descended from both the Puritan leader John Winthrop and the fiery theologian Jonathan Edwards, Winthrop, like J. N. Reynolds, portrayed Humboldtian exploration as an affront to New England complacency, as something undertaken by "hearts undebased by sense and unbewildered by mammon." About *The North*, Winthrop noted that Church had left his viewers without any grounding, that the picture daringly confined itself to the forms of "air, light, and water." Most of the picture was craggy and fragmented, and the only smooth surface, on the face of the iceberg, was clearly marked as an avalanche field, the site of recent chaos and violence. It blazed with whiteness. If there was a unity here, it was a unity without diversity.[5]

Less than two months later, Winthrop became one of the first Union officers killed in the Civil War. On June 10, 1861, ten days before Bull Run, the thirty-two-year-old major took a bullet in his head during the brief and confused skirmish at Great Bethel.[6] An abolitionist and a devoted Humboldtian, he died in the midst of a brutal Darwinian conflict.

In 1862, several of his writings received posthumous publication, and Winthrop would eventually become well known, on both sides of the Atlantic, for four novels, a book of poetry, and a travel narrative. Apparently, like many other Americans, Winthrop had become fascinated with the West in the 1850s, as pioneers poured in and violence erupted over whether the new settlements would permit slavery. Unlike most others, Winthrop actually set out to explore the frontier on his own, riding and paddling his way through Oregon and Washington Territories, consorting with Indians, and writing up what would eventually become *The Canoe and the Saddle: Adventures among the Northwestern Rivers and Forests*, which was in its seventh edition by 1864. Winthrop was well aware of the dangers of exploration, but, on the last page of his narrative, he reaffirmed his belief in its redemptive power, in the importance of wandering through borderlands: "And in all that period while I was so near to Nature, the great lessons of the wilderness deepened into my heart day by day, the hedges of conventionalism withered away from my horizon, and all the pedantries of scholastic thought perished out of my mind forever."7 ■

PART THREE

WEST

Clarence King's Experience of the Frontier

CHAPTER 6

Mountaineering in the Sierra Nevada

The Art of Self-Exposure

When I would recreate myself, I seek the darkest wood, the thickest and most interminable and, to the citizen, most dismal swamp. I enter a swamp as a sacred place,—a sanctum sanctorum. There is the strength, the marrow of Nature. . . . How little appreciation of the beauty of the landscape there is among us! We have to be told that the Greeks called the world Kosmos, Beauty, or Order, but we do not see clearly why they did so, and we esteem it at best only a curious philological fact. For my part, I feel that with regard to Nature I live a sort of border life. . . . Unto a life which I call natural I would gladly follow even a will-o'-the-wisp through bogs and sloughs unimaginable.
—Henry David Thoreau, "Walking" (1862)

If one loves to gather the material for traveller's stories, he may find here and there a hollow fallen trunk through whose heart he may ride for many feet without bowing the head. But if he love the tree for its grand own nature, he may lie in silence upon the soft forest floor, in shadow or sunny warmth, if he please, and spend many days in wonder.
—Clarence King, *Mountaineering in the Sierra Nevada* (1872)

Theodore Winthrop's body was brought back to his hometown, New Haven, Connecticut, at the end of June 1861. You can still see the granite cross he requested for his grave in the Grove Street Cemetery, just across the street from Yale's Sheffield Scientific School—where Clarence King had begun his studies a year earlier. The base of the tomb was engraved only with Winthrop's name and the date and place of his death, but the cross itself was "sculptured with the endless cord, the emblem of eternity": it looks as though it is draped in impossibly tangled roots. A beloved son of Yale, Major Winthrop was given a hero's funeral, with bugles and muskets and eulogies and a team of students to carry his coffin across campus. According to firsthand accounts, "the whole town was deeply and sincerely moved," and much of the university's population came out to watch, contemplating Winthrop's lineage, his wandering spirit, his youthful energy, his commitment to a unified America.[1]

Within a few months, as Winthrop's posthumous writings poured from the presses, Clarence King began to see the fallen soldier as something of a personal hero. He even fashioned his appearance to match the major's, growing thick whiskers down the sides of his face.[2] But he did not join the Union army. Rather, once he finished his geological studies in the summer of 1862, he turned his thoughts westward, in the spirit of *The Canoe and the Saddle.*[3]

Perhaps even more influential than Winthrop's funeral were King's classes with the geologist James Dwight Dana—friend of J. N. Reynolds and veteran of the Ex Ex—and, especially, those with George Brush, the young professor who ten years earlier had traveled to Berlin with Benjamin Silliman to visit Humboldt. King often spent evenings at Brush's house even after graduating, and one night, in October 1862, his mentor read aloud to him from a particularly memorable letter. Written by another of Brush's former students, William H. Brewer, now a field geologist in California, it told of a moonlight ascent of the volcano considered at that time to be the highest peak in the United States: Mount Shasta. The New England wilderness was quite pleasant—but there were no volcanoes in the East.

The Shasta climbing party, according to Brewer, had to stumble over long patches of loose, gravelly lava and slog through "a desolate stretch of dark rock and glistening snow." Along the way, the mountaineers also noted that "sulphur water, boiling hot, issues from several orifices, and steam, charged with sulphurous gases . . . , issues from many vents." The chill of the night made the snow easier to walk on, but, combined with the "rarified air," it also left the climbers with bloodshot eyes, blue lips, headaches, nausea, and a powerful desire to curl up and go to sleep. Finally, at noon the next day, they gained the summit, only to find themselves surrounded by a thick haze. Yet Brewer was still able to see Lassen's Butte, about eighty miles to the south, "an isolated island peak of black rock and white snow rising from this sea of smoke," and despite his discomfort he found himself filled "with awe as well as admiration." Earth, air, water, fire were all captured in a moment of vision atop a volcano. Winthrop, a little farther north, had described other fiery, snowcapped peaks in the Cascade Range—Mount St. Helens, Mount Rainier, Mount Baker—but Shasta, with its mysterious Indian name, surpassed them all. Perhaps the West, in its endlessness, could offer many more such wonders and epiphanies. On that autumn night in Professor Brush's living room, King decided he had to find out.[4]

He went to the frontier to see, and to experience, nature. He would follow Humboldt's lead, perhaps more devotedly than any other American, in subjecting himself to the cosmos. His goal in the wilderness was to "expose myself, as one uncovers a sensitized photographic plate, to be influenced."[5] Emerson taught his readers to "see the world to be the mirror of the soul," but King believed that the modern soul, in rapid coevolution with the camera, ought to mirror the world.[6] He recognized the difficulty of leaving his baggage, his heritage, his New England assumptions, behind. But he insisted on making the effort.

Within a few years, King's soul would be imprinted with the geography of the West. "I never tire of overlooking these great wide fields," he wrote from the top of Mount Shasta itself, "studying their rich variety, and giving myself up to the expansion which is the instant and lasting reward."[7]

A different kind of expansion had driven many Americans for the previous two decades—and driven the United States toward disunion. Certain of Humboldt's disciples—Frémont, Abert, Möllhausen—had conducted explorations in the West in the 1840s and '50s, but even they had little time to "give themselves up" or "expose themselves" to anything. Their expeditions, far from being Humboldtian exercises in humility, curiosity, and mutuality, were designed primarily to facilitate the passage of the railroad and the waging of Indian wars—designed, in short, to encourage the conquest of the frontier by restless white pioneers.[8] And they were effective. By 1860, some 250,000 settlers had passed along the newly opened westward routes, such as the Humboldt Trail through Nevada.[9] Clarence King would later refer to their achievement as a "vast ACT OF POSSESSION," as "this great sweeping campaign against nature, this prodigious advance of a horde of homemakers." He admired their hardiness, but he was also suspicious of them—especially those who showed no respect for the land and, most important, those who intended to practice slavery in the West.[10] As the country expanded, the great question confronting its people, the question that, in the minds of many, had necessitated the war, was: would the frontier be free?

The West remained a cipher, even to those who settled there. It is not an easy land to read. In the late 1840s, the prime era of Manifest Destiny, confidence had run high in the pioneers' ability to transplant the farms of the humid Midwest to the arid plains surrounding the Rockies. But blood, rather than rain, followed the plow. Migrants on the Overland Trail had never known

16 Clarence King—"Mountain Climbing," c. 1869.
Timothy H. O'Sullivan.

what to expect from different groups of Native Americans—most of whom looked equally threatening to the settlers—but the one discernible trend in the mid-1850s was an increase in violent confrontations, usually spurred by the federal government's attempts to squeeze tribes onto smaller and smaller reservations. Meanwhile, especially after the Kansas-Nebraska Act of 1854, pioneers were killing *each other*. The new law held that territories would become free or slaveholding states depending on the will of the people who settled them, so Northerners and Southerners swarmed onto the plains toting shotguns and broadswords, and almost immediately formed militias to attack each other's outposts.[11]

In 1856, the Republicans established themselves as a new antislavery party and ran John Charles Frémont as their first presidential candidate, under the slogan "Free Land, Free Men, Free Labor—Frémont." The hope was that the explorer's status as a western hero would help unite the frontier with the mercantile North in opposition to the plantation South—but James Buchanan, by refusing to take any stance at all on the extension of slavery, won the White

House for the Democrats, whom Humboldt referred to that November as "the disgraceful party which sells negro children." Watching from overseas, Humboldt thought it a "crime" that "Buchanan will be the next President and not Frémont . . . , to whom it is owing that California did become a free state."[12] California, like much of the West, had boomed in the early 1850s, but now, with Kansas bleeding, with Indian raids on the rise, with the cracked earth refusing to yield crops, with the gold rush finally petering out, even the optimistic settlements around Frémont's Golden Gate seemed vulnerable, contingent. Word gradually trickled back East, and migration slowed considerably. When Mark Twain deserted his Confederate troop in 1861 and lit out for the Territory—as described in his popular book *Roughing It*—he was looking for escape, for the exhilarating freedom of wide-open spaces. But what he found was a desert frontier "whose concentrated hideousness shames the diffused and diluted horrors of the Sahara." To Twain, the country between Omaha and San Francisco seemed to consist of a few bedraggled mining communities surrounded by dust, ashes, and the scattered bones of migrants and their oxen. The American West was "one prodigious graveyard."[13]

Abraham Lincoln won the election of 1860 in part by insisting that the West could be reclaimed and rehabilitated as a unifying force, as the future of the nation. And even in the midst of the Civil War, he pursued the agenda he had laid out in his campaign, proclaiming the reopening of the frontier in 1862 when Congress passed the Homestead Act, which offered individual yeomen farmers 160 acres of government land for the token fee of ten dollars. It was the Civil War Congress, too, that granted huge tracts of federal property to the newly incorporated Union Pacific Railroad, in the hope that a transcontinental line could establish a permanent, ironclad link between the ravenous industries of the North and the seemingly boundless natural resources of the West. Meanwhile, these civil investments were backed by a military buildup. Though the most famous battles of the Civil War were fought in the East, the number of soldiers in the West actually doubled between 1860 and 1865, leading to increased tensions with Native Americans and one of the most infamous army atrocities ever, the 1864 massacre at Colorado's Sand Creek Reservation, which in turn launched the bloodiest, most brutal period of war between whites and Indians in American history. At Sand Creek, just before sunrise on November 29, some seven hundred men of the Third Colorado Regiment surprised a Cheyenne camp whose young warriors were away on a hunt. The savage, carefully planned attack left more than two hundred corpses sprawled on the plain. Some three-quarters of the dead were women and children, and the rest were Cheyenne elders—including Lieutenant James W. Abert's dear friend, the eighty-five-year-old Yellow Wolf.[14]

Clarence King's decision to head west, then, was not only an escape into nature but also a plunge into politics. The young geologist had considered joining the war effort and going south instead. He was known at Yale, after all, as an "enthusiastic abolitionist," and he acknowledged in a letter that his "heart and soul" were aligned with "radicals" like the militant antislavery advocate Wendell Phillips. From early childhood, he had been deeply influenced by his grandmother, Sophia Little, who, he later explained, "ate no sugar but free-soil maple and refused Southern oranges, as they were to her mind 'full of the blood of slaves.'"[15] Yet King also considered himself a pacifist. In the late months of 1861, he had burned with Union fire, but as he approached graduation in the spring of 1862, he renounced his violent rage against the Confederates: "When I said I wanted to 'push a bayonet' I was wrong," he wrote to his best friend, Jim Gardiner. "I was hot with passion and I was excited by the outrage of one of my pet ideas, 'freedom.' God knows that for my country I *would* 'push a bayonet' and that I would not quail before death for my land, but the act would crucify in me many of my noblest impulses. It is like tearing my soul in sunder."[16] Instead of killing his Confederate brothers, King would fight for national unity by helping to reopen the West for pilgrims from the North, by mapping and surveying the territories in the tradition of Frémont. Of course, securing the frontier for whites—even idealistic abolitionist whites—generally meant dispossessing Indians. But that just made the West even more intriguing to King as an intellectual and moral puzzle. A committed Humboldtian, King espoused the unity of races, upheld the rights of blacks, and yet his "convictions concerning Indians," as he later admitted, "were formerly tinged with the most sanguinary Caucasian prejudice."[17] By immersing himself in the confused wilderness of the West, in the cloudy American future, King would have the opportunity to see what Humboldt's theories meant on the ground, to discover what kind of unity might actually be possible in the harsh, brittle frontier world to which red, white, and black Americans would all have to adapt. Rather than go to war over his convictions, he went west to put his convictions to the test.

King rarely unified the diversity of those convictions, rarely overcame all his prejudices, rarely resolved his inner conflicts. In 1893, his unending turmoil would land him in Manhattan's Bloomingdale Asylum. But in the 1860s and '70s his struggles to see in new ways at least led him to a kind of Humboldtianism, to a compellingly complex view of the cosmos. "He loved a paradox," noted his friend Henry Adams, "a thing, he said, that alone excused thought. No one, in our time, ever talked paradox so brilliant." Adams, the tormented conscience of Victorian America, is often cited as the representative intellectual of the postbellum period, but Adams himself saw King as the most typical American of his day, as the man who most fully internalized the tensions of his rapidly expanding society, waging a civil war within his own body, mind, and soul.[18] Eloquent equivocations flowed from his pen as he pondered questions relating to Indians and the settlement of the West, the preservation versus the development of natural resources, moral restraint versus the masculine rush toward prosperity, tradition versus modernity. Perhaps his most productive paradox, though, was the one that arose from his confusion over how best to submit himself to nature: as a scientist or an artist.

King actually put himself through a rigorous training program in both science and art, reason and intuition, analysis and absorption, before he headed to the frontier in 1863. And his instinct not to value either of these perspectives over the other would turn into one of his deepest and longest-standing Humboldtian commitments. In the fall of 1862, besides continuing to consult with Professor Brush, King spent long hours with Professor Dana's brand-new *Manual of Geology* and traveled up to Harvard to hear Louis Agassiz give a series of lectures on his specialty, glaciology.[19] "Look at your fish!" Agassiz famously screamed at his students in dissection labs, and even in glaciology lectures his goal was to make people into more careful observers of nature's details. You want proof of his radical theory of the Ice Age? You want to understand his insistence that much of Europe and America had been covered by vast sheets of ice whose grating and polishing power had largely determined present-day geologic formations? Just take a walk through Massachusetts, but this time note the erratic boulders like those described by James Eights in the Antarctic; note the smooth, flat-topped rocks that former generations had attributed to Noah's Flood; note the gravel deposits and the striated granite. Or, better yet—and here Agassiz seemed to be speaking directly to King—go climb the snowcapped mountains of the West, and compare their geologic surroundings to the U-shaped valleys and terminal moraines Agassiz had found in the Alps, where significant numbers of glaciers were still active. Yet even this brand of minute scientific observation entailed wild leaps of imagination: Agassiz was asking his students to peer at plain surfaces of polished stone and see thousands of years into the fantastic and dramatic past.[20]

It makes perfect sense, then, that King, upon settling in New York City in January 1863, should have immediately helped to form an aesthetics society. Besides Dana's textbook, he was also reading *Cosmos*, and perhaps Theodore Winthrop's pamphlet on *The Heart of the Andes*, and bits of Thoreau as well, and he had begun to see art as a necessary counterweight to science. Careful study led to a deeper understanding of nature, but no scientist could quite match the appreciation Frederic Church had gained for the physical world when he set up his easel amid the vibrant blossoms of the Andean jungle. As a budding geologist, King had to step back from nature, disentangle himself from it, get his bearings, try to be objective. Yet what he really wanted to do was follow Thoreau's dictum that "there is a subtle magnetism in Nature, which, if we unconsciously yield to it, will direct us aright." He yearned to saunter westward, to learn how to trust his instincts as guided by occult forces, to produce something equal to Thoreau's writings or Church's paintings— for, as Thoreau noted, "true art" is ultimately "the expression of our love of nature."[21] True science could never aspire to such a goal.

At the same time, though, the members of the Association for the Advancement of Truth in Art, whom King's friend Jim Gardiner dubbed "the practical Ruskinites of the city," felt that all serious artists had to engage in a near-scientific study of nature if they harbored any aspiration to produce true art. Their literary hero, John Ruskin, in volume 2 (titled *Of Truth*) of his popular five-volume work, *Modern Painters*, had emphasized that the highest art involved capturing "the great verities of the material world."[22] So, when this radical group began publishing its own journal, called *The New Path*, the leaders of the association expressed contempt for any past picture that had not treated nature "in such a way that the poet, the naturalist, *and the geologist* might have taken large pleasure from it." The American Pre-Raphaelites, as they came to be known, for their resemblance to the defiant British Brotherhood of the same name, explained that they had gleaned from Thoreau not just inspiration and a fine-tuning of their intuition, but also the skills necessary to observe and depict nature accurately: they had learned the "science of art." If you were a theorist and a critic, you had to think of yourself as a "botanist with pictures for plants"; if you were one of the few who actually painted for a living, you might spend hours "trying to get acquainted with a fragment of flint rock." Only this approach could lead to a national art that was appropriate to this rough-hewn nation, blessedly free of European embellishments. "Perceptions naturally direct and true," claimed the "Introductory" letter of *The New Path*'s first issue, "are nowhere to be found today as pure as they are in America." Of course, "cold, remorseless fidelity" to the material world could not be the *only* contribution of the American artistic community. As one of King's associates put it, "We do not believe that mere faithful transcript from nature can ever be the greatest art. But we believe and positively affirm, that there can never be any degree of greatness without this as a basis. . . . Naturalism is not all we believe in, but we know it must come first."

King, meanwhile, honed his vision; as a friend later explained, "his observations of natural objects, plants, animals, or rocks, were so vivid that they seemed to photograph themselves upon his memory, so that he could recall the picture at will."[23]

King himself had a complicated relationship to Ruskin, which mirrored his own struggle to balance science and art. (The two men actually met in a London art gallery when Ruskin walked past King as the geologist was commenting on a painting. The professional art critic was so impressed by this amateur that he invited him back to his estate and offered him a landscape by his favorite painter, J. M. W. Turner. Presented with two from which to choose, King remarked, "One good Turner deserves another," and requested both.) At times, Ruskin seemed to King too obsessed with the anthropocentric view, too scornful of plain, wild American scenery, which, the Englishman once commented, "cannot acquire picturesque significance, or rightfully claim to excite human sympathies, till man has consecrated it." What of the simple rocks and plants? King could get fed up with the famous aesthetician: "The varying hues which mood and emotion forever pass before his own mental vision mask with their illusive mystery the simple realities of nature, until mountains and their bold, natural facts are lost behind the cloudy poetry of the writer. Ruskin helps us to know himself, not the Alps." Yet Ruskin, for all his humanness, for all his love of the pastoral, could also appreciate the geological sublime, could also find inspiration in a mountain's wild summit. And he saved his reputation in King's eyes when he preached "truth to nature." The final volume of *Modern Painters*, published in 1860, was called *Mountain Beauty*, and it featured chapters on "Gloom," "Glory," "Crests," and "Precipices," in which Ruskin reverted to the radicalism of his second volume and celebrated realistic artworks depicting the raw power of landscapes beyond the reach of human ingenuity. It was surely this final volume that King had in mind when he finally reached California and, catching his very first glimpse of Mount Shasta, shouted out: "What would Ruskin have said if he had seen *this!*"[24]

Ultimately, the five volumes of *Modern Painters* seemed to provide King with a perfect supplement to the five volumes of *Cosmos*: Ruskin, who a few times noted in letters or journals that he had spent his teatime reading Humboldt, elaborated for artists what it meant to humble oneself before nature in the Humboldtian manner. "However great a man may be," Ruskin explained, "there are always some subjects which *ought* to throw him off his balance; some by which his poor human capacity of thought should be conquered, and brought into the inaccurate and vague state of perception, so that the language of the highest inspiration becomes broken, obscure, and wild in metaphor."25 To know nature, then, was to question your beliefs, to upend your identity; you had to seek out dizzying precipices where it was hard to breathe.

Groping for holds, King, a much more pious man than Humboldt had ever been, clung to his religious faith, to the "Law of sympathy" and the "Law of sacrifice." His first journal entries in the West suggest a humbling search for "God's love" and "His brightness" and "Glory." But as he allowed his pencil to scratch out the contours of his thought, he began to focus not so much on radiance as on "dark mirrors," on "occult laws" and "hidden veins of truth"— not on God but on nature. Both art and science first served as conduits, aids in our "journeyings" into nature and its "intricate mazes of fact," where we had to "strain our vision . . . to learn its building, its form, its color, and its texture," and ultimately its "harmonies of structure." The only society worth constructing was one modeled on these harmonies, which were in turn capable of rejuvenating our arts: "Nature is a solemn fact, a glorious reality, which ought to move us to higher thought and true nobility." If nature was the ultimate school, though, it would have to be protected. Its higher purpose meant that human beings had to look beyond whatever immediate utility certain resources might seem to offer. We would always have to approach the physical world with a spiritual awe, "with gentleness and humble adoration. You Clarence King, never dare to look or speak of nature save with respect and the admiration you are capable of." King spent much of his life wavering between what he called the "two modes of studying nature," the "receptive" and the "analytical"—trying to decide, in a society whose professions were becoming increasingly specialized, whether to be an artist or a scientist. But he knew from the age of twenty-one, when he first reached the frontier, that nature itself was "the key to Art and Science."26

"No one who has not witnessed the autumn phases of California," King wrote in his journal on September 7, 1863, "can appreciate the glories of the contrast of Blue and Gold in vast extent." Having successfully crossed the plains and the Great Basin and caught up with Professor Brewer and the California Geological Survey, King was now a professional scientist, but he still appreciated nature most deeply for its colors. In New England, he had grown up with autumnal reds and yellows, oranges and browns—just as I did, more than a century later. When I experienced my first California fall, it was a letdown, a study in scorched, sere dullness. For King, though, the sallow hills were indescribably brilliant, throbbing with novelty. When he first met Brewer, by chance, on a steamboat headed down the Sacramento River toward San Francisco, he immediately asked to join his crew, to be given an opportunity to explore those hills and immerse himself in the golden sheen of the West.[27]

By the end of September, Brewer had led the survey back to Mount Shasta, and King had fallen in love with America's frontier landscapes—in large part because they proved so stimulating to both his "receptive" and his "analytical" faculties, in equal measure. "These valleys are new and strange to me . . . ," he raved in his first California journal; "today I saw . . . sage brush again! Oh!" In a culture saturated with images of the entire world, it is virtually impossible for us to recapture King's rapture upon experiencing the sights of the West, upon gazing for the first time at lava fields and canyons and gigantic trees the like of which he had scarcely been able to imagine while living in New England. For us, it would be like suddenly walking on Jupiter (one could have said Mars until quite recently). King filled his field notebook with unthinkable colors—the subtle gradations of blue in the sky, a "red purple lava peak," the yellow beams of sunlight on the blue-green pines and firs—but also with fossils and "crystalline masses" and the precise depth of ore deposits and morning thermometer readings and "flows of dark basalt showing among its fractured blocks a beautiful concentric structure." Art and science finally flowed together into a powerful current of naturalistic perception and depiction, allowing King to swim with nature's tide. Ultimately, though, it was the mountains that hooked him, that led him most directly to the sense of awe and humility that he felt should characterize the human relationship with the natural world. "Shasta," he marveled, "rises in a clear refined curve . . . , the summit thousands of feet above the floating clouds. A gold glow, faint, rests on the patches of snow on the grand old volcano."[28] King had never before seen a piece of the earth so high above his head.

King's journal of his westward journey across the continent was destroyed by a Nevada fire, from which King barely escaped with his life, but it seems clear that his fascination with the frontier had been confirmed almost as soon as he crossed the Mississippi. Everything was bigger here, from the prairie grasses to the herds of buffalo to the Rockies suddenly rising from the flatlands. And the Indians were wild, free, unpredictable. Having joined up with a mule trader and his wagon train in Missouri, King, traveling with his dear friend Jim Gardiner, had virtually all of the classic migrant experiences as the caravan made its way across the plains, past the Great Salt Lake and the new Mormon settlements, and finally along the Humboldt Trail through Nevada. King even took an extra day to go on a buffalo hunt, during which he wounded a particularly tough male and, failing to kill it, suffered a severe leg sprain when the animal turned on him, charged, and toppled his mount, leaving King temporarily pinned while his horse flailed in agony.²⁹ Several weeks later, according to the journal of J. T. Redman, one of the mule drivers in King's convoy, the group got to meet Brigham Young, leader of the Mormons, who memorably "made a short speech to us with a warning that they wanted to be let alone as they had been driven out from the States and settled on a desert and advised us to give an Indian a biscuit instead of a bullet."³⁰

The wagon train never had any violent confrontations with Native Americans, but Redman did report seeing "about 900 Indians of the Snake Tribe held as prisoners at [Fort Bridger]," the Snakes "being one of the hostile tribes that year." Of course, on the trail, the group had no way of knowing which Indians were which, let alone which ones meant to be friendly, so Redman and several others felt "compelled to stand guard every night to keep the Indians from attacking us and running off our stock (horses and mules)." At the same time, though, Redman noted that certain Indians actually helped them find grass for the livestock, and when prowling coyotes made the camp restless, "the Indians . . . would begin to whistle and the stock would quiet down." Throughout the journey, Indians remained mysterious to the westward travelers, depicted more as symbols than as human beings, seeming to represent the inscrutability of the frontier: "they came into our camp," Redman says of some Missouri Indians, probably Pottawatomies, "and stood gazing at us"— and our narrator leaves it at that. It seems understood that such silent standoffs were the best exchanges either side could expect, given the highly charged circumstances.³¹

All along the frontier, as Jim Gardiner recalled, King "showed his wonderful power of entering the lives and sympathies of every human being in the train"—even "the half-breed Indian hunter." But both Gardiner and King seemed to open their hearts and minds widest to the landscapes. "We became so fascinated with the life and so interested in the vast loneliness of those deserts," Gardiner wrote, "that I would gladly have turned around and traveled right back over the same road." As for King, he had already decided to spend his "life as an explorer in the unknown heights of the greatest of American ranges."³²

———

King and Gardiner would pass the next ten years together in the deserts and mountains of the West, first with the California Geological Survey, and then on King's own Geological Exploration of the 40th Parallel. The two young men had been bosom companions since boyhood, and for King, in particular, part of the appeal of the frontier was the opportunity it afforded him to spend long blocks of time in the intimate company of the man he called his "Dear

Brother" or his "darling fellow." On occasion, though, Jim was apparently distracted from his beloved "Clare" by eligible young women: as J. T. Redman reported, when their wagon train was joined by a new group of migrants, among whom was a certain "Miss Amanda," sporting fashionable bloomers, Jim "seemed quite fascinated with her charms and rode horseback with her as long as they traveled with us, and King was uneasy about him."[33] In the summer of 1868, Gardiner even took the step of getting married, forcing King at times to look elsewhere for the close male connections he craved. King found them, with a climbing partner named Dick Cotter, and even with Henry Adams, and later with the Washington insider John Hay. But no one ever knew King the way Jim Gardiner did.

It would be tempting to speak of Clarence King, in terms of personality, as the reincarnation of Alexander von Humboldt—if Humboldt had not still been very much alive when King was born (in 1842). Like Humboldt, King was known as a charismatic raconteur, the kind of ebullient and intelligent charmer around whom crowds gathered both at social clubs in New York City and at mess tents in the desert. "His wit and humor," Adams gushed, "his bubbling energy which swept everyone into the current of his interest; his personal charm of youth and manners . . . marked him almost alone among Americans."[34] Yet, as only Gardiner fully realized (until King was institutionalized), this font of knowledge and cleverness had a fractured and perturbable soul.

In his early letters to Gardiner, King clearly lacked confidence, wallowed in melancholy, brooded over his sense of isolation and disconnection. Though he possessed more ambition and initiative than Gardiner, he confessed that he never felt good enough for his friend: he thought himself "below" Gardiner "intellectually" and worried that although he "gained so much" from the relationship, he was hopelessly "incompetent to give you anything but silent sympathy and love." But love, perhaps, could make all the difference. "Do you remember our old Hartford walks," he asked wistfully, "how we used to talk whole evenings and tell each other all our feelings and sympathize as to life's trials and hope and wonder for the future, and how we declared our love and resolved never to lose our confidence in each other and never get the world's bashfulness of saying 'love.' Yes my brother those were happy days and when life seems to us as a desert land we can look back on them as a bright oasis of shining hours."[35]

Even as a very young man, then, it was important for King to *walk*, to get away from the centers of society, in order to gain the freedom of true intimacy. His career eventually allowed him to explore himself and his sexuality in ways that would have been impossible anywhere near the conventional, urbane

social enclaves of his mother's New England. Having been raised in Rhode Island exclusively by his mother and grandmother—his father had died in the
China trade when Clarence was only six—King remained extremely close to
the women in his family but also relished his opportunities to experience the
male culture of exploring expeditions.[36] Of course, as with Humboldt, we will
probably never know all the secrets of King's love life, probably never be certain whether his friendships with men blossomed into full-fledged relationships. We do know, however, that King never settled down, never formed strong
attachments to any women of his own social class—and was never strong
enough himself to acknowledge in public that he *had* formed attachments to
women who might have caused him a great deal of embarrassment in his
elite Anglo-Saxon circle. Perhaps he suffered all his life from a subconscious
fear that his doting mother would see a marriage partner as competition. Perhaps he suffered because he lived in a society where a healthy bisexuality was
simply unthinkable.[37]

Regardless, he clearly suffered—except when he was with Jim Gardiner.
And when the two friends were not together, Clare would write, and ask Jim
to "consider yourself kissed goodnight." "I turn to you then, my brother, as the
one friend of heart and soul whom I can enfold in *me* and to whom as far as
language conveys the soul's meaning nay farther, to where true hearts understand the unspeakable part of us, I open."[38] For Jim, the relationship clearly
petered out after a while, making him fairly typical of his time: countless
young men from elite, eastern families formed intensely romantic ties with
male friends in the nineteenth century, only to abandon them for conventional heterosexual unions a few years later.[39] For Clarence, though, who continually used the same language of communion to describe his bond both to
Jim and to nature, deep male love was something to cling to. Later, in the
1880s and '90s, he would write countless wistful letters to John Hay, revealing "how in my heart of hearts I long to see you. . . . Life would be far more
worth living were I your next door neighbor for all time."[40] Meanwhile, Clare
journeyed with Jim into the region of true hearts and unspeakable parts: "I
trust you wholly, I pour my soul into yours and am one with you."[41] This secret connection—many of his letters were marked "PRIVATE"—this forbidden, cosmic tie, gave King the faith he needed to continue with his sometimes
lonely work—not to penetrate the natural world, but to open himself to it, to
flow into its currents, to become receptive enough to make a real connection.

At the same time, King was not as successful as Humboldt at sublimating the
bitterness he felt over his isolation, and he occasionally adopted a much more
aggressive, masculine stance toward nature—and women. Late in life, King

was part of a small circle of friends known as the "Five of Hearts," with Henry Adams and John Hay (who had been President Lincoln's personal secretary) and their wives Clover and Clara. The group even made up their own stationery, and the three men often professed their love for each other, but King was clearly the odd man out. On Hearts notepaper, in May 1885, King told Hay he was making "conscientious efforts . . . to find the sixth heart," yet in the same letter, characteristically, he described the refined women of his social circle in New York as having "macaw voices" which left him "in a state of irritated daze."42 The only women he ever seems to have admired were either dark-skinned or working-class—Indians, African Americans, Polynesian dancers, barmaids, nurses—and he generally dehumanized these acquaintances by portraying them simply as "primaeval" or "archaic" or "natural" or "grandly barbaric," as possessing an "oceanic fullness of blood and warmth"— in other words, as fulfilling his primitivistic fantasies without challenging him intellectually.43 When, in turn, he portrayed nature as female, he sometimes lost his capacity for fellow-feeling, for love and communion, and adopted a stance of scornful superiority. The curves of sand dunes became "strange forbidding hill-forms," amid "leagues of desert, from which no familiarity can ever banish suggestions of death." Alternatively, certain curvaceous landforms could offer primeval challenges, in which cases King would grow excited, spurred by the prospect of conquest, "animated by a faith that the mountains could not defy us."44

This was the attitude with which King climbed Mount Tyndall, for instance, in the Southern Sierra, in 1864, whose apex revealed to him nothing but "the highest, most acute, aspects of desolation."45 And it was precisely this sort of description of his frontier adventures that has led several scholars, understandably, to attack King for his embrace of a "masculine primitive ethos," for his "blustering style," his overly "romantic ideal of self-reliant heroism," his gratuitous "risk-taking as a form of existential encounter," his "adversarial" relationship to nature.46 Without doubt, King's rhetoric did significantly more violence both to people and to nature than that of Humboldt or even J. N. Reynolds. His prose can be hard to read in the culture of the twenty-first century; there are times when it makes me squirm. What keeps me coming back to King's writing, though, and his life, is his commitment to challenge himself, to question the validity even of the masculine posturing that seems to have grounded his identity—and that of countless other white men in Victorian America47—in the midst of his sexual confusion and isolation. King started up Mount Tyndall with defiance in his heart, and he expressed a somewhat dismissive contempt at the peak's summit, but during the actual climb, his Romantic risk taking produced moments of intensely stimulating strain.

From the valley floor, King peered up to observe the "great pyramidal peak which swelled up against the eastern sky," the "straight, isolated spires of rock" "piercing thick beds of snow"; meanwhile, the deep blue lakes were "embosomed in depressions of ice." He gathered his energy for a burst of manly daring—but in fact his climb was repeatedly halted by the vagaries of weather and nature. Stuck on a ledge as night fell, no longer able to keep his blood flowing, King began to absorb the chill of his surroundings, despite his blanket: "the longer I lay the less I liked that shelf of granite; it grew hard in time, and cold also, my bones seeming to approach actual contact with the chilled rock."48

From the moment King first published such climbing tales, in his reputation-making book, *Mountaineering in the Sierra Nevada* (1872), he was criticized, even by his friend Henry Adams, for "carrying sensationalism too far for effect."49 Indeed, John Muir, King's foremost climbing rival, commented, upon reading King's description of his ascent of Mount Tyndall: "He must have given himself a lot of trouble. When I climbed Tyndall, I ran up and back before breakfast."50 But Muir arrived in California a full five years after King, and that half decade marked the crucial last stage of Sierran exploration. The danger was palpably real for the survey climbers of 1863 and 1864, who had no maps or trails, who had not read anyone else's descriptions of the area, who could never tell when the next abyss would cleave the earth. Professor Brewer's pithy response to Muir's braggadocio revealed a pride of first discovery: "It's easy enough to climb a mountain when you know where to go."51

What perhaps matters most in King's writing is his willingness to reveal his acts of self-exposure, to let us see him "hanging between heaven and earth," to help us trace step by precarious step the process by which his experiences led him to his particular, subjective reading of the land. King always "looked for lessons" in nature, believing "that God created all with design," that "lessons were taught in nature which were not elsewhere"—but that not all the lessons were comforting. Sometimes wilderness experiences meant confronting the fatal force of the inhuman, the harsh actuality of the material world; sometimes you wound up seeing too clearly what you were up against. Even when writing about other peaks, King recalled the lessons he learned "on Mount Tyndall, where an unrelenting clearness discovered every object in all its power and reality. Then we saw only unburied wreck of geologic struggles, black with sudden shadow or white under searching focus, as if the sun were a great burning-glass, gathering light from all space, and hurling its fierce shafts upon spire and wall."52 It was a moment comparable to

Thoreau's experience atop Mount Ktaadn, in Maine—perhaps the most intense wilderness experience anyone has ever described in the gentle Northeast: "Think of our life in nature,—daily to be shown matter, to come in contact with it,—rocks, trees, wind on our cheeks! the *solid* earth! the *actual* world! the *common sense! Contact! Contact! Who* are we? *Where* are we?"⁵³ These were men both terrified and thrilled.

In the end, King, like Thoreau, expressed far more awe and respect than scorn for the landscape. And though there were times when he did play the Romantically self-reliant hero who made magnificent escapes from the tightest spots, he more often played the Romantic climbing partner whose love for his close companion could redeem any doubts he might have had about the worthiness of the world around him. Since Jim Gardiner was away on another expedition in 1864, King climbed Mount Tyndall with his friend Dick Cotter, a mule packer from their trans-Mississippi wagon train who had grown so fond of Clare and Jim that he joined them in asking Brewer for jobs with the California Geological Survey. (Brewer, by the way, had discouraged them from making the attempt on Tyndall, for he "had seen the difficulties" involved—but, as the professor reported, "the intrepid Clarence King earnestly begged to try with Cotter to reach the crest.") So, on that night when King felt as though his bones were starting to freeze, he and Dick simply "rolled ourselves together. . . . How I loved Cotter! How I hugged him and got warm, while our backs gradually petrified, till we whirled over and thawed them out together!" King clearly relished his relationship with this brave, robust, hardworking drover, who always "came up the rope in his very muscular way," whose background was so far removed from the Newport mansions and drawing rooms and lecture halls in which King had grown up. Of course, such a connection was possible only out West, in the wild, on the frontier—at least, that's how it seemed to King. While King may well have approached mountaineering with what one scholar has called "a masculine exhilaration in achievement," he also climbed mountains in order to escape the masculine competitiveness of cities and to experience instead the intense cooperation necessitated by extreme conditions. And this kind of love—whether King thought of it as deliciously transgressive with regard only to social class or also to sexual mores—could even negate the most frightful powers of inhuman nature: "In this pleasant position we got dozing again, and there stole over me a most comfortable ease. The granite softened perceptibly."⁵⁴

Later in their climb, King and Cotter found that they had to go down some sheer, "glacially polished" cliffs in order ultimately to continue going up. So, "as the chances seemed rather desperate, we concluded to tie ourselves together, in order to share a common fate." Each man lowers the other down the walls in turn, in an image that almost perfectly mirrors a chapter of *Moby-Dick* called "The Monkey-Rope," during which Ishmael, "united" with Queequeg by "an elongated Siamese ligature" while the harpooner dangles above a whale, realizes for the first time the extent to which mutual dependence is a fundamental law of the cosmos. To Ishmael, this link with "his dear comrade and twin-brother" seemed a symbol of the connection between "every mortal that breathes" and "a plurality of other mortals." King's vision of sympathy and interconnection, like Melville's, is ultimately unrealistic, since it occurs outside of society. But it reveals something important about his belief in an ecologically structured world. For all the classic markers of his status as a member of Victorian America's white male elite, King at least expressed a desire to cross the uncrossable boundaries of his class, to break free from the hubris of refinement and artifice and greed and urbanity, to understand the ways in which he depended on the principle of unity in diversity—a principle that was visible to the naked eye only in the wilderness of the West. Through the conjoining love of men like Dick Cotter, King could come to appreciate "the connection of the material with the human, the esthetic with the eternal, the cosmical relation of God's earthly planes."[55]

"I think that the present generation are perhaps too much inclined to overestimate the works of man," King wrote to Jim Gardiner back in 1860. "Cold calculating *gainers* form a large part of my acquaintances," railed another letter. And on yet another "very dark" day of "blank uncertainty," King complained about the number of "men whose only end in life is to get rich and retire."[56] So how did he end up, in January 1864, wielding his geologist's hammer in service to the California gold rush?

When the members of the survey had gathered in San Francisco for Christmas, the previous month, King had waxed poetic on the topic of alpine landscapes. He raved in particular about Shasta, "that huge spectral mountain, brilliant yet wrapped in aerial mystery, surging up far above all earthly height."[57] And he also reported having caught a glimpse, on a transcendentally clear day during his November fieldwork, of the sublime, towering peaks of the Southern Sierra. Brewer certainly shared some of King's enthusiasm: "Although I have often reached great altitudes," he wrote, contemplating his experience with King on Mount Shasta, "that day stands out in my memory as one of the most impressive of my life." Yet he also worried, in a letter to George Brush, that King might be having too difficult a time with the "unpleasant work" of science—that he might be too much of a dreamer and mountaineer, and not enough of a geologist.[58] Josiah Dwight Whitney,

Brewer's superior and the actual chief of the California Geological Survey, shared these concerns. Whitney seemed of "even heavier caliber than Agassiz," and the chief certainly had no use for visionaries.[59] Though a devoted Humboldtian—he had made the requisite pilgrimage to Berlin in 1844—Whitney tended to downplay Humboldt's embrace of subjectivity, generally portraying the author of *Cosmos* as the consummate empiricist. Upon Humboldt's death in 1859, just before Whitney left Harvard to launch the California Survey, he had given a talk entitled "Science, and Humboldt as Its Representative Man."[60] Now he wanted to make sure that Clarence King, too, understood Humboldt in an appropriately rigid way, that he had the discipline to carry out Whitney's orders with adequate precision, the endurance to do the necessary hammering and collecting and the endlessly elaborate calculations. As the survey's paleontologist, William Gabb, commented at Christmas, upon hearing King burst yet again into his "rhapsodies of admiration": "I believe that fellow had rather sit on a peak all day, and stare at those snow-mountains, than find a fossil in the metamorphic Sierra."[61] So, just a few days before King's twenty-second birthday, Whitney sent the young Romantic to a decidedly modest peak called Mount Bullion, with the assignment of whipping himself into shape and producing a comprehensive geologic and topographic survey of the surrounding gold-mining area.

"Can it be?" King asked himself. "Has a student of geology so far forgotten his devotion to science? Am I really fallen to the level of a mere nature-lover?" If he had spent Christmas with his friends in New York, he surely would not have been racked with such self-doubt. But this was California, and his colleagues, having risked their reputations by shipping out to the uncultured frontier, could not rest their professional credibility on a dilettante and an aesthete. As King made his way to Mount Bullion, sitting alone—for Whitney had deliberately chosen to give him a solo mission—in the stagecoach, which was beginning "to rumble over upturned edges of Sierra slate," he kept recalling Gabb's stone-hearted comment. The paleontologist's words seemed to have become the words of God, and the coach's "every jolt seemed aimed at me, every thin sharp outcrop appeared risen up to preach a sermon on my friend's text." King, repenting, was determined to change his instincts, to feel the excitement of fossils, to succeed as a scientist—to bring back to Whitney a piece of definitive evidence that would reveal which geologic era had produced the gold-bearing regions of California. But nature had a way of undermining his scientific vows:[62]

I re-dedicated myself to geology, and was framing a resolution to delve for that greatly important but missing link of evidence, the fossil which should

clear up an old unsolved riddle of upheaval age, when over to eastward a fer-
vid crimson light smote the vapor-bank and cleared a bright pathway
through to the peaks, and on to a pale sea-green sky. Through this gateway of rolling gold and red cloud the summits seemed infinitely high and
far, . . . as gloriously above words as beyond art. Obsolete shell-fishes in the
metamorphic were promptly forgotten.

This image of glorious light, "above words" and "beyond art," was clearly
not the positivist light of factual knowledge, not a confirmation of scientific
progress and penetration, but rather a divine, Romantic vision of transcen-
dence. It was a reminder that words and art—and the natural world itself—
were what King cared about most. Back at Yale, he had told Gardiner that he
was "happy in my studies," that "I love science and language dearly." But even
then, despite the fact that he was pursuing a scientific degree, he admitted
that he preferred the study of "Language, as an enriching cultivation,"
whereas "I don't love the practical minutiae or lower details of science, although I work at these for discipline." Even upon achieving success as a geologist after several years in the field, King wrote in his journal of "the intense
yearning I feel to get through my analytical study of nature and drink in the
sympathetic side."[63] Science had a way of dehumanizing the scientist and
stripping nature to its barest bones.

Ironically, in January of 1864, King did in fact find that elusive fossil. His
professionalism and self-discipline seemed to triumph over his instincts.
Still, though, no glorious light led King to his startling discovery. Gray gloom
reigned in the realm of geology:[64]

> For many days thereafter I did search and hope, leaving no stone unturned,
> and usually going so far as to break them open. Indeed, my third hammer
> and I were losing temper together, when one noon I was tired and sat down
> to rest and lunch in the bottom of Hell's Hollow, a cañon whose profound
> uninterestingness is quite beyond portrayal. Shut in by great monotonous
> slopes and innumerable spurs, each the exact fac-similie of the other; with
> no distance, no faintest suggestion of a snow-peak, only a lofty chaparral
> ridge sweeping around, cutting off all eastern lookout; with a few disor-
> dered boulders tumbled pell-mell into the bed of a feeble brooklet of bitter
> water,—it seemed to me the place of places for a fossil. . . .
>
> Jagged outcrops of slate cut through vulgar gold-dirt at my feet. Picking
> up my hammer to turn homeward, I noticed in the rock an object about the
> size and shape of a small cigar. It was the fossil, the object for which science
> had searched and yearned and despaired . . . ! The age of the gold-belt was
> discovered . . . !

Down the perspective of years I could see before me spectacled wise men of some scientific society, and one who pronounced my obituary, ending thus: "In summing up the character and labors of this fallen follower of science, let it never be forgotten that he discovered the cephalopoda"; and perhaps, I mused, they will put over me a slab of fossil rain-drops, those eternally embalmed tears of nature.

But all this came and went without the longed-for elation. There was no doubt I was not so happy as I thought I should be.

Embosomed in this evocatively dreary description, of course, is a denunciation of King's social equals, the dominant players in the era just then becoming known, in the phrase of Mark Twain and Charles Dudley Warner, as the "Gilded Age." As King well realized, Whitney had sent him not just on a scientific mission but on a capitalist quest. Dating the gilded soil meant being able much more readily to identify *other* soil and rock that might bear golden fruit: the rush could be extended. Whitney did intend to use King's survey in his scientific reports, but he was also employing King on this particular mission as a favor to his friend Frederick Law Olmsted, fresh from designing Central Park in New York City and now serving as superintendent of the infamous Mariposa Estate, a large tract of land surrounding Mount Bullion that had originally been owned by none other than John Charles Frémont. Olmsted's straightforward goal was to figure out how to expand operations so as to increase output at the slumping Mariposa Mining Company.[65]

Because of King's complicity in this and many other mining and investment schemes, the literary critic David Mazel has suggested—and this is the scholarly consensus—that "the only consistent thread running through the widely varied activities of King's career" was probably "his reading and writing of the West in ways that served an ideology of capitalist expansion."[66] Yet King's description of his experiences in Hell's Hollow clearly equates scientific and materialistic progress, and neither pursuit comes off as particularly exciting or fulfilling. Indeed, his quest had left him empty, even bankrupt. King tried to imagine himself in the future as a hero among scientists, those bringers of progress—yet what could the opinion of future paleontologists do for him in the present? He deplored "the practical mundane spirit of this age of science and business." While the cephalopoda might take on the shape of a celebratory "cigar" to be smoked by some newly minted mining or railroad magnate, the fossil had meant merely tedious research for King. Trusting to science, worshipping the "vulgar gold-dirt," Americans of the late nineteenth century distanced themselves from any immediate experience of nature, from the possibility of communion. One of the most vicious portraits in King's first western journal was that of the "miser pilgrim," who had come to California with no attachments and with no other purpose than to live in the hills accumulating nuggets—the way a scientist accumulates fossils.[67]

In the end, it was far more important to King to use Hell's Hollow to attack the values of his social class than to portray the dark canyon in any sort of realistic way. He had chosen art, the human, the subjective, over scientific truth. His conclusion was that paleontologists were "scientific autocrats," anyway, and rather "light-minded" in going about their dull empirical business.[68] The bright beams of scientific progress would have left him blinded, without any perspective he could call his own. King's frontier experiences had, after only a few months, led him to the same conclusion that Friedrich Nietzsche would reach a few years later, that humanity's "will to truth" was really nothing but our "will to self-belittlement": "all science . . . has at present the object of dissuading man from his former respect for himself." If King had read Nietzsche while climbing in the Sierras, he perhaps better than anyone else would have understood the philosopher's assertion that "the scientific conscience is an abyss."[69] What King concluded on his own is that "the purely scientific brain is miserably mechanical; it seems to have become a splendid sort of self-directed machine, an incredible automaton, grinding on with its analyses and constructions"—that in fact it might have the power to kill the aesthetic perspective in his society, that the current "avalanche of materialistic and scientific activity" perhaps represented the "perennial artifice which wellnigh always chokes with its weedy growth the rare, fine flowers of art."[70] So King, for the next decade, would be primarily an artist. He would continue with his field training, with the drudgery of analysis and calculation, but he would also break free of Whitney's control and find plenty of time to enjoy wild, undisciplined scenery and the company of like-minded souls. From his arrival in the West in 1863 until the publication of the fourth edition of Mountaineering in the Sierra Nevada in 1874, King focused on his love for men and mountains, embraced the freedom of the frontier—gave himself over to a life of rambling.

King's successful discovery and analysis of the cephalopoda earned him some breathing room, got him out of Hell's Hollow. Whitney, now satisfied that the young apprentice had paid his scientific dues, even allowed him to explore the snow peaks of the Southern Sierra a few months later—thus giving him the opportunity to climb Mount Tyndall—and also assigned King and Jim Gardiner to do the first official mapping of the already celebrated Yosemite Valley. Then, after funding for the California Geological Survey ran out in 1866, King managed to become the director of his own expedition, along the

40th Parallel. His fieldwork over the next several years was concentrated in the desert regions of the Great Basin, and his commitment to professionalism still required that he consistently produce scientific analyses, but he also had the liberty to revisit the High Sierras quite frequently and make sure he was still in sympathy with nature. Later in life, he would consistently claim that "my years in the Sierras and the plains of California, Oregon, and Nevada were the happiest I have ever known or ever expect to know," and it was the exhilaration, the breathlessness, of his experiences here that afforded the success of *Mountaineering in the Sierra Nevada*.[71]

The book contained precisely what Walt Whitman had asked for in the exhortatory essays he called *Democratic Vistas*: "a little healthy rudeness, savage virtue"—"the genius of our land" as experienced by "stalwart Western men." To William Dean Howells, "Dean of American Letters," *Mountaineering* was a fit capstone to the frontier tradition founded by Washington Irving and continued by Francis Parkman (in *The Oregon Trail*) and Richard Henry Dana (in *Two Years before the Mast*). King, having abandoned science and granted his imagination more leeway, "brilliantly . . . portrayed a sublime mood of nature, with all those varying moods of human nature which best give it relief. The picture is none the less striking for being of a panoramic virtue; that is the American virtue, as far as we have yet got at it in our literature." *Mountaineering* hung balanced on a precipice between Romanticism and Realism: its tales of adventure provided King's predominantly East Coast readership with escapes from the hard, biting social truths of their modernizing cities (and of Howells's novels)—but escapes full of scenery rendered with "map-like distinctness," grounded in what Whitman called "real mental and physical facts."[72] King's many moods and prejudices distorted his descriptions and at times caused a kind of blindness; there are some outright fictions in the book. At other times, though, the writing achieves a delightfully slippery exactness, something King considered far superior to anything a scientific perspective could have produced.

King expanded the horizons only of those readers willing to absorb the dizziness caused by towering trees along with the soothing calm of alpenglow and the rush of climbing conquests. He would not be simply another tourist, arriving with all due pomp and circumstance at Yosemite's world-famous Inspiration Point, only to "dismount and inflate," and then "burst into rhetoric." And he had no desire to prod his readers into joining the "army of literary travellers" who were "already shooting our buffaloes" and were even starting to "cause themselves to be honorably dragged up and down our Sierras, with perennial yellow gaiter, and ostentation of bathtub." By using the possessive pronoun "our," King seemed to create an elite—but public—conglomeration of more deliberate travelers, people who could responsibly share ownership of the land and animals with Native Americans: you had to come west with a certain attitude.[73]

When everything came together in his writing, King, like Humboldt in his *Personal Narrative*, displayed a painter's sensitivity to light and color and a naturalist's sensitivity to biological detail, as well as a social critic's determination to condemn his compatriots' expansive ambition, to expose it as merely another expression of "the smothering struggle of civilization." Americans, living in cramped, urban spaces, had become overly "sophisticated, *blasé*, indifferent to nature, and conventional to the last degree." In *Mountaineering*, King lamented the typical American's "passing from a life of nature to one artificial," which resulted in "the fatal blunting of all his senses"; perhaps he ought to go back to his "true home" beneath the "trunks of sheltering pine . . . in the silent darkness of the primeval forest." The wilderness of high mountains provided a kind of higher education, taught people how they could build a society whose fundamental structure would express respect for the world beyond themselves, would provide opportunities for contact in the Thoreauvian sense, for direct experiences. Let the young American make "the constellations his tent, the horse his brother," rather than have "life, originality, and the bounding spirit of youthful imagination stamped out of him" in the civilized cities. King's Romantic sensibility brought him dangerously close to an endorsement of the frontier as merely a temporary, escapist, solipsistic fantasy—and of course King himself never completely let go of his own involvement in the centers of civilization. Generally, though, he upheld nature as a means of helping individuals transcend their egotism, by focusing their gaze outward, by keeping them moving, by forcing them into confrontations with the complex realities of the cosmos. You can never completely erase your mental filters, never have a thoroughly unmediated experience of anything. But you can try. In countless passages of *Mountaineering*, past-tense narrative spontaneously becomes present-tense visual excitement: "Our eyes often ranged upward, the long shafts leading the vision up to green, lighted spires, and on to the clouds. All that is dark and cool and grave in color, the beauty of the blue umbrageous distance, all the sudden brilliance of strong local lights tinted upon green boughs or red and fluted shafts, surround us in everchanging combination as we ride along these winding roadways of the Sierra."[74] Let the trees themselves guide your vision, and you'll truly start to see.

———

Yosemite Valley became a tourist destination, a place that was almost impossible to see except from the perspective of Inspiration Point, in the summer of

1864, when Congress granted it to the state of California to be forever preserved "for public use, resort and recreation." A few months earlier, the valley's fate had been uncertain, but in March Frederick Law Olmsted had come over from the Mariposa Estate to help Professor Whitney resolve the situation, and the two friends had sent a detailed report and some splendid stereoscopic photographs, taken in 1861 by the San Francisco photographer Carleton Watkins, to Senator John Conness. The California senator handed the photos around Washington, the bill quickly passed, and the precedent was set for America's system of national parks. King and Gardiner got to establish the park's boundaries with their autumn survey; their map, despite a few small errors, is itself a work of art. Watkins's photographs, meanwhile, represented a powerful opportunity, as Humboldt had pointed out, to enhance people's appreciation of nature—but, for King, they also represented a challenge. When he wrote about Yosemite, he vowed to be as graphically accurate as Watkins's photographs, but also to add something to them, to give his readers a chance to see beyond the valley's pretty, picturesque qualities.[75]

Wilderness preservation, from the very beginning, was both a radical and a conservative idea. On the one hand, people like Whitney and Olmsted believed they were taking land permanently out of production, setting a limit on industrial capitalism. Their highest goals were Humboldtian: to save pockets of nature that could be studied by scientists and enjoyed by the general public, so that ultimately all of us would feel more deeply the physical world's beauty, power, and significance to our lives. On the other hand, places like Yosemite seemed to have little industrial value in the first place—you can be sure Olmsted confirmed that there was no gold in the valley—and making them into tourist resorts was potentially just another form of capitalist exploitation. Yosemite was also far enough off the beaten track that, for the time being, it would be accessible only to elite travelers: this was not a democratic public space.[76]

Moreover, starting with Yosemite, in 1864, the preservationist movement committed itself to a vision of a pristine nature that erased all human influence. Suddenly, we were detached from the cosmos: we lived in cities, purely artificial constructions; when we experienced nature, it was merely as silent, invisible visitors, who never lingered long enough to be captured in a photograph. In truth, the picturesque views and contours of Yosemite Valley had been created by centuries of land and fire management: like all landscapes, it was both natural and artificial. But preservationists did not concern themselves with the people who had inhabited the land before the arrival of white settlers and miners. Just thirteen years previously, a group of soldiers called the Mariposa Battalion had entered Yosemite Valley for the explicit purpose of

17 Map of the Yosemite Valley, 1865. Clarence King and James T. Gardner (usually spelled "Gardiner").

exterminating or removing the local Indians. By the time Olmsted and Whitney filed their report and sent off Watkins's photographs, this violent history had been obliterated; indeed, their vision of wilderness preservation helped obliterate it. As essayist Rebecca Solnit has observed, "Nothing in any of these images or any of these agendas suggested that Yosemite was a battleground before it was a vacation destination."[77]

Clarence King, in characteristic fashion, absorbed and extended these paradoxes. He saw Yosemite as an answer to Hell's Hollow, as a place where greed would not interfere with direct experiences of nature. Yet he was not enough of a Humboldtian to correct Olmsted and Whitney's erasure of Native Americans from the valley's landscape. In the chapters of *Mountaineering* that deal with Yosemite, King focuses almost exclusively on nature, on domed mountains and ice fields and waterfalls and dark pines. At one of his campfire breakfasts, though, he at least gives us a fleeting glimpse of "a family of Indians," who "sat silently waiting for us to commence [eating,] and, after we had begun, watched every mouthful from the moment we got it successfully impaled upon the camp forks, a cloud darkening their faces as it disappeared forever down our throats."[78] This isn't much, and King seems distant, noncommittal—but it's more than we get in Watkins's people-less, ahistorical photographs. The Indians are possibly pathetic, passive, unable to fend for themselves, but their posture of watchfulness is also an assertion that they understand what the whites are doing to them; perhaps it is even a condemnation of the invaders' ravenous cruelty and selfishness. Since so much of King's prose involves his own watchfulness, his careful viewing of the world, it is striking, and disquieting, when he is the one being watched.

What we can count on from King is a more diverse set of views than we get from Watkins; King's work adds up to a grittier, more challenging picture of Yosemite. His black-and-white literary snapshots use certain obvious photographic techniques, but they sometimes come out more grotesque than picturesque: "A few weather-beaten, battle-twisted, and black pines cling in clefts, contrasting in force with the solid white stone." If King fails to recount the Mariposa Battalion's war against the valley's native population, he at least offers evidence of some sort of past violence. In his hands, Yosemite seems a place not only of "misty brightness, glow of cliff and sparkle of foam . . . , charm of pearl and emerald," but also of "stern sublimity," and "geological terribleness," and "confused wreck of rock and tree-trunk thrown rudely in piles by avalanches." He shows us eruptions and torrential floods and gusts of wind and metamorphic uplift and plunging precipices and glacial grinding: you never know what to expect. The physical world is a "stern, powerful reality." You might come here in some seasons to find a "lovely park" or "pleasure-ground," but at other times you'll confront catastrophes. "Nature in her present aspects, as well as in the records of her past, here constantly offers the most vivid and terrible contrasts." Olmsted and Whitney had sold Yosemite as a place to which civilized travelers could escape, as a garden of inspirational delights. But King wrote of it as a landscape to which you had to open yourself: you were not truly experiencing Yosemite unless you recognized your vulnerability here. You came not to forget about where you lived, but to learn lessons you could bring back home. You came for the experience of disorientation: "I found it extremest pleasure to lie there alone on the dizzy brink."[79]

In 1866, after two more years of field experience in California, King got to return to Yosemite, this time in the company of Carleton Watkins. The Geological Survey was about to close up shop, and Whitney wanted a complete scientific and artistic record of what he and his crew had accomplished, in the form of a volume, to be called simply *The Yosemite Book*, explaining the technical facts about the valley and its geological history through both word pictures and lavish photographs. The results were striking, and Watkins's photos, drained of their 1861 picturesqueness, captured King's desire to make contact with physical reality. The scenes were less balanced, more overwhelming. Instead of careful mixes of small rocks and trees and mountains and rivers, Watkins offered full-frontal views of stark cliffs, seen from below so that the barren walls looked as if they were about to fall forward onto the viewer. Other photos showed sharp drop-offs from dizzying heights, and narrow precipices, with no clear perspective, no flat ground where the viewer can secure his footing. And everywhere there are cropped, coarse, crooked trees—trees that seem to stare back at you. Long, bare trunks block the classic view of Yosemite from Inspiration Point. To arrive at the repose hanging in a misty glow at the far end of the valley, you must first come to grips with the unsettling violence of upthrust timber.[80]

In October of 1866, just after Watkins and King finished their work in Yosemite, the Union Pacific Railroad crossed the hundredth meridian. Its spiked tracks now covered 250 miles of prairie, stretching westward out of Omaha. Most people in America were thinking not about wilderness preservation but about technological unification.

Thomas C. Durant, vice president of the line, was thinking about celebration. Hoping to stimulate investment, he planned a three-day excursion to the end of the tracks, a glamorous frontier camping trip. Durant invited "two hundred and fifty of the most distinguished citizens of America," as well as a staff of highly acclaimed chefs, a couple of brass bands, several journalists—from all the major cities back east—and a photographer. As reported in two Omaha newspapers, the *Weekly Republican* and the *Weekly Herald*, the railroad magnate went all out to impress his guests. Beforehand, Durant had contacted Peter LaCherre, the chief of a local group of Pawnee Indians who so far had maintained friendly relations with the white people gradually encroaching upon their territory. LaCherre and some of his men, having received careful instructions (as well as some valuable gifts), welcomed Durant's party to the prairie with an elaborate war dance. Then, at 4:00 a.m., as the excursionists snored peacefully in their tents, the Pawnee staged a mock raid against the camp. Durant's guests were a little sleepy the next morning, but still thoroughly excited. They spent their day posing for photographs at the hundredth meridian marker and watching as the Pawnee pretended to wage war among themselves. That night, there was a festive dinner, followed by music and vaudeville acts. On their last day of "roughing it," the excursionists went for short strolls, taking in the natural scenery and observing the Irish workmen as they laid rails and drove spikes. Finally, in the early-evening darkness, Durant gathered his guests back into the railway cars. He called out to them, to get their attention. And then, at his command, the prairie suddenly burst into flames. The excursionists stared out their windows in thrilled amazement, as twenty miles of tall grassland slowly burned to the ground.[81]

Three months later, in January 1867, just after turning twenty-five,

Clarence King strolled into the Washington office of Edwin McMasters Stanton, President Andrew Johnson's secretary of war, and began lecturing him about the Union Pacific Railroad. Stanton, with a reputation for great administrative efficacy and even greater impatience, very nearly tossed the young geologist out on his shirttails. After just a few minutes, though, the secretary's interest was piqued. King was proposing a National Geological Survey, whose object would be to explore the lands surrounding the proposed route of the transcontinental railroad, roughly following the 40th Parallel from the Rockies to the Sierras, straight through the Great American Desert. Given the Union Pacific's track record, so to speak, Stanton was already convinced that he would soon be able to expand the Indian Wars and settle farmers on the plains of western Nebraska and eastern Colorado. But the future of Utah and Nevada was still an open question.

Following Thomas Durant's example, King both reinforced and exploited newly optimistic American attitudes toward western nature: he asserted that the land along the 40th Parallel was in fact not barren, "but full of wealth; the deserts are not all desert." No one knew the extent of the gold, silver, and coal deposits that might be found there, since, after the establishment of a few well-known mining towns, no one had ever bothered to look any further. Moreover, with an increase of settlement in the region, "the vast plains will produce something better than buffalo, namely beef; there is water for irrigation, and land fit to receive it. All that is needed is to explore and declare the nature of the national domain." Stanton decided to confer with Brigadier General A. A. Humphreys, chief of engineers, and the general responded to King's proposal with unreserved enthusiasm. The geology of the region in question, he agreed, was "unknown, and its determination cannot fail to prove of material value to the interests of the government and country." In March, President Johnson signed a bill establishing the Geological Exploration of the 40th Parallel, with Clarence King, U.S. geologist, in charge. King would report to Humphreys, who enjoined him, in the name of the material interests of his country, "to examine and describe the geological structure, geographical condition, and natural resources" of the far western frontier, from Colorado to California. As for Stanton, he provided King with an avuncular warning: "the sooner you get out of Washington the better—you are entirely too young to be seen about town with this appointment in your pocket—there are four major generals who want your place."[82]

King had pulled a J. N. Reynolds—and even managed to avoid getting kicked off his own expedition. By insisting on the practical, national benefits of his proposed exploration, by playing the economy card, he had paved the

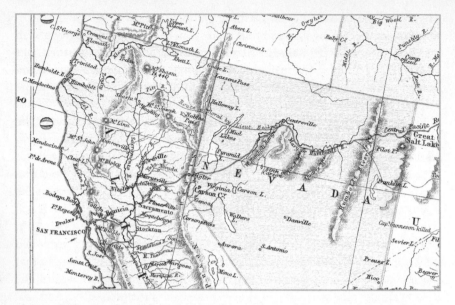

18 Detail of a U.S. map, 1870, showing the area covered by King's 40th Parallel Survey. Note the Humboldt toponyms.

way for a more thoroughly scientific survey of the West than had ever been attempted. As with Hell's Hollow, King's work in the Great Basin would eventually open more natural resources to exploitation, but it would also lead to more Yosemites—to the preservation of lands and to more direct, Humboldtian confrontations with the various meanings of wilderness. Although overseen by the War Department, the 40th Parallel Exploration, unlike the Ex Ex and every other previous federal expedition, was conducted entirely by civilian scientists: there was no Wilkes to sabotage King's careful selection of professional intellectuals. Jim Gardiner had already made enough of a name for himself to get offered a position in geodesy at Harvard's Lawrence Scientific School, but he quickly turned it down as soon as King came calling: Gardiner became the chief's second in command and the principal topographer. The two friends then proceeded to hire a team of geologists, a zoologist, a botanist, and, perhaps most important, a photographer—Timothy H. O'Sullivan, fresh from the Civil War. By May, the *New York Times* was announcing its support for this in-depth exploration of the West by asserting that "all the work of nature in that wild and unknown region is to be scanned by shrewd and highly educated observers." And Henry Adams claimed that King, by convincing Stanton to throw his weight behind the 40th Parallel Survey, "had managed to induce Congress to adopt almost its first modern act of legislation."[83]

Finally leading his own expedition, taking notes in Nevada, surrounded by Humboldt place-names—"fossil shells in Tufa at Antelope district near Humboldt"—King often considered the example of Humboldt's career and pondered his various theories. "Humboldt finds strata of warm air near the ground. It is an interesting question in Nevada to experiment on this!" "Humboldt on the plumbago polish of granitic rocks, Vol 2 Travels page 243." "Antithesis of *Cosmos* and *Divina Comedia*—the most striking opposition in European literature." Even back in 1864, in California, King had tested out various Humboldtian theories in his journals: "The vegetation of the Sierra Nevadas is plainly divisible into four well defined belts or groups. These belts follow climate rather than geographical lines."[84]

There is no simple, direct line of influence from Humboldt to King. Even King's juxtaposition of Humboldt and Dante is characteristically slippery and paradoxical. Was he reading *Inferno* or *Paradiso* at the time? Did he mean that the *Divine Comedy* represented humanity's religious and spiritual side, while *Cosmos* stood for all that was modern and scientific about the nineteenth century? Did he think of Dante as searching for the one true path through the dark wilderness, while Humboldt exhilarated in the experience of being lost, of staring into fiery, infernal abysses? Certainly, when King referred to Dante explicitly in his published writings, the images were of hell, and King seemed to take great delight in expressing a Humboldtian appreciation for landscape features that Dante described in order to evoke horror: "I rose and went out upon the open rocks, allowing myself to be deeply impressed by the weird Dantesque surroundings;—darkness, out of which to the sky towered stern, shaggy bodies of rock; snow, uncertainly moonlit with cold pallor; and at my feet the basin of the lake, still, black, and gemmed with reflected stars, like the void into which Dante looked through the bottomless gulf of Dis."[85] There is an icy foreboding here, but also a comforting peacefulness, an ease with which King exposes himself to the hard reality of the rock. While Humboldt's influence on King may have been too organic to be easily defined, it seems to me that perhaps the most important lesson King drew from reading *Cosmos* and the *Personal Narrative* was simply his determination to be a conscientious viewer, to go off the beaten track and observe every different kind of landscape, no matter what value government bureaucrats or classic writers might attach to them—to see more than his own reflection when he peered into a pool of water, and to be willing even to stare into the void of a bottomless gulf. He learned from Humboldt to see photographically, and that vision truly came into its own when King started collaborating with Timothy O'Sullivan. In the 1870s, King would produce detailed maps and scholarly articles and heavy tomes full of Humboldtian science, but in the late 1860s, his most important contributions, besides the sketches that would become *Mountaineering in the Sierra Nevada*, were the mammoth wet-collodion prints of the 40th Parallel Survey.

It is fitting that several of O'Sullivan's pictures show King—or someone who looks a lot like King—staring down a dubious snowfield or a blinding white cataract, or into a geyser mouth or sulfur spring. These are characteristically western landscapes, but the images of dazzling whiteness and endless blackness could easily have come from the South American or South Pacific writings of Humboldt, Reynolds, Poe, or Melville. In "Hot Sulphur Spring, Ruby Valley, Nevada," a grimy, disheveled Narcissus figure becomes entranced, looking deep into something very like what Poe called, in a short story, "the rank miasma of the tarn." In "The Falls of the Shoshone," a landscape that King thought provided "a frightful glimpse of the Inferno" and whose "ramparts of lava" and "ghostly, formless mist" he described in exquisite detail, O'Sullivan's viewer stares down a cascade of frothing, white water that eventually drops into the blur of a depthless, blackening chasm. O'Sullivan's negative for "Geyser, Ruby Valley," shows an entire person lying on his side and peering down into a pit, but in every print the man's head is cropped off.[86]

O'Sullivan wanted to portray Man Viewing. His figures are dirty, tired, believable. They are usually dwarfed by the scenery, but this limitless wilderness isn't pure or pristine, because a man has obviously entered it. The landscapes may occasionally seem inhuman, but the man gives you a way into them: they are starkly beautiful, real, direct, sometimes panoramic, sometimes narrowly confined. O'Sullivan's camera can't quite lie, yet the presence of the viewer reminds us that these are all interpretations, representations, dependent on a particular mental perspective and physical angle. And they took immense effort: the summit views, despite their terrible calm, suggest all the struggles of the ascent. As O'Sullivan himself wrote, of the Rockies: "It was a pretty location to work in, and *viewing* there was as pleasant work as could be desired; the only drawback was an unlimited number of the most voracious and particularly poisonous mosquitoes that we met with during our entire trip. Add to this the entire impossibility to save one's precious body from frequent attacks of that most enervating of fevers, known as the 'mountain ail,' and you will see why we did not work up more of that country." There was also the problem of crossing "snow-drifts from 30 to 40 feet in depth": the men and mules, "burdened with their heavy packs . . . , were frequently lost from sight." Even in the absence of snow, the surveyors found that their footholds were "very insecure, and danger from fragments of rock that are frequently

19 "Hot Sulphur Spring, Ruby Valley, Nevada," 1868. Timothy H. O'Sullivan.

20 "The Falls of the Shoshone, Snake River, Idaho," 1868. Timothy H. O'Sullivan.

21 "Geyser, Ruby Valley, Nevada," 1868. Timothy H. O'Sullivan.

dislodged by those who are in advance is continually experienced by the climbers in the rear." O'Sullivan's pictures were challenges to Americans to engage with "a country absolutely wild and unexplored," to see what they were actually dealing with and develop an honest relationship with these rugged territories.[87]

Of course, the character of O'Sullivan's photos comes as much from his own sensibility and experiences as from his collaboration with King. O'Sullivan took the famous Civil War picture called "A Harvest of Death," and this kind of bleak, devastating imagery haunted him on the frontier.[88] But King's field papers reveal that whenever he split up his surveying crew, he almost always paired himself with O'Sullivan, and as the photo historian Weston Naef has commented, no members of the survey's inner circle could escape "the outspoken esthetics of Clarence King."[89] These two men clearly shared a desire to offer realistic depictions of the West and of human interactions with the West, to do more than simply illustrate geological concepts or celebrate mountain glory, and King was so smitten with O'Sullivan's initial efforts along the 40th Parallel in 1867 that he invited him back for each of his field campaigns thereafter. Carleton Watkins was a decent second choice in 1870, and in 1871, when neither photographer was available, King fell back on the painter Albert Bierstadt. But Bierstadt, whose grandiose canvases could compete with Frederic Church's in size and popularity but not in Humboldt-ian complexity, was a bitter disappointment to King, who found his work to be hopelessly Romantic. "It's all Bierstadt and Bierstadt and Bierstadt nowa-days!" says one of King's frontier acquaintances in *Mountaineering*. "What has he done but twist and skew and distort and discolor and belittle and be-pretty this whole doggoned country? Why, his mountains are too high and too slim; they'd blow over in one of our fall winds. . . . He hasn't what old Ruskin calls for."⁹⁰

Neither did most of the other photographers working on the various west-ern surveys of the 1860s and '70s. The most famous one, William Henry Jackson, once climbed for ten hours in the Rocky Mountains to photograph the pristine Mount of the Holy Cross from the perfect angle. The cross of snow seemed to stand out beautifully against the black rock. When Jackson got back to Washington, though, he found "that the right arm of the cross was stunted and not quite straight." In order to save his image of nature's mani-fest approval of man's holy journey into the wilderness, he retouched his negative. Later, a perfect print of the Mount of the Holy Cross fell into the hands of Henry Wadsworth Longfellow, and the poet paid tribute to this sa-cred image in Romantic verse: "There is a mountain in the distant West / That, sun-defying, in its deep ravines / Displays a cross of snow upon its side."⁹¹ It was precisely this tradition of sentimentalism against which King and O'Sullivan rebelled; this was why they produced dark images of ex-hausted men confronting all sorts of dark, exhausted landscapes.

Sadly, in the twenty-first century, the dominant tradition in nature photog-raphy is the one represented by Jackson's print of the Mount of the Holy Cross and by Watkins's earliest Yosemite pictures. You have to look hard to get be-yond the pristine, airbrushed lakes of Sierra Club books and the glowing rock faces of Ansel Adams calendars—in the same way that you have to look hard to get beyond the wilderness preservation tradition that dominates today's en-vironmental movement. Preservationism is in many ways an admirable doc-trine, and photographs like those taken by Ansel Adams are gorgeous paeans to the nonhuman world, powerful assertions that wilderness is a good unto it-self. Yet such images also paper over the history of our interactions with the land and with each other, and for that reason it is tempting to imagine a slightly different tradition of nature photography, one not so concerned with wilderness preservation as with contact, connection, communion, with bal-ance and humility, with the real decisions confronted by the people laboring in the landscape. What if our coffee-table books showed O'Sullivan's images of ragged men overlooking borax lakes, or smokestacks rising from mountain

22 "Gould and Curry Works, Nevada," 1867–68. Timothy H. O'Sullivan.

23 "Gold Hill, Nevada," 1867–68. Timothy H. O'Sullivan.

24 "Gould and Curry Mine, Cave-in, Nevada," 1867–68. Timothy H. O'Sullivan.

25 "Savage Shaft, Miners and Mine Car, Nevada," 1867–68. Timothy H. O'Sullivan.

26 "Indians: Group (Shoshones), Nevada," 1868? Timothy H. O'Sullivan.

27 "Shoshone Indians: Group (Buck and Staff), Nevada," 1868? Timothy H. O'Sullivan.

mining towns, or even the cropped bodies of dark-faced miners attempting to cope with a cave-in? What if they stopped offering us perfect, magical escapes?[92]

Of course, there are also omissions in O'Sullivan's photos. Perhaps the most radical western pictures of the early 1870s came from the camera of Eadweard Muybridge, who actually showed the Miwok-Paiute Indians still surviving in Yosemite Valley; he even photographed the Native American communities that fought the U.S. Army in California's Modoc War of 1873. Muybridge documented the Indians' intimate connection to California's monumental landscapes—not as hopelessly backward Romantic savages but as human beings inhabiting a homeland—as well as white attempts to erase Indians from those landscapes.[93] O'Sullivan and King never went quite that far, never had the same kind of Humboldtian epiphany that J. N. Reynolds had in the land of the Araucanians. But in the late 1860s King did hire several Indians to help with his surveying, and he and O'Sullivan carefully arranged a few group portraits of some Shoshones in Nevada. The pictures are artificial, posed, but also direct and serious: men, women, and children stare back at the camera, well-dressed and confident yet solemn at the same time, perhaps slightly threatened or suspicious. Sometimes a couple of survey workers share the frame, and they clearly own the power, clearly represent the forces of change—but the Indians come off as their equals. All the subjects of the photograph are subjecting themselves not only to the lens but to the land, bracing themselves for shocks, struggling to understand the shifting shape of the frontier.

———

One of O'Sullivan's most striking portraits of Clarence King puts him somewhere in the East Humboldt Mountains, in 1868, staring down the shady side of a snowfield. It is unclear how he managed to attain this narrow ledge of a summit, and O'Sullivan seems to have set up his camera in midair. Of course, the other question hanging in the air is: how will they get down? King appears calm, settled, stable, but his right foot is just slightly forward, as if testing the snow, and his head and shoulders tilt longingly toward the earth. Perhaps, after a few more seconds of contemplation, he'll hurl himself off the cliff and go sliding down the mountain on the soles of his shoes.

Why not? This was how he had descended Shasta in the fall of 1863: "Now there was a slope of snow . . . ," he wrote in his journal, "at an angle of 60 degrees (mighty steep) down which I had a desire to 'glissade,' so Brewer took the barometer, and . . . with a little trace of scare I jumped . . . and began to slide with an awful rapidity." Inevitably, King soon approached a crevasse, which might well have killed him had it been a little wider. But his momentum was

28 "Summit, East Humboldt Mountains, Nevada," 1868. Timothy H. O'Sullivan.

enough to send him airborne, and he landed safely on the far side, his heels scraping against the edge. "The snow flew in a cloud from my feet. Ah! It was splendid."94 Perhaps, even in his private journal, King showed a tendency toward sensationalist bravado, but for all the macho overexuberance of his risk taking, I think it ultimately was meant to reinforce a sense of humility. In a moment of pure exhilaration, he had captured his purpose in coming west in the first place, the spirit with which he meant to abandon himself to the currents of nature, to get thrown off balance and exposed to the world's power.

King arranged to go back to Shasta in 1870. From the moment the grand peak came into view, King "dated a new life," and he and his survey companions, including Carleton Watkins, "let our hearts open to the gentle influence of our new world." Their intention was to study the line of dormant volcanoes stretching up from Shasta to Rainier, and perhaps to compare these to the much more active Andes. Indeed, in the cone of Shasta King saw a much more threatening image: a "broad broken summit singularly like Cotopaxi," the Andean peak whose frequent eruptions Humboldt had witnessed and Frederic Church had painted. King, for one, would not have been surprised when Mount St. Helens blew its top in 1980. Indeed, the whole landscape appeared to him "unstable . . . , in every state of uncertain equilibrium," and full of "tremendous contrast." "I found it strange and suggestive," he wrote in *Mountaineering*, "that fields of perpetual snow should mantle the slopes of an old lava caldron, that the very volcano's throat should be choked with a pure little lakelet, and sealed with unmelting ice."95 Eventually, since there was less heat here than in the South, King turned his attention to the cold. Humboldt and Reynolds got to peer into smoking craters; King got to stare down rifts in the ice and snow. And, in 1870, on what he called the "flanks" of Mount Shasta, he got to make a scientific discovery far more intriguing to him than any fossil.

The morning of September 11, when King and his climbing partners started their ascent, "was cool and clear with a fresh north wind sweeping around the volcano and bringing in its descent invigorating cold of the snow region." Assisted by a local guide who had opened a way station at the base of the mountain, King tried a new approach, from the west. As usual, he was aided in achieving mental openness by the very "dislocation" of the landscape. "The footing" on the western slope "at times gave way provokingly, and threw us out of balance." Eventually, the party made it to the top of a secondary cone, called Shastina, where today you can find Lake Clarence King and where King himself beheld "a scene of wild power." Below the north rim of the crater that divides Shastina from Shasta lay not your typical white snowfield, inviting glissaders to kick up its powder, but a sheet of ice about three miles long, riven with yawning blue crevasses—an active glacier.

Science suddenly recaptured King's spirit. As the still-young geologist explained, in an article for the journal founded by Benjamin Silliman (and now edited by James Dwight Dana), while much of the high country in the American West had clearly been carved out by glaciers, no scientist up to this point had been able to study the actual process of glaciation on American soil. The shock King experienced upon viewing Shasta's glacier had been profound; he had immediately written a letter to his supervisor, General Humphreys: "The labor has been extremely severe: our camps were frequently up to 11,000 feet above sea level, and on one occasion over 14,000 feet. We completely upset the ideas of Humboldt and Fremont concerning the mountain itself, and have made the somewhat startling discovery of immense existing glaciers. This is the more surprising when we consider that Whitney, Brewer, Dana, and Fremont all visited the peak without observing them, and that Whitney, Dana, and Agassiz have all published the statement that no true glaciers exist in the United States. May I ask that this be kept private until my return."96 King's new climbing route, his fresh approach, had made all the difference. Now he

29 "Mount Shasta, Siskiyou County, California," 1870. Carleton Watkins.

30 "Commencement of the Whitney Glacier, Mount Shasta, Siskiyou County, California," 1870. Carleton Watkins.

could look into the past and imagine the continent covered in booming, grinding fields of ice, could investigate the percolation of streams through crevasses, the causes of avalanches, the construction of moraines. Perhaps he could even settle the question of whether valleys like Yosemite resulted primarily from glacial activity or from the inner fire of vulcanism.

King embraced these scientific opportunities and spent the rest of the decade working on them. In 1870, though, the most important thing was still the simple act of experiencing nature—the act of "going to the edges of some crevasses and looking over into their blue vaults, where icicles overhang and a whispered sound of water-flow comes up faintly from beneath." King balanced his glacier article in Dana's scientific journal with an article on the same topic in the popular *Atlantic Monthly*. And in *Mountaineering*'s Shasta chapters, King once again comes across as a character created by Poe or Melville, an aesthete in the wilderness, eager for new experiences. Like Humboldt and Reynolds, King does continue taking all the necessary readings and measurements, even under the most trying conditions. Meanwhile, though, he is constantly chasing avalanches, inching along narrow brinks, losing his way in the "labyrinth" of ice cascades and gaping cracks, climbing from erratic boulder to erratic boulder over debris fields that are caving in on themselves, seeking out "views down into the fractured regions below": "We were charmed to enter this wild region, and hurried to the edge of an immense chasm." At one point, King admits that he "had the greatest desire to be let down with a line and make my way among these pillars and bridges of ice." But his partner's monkey rope is, fortunately, too short, so the companions simply "pushed on, tied together with our short line, jumping over pits and chasms, holding our breath over slender snow-ridges." The lure of the crevasse, King ultimately realizes, is parallel to the lure of the flame—and he is the moth. "To the siren babble of mountain brook is added all the tragic nearness of death."[97]

In the end, King not only survived on Shasta but even achieved the kind of fulfillment that consistently eluded him in more civilized settings. The lone, towering volcano became a symbol for him of independence and confidence, of direct confrontation with the dark and dizzying reality of the cosmos. On Mount Shasta, science gave him clarity, brilliance, distinctness, transparency, while art preserved his sense of mystery, humility, color, connection. From a high enough vantage point, in the West, you can see almost all of nature's diversity: "from rich, warm hues of rocky slope, or plain overspread with ripened vegetation, out to the high pale key of the desert." If some mountains, like Tyndall, suggested nothing but desolation, Shasta encompassed the full range of human experience. King's climb went beyond Romantic escape, beyond professional accomplishment, beyond capitalist striving, beyond masculine conquest. In *Mountaineering in the Sierra Nevada*, King summed up his perspective on Shasta with a characteristic combination of precision and suggestiveness: the overriding "sensations" he felt at the summit, he explained, were "geography, shadows, and uplifted isolation."[98]

CHAPTER 7

"Catastrophism and the
Evolution of Environment"

A Science of Humility

The West of which I speak is but another name for the Wild; and . . . in
Wildness is the preservation of the World.
—Henry David Thoreau, "Walking" (1862)

The earliest geological induction of primeval man is the doctrine of terrestrial catastrophe. This ancient belief has its roots in the actual experience of man, who himself has been witness of certain terrible and
destructive exhibitions of sudden, unusual, telluric energy. . . . Catastrophism is therefore the survival of a terrible impression burned in
upon the very substance of human memory.
—Clarence King, "Catastrophism and the Evolution of Environment" (1877)

Clarence King liked to be known as a reasonable man, and in 1870 he had offered a reasonable justification for his desire to revisit Mount Shasta. The volcano was relevant to his survey of the Great Basin because of "the lava systems
which flow eastward from the peak, and connect themselves with the basalt
regions of the Nevada Desert." Indeed, as Timothy O'Sullivan put it, the area
between Shasta and the Humboldt Sink provided "picturesque evidence of
great volcanic convulsions that have occurred in years long since passed
away." For all of King's interest in glacial ice, his reading of Humboldt had
kept him focused on the significance of vulcanism. Indeed, both the practical
and the theoretical value of his work in the Shasta region were undeniable:
"much of the fertility of the northern portions of California and Nevada, many
of the most important silver deposits, and the most grave of the unresolved
questions of Cordillera geology lie in this cordon of volcanic rents."[1] But, the
following summer, General Humphreys insisted that King take his fieldwork

back to the desert between the Rockies and the Sierras. The general could see no further reason to let King pursue his California dreams. By 1871, there was mounting pressure to finish up the 40th Parallel fieldwork and continue issuing reports like the immensely useful *Mining Industry* of 1870, the survey's first official publication. While the scientists pondered their grave geological questions, congressmen and investors demanded more data.

One of the charms of fieldwork, though, is freedom from supervision. King, knowing that his time on the frontier was almost up—even in 1870, he was already complaining of "the increasing complexity and conflicting nature of my duties in connection with the preparation and publication of my reports"[2]—arranged for his crew to start surveying without him in Wyoming Territory in the late spring of 1871, while he went on to California by himself, to buy supplies for the rest of the season—and to do a little climbing. Back in 1862, Shasta had been considered the tallest mountain in the West, but King's own observations and climbs in the Sierra in 1863 and '64 revealed that the grand volcano was actually shorter than some of the high peaks clustered in the southern part of the chain. And the highest of these, Mount Whitney, had never actually been climbed by anyone equipped with the tools to measure it. Volcanoes were fascinating, but if King had time for only one more jaunt in the California mountains, he was going to embrace the exhilaration of altitude. Humboldt had had similar inclinations: he had taken great pride, after all, in setting an altitude record on his favorite Ecuadoran peak, Mount Chimborazo, despite being halted by an impassable crevasse just a few hundred feet from the top. If Shasta was King's Cotopaxi, Mount Whitney, though not volcanic, was his Chimborazo.

King had actually made an attempt on Whitney in 1864 that almost perfectly recapitulated Humboldt's experience on Chimborazo, though it was a sheer granite cliff that finally turned him back. The climb had left him blinded and hypothermic: "When I hung up the barometer, a bitter wind was blowing; it hailed and snowed and the clouds closed in around me shutting off all view. I had been all day abroad and the fatigue and excitement exhausted me greatly, so much that although chilled through I could scarcely prevent myself from lying down and sleeping. A strange carelessness came over me."[3] So when King left Lone Pine, California, in late June of 1871, for another assault on "the loftiest peak in the Union," he was after not only freedom but redemption. A successful ascent would cap his nine-year career in the field and provide the climactic chapter of *Mountaineering in the Sierra Nevada*, scheduled for publication the next February.[4]

King had been pegged by his coach driver, who "scanned me from boots to barometer," as a scientist come to measure a mountain, but, typically, he saw this climb as transcending the pragmatic concerns of science, as achieving almost the status of a great artwork, undertaken purely for its own sake: he was climbing Mount Whitney because it was there. He had nearly died on this peak, yet here he was reimmersing himself in its dangers, braving disorientation, perhaps to prove his manhood or his power over nature, or perhaps simply to reexperience that rare feeling of true immersion. Even when he attained the summit and was rewarded with a view, it was far from clear. The sharpminded, barometer-toting scientist might have conquered the mountain, but, in dim swirls of fog, the atmosphere had reasserted nature's mystery: "When all else was buried in cloud we watched the great west range. Weird and strange, it seemed shaded by some dark eclipse. Here and there through its gaps and passes, serpent-like streams of mist floated in and crept slowly down the canons . . . , and in a moment a vast gray wave reared high, and broke, overwhelming all." The mountainscape, as vast and humbling as the ocean, had transformed King's redemptive ascent into a confirmation of his blindness.

As usual, King's companion, in this case "Paul Pinson, as tough and plucky a mountaineer as France ever sent us," provided comfort, a firm helping hand, and both spiritual and physical warmth. But even Pinson's extra pair of eyes and expert opinion could not prevent the two climbers from becoming hopelessly befogged on their descent, "until suddenly we came upon the brink of a precipice and strained our eyes off into the mist. I threw a stone over and listened in vain for the sound of its fall. Pinson and I both thought we had deviated too far to the north . . . , so we turned in the opposite direction . . . , but in a few moments we again found ourselves upon the verge. . . . We were evidently upon a narrow blade." The companions could do nothing, for a while, but sit down, huddle together, and try to puzzle out their route, during which time, of course, they risked falling victim to the even more dangerous fog of hypothermia. Fortunately, they did not pause too long, did not get caught in one of the dead calms J. N. Reynolds so rightly feared. Several hours of rope work, careful crawling, and a kind of prolonged stumbling brought them safely back to their camp.

These were certainly interesting experiences, but King's embrace of them couldn't cover up his overriding sense of defeat. Explorers like to be first. Indeed, their reputations often rest on claims to precedence. What King and Pinson had found at the top of Mount Whitney, however, was that "a small mound of rock was piled upon the peak, and solidly built into it an Indian arrow-shaft, pointing due West." King did his best to avoid holding a "grudge" against his "Indian predecessor," who clearly had earned "the honor of first finding that one pathway to the summit of the United States, fifteen thousand feet above two oceans." At least King would be the first to *know* that he had reached his nation's apex, and the first to *measure* those fifteen thousand feet. So he cheerily hung up his barometer on the Indian's cairn—only to get a reading that, when all the computations were complete, seemed to make Mount Whitney significantly shorter than several of its closest neighbors. Well, science was not infallible. "Realizing at once that this must be an error," King later explained, "I attributed it to some great abnormal oscillation of pressure due to storm, and decided not to publish the measurement"—sticking, instead, to his round estimate of fifteen thousand feet. *Mountaineering in the Sierra Nevada*, like King's ascent itself, was a work of art, after all—not science. He could let an associate publish the correct figure in some obscure journal in a couple of months. Besides, "on carefully studying the map it was reassuring to establish beyond doubt" that he must have climbed Mount Whitney. But King did acknowledge, toward the end of his Mount Whitney chapter, that the whole escapade had been rather disconcerting, that he felt almost as though it had been some sort of delusory dream. From his base at Lone Pine he found himself staring persistently back at the high peaks, searching for some confirming sign of his successful climb. The image he leaves us with is one of doubt and brooding: "I constantly looked back and up into the storm, hoping to catch one more glimpse of Mount Whitney; but all the range lay submerged in dark rolling cloud, from which now and then a sullen mutter of thunder reverberated." Was it really 1871, or was he stuck in 1864?⁵

Perhaps, as he wrote that final description, King had a premonition of the moment, two years hence, when a former assistant to the California Geological Survey would gleefully inform the world that, in fact, on that dark 1871 afternoon, King had ascended the wrong mountain.

———

Immediately after the climb, shaken but ultimately oblivious of his mistake, King hustled back to Wyoming, where a stranger had come to visit his 40th Parallel camp, a stranger who would soon become another of King's bosom companions. The way Henry Adams tells it, narrating his own life in the third person, "Adams fell into his arms. As with most friendships, it was never a matter of growth or doubt." On the night of their first encounter, they shared a bed in camp, "and talked till far towards dawn." For Adams, anyway, bearing the burden of his overcivilized background and his descent from two past presidents, this connection was life altering, and this night on the rugged frontier took on heavy symbolic weight. "King, like Adams," he later wrote, "and all their generation, was at that moment passing the critical point of his career." Things would turn out badly. There is a huge gap in Adams's famous autobiography, *The Education of Henry Adams*, and it comes immediately after this moment in the Wyoming desert: the story jumps twenty-one years, to

1892, a time when, in Adams's view, the entire nation had become lost in the desert. King was bankrupt, and on the verge of a nervous breakdown, and Adams was chronically depressed. Neither one "knew whether they had attained success, or how to estimate it, or what to call it, and the American people seemed to have no clearer idea than they. . . . They were wandering in a wilderness much more sandy than the Hebrews had ever trodden about Sinai." The desert of 1871, though, had held immense promise. Adams felt powerfully inspired in King's presence and immediately judged this explorer to be "the ideal American": "King's abnormal energy had already won him great success. None of his contemporaries had done so much, single-handed, or were likely to leave so deep a trail. . . . Whatever prize he wanted lay ready for him—scientific, social, literary, political—and he knew how to take them in turn. With ordinary luck he would die at eighty the richest and most many-sided genius of his day." So: King was a self-reliant, complicated, discerning dynamo. He "had everything to interest and delight Adams. He knew more than Adams did of art and poetry." And, perhaps most important, King "knew America, especially west of the hundredth meridian, better than anyone."[6] He was quintessentially American because he was directly in touch with the American landscape. Yet this renowned geographer and geologist, this firmly grounded, thoroughly modern, eminently reasonable man, this field-tested cartographer of the American West, had just convinced himself that he had climbed to the nation's summit when in fact he had gone up and down a humble spur to which he himself, back in 1864, had affixed the tender name "Sheep Rock."

So much for maps.

Adams had been correct in his assessment of King's professional status, but all of the scientist's accomplishments were suddenly cast in doubt on August 4, 1873, when W. A. Goodyear presented his definitive, disabusing report to the California Academy of Sciences.[7] Goodyear himself, however, had stopped at the top of Sheep Rock; since he had made the climb on a clear day, he immediately saw Whitney looming over him, but he was so stunned by his discovery that he felt compelled to rush back home and write up his experience, leaving the true Mount Whitney unmeasured. A fellow graduate of Yale's Sheffield Scientific School, Goodyear also wrote a rather haughty letter to Professor George Brush suggesting that King saw "many things through a strange magnifying glass."[8]

King got wind of Goodyear's findings a month later, while already on his way to the West Coast, and "of course I lost no time in directing my steps toward Mount Whitney, animated with a lively delight which was quite unclouded by the fact that two parties, who had three thousand miles the start of me, were already en route and certain to reach the goal before me."9 This time, King managed to climb the real peak, and, seeming truly untroubled by his failure to be the *first* scientist to sit atop America, he quickly scribbled a second Mount Whitney chapter for a revised fourth edition of *Mountaineering in the Sierra Nevada* (the first three printings had sold out as soon as they came off the presses). In the book's new preface, he claimed to have learned a profound lesson—he could have avoided this humiliation if only he had trusted his barometer—and thus to have reached a watershed in his career: "There are turning-points in all men's lives which must give them both pause and retrospect. . . . It is thus with me about mountaineering; the pass which divides youth from manhood is traversed, and the serious service of science must hereafter claim me."10

King did complete his fieldwork in December 1873, and he then retreated to the lab and the office for the rest of the decade. In 1876, he would become the youngest person ever elected to the American Academy of Sciences (and, following in Reynolds's footsteps, he was named a corresponding member of the Boston Society of Natural History a few years later);11 in 1878, he would publish his magisterial *Systematic Geology*; and in 1879 he would take on the role of director of the just-formed U.S. Geological Survey. Even during these years of supposedly "serious" study, however, King maintained his same sense of skepticism about what he was doing, his suspicion that the scientific perspective, devoid of art, was probably just as skewed as that of a mountaineer lost in a fog bank. He insistently defined himself as a scientist, after 1873, but his science continued to be marked by his intense awareness of human limits and fallibility. His geology was certainly systematic; at the same time, though, it was humanized by humility, sympathy, and imagination.

King took great pains to make his readers realize that the blinders were off for his 1873 ascent of Mount Whitney, that this time the powerful light of science would pierce any storm clouds. Hailing "the geological history and hard, materialistic reality of Mount Whitney, its mineral nature, its chemistry," he swept aside his artistic inclinations, breathing deep of "the liberating power of modern culture which unfetters us from the more than iron bands of self-made myths." As King stared up at the range, an old Paiute Indian stepped out of the forest to join him in his reverie. But this poor "savage" could behold merely a legendary mountain, for he was burdened with "a hundred dark and gloomy superstitions." The man's stories were intriguing—but outdated. It fell to the cutting-edge scientist to see "the great peak only as it really is, a splendid mass of granite, 14,887 feet high, ice-chiselled and storm-tinted, a great monolith left standing amid the ruins of a bygone geological empire."12

Yet science's conquest remains unconvincing. Ultimately, the Mount

Whitney climb turned out to be one of the most demanding of King's career, requiring more than a week and several different approaches. King had to stop at one point because he was overcome by "a three days' painful illness," and that's not even to mention "the days of snow and sleet we spent under a temporary shelter constructed of blankets." Moreover, his explanation of why his own map had led him astray in 1871 leaves open the possibility that we might never get the maps right: cartography depends on the accurate use of a compass, and it takes just a small deposit of magnetic rock to skew a compass reading. King wasn't even sure that a perfect map would have helped: "infallibility in retrospect is one of the easiest conditions imaginable," he suggested; "yet when the ever-fresh memory of those seething cloud-forms comes back to me . . . , I am free to confess that I should make the same mistake again."[13] The fogs keep rolling in: nature is ever-changing, and all we have to go on is our experience of it.

King also used passages in this chapter to attack "the crushing juggernaut-car of modern life and the smothering struggle of civilization." So his concluding embrace of modernity and "materialistic reality" seems somewhat forced, given his scorn for "the shallowness of society." Indeed, despite the difficulties of his ascent, King found himself once again thoroughly relieved to be back in the wilderness: he could feel settled only when he was far away from the settlements. "Perhaps there is no element in the varied life of an explorer so full of contemplative pleasure," he wrote, "as the frequent and rapid passages he makes between city life and home; by that I mean his true home, where the flames of his bivouac fire light up trunks of sheltering pine and make an island of light in the silent darkness of the primeval forest." Such transitions into older worlds helped him to appreciate the "unperishing germ of primitive manhood which is buried within us all under so much culture and science." It seems particularly important to notice, then, that before King dismissed as archaic and superstitious the bowed Indian who had shared his fascination with Mount Whitney, he had listened to what the "gaunt, gray" man had told him, "in strange fragments of language." From his perspective, the Paiute elder explained, "the peak was an old, old man who watched this valley and cared for the Indians, but who shook the country with earthquakes to punish the whites for injustice toward his tribe." King in the end felt compelled to uphold science over superstition, but his rhetoric reveals just how sad he was that "the myth-maker has been extinguished in modern students of mountains" and just how much he sympathized with the old Indian's condemnation of modern white society.

The final view King described from the summit of Mount Whitney also suggests the extent to which he still believed that modern science needed to be tempered by art, just as "civilization" needed to be tempered by wilderness. His panoramic vision reads as though he had just been studying a painting by Frederic Church: he saw every detail, clearly enough, but not too clearly; he saw the real and the ideal, the blending of two seemingly irreconcilable perspectives; he saw the truth of the brutal desert and the jutting rocks—but even such inhuman wildness somehow came across as appropriate, as comforting. He had seized an opportunity to be true both to the landscape and to himself. King claimed to enjoy his trips between the city and the country, between complex analysis and simple sympathy, but in fact he was exhausted, and years of cognitive dissonance, spent bouncing back and forth between geology and mountaineering sketches, had created a bifurcation in his very being. In part, King climbed Whitney just because he felt a mysterious impulse to do so; in part, he was consciously searching for peace and unification. Finally, at the summit, he succeeded in opening himself to the harmony of the cosmos:

> Around the wide panorama, half low desert, half rugged granite mountains, each detail was observable, but a uniform, luminous medium toned, without obscuring, the field of vision. . . .
>
> It was like an opal world, submerged in a sea of dreamy light, down through whose motionless, transparent depths I became conscious of sunken ranges, great hollows of undiscernible depth, reefs of pearly granite as clear and delicate as the coral banks in a tropical ocean. . . . There was no mist, no vagueness, no loss of form nor fading of outline—only a strange harmonizing of earth and air. Shadows were faint, yet defined, lights visible, but most exquisitely modulated. . . . I do not permit myself to describe details, for they have left no enduring impression, nor am I insensible of how vain any attempt must be to reproduce the harmony of such subtle aspects of nature.

Aspects of Nature—Views of Nature—*Ansichten der Natur:* these visions may have seemed irreproducible to King, but they would have been immediately recognizable to Humboldt.[14]

———

Unfortunately, Humboldtian science was undergoing something of a crisis in 1870s America. On the one hand, King's very career illustrated the extent to which Humboldtianism had flourished in the United States in the mid-nineteenth century: the countless exploring expeditions of the 1840s, '50s, and '60s, whatever their pragmatic goals, had created significant opportunities for young, broad-minded scientists to develop a deep appreciation for the way natural and social forces interrelated and interacted. J. N. Reynolds's vision of scientific exploration had in large part been realized. On the other hand, though, the expansion of science's role in American society brought with it a great increase in specialization. Fieldwork lost its appeal as laboratories proliferated and microscopes became more powerful. Eventually, Theodore Roosevelt would feel compelled to lament that "the average unfortunate student who has taken up scientific work in the colleges [has] been carefully trained not to do the field work which in the past has aided in producing men like Humboldt." Spectroscopists and entomologists, histologists and embryologists now garnered all the renown; naturalists in the Reynolds mold were considered rank amateurs. No one was looking at the Big Picture anymore. At the same time, as science became professionalized and institutionalized, its prestige grew to be unsurpassed in American culture. Whenever you needed an unassailable justification for a social policy, you invoked science, the way earlier generations had invoked religion. Humboldtian humility and uncertainty had been superseded by the confidence of positivism. Many scientists would have scoffed at King's subtlety and mystical luminescence; for them, empirical inquiry led not to modulated views but to the clear, deterministic laws that governed the universe. There was a new way of seeing in the Western world.[15]

American scientists had in effect taken credit for the wave of technological and industrial change that had overwhelmed the United States in the previous decade or so. This was the era of the phonograph and the electric lightbulb, the typewriter and the telephone. The displays at the 1876 Centennial Exposition in Philadelphia had included not only photographs by Timothy O'Sullivan but also sewing machines and power looms, as well as the delicate wire cables of Edison's multiplex telegraph. Old conceptions of space and time were fracturing and fading. In 1869, the year of Humboldt's centennial, Thomas Durant's Union Pacific was back in the news, joining with the Central Pacific in Utah to complete the transcontinental railroad. As a result, the natural resources of the West came under vastly increased pressure, and the federal government did whatever it could, in many cases, to encourage their rapid exploitation. A new 1872 mining law, for instance, essentially granted public lands to anyone who claimed to find valuable minerals on them. Moreover, communities and even entire regions could now begin to specialize in particular cash crops or manufactures for export to any other part of the nation. Science had proved that the narrow approach, the microscopic perspective, served to maximize one's returns.[16]

According to the scientists themselves, however, the true measure of their professional worth could never be merely their work's practical applicability. As soon as their reputation peaked, they began upholding "pure," abstract science as the key to *real* human progress. The first graduate schools in the United States, just now opening their doors, claimed that their science programs, unlike those of government bureaus or technical colleges, were beyond the reach of mundane, political concerns. They addressed quandaries internal to each field of study, problems that arose in the vacuum of the lab: only this type of uncorrupted research, they insisted, could help humanity truly understand the way the world works. Under the right circumstances, such an approach—science for science's sake—could actually serve Humboldtian ends, by relieving science of its long-standing role as servant to capitalist expansion. In reality, though, increased specialization usually made it easier for capitalists to exploit scientific developments, and scientists' insistence that they could be held accountable only by their highly trained peers pushed their various fields further and further toward elitism.[17] Back in the 1830s, J. N. Reynolds had acknowledged that a certain amount of specialization had become necessary, but he also knew that any kind of Humboldtian science depended on the specialists' close cooperation and their responsiveness to the changing needs of different social groups. The idea was that all peoples had to learn how best to adjust to their complex, tangled, multifaceted environment. Now, half a century later, committed Humboldtians like the government naturalist C. Hart Merriam, author of *Results of a Biological Survey of Mount Shasta*, worried that their narrowly trained colleagues were becoming "blind to the principal facts and harmonies of nature," because they were no longer "studying the forms, habits, and relationships of animals and plants." Ernst Haeckel coined the word "oecologie" in 1866, but in practice fewer and fewer scientists were thinking ecologically. Breadth and connectivity had been sacrificed for depth: each physical fact was now understood only in isolation. And, even worse, this kind of science, bereft of relationships, tended to isolate the scientists themselves, in far from inspiring settings. Leading intellectuals like the physicist Henry Rowland made "A Plea for Pure Science" and then withdrew from public life. As Merriam put it, scientists no longer experienced or conferred on others the basic feeling that motivated Humboldt and all his disciples: their simple "delight in contemplating the aspects of nature."[18]

Clarence King had been resisting this trend even before his watershed experiences atop Mount Whitney and Sheep Rock. "Ironize and ridicule specialism," he wrote in an 1868 memorandum to himself.[19] To an explorer in the field who was confronted on a daily basis with organic interrelatedness, the attempt to understand facts in isolation seemed ridiculous. Of course, King also realized that some of his own scientific interests were rather arcane: while working on his magnum opus, *Systematic Geology*, he liked to joke that "he was writing a book which just three people in the United States would care to read."[20] But the actual tome, once it finally appeared, had the scope of the gigantic, wide-open landscape that King had surveyed. As Henry Adams put it, reviewing his friend's book: "the most satisfactory part of Mr. King's work, next to its scientific thoroughness, is the breadth of view . . . , the vigor and grasp with which the author handles so large a subject without allowing himself to be crushed by details."[21] Indeed, King had always worried more about being crushed by science and society than by boulders or avalanches: in the catastrophic wreck of western landscapes, he saw an impending human catastrophe. In 1873, in his new Mount Whitney chapter, his reference to "the crushing juggernaut-car of modern life" seems to have been a direct reflection of his private doubts about his geological career. Just a few months earlier, in a letter to Jim Gardiner, King had declared himself "convinced that science goes on and progresses at the *expense* of those absorbed in her pursuit. That men's souls are burned as fuel for the enginery of scientific progress. And that in this busy materialistic age the greatest danger is that of total absorption in our profession. . . . We give ourselves to the *Juggernaut of the intellect.*"[22]

Both science's new professionalism and society's new scientism, then, tended to separate human minds from their accompanying bodies and souls, leaving men like King drained and exhausted. In an 1870 book review that made several explicit and admiring references to Humboldt, King decried "the cold and bloodless accuracy of science."[23] And he playfully elaborated on the point in *Mountaineering in the Sierra Nevada*, referring to a particularly cold night on Mount Shasta: "I abominate thermometers at such times. Not one of my set ever owned up the real state of things. Whenever I am nearly frozen and conscious of every indurated bone, that bland little instrument is sure to read twenty or thirty degrees above any unprejudiced estimate."[24] His barometer may have been right atop Sheep Rock, but his compass had been wrong when he'd drawn the first map of the Southern Sierra; ultimately, it made more sense to valorize human experience than to trust even the most precise, modern instruments. Besides, once you started pondering the larger problems of the earth's history and of humanity's changing relationship to the environment, the significance of particular measurements started to fade, and technical accuracy seemed almost irrelevant. "It is eminently the character of geological inquiry," King wrote to General Humphreys in 1873, "that the deeper one penetrates the subject, the more problems appear."[25] A sense of humility, then, was the scientist's first requirement; and after that, in the Humboldtian field guide, came questions of subjectivity and vision. On the verge of completing his scientific writings, in 1876, King arrived at the conclusion that "geology itself is chiefly a matter of the imagination—one man can actually *see* into the ground as far as another."[26] He put his faith in a science not of the mind but of the soul.

———

Professional science, meanwhile, was putting faith and souls into a brand-new context: lurking behind the general trend toward specialization, buttressing and justifying it, was the development of Darwinism. Evolution was absorbed into American intellectual culture with remarkable ease: *On the Origin of Species* had laid the groundwork in 1859, and 1871's *The Descent of Man* created a legion of scientific converts, many of whom retained their Christian beliefs without the slightest discomfort. (The drag-down battles between evolution and religion in the United States wouldn't occur until the 1920s.)[27] The ascent of Darwinism meant that white Christian Europeans and Americans could no longer believe in a divinely sanctioned separation of humanity from lowly animals. But, thanks to creative thinkers who quickly converted evolutionary science into social theory, they did manage to retain their faith in their own status as the highest living beings. After all, they were the ones who had become the most specialized in their knowledge and their societal roles, and a close reading of Darwin clearly revealed that the most successful animals were those perfectly designed to perform one particular function or adapt to one particular environment. Darwin himself agonized over whether or not his theory of evolutionary change, which resulted from "the struggle for existence," from competition between organisms for limited resources, represented progress.[28] Mostly, he tried to characterize natural selection as a process of blind, random chance: narrow-beaked finches are more likely to survive under certain environmental circumstances, but broad-beaked finches have the advantage under other circumstances. "Never use the word[s] higher and lower," Darwin told himself in 1847. By the end of his life, though, he had written publicly of his suspicion that an "endless number of lower races" had to be "beaten and supplanted" by "the higher civilized races."[29] Ultimately, it was this racist strand of Darwinism combined with the social trend toward specialization that forced the Humboldt current out of the American mainstream.

More than a century later, one might be tempted to see Darwinian strife and chaos as simply having beaten and supplanted Humboldtian harmony and cosmos. But major historical shifts, like natural selection itself, tend to play out in much more complicated ways: there are always tides or eddies to contend with. The closest readers of Darwin realized that large parts of his theories were in fact derived directly from Humboldt, for evolution was essentially ecological. "We may all be netted together," Darwin mused in the late 1830s, and web-of-life imagery appears in much of his work. So when Haeckel first started writing about "oecologie," he defined it as "the study of all those complex interrelations referred to by Darwin."[30] Indeed, when Darwin published *The Descent of Man*, the double entendre so blatant in his title should have suggested humanity's ultimate comeuppance, should have provided a climactic lesson in Humboldtian humility. And some observers, like William Henry Hudson, did actually identify this perspective as forming the very core of Darwin's work: "we are no longer isolated, standing like starry visitors on a mountain-top, surveying life from the outside; but are on a level with and part and parcel of it."[31] The themes of Darwinism that meshed best with other social doctrines, however, that more easily became social Darwinism, were the ones emphasizing competition, the ones putting Anglo-Saxons back on the mountaintop. It is of course also tempting to believe that Darwinism actually *spurred* these other trends, since Darwin himself probably coined the word specialization in the 1840s, and since he did in fact use the terms "higher" and "lower." But science has never had *that* kind of power in society. Certain aspects of Darwinian theory became the new gospel because they validated already-developing shifts toward corporate capitalism, racism, and imperialism: they survived because they fit.[32]

Powerful intellectual currents affirming the progressive march of white civilization had been stirred by the British writer Herbert Spencer back in the early 1850s, a few years before Darwin published his theory of evolution. It was Spencer who coined the phrase "the survival of the fittest" and who thus became the hero of most social Darwinists, those Panglossian defenders of social and economic hierarchies as being the natural results of our natural urge to compete on the free market. The doctrine of Spencerian progress, in other words, assumed that everything was already as it should be, or was at least on the verge of being so: the fact that certain people rose to the top simply confirmed that they deserved to be there. While this law of competition "may be sometimes hard for the individual," as the robber baron Andrew Carnegie put it, "it is best for the race." In short: leave the laissez-faire system alone.[33]

Spencer's evolutionary progressivism fit particularly well in the United States not only because Gilded Age culture so eagerly embraced materialistic acquisitiveness, but also because the so-called American school of anthropology had for several decades been preparing the ground for social theories that would prop up the existing race regime. The eminent craniologist Samuel George Morton, who had initially garnered some fame for his analysis of the fossils sent back to Peale's Museum in Philadelphia by the Lewis and Clark expedition, became one of America's best-known intellectuals in the 1840s by practicing a bizarre form of pseudoscience that allowed him to declare the United States' brutal treatment of blacks and Indians to be entirely justified, because these lower races simply had less brain capacity than whites did. In ill health and disdainful of fieldwork, Morton led a seemingly morbid life in his laboratory, soliciting and then measuring skulls from all over the world, often without considering factors like body size or sex. Both his specimens and his conclusions were utterly unreliable, but his many folios were well received by the scientific community, and his ranking of world races—white at the top, Native American and black at the bottom—was quickly accepted in the United States. Morton, directly contradicting Humboldt, explained this racial hierarchy by embracing polygenism, the theory that humanity did not in fact share a common origin but rather that each race was created separately and granted discrete levels of potential intelligence. In other words, the political and historical analysis so crucial to Humboldt's position, the belief that colonial oppression accounted for any seeming differences in racial capacities, was traded in for a commitment to positivism and a pair of calipers.

Morton's polygenism dominated American science in the 1850s and '60s, thanks in large part to the conversion of Humboldt's disciple Louis Agassiz, who had been a monogenist while still living in Europe but who came under Morton's influence almost as soon as he moved to the United States in 1848. Agassiz, a more successful popularizer than Morton himself, even developed an influential response to Darwin's assertion of the unity of the human species. The theory of evolution might well indicate the common descent of all humanity, Agassiz argued, but certain groups of people had clearly progressed further from the animal kingdom than others. Agassiz remained a Humboldtian glaciologist, but his new stance on anthropology suggested the deepening fault lines in Humboldtianism in general. By the time Clarence King was listening to Agassiz lecture at Harvard, in 1862, even most abolitionists were convinced of the innate inferiority of colored people. Scientific racism ruled the day.[34]

———

No single American ever approached Humboldt's multifaceted radicalism, which, in its sweeping political critique, combined a scientific justification of human equality with a profound understanding of natural resources and their use. But it was significantly easier to be *more* Humboldtian in the day of J. N. Reynolds and James Abert than in the catastrophic era of the Civil War and Reconstruction. Early American abolitionists and Indian advocates, following Humboldt, were strict monogenists, who still may have thought of nonwhite races as being degraded but who were generally open to acknowledging their equal capacities. Later abolitionists tended to be polygenists who hated slavery simply because it was a black mark against the higher white races.[35] Similarly, most "friends of the Indian" in the 1870s and '80s insisted on a policy of strict assimilation—that is, a repudiation of native cultures— not because they imagined that Indians could be full members of white American society, but because they had become converted to the belief that all native cultures were savage and that it was the white man's burden to help the red man develop.[36] The culture of scientific racism made it that much harder for someone like Clarence King to validate blacks and Indians. Moreover, coming of age during both the Civil War and the bloodiest Indian Wars, King faced a more charged political situation than had earlier explorers.

Unlike Reynolds, King actually spent a great deal of time on the far frontiers of American settlement, and once, while doing some extra survey work in Arizona Territory in 1866, he and Jim Gardiner were captured and almost burned at the stake by a group of young Hualapai warriors upset about white invasions of their traditional homeland. The Hualapais had actually gotten the fire going before the survey's military escort finally rode into view and saved the two scientists. Given such experiences, it is not entirely surprising that King generally had a hard time recognizing the injustices done to Indians by white Americans, that he fell back on the prejudices of his time to assert that the native of the Southwest had "but one passion in life—assassination; one bequest from father to son—the tiger love of human blood; one mental activity—treachery." King was also too much a man of his own culture to make any serious effort to understand native customs on their own terms. He did have some friends among various tribes during the time of his fieldwork, but when he witnessed a particular chief called Buck taking a new spouse just one day after his former mate had passed away, he was appalled to the point of absolute scorn. There had been times when he "felt as if any policy toward the Indians based upon the assumption of their being brutes or devils was nothing short of a blot on this Christian century." But now he joked about extermination and eventually decided simply to "avoid all discussion of the 'Indian question.'"[37]

In the political realm, King's colleague John Wesley Powell was much more committed to the project of using his "intimate knowledge of Indian character and life" to uphold the natives' "virtues" and "fidelity."[38] In private, though, King did tell a friend that "some of the hours dearest to him were those spent . . . idly dreaming about a Paiute village." He also expressed his approval for the resoundingly positive ways in which the committed Indian advocate Helen Hunt Jackson had portrayed natives in her books *Ramona* and *A Century of Dishonor*.[39]

In his official capacity as a surveyor and cartographer, meanwhile, King had a crucial opportunity to honor Native Americans' attachments to particular places by adopting their favored toponyms—and he did sometimes seek out Indians' contributions for his maps. A letter to General Humphreys of February 1873 reveals that King was planning "a trip to the Pah Ute and Shoshone Indian camps of Nevada in order to get from them accurately the names of such mountain ranges and groups as are now without recognized names." Of course, this was poor compensation to the Paiutes and Shoshones, given the amount of land they lost to white settlement and resource exploitation, but even today it serves to remind people of the West's multicultural history. Staring at a map, you might not immediately know why a particular place in Nevada is called Winnemucca (home to the Humboldt Museum and the Humboldt County Library), but, like King himself, you might be pleased at least to realize that some of America's most beautiful places— King was thinking in particular of the Shoshone and Yosemite waterfalls— were still "happily bearing Indian names." Some scholars would argue that scientists like King used native toponyms merely to ease the national conscience and erase the government's acts of violence. But King's actual intentions may be better reflected by the fact that Chief Winnemucca himself saw fit to ask King to serve as his official emissary to the local Indian agent.[40]

King's efforts to use Indian place-names may even begin to seem rather noble when one considers the alternative approach of simply ignoring local traditions and seizing control of the cultural landscape the same way whites had seized control of the physical landscape, incorporating it into a rubric of federal management and a rhetoric of white scientific imperialism. Indeed, King himself often fell back on this much less respectful naming method: when he christened Mount Whitney, for instance, he was quite deliberately casting aside the peak's well-known Indian name—Waijau, meaning "pine mountain." Clearly his meeting with the Paiute elder in 1873 at the mountain's base had not inspired him to reconsider.[41] He was not even above using topographical names as political plums: when Senator John Conness of California pushed the Yosemite preservation bill through Congress, King promptly named a peak for him (and Conness later helped King renew the funding for his 40th Parallel Survey).[42] When King added Mount Agassiz to his map, though, it was at least clear that he was honoring the glaciologist and not the raciologist. Henry Adams seems essentially correct in his remark that King was loyal to Lincoln's Republican Party primarily because of his "love of archaic races," his "sympathy with the negro and Indian and corresponding dislike of their enemies."43 Like Spencer, and to some extent Darwin— indeed, like most of his contemporaries—King did believe in the "archaism" of African Americans and Native Americans, in their impeded development, but he at least defended them as transcending most of the stereotypes held by white society.

King's views of African Americans mirrored his views of Indians in that they seemed based more on personal experience than on broad political principles. Also, they tended to get expressed more often in private letters and conversations than in publications, and they ran the gamut from profound respect and even intense attraction to the standard condescension of his day. According to his colleague James Hague, King "had many friends among the negro people, and often sought their companionship when opportunity offered"—an assertion apparently borne out by King's close relationship with one of his assistants on the 40th Parallel Survey, Jim Marryatt, and with his personal valet later in life, Alexander Lancaster. There is also evidence in King's papers showing that he went out of his way to support a black church in Georgia, sponsor an African American orphan in New York, and help an aging "quadroon" receive the money she had been promised by her dead father, a prominent Southern politician. Once, in the late 1880s, after being forced by his doctor to spend two months at some hot springs in the South, King explained to Hay that "I have been interviewing a lot of Negroes about the recent notices to leave, in various parts of Arkansas. The rebs usually give them about 12, sometimes 24, hours to leave the country. A delegation of these outraged blacks went to the governor and formally demanded protection but were laughed out of town. Arkansas is to me the most barbarous and terrible place I ever was in." And occasionally King's feelings about race did in fact bubble over into political generalizations: he vehemently supported Cuban independence, for instance, at a time when many Americans insisted that the colored islanders were incapable of self-rule.44

When King expressed his general opinions about the qualities of African Americans, however, he tended to come across as a typical promoter of demeaning stereotypes. Ultimately, blackness for him was a "grandly barbaric Congo woman . . . , very black and very silent . . . , gay and loose with a heredity of sweetly dissolute centuries." As Henry Adams put it, King took an interest in "the archaic female, with instincts and without intellect"; he seems to have fetishized women of color as animalistic sex objects. Yet it is significant that whenever King spoke of "primeval" black women, he was invariably differentiating them from the boring, inhibited, intellectual white men of his own social class—men like Adams himself, for instance, whom King once described as "a mere cerebral ganglion," a hopelessly rational highbrow who had progressed so far on the evolutionary scale that he had completely lost touch with his instincts and now refused to "avail himself of the glorious privilege of drifting with the ebb and flow of his emotions." King's own instinct, then, was to celebrate the way his mother's new black servants had transformed the "grim calvinistic scotch propriety" of her housekeeping: "Civilization so narrows the gamut! Respectability lets the human pendulum swing over such a pitiful little arc that it is worthwhile now and again to see human beings whose feelings have no inflexible bar of metal restraining their swing to the limits set by civilized experience and moral law."[45] Again, King's primitivistic attitude has something of the voyeur in it: he dehumanized the servants as purely physical and emotional beings and never really got to know them. But, late in his life, King in fact did get to know an African American woman named Ada Copeland quite intimately, falling deeply in love with her over the course of several years. We'll never discover the extent to which King actually respected Ada, appreciated her as more than a stimulus to his less inhibited half, the wilderness-seeking side of his soul. Perhaps, even at the time of his death in 1901, he retained his prejudice that, in evolutionary terms, whites were modern and blacks archaic. His intense relationship with Ada, however, forces us at least to consider the seriousness of the joke he often made to his friends, "that miscegenation was the hope of the white race."[46] Many of King's contemporaries, of course, considered racial mixing to be nothing short of an evolutionary catastrophe.

———

On June 26, 1877, Clarence King stood before the brand-new graduates of Yale's Sheffield Scientific School and delivered a lecture entitled "Catastrophism and the Evolution of Environment." It was intended simply as a distillation of the argument he would make in *Systematic Geology;* but, thus concentrated, it became, in King's own words, "nothing less than an ignited bomb-shell thrown into the camp of the biologists." From the moment King started talking, the professors in the audience knew just how radical the repercussions would turn out to be, how drastically currents would shift. The famous paleontologist O. C. Marsh, for instance, despite his friendship with the speaker, "got very warm before King finished and used his fan vigorously."[47] King was attempting to combine Humboldtian geology with Darwinian evolution, was tracking through history the most important ways in which humanity and the earth had been actively interlinked. Back in 1799, as Humboldt set out on his American expedition, he had expressed his intention "to find out how nature's forces act upon one another, and in what manner the geographic environment exerts its influence on animals and plants."[48] Now,

King was not only exploring the idea of a human ecology—that "series of complicated relationships with contemporaneous life, but, besides, with the general inorganic surrounding, involving climate and position on the globe"—but also mixing in the element of deep, evolutionary time. Out West, King had examined lava flows and grinding glaciers, had "read the record of upheaval and subsidence, of corrugation and crumpling of the great mountain chains," and now he wanted to understand the impact of rock and ice, of geological catastrophes, on the development of human cultures.[49]

More than fifty years later, the pioneering environmental ethicist Aldo Leopold would comment that "the two great cultural advances of the past century were the Darwinian theory and the development of geology."[50] Leopold wasn't actually referring to King's Sheffield lecture, but he could have been, for it was here that King injected into American culture the sense of scientific humility that today is the linchpin of America's environmental movement. Indeed, the greatest contributions of science have generally been those that deprived us of our place at the center of the universe, or on top of the chain of being. Copernicus and Galileo forever banished geocentrism, showing us that we actually lived upon a tiny grain of sand in an infinitely immense desert. More recently, Freud taught us not to be so proud of our supposed rationality. In the nineteenth century, Darwin, when interpreted through a Humboldtian lens, brought us into much closer contact with the animals, and geology established the fact that humanity was utterly insignificant in terms of the earth's history.[51]

King, in his lecture, did celebrate the development of science, did praise humanity in general and the Sheffield School in particular for inventing "means of clearing away the endless rubbish of false ideas from the human intellect, for the lifting of man out of the dominion of ignorance." But ultimately his lesson constituted a warning about the dangers of scientific arrogance. "Man's enthusiastic hand," King told the Sheffield students and faculty, "may clear away the shallow dust or rubbish from an Oriental city, and lay bare the stratified graves of perished communities: it is only a mountain torrent which can dig through thousands of feet of solid rock and let in the light of day on the time-stained features of a long-buried continent." Moreover, the light of science was just barely starting to shine on the secrets of torrents and solid rock and buried continents: "We pour out our passionate questionings, and hearken lest mute nature may this time answer. But nature yields only one syllable of reply at a time."[52]

What was most significant, and most avant-garde, about the way King deployed the lesson of Humboldtian humility was his emphasis on nature's catastrophes. The geological fashion, in 1877, was to frown upon catastrophism, a theory that many scientists associated with devout Christians who had not fully grasped Darwin and who clung to the story of Noah's flood as proof of God's power to punish humanity's sinfulness. Indeed, catastrophism had been around for a long time, and old-school believers had tried to argue that life on earth, through the intervention of supernatural forces, had been repeatedly destroyed in cataclysms and then repeatedly re-created. But modern, scientific catastrophists like King were merely asserting their faith that the rate of geologic change had varied over the ages. The fossil record does suggest moments of rapid, wide-ranging, overwhelming destruction: entire faunas have clearly been extinguished. Indeed, Humboldt himself had argued that "the whole globe appears to have undergone the same catastrophes. At a height superior to Mount Blanc, on the summit of the Andes, we find petrified sea-shells; fossil bones of elephants are spread over the equinoctial regions. . . . In the New World, as well as in the Old, generations of species long extinct have preceded those, which now people the earth, the waters, and the air." Adherents of the dominant uniformitarian school, however, insisted that while the evidence was obviously catastrophic in scale, the events indicated might nevertheless have taken a considerable amount of time to occur.

Uniformitarianism, a theory originally developed by Darwin's friend Charles Lyell, was all about gradual change and long time spans. A slowly evolving environment and slowly evolving species were mirror images of each other. In other words, uniformitarians embraced the humility of deep time, but not of the environment's omnipotence: random mutations within species mattered much more to them than climatic shifts or volcanic activity. They insisted that the present was the key to the past: all geological change had occurred at uniform rates, namely the ones still observable today. King looked at canyons and saw prehistoric torrents; his colleagues saw long-lived rivulets. "Give our present rivers time, plenty of time," the uniformitarians declared, "and they can perform the feats of the past." But King thought there were many things, both cosmic and chaotic, that lay beyond our field of vision. He attacked his colleagues for their "positive"—or positivistic—"refusal to look further than the present, or to conceive conditions which their senses have never reported." The uniformitarians were mere empiricists. They could never be true geologists, King insisted, pronouncing his final, dooming judgment, because "they lack the very mechanism of imagination." When you look at your fish, you must also see beyond it. But Lyell had "put blinders on the profession."[53]

One reason the uniformitarians remained blinded was that they simply refused to trust Louis Agassiz. In developing his theory of the Ice Age—a climatic catastrophe—Agassiz had compiled compelling evidence of drastic, sudden changes in flora and fauna (some animals became extinct simply because they weren't able to migrate southward quickly enough), but his stubborn anti-Darwinism wound up casting doubt on even his best-grounded glaciological ideas. To Lyell and Darwin, catastrophes seemed to break evolution's gradual linear progression. So, as King explained in his lecture, "biology, as a whole, denies catastrophism in order to save evolution." But King sought "a dovetailing together of the two ideas": he defended "suddenly-destructive change" without endorsing the kind of "all-destructive change" that would snap the evolutionary chain. In short, he argued for an understanding of human-environment interactions that accepted Darwin's model of speciation but reinstated a Humboldtian model of respect for nature's determinative power. And, just as the vast jungles and towering volcanoes of the equatorial region inspired Humboldt with more awe than Europe ever had, so too did the "grander features" of the American West help determine King's ultimate interpretation of environmental change. "If the uniformitarians can derive any comfort from Eastern America,—and I suppose they justly may,—they are welcome to it," King wrote. "The rate of subsidence in the east . . . may be called uniformitarian." But King had spent his career exploring the deserts and canyons of the frontier, where humanity seemed smaller and less significant and where the rate of geological change was "distinctly catastrophic in the widest dynamic sense": this was the lesson he had learned while "looking back over a trail of thirty thousand miles of geological travel." One reason King's lecture seemed so radical to the uniformitarian "fashion followers," as King himself called them, was that it was simply one of the first assessments of the new geological evidence provided by the scientific surveying of western territories—whose very landforms often seemed beyond belief.[54]

King wanted his audience to engage in "geological induction," in "physical conjecture"; he wanted them to see different sections of the earth's crust breaking apart to form huge abysses or coming together to fold up mountain ranges. Thanks to Timothy O'Sullivan's photographs and his own word paintings, he could actually reveal to them the new "out-door facts of American geology," the contours of places like Yosemite Valley—could give them "a grand view down the cliff" and point out the "unmistakable ice-striae, showing that the glacier of Mount Hoffmann had actually poured over the brink." But what he truly desired was for them to develop the skill of "restoring in imagination pictures of the past": "how immeasurably grander must it have been when the great, living, moving glacier, with slow invisible motion, crowded its huge body over the brink, and launched blue ice-blocks down through the foam of the cataract into that gulf of wild rocks and eddying mist!" King felt that they should approach nature not as narrow fact finders but with "the open-mindedness of a . . . man searching after immemorial knowledge." They had to be scientific poets, for "only to poets is it given to plunge their souls into the sensitizing solution of the imagination, to hold them up to the invisible actinic light of other days, and to develop a true picture of a forgotten age."⁵⁵ (Historians, then, might learn a thing or two from poetic geologists.)

Through his own process of Humboldtian submission, exposure, and immersion, King had come to a particular vision of geology, and now he wished to share not only that vision but the experience behind it. He himself had been trained at Sheffield to envisage the gradual deposition and blending of detritus and sediment, the slow erosion of gorges by persistent trickles. But what could explain the gigantic canyons he had seen out West, which were not being eroded at all anymore? "There are left hundreds of dry river-beds, within whose broad valleys, flanked by old steep banks . . . , there is not water enough to quench the thirst of even a uniformitarian." Those "once-powerful streams," now "dead from drought," proved that water erosion depended on the "cycle of climate," which itself had drastic peaks and valleys. Take the impressive canyon of the Green River, along the 40th Parallel. Using standard methods of paleontological analysis, King determined that it had been formed in the Quarternary Age—during much of which, according to Agassiz, this part of America was under ice. So King had his listeners imagine the river gorge being carved out not by the Green River as it flowed in 1877, but first by a massive retreating glacier, and then by glacial floods in "an age of water catastrophe whose destructive power we only now begin distantly to suspect." There had to be more to that canyon than met the eye; it was always safest, and most reasonable, to assume that nature was far more powerful than we could ever imagine based on observable facts. "It is only lesser men," King insisted at Yale, "who bang all the doors, shut out all doubts, and flaunt their little sign, 'Omniscience on draught here.'"

Again, the key lesson turned out to be one of humility—and caution. The subtext of King's lecture on environmental catastrophes was about the care with which we ought to treat the environment, given its power over us: "You may divide the race into imaginative people who believe in all sorts of impending crises,—physical, social, political,—and others who anchor their very souls *in statu quo*. There are men who build arks straight through their natural lives, ready for the first sprinkle, and there are others who do not watch Old Probabilities or even own an umbrella. This fundamental difference expresses itself in geology by means of the two historic sects of catastrophists and uniformitarians. Catastrophism, I doubt not, was the only school among the Pliocene Californians after their families and the familiar fauna and flora of their environment had been swept out of existence by basalts and floods."

King, clearly, was an imaginative soul who believed in impending crises, and he knew all too well the danger associated with rapid rates of change. In 1877, he worried about society's thoughtless embrace of scientific progress, and, specifically, about the mounting tension surrounding the railroads and their angry workers, who were about to rise up in a series of strikes that would spark a bloody eruption of social violence. The breakneck advance of technology, King warned, did not guarantee "evolution"; the supposedly unifying transcontinental railroad had in fact torn the Union apart again, and it had also helped to deepen the "rude scars of mining" on the land. To align yourself with science and its supposed control over nature was to forget nature's ultimate power. The ground can shift, and tracks can crack; mines, as Timothy O'Sullivan's avant-garde underground photographs suggested, have a tendency to collapse. Groping for an illustration of his catastrophic theory, King told his listeners in New Haven that they could understand the "geologic struggles" he had examined in the American West by imagining a scene wherein a "train at full speed dashes against a bridge pier and is utterly wrecked."[56]

King felt he had been overtaken by the "avalanche of materialistic and scientific activity" that now defined his society. He was approaching another watershed moment in his life. Later on, he would be firm in his condemnation of the Gilded Age: "Ours is a vulgar, but remarkably active civilization, given over for the most part to the energetic pursuit of personal prosperity and the struggle for material good. Of all ages and all lands this is the one where for the mind's and the soul's sake a brilliant struggle must be made to stem the almost irresistible current of sodden materialism."[57] But King had self-consciously committed himself to science in the 1870s, and that commitment, as he well recognized, went hand in hand with an embrace of progress and the future-oriented mechanism of capitalistic investment. While he pored over his specimens in his laboratory in New York City (just around the corner from where J. N. Reynolds had lived in the 1830s), trying to focus his imagination on the geologic past, he found himself worrying about his financial future. His hundreds of wafer-thin cross-sections, specially prepared for microstructural analysis, simply couldn't hold all of his attention. What would he do next, now that there was no more funding available for fieldwork and the reports of the 40th Parallel Survey were almost completed? His expertise in mining and western lands seemed to direct him toward investments in natural resources and their exploitation, but that option seemed, well, crass. As it turned out, he would struggle for the rest of his life with the problem of how best to use his scientific training. He now had to balance not only science and art but also two different kinds of science: the pure, abstract, prestigious kind, and the pragmatic, applied, much-more-remunerative kind. The tide was turning against many-sided men, as it had for J. N. Reynolds before him, and the most talented people in America were expected to earn huge amounts of money.[58] King was resisting the current of sodden materialism, but things were starting to get a little damp.

In 1893, he would publish a highly specialized, and highly muddled, article about "The Age of the Earth,"[59] but 1878's *Systematic Geology* was his last significant contribution to the professional scientific community. And even while he was writing *Systematic Geology*, King had begun investing in cattle and some of the best grassland in Wyoming—he had seen it himself—and inviting the other scientists of the 40th Parallel Survey to join him. They made a killing. Of course, they immediately felt guilty about buying into what Reynolds would have called their society's tendency toward engrossment—though not so guilty as not to be able to rationalize their actions. "Clare and I have had to face this matter in a practical form," Jim Gardiner wrote to his mother in 1879, "passing as we have done from the humble pecuniary inheritance of science to cattle kingdoms in the West: and at our last meeting in New York we pledged to one another the promise that however rich we may become we will by simple living, and use of our money for others, do our part to stem the tide of selfish extravagance."[60]

Wading in that gentle tide, King allowed business associates to handle his ranching affairs on the frontier while he remained in the East to popularize his geological theories and talk to politicians about consolidating the federal government's surveying efforts. *Systematic Geology*, if not exactly the explosive publication King hoped it would be, did make a splash. "Right or wrong in his theories," Henry Adams wrote, in his review of the book, "Mr. King could hardly fail to stimulate his students."[61] At the very least, the 40th Parallel Survey, following in Humboldt's footsteps, had shed new light on the continuity of New World geology. King made this point explicit on page 5 of his book when he insisted that his countrymen not refer to the peaks of the 40th Parallel as the Rockies—he found "so loose and meaningless a name" to be "simply abominable"—but rather that they adopt "that general name, Cordilleras, which Humboldt applied to the whole series of chains that border the Pacific front of the two Americas."[62] The term "cordilleras," after all, was hardly loose and meaningless, for it refers to cords, lines of linkage—chains of connection.

As it turned out, beyond stimulating his students—by penning a catastrophic story of submersion, collision, and eruption that the historian William

Goetzmann has characterized as "only a trifle less dramatic than Genesis"[63]— King also happened to be essentially right in his theories. Stephen Jay Gould, cocreator of the theory of "punctuated equilibrium," has argued that "modern geology is really an even mixture of two scientific schools: Lyell's original, rigid uniformitarianism and the scientific catastrophism of Cuvier and Agassiz"—and, we might add, of Humboldt and King. Usually, geologic processes operate with great deliberateness, but there are always surprises: we live, as Gould put it, in "a world composed of quasi-stable systems that re-sist stress to a breaking point and then flip rapidly to a new equilibrium." Or, in the words of a current geology textbook, "The history of any one part of the earth, like the life of a soldier, consists of long periods of boredom and short periods of terror." When King examined western evidence of sedimentation during the Carboniferous period and found that there was a drastic split be-tween land-detritus and limestone layers, rather than a gradual blending to-gether, he was essentially right to say that "the sudden change of sediment simply means a sudden physical change."[64]

Systematic Geology did not cause a sudden about-face in American science—Lyell would remain dominant for another century—but the book was at least impressive enough to help King, after a lobbying campaign of Reynoldsian proportions, to secure the job of director of the brand-new U.S. Geological Survey. The struggle to create the survey, meanwhile, had been thoroughly uniformitarian in nature. King's success in launching the Geo-logical Exploration of the 40th Parallel back in 1867 had spurred the creation of three other frontier expeditions: the Geographical and Geological Survey of the Rocky Mountain Region, under John Wesley Powell; the Geological and Geographical Survey of the Territories, under Ferdinand Vandiveer Hayden; and the Geographical Surveys West of the Hundredth Meridian, under George Montague Wheeler. A new era of scientific exploration had erupted onto the scene.[65] As their names suggest, though, these four operations over-lapped considerably in purpose, method, and even physical location, which resulted in a chaos of inefficiency and bickering. A few concerned scientists and politicians attempted to consolidate and reform the surveys starting in the early 1870s, but their efforts were effectively blocked by western congress-men. A better-coordinated scientific assessment of frontier territories, after all, might deprive these politicians of their influence over the General Land Office, which was notorious for handing out huge "homesteads" to well-connected businessmen seeking monopolies in all the industries dependent on good pasture, mineral beds, and water sources. So it was not until 1877, when Rutherford B. Hayes replaced Ulysses S. Grant as president and brought with him a more liberal group of advisers, that the trickle of reform-minded

activity finally started to have some erosive effect on the corruption that dominated the government's administration of western lands.[66]

A lobbying effort spearheaded by King and Powell, and propelled by
several of their friends, including the new secretary of the interior, Carl
Schurz, as well as Congressman Abram S. Hewitt, resulted in a fierce legisla-
tive debate in 1878 and '79. One highlight was a harangue by Hewitt that referred to the country's western lands as its "heritage of the future" and
portrayed that portion of the public domain as being rapidly "mortgaged to
grasping corporations or to overpowering capitalists." As King noted, "the
contest here is a tremendous one, having waked up no end of bitterness and
consolidated a powerful lobby," which openly referred to itself as representing
the West's "landed interests" and which referred to the scientists as "revolu-
tionists" and "charlatans." Of course, even most of the reformers had made
investments in resource-rich western lands; their goal had never been to cut
off the public domain from use, but rather to develop it sustainably and equi-
tably, in the public interest, rather than for the benefit of a few powerful land
barons. From our perspective, 125 years later, the ultimate passage of a law
that consolidated the different western explorations into one U.S. Geological
Survey (and also created a new Public Lands Commission) may seem like a
minor accomplishment, especially since the law did not contain any explicit
environmental restrictions. But, in April 1879, when King was appointed as
the first director of the USGS, and he announced that the survey would immediately "begin a rigid scientific classification of the lands of the national domain . . . , for the general information of the people of the country," he
marked a shift in awareness of land use in the United States even more significant, perhaps, than the one signaled by the preservation of Yosemite Valley as a public park in 1864.[67]

Suddenly, the opinion of civilian scientists about the appropriate use of
federal lands was deemed more important than the opinion of army men or
business leaders. Development would no longer just follow the gold and the
rifles. As the scientist and politician George Perkins Marsh explained, in language showing a clear affinity with both Humboldt and King: "The . . . composition of the great masses of terrestrial surface, and the relative extent and
distribution of land and water, are determined by geological influences . . .
remote from our jurisdiction. It would hence seem that the physical adaptation of different portions of the earth to the use and enjoyment of man is
a matter so strictly belonging to mightier than human powers, that we can
only accept geographical nature as we find her, and be content with such soils
and skies as she spontaneously offers."[68] It was time to start adapting to the
environment.

In some ways, John Wesley Powell had been even more instrumental than King in lobbying for the creation of the USGS, and the book he published in 1878, a *Report on the Lands of the Arid Region of the United States*, was, from a political standpoint, a more radically Humboldtian text than *Systematic Geology*. (That same year, Powell founded an intellectual society in Washington called the Cosmos Club.) And it was Powell who truly crushed the myth that "rain follows the plow," who insisted most directly that, especially in a region where water was scarce, "the regulations and the national government" must be "adapted to the physical conditions of the country" rather than the other way around. Powell even politicized his ecological analyses by insisting that divisions of land based on federal surveys should be "so made as to give the greatest number of water fronts. For example, a brook carrying water sufficient for the irrigation of 200 acres of land might be made to serve for the irrigation of 20 acres to each of ten farms. . . . But if the water was owned by one man, nine would be excluded from its benefits."[69]

This emphasis on adapting regulations to the pattern of the land, though, was ultimately just a practical application of King's theory of "the survival of the plastic," his assertion that species must constantly adjust to a variable environment—"a widely different principle from the survival of the fittest," which he rightly saw as little more than an excuse for more powerful social groups to wage an ongoing war against less powerful groups. Moreover, Powell became increasingly optimistic in his old age about humanity's ability to make drastic changes in local ecosystems, especially through the power of irrigation: in the 1880s, he started talking about "redeeming" and "reclaiming" the arid lands of the West rather than simply adapting to them. But King retained his deep Humboldtian conviction about the terrifying power of the inorganic environment, about the determinative force of climate, of fire and ice and rock. Even in 1892, he was still writing about the necessity of submitting to the "supreme dominion of natural laws."[70] His creed would always be tinged with catastrophe.

On October 29, 1893, a mild Sunday glowing with bright northeastern foliage, Clarence King was arrested in Central Park. The eyewitness accounts differ as to exactly what happened, but everyone agreed that King appeared gaunt, ragged, and disoriented. Somewhere in the vicinity of the lion cages there erupted a chaos of bumping and shouting, and eventually King was booked by two detectives for having "acted in a disorderly manner . . . in the presence of a large crowd." He got off with a fine of just ten dollars, and King's lawyer insisted in the newspaper that his client's arrest had been an act of harassment. King's doctor, however, admitted that the fifty-one-year-old geologist was suffering from "nervous depression." Indeed, King had been grappling with some form of this ailment since 1881, when he resigned from the directorship of the USGS after serving only two years. And in the weeks just before his arrest he had been feeling an intense pain in his spine and noticing gaps in his memory. Some nights, he had found himself wandering in dark, dirty neighborhoods of the city, "without any notion," as a friend explained, "of how he came there." Now he decided to leave his fate up to his physician, and two days after his arrest, on Halloween, he was committed to the Bloomingdale Asylum. On November 3, the *New York Sun* ran a front-page headline asking: "IS CLARENCE KING INSANE?"[71]

People were curious, because in the previous three decades King had earned a reputation as one of the most reasonable men in America. Could the faculties of a cutting-edge scientist, in an era defined by science, really have deteriorated this far? Friends within the intelligentsia may have appreciated the extent to which King's accomplishments had always depended on "that essential quality of an investigator—scientific imagination," but in the newspapers King was known almost exclusively as "a cool-headed man of scientific education."[72] After all, he had rationalized the federal government's haphazard exploration of its territories and contributed a systematic vision not only of geological history but also of the administration of the USGS, whose methods and scope even today bear the mark of his careful planning.[73]

The crest of King's wave of scientific fame, especially in popular circles, had come back in 1872, when newspapers all across the country celebrated him for having exposed the Great Diamond Hoax. That October, King and his men had just arrived in San Francisco, to write up the results of another long season of fieldwork, when they heard that a new mining company had recently been capitalized at ten million dollars, based on a huge diamond find at an undisclosed location in the West. The value of the gem-studded fields had been verified by Henry Janin, a highly respected mining engineer whom King knew to be both skeptical and careful. At the time, though, speculation was rampant in California society, and mineral swindles were a dime a dozen. Moreover, King was about to issue a report stating that, if the belt covered by the 40th Parallel Survey was any indication, the West as a whole contained precious few precious gems. So he sought out Henry Janin—who immediately urged King to invest. When pressed, however, Janin admitted that he had been allowed only one hour for his investigation of the diamond fields and that he had been brought there blindfolded and so had no idea of the exact location. All he could say was that "the discoveries had been made upon a mesa near pine timber." King, of course, knew a hundred such tablelands, but in a feat of memory and comparative geography worthy of Humboldt himself, King used this bare description, in combination with a few other salient facts—the amount of time it had taken Janin, by train and on horseback, to reach the mesa from San Francisco; the "unfordable" conditions of certain rivers that summer; the geological characteristics that the surrounding area must have displayed in order to convince an expert like Janin that diamonds could actually be produced there—to determine the exact spot in question. He conferred with his colleagues, and they left for Wyoming Territory the next day.74

It was a harrowing journey, for snow had already started falling in the mountains—and there was a lot of money at stake. But King found notices posted by the diamond company exactly where he expected to. As he later explained in an official deposition, he and his crew "lay down upon our faces, and got out our magnifying glasses and went to work, systematically examining the positions of the stones and their relation to the natural gravels." Their result, after three days of research, pointed to "an unparalleled fraud." They had found combinations of diamonds, rubies, emeralds, and sapphires that seemed to constitute an "impossible occurrence in nature." No human being could know exactly what nature might be capable of, and King admitted that the swindlers had found a location "where every geological parallelism added a fresh probability of honesty." But the evidence of tampering—the never-varying ratio of a dozen rubies for each diamond, for instance—left the scientists with little doubt as to what was going on. Indeed, after King got back to San Francisco and made his findings public, the original "discoverers" of the diamond fields quickly disappeared, and the company collapsed. In the newspapers, the central story—a story repeated in many of King's obituaries almost thirty years later—focused on how pure science had triumphed over capitalist greed.75

Ironically, King himself had abandoned his scientific post at the USGS in the hope of striking it rich. Between 1881 and 1893, he opened his own bank in Texas and staked at least six claims in Mexico, Cuba, and the United States, prospecting for silver, gold, quartz, iron, coal, manganese, and phosphates.76 He spent almost all of his time during this period either running his various mines or trying to convince wealthy friends and acquaintances to invest in them. Henry Adams complained that his friend "was passing the best years of his life underground."77 On occasion King published magazine and journal articles and even did a bit of lab work, but, by the age of forty, he had essentially abandoned both art and science in favor of business. Yet he never enjoyed what he was doing, and he never made his fortune. By 1889, King felt sure that he had failed on every front: "Now in middle age," he wrote to Adams, "I am poor, and what is worse, so absorbed in the hand-to-mouth struggle for income that I see the effective literary and scientific years drifting by empty and blank, when I am painfully conscious of the power to do something had I the chance."[78]

It's hard to know what King really wanted. Even while he clung to the elusive goal of making a fortune, he continued to rail against the capitalistic values of his society. His writings of the 1880s and early '90s overflowed with scorn for America's "plutocrats, with their yachts and drags and squanderings," and for the "back-breaking weight of our dull zeitgeist," which was based on nothing more than a "debauched materialism."[79] He seems to have thrown himself into the debauchery partly because he had a taste for luxury and dreamed of being a patron of the arts, and partly because he felt a financial obligation toward his mother's family. When his stepfather died suddenly during his first stint in California, as King put it, "I found myself at 24 years of age with 11 people dependent on me alone."[80] Science had never paid particularly well: all too often, on the 40th Parallel, he had found himself dipping into his private funds just to make sure his men had adequate supplies. And whenever King did make a little money—on his ranching operations and a couple of his mines, for instance—he was generous with it. As his field colleague S. F. Emmons commented, King's "tender affection and solicitude for [his mother's] welfare was one of the most marked traits of his character, and through all the many vicissitudes of his checkered life his first thought and duty was to provide for her comfort and happiness." His friends, meanwhile, often commented on his "many acts of loving kindness," and Emmons started off a memorial essay for King by asserting that "the recipients of his beneficence, whether of mind or purse, alone can realize with what grace and freedom he gave."[81]

Perhaps the most remarkable aspect of these difficult years was that King remained utterly beloved of his friends—even those who had sunk thousands of dollars into his ill-fated mining endeavors. Most scholars, trying to evaluate King's career, have concluded that this latter part of his life simply reveals the shallowness of the current that had been driving him all along. Even the normally judicious Wallace Stegner asserted that "Clarence King failed for lack of character, persistence, devotion, wholeness."[82] But if one evaluates people by the connections they created and maintained with others, then King was phenomenally successful. John Hay admitted to the novelist Henry James that he felt more settled around King the wanderer than around any of his other friends: "where *he* is, there is my true country, my real home." And James agreed that King seemed to be possessed of "magic spells," that despite being "a queer, incomplete, unsatisfactory creature," not to mention "slippery and elusive," he was nevertheless "the most delightful man in the world."[83] During his time in the field, King inspired the love of his coworkers by displaying the determined optimism without which Humboldt and his disciples never could have survived, let alone developed an appreciation for the cosmos. As Emmons explained, "there were many times, especially in the first two seasons' campaigning in the deserts of Nevada, when, through weakness resulting from malarial fever contracted in the Sacramento Valley bottoms, the impossibility of obtaining potable water, a shortness of food supplies, or delays from storms or other causes, discouragement took possession of different members of the party. But King's abundant courage and energy never failed and his fertility in expedients was equal to every emergency; so that he gradually impressed upon every member of his corps such confidence in his ability as a leader that their personal devotion to him and their faith in the complete success of the undertaking knew no bounds."[84]

King spurred an extravagant devotion from friends like Gardiner, Adams, and Hay, and, remarkably, even from several other acquaintances who could not possibly have been as close. After he died, eleven friends contributed appreciative essays to a volume called *Clarence King Memoirs*, one of the most remarkable books I have ever read. The four-hundred-page tome contains a certain amount of fluff, of course, but the depth of understanding is palpable: these people clearly spent hours, days, trying to come to grips with the meaning of King's life. Hay noted that King "possessed to an extraordinary degree the power of attracting and attaching to himself friends of every sort," while William Dean Howells confirmed that King had a deep "feeling for those who do the hard work of the world, that others may enjoy their ease," that he "felt his unity with all men." Yet his constant—and admirably broad—quest for connection makes one wonder whether he himself ever managed to experience what he sought.[85] "Save for my family and you and yours," King wrote to Hay in 1884, "I am hopelessly 'out of it' and 'out of it' forever."[86]

King seemed to bubble over with vitality, but in fact his later years, like Humboldt's, were defined by repression. He decided that in order to survive the Gilded Age with his faculties intact, he would have to "smother and hide my pessimistic hate of civilization and be as straightlaced and wooden and fatuously American as anybody."[87] The illnesses he picked up in the field, meanwhile, from rheumatism to hernias to random fevers to a spinal polyp, consistently came back to plague him—as was the case for virtually every nineteenth-century explorer who didn't actually die on the job. In 1881, just a couple of months after King left the USGS, the malaria he had probably caught in Nevada's Humboldt Sink began to flare up again. "My physicians shake their heads," he wrote to Emmons, "and tell me that it will take several

31 "Survey Company: Group, 'The immortal few who were not born to die,' " 1867? Timothy H. O'Sullivan. Clarence King is at center just in front of the flagpole, with Dick Cotter atop the pole.

months yet to tell whether this outbreak of the malady is going to be curable or fatal."[88] That same year, the physician George M. Beard published a treatise called *American Nervousness*, describing the neurasthenic ailment prostrating highly educated men and women throughout the United States. King, after toiling for a couple of years as a Washington bureaucrat, had developed a classic case. As Beard explained, in terms King surely would have understood, "modern nervousness is the cry of the system struggling with its environment."[89] King's experiences on the frontier, as exhilarating as they might have been, had left him ill-suited for life in society. He was a catastrophe waiting to happen.

———

King told his close friends about many of his torments and dilemmas, but there was one secret he never revealed to them: in September 1888, he married Ada Copeland, the African American nursemaid with whom he had fallen in love over the course of that year. He also kept his own true identity secret from Ada until just before his death, having introduced himself to her as

James Todd, a humble railroad porter. Here was King's final paradox. "Love" is not too strong a word for what he seems to have felt toward Ada: "Ah, dearest, I have lain in my bed and thought of you and felt my whole heart full of love for you. It seems to me often that no one ever loved a woman as I do you."[90] And yet he deceived her, and all his friends and relatives, for more than a decade, because he could not bear the idea of anyone finding out that he had so blatantly flouted the conventions of his society. Ultimately, this was neither a radical rebellion nor a simple act of exploitation: it was a characteristic attempt to seek connection in a way that was guaranteed to fail. Just like the forbidden attraction King felt to certain men, his relationship with Ada ignited his passions and his desire for unity, but in the end it merely reconfirmed his alienation, his disgust, his utter aloneness.

Ada was twenty-six when they met, almost twenty years younger than King: it is hard to get around the idea that the middle-aged white man was using the inexperienced black woman merely for his pleasure. Yet one must remember just how scandalized his friends and family would have been if the relationship had come to light. Most of his acquaintances probably would have disowned him. King often complained of "family cares" and "those family situations which with frightful certainty avalanche themselves down upon me," yet he simply refused to upset his mother.[91] His attachment to an African American, he figured, would surely kill the fragile old woman with whom he shared a "close intellectual companionship that never weakened during his lifetime" (she outlived him, as it turned out).[92] We still may fault King for his lack of courage. But there was also a practical consideration: by 1888, he was tens of thousands of dollars in debt to John Hay, and he actually took his financial obligation toward Ada—and the five children she had with him—quite seriously, so he wanted to be sure that he would inherit his mother's house when she did finally pass away. He quite literally could not afford a scandal. Moreover, he did provide Ada with a place to live in New York City, and with a cook, a laundress, a nurse, and even a gardener. He showed their children how to play the piano, and he brought them presents when he came back from trips. "My first duty now in these hard times," he wrote to Ada, "is to make money enough for your expenses, and on that I use all my strength." Their marriage may have been clandestine and unlicensed, but "ever since I put that ring on your finger I have worked and prayed for you and will do so till God parts us by death. God bless you my own, my only one."[93]

All we can know for sure about their relationship is that King did rearrange his life for Ada's benefit in the 1890s. To John Hay, King's unpredictable trips and his bizarrely opaque requests for money suggested that his friend "must have joined some oath-bound order, which pledged him, under fearful sanctions, never to tell anybody anything."[94] What Hay didn't realize, of course, was that when King's stress overwhelmed him in 1893, his financial obligations had approximately doubled. It would have been a stretch for almost anyone to support two families in 1893, a year that saw the most profound market collapse the United States had ever experienced and "the most drastic deflation in the memory of man." Henry Adams later remembered that "men died like flies under the strain." As for King, the pain in his spine was keeping him up all night, and, shortly before his own collapse, his El Paso bank folded, the victim of both financial panic and managerial graft. "I have lost everything," King wrote.[95]

Paradoxes held King together, and they also tore him apart. Already an artist and a scientist, he also became both a rabid anticapitalist and an aspiring businessman, an exploder of aristocratic conventions and a conventional aristocrat, a friend of the Indian and an Indian hater, a lover of a black woman and a racist, a lover of women and a lover of men, a stubborn optimist and a tortured depressive, a socialite and an isolate. At least he could never be accused of overspecialization. Unfortunately, though, he soon began to realize that even his closest friends were finding it impossible to understand his contradictions. In 1897, he told John Hay that he was spending most of his days "in the silent spaces of the desert," playing "the primitive like Thoreau," purposefully remaining "as lonely and isolated as an anchorite. Months go by with no one to talk to. . . . And perhaps I oughtn't to admit it, but I have grown to love the uncomplicatedness of it all. You have always thought my alleged savagery of soul a mere attitudinizing but you were wrong."[96] King was mired in some sort of borderland, paralyzed by the swirling sands of a perpetual frontier.

In July 1893, just three months before King's breakdown, Frederick Jackson Turner delivered a lecture that would resonate across the silent spaces of the continent for decades to come, on "The Significance of the Frontier in American History." Turner, casting himself as a scientific-minded student of historical processes, first announced that Americans clearly owed all of their best national characteristics—from their embrace of participatory democracy to their rugged individualism to their spirit of independence—to the existence of the American frontier. Then he explained that his careful analysis of certain statistics given in the 1890 census indicated that the frontier had disappeared. The American pioneer had always struggled with the wilderness, carved out a niche for civilization, defined himself through his relationship with free, open land—but now the data showed that, on average, there were at least two people settled on every square mile of the West, a density level no longer compatible with true self-realization.

Turner's countrymen would have to look to new frontiers beyond their own borders.[97]

Much of the infamous "Turner Thesis" now sounds moderately ridiculous. Americans never encountered open land, after all—first they had to dispossess the Indians. If the frontier did perhaps encourage a certain ruggedness, then it perhaps also fostered violence, the overexploitation of natural resources, a general sense of irresponsibility, and an antidemocratic anarchism. Moreover, whatever the actual spread of settlement might have been in the 1890s, the federal government still owned about half the land in the West: a considerable "safety valve" existed for the release of population pressure. Yet Turner had tapped into a kind of cultural panic in the United States, based in part on the failure of the economy but also on a perception that settlement and industrialization were legitimate threats to American manliness and to the country's Manifest Destiny to achieve international dominance. Turner's speech was heard only by members of the American Historical Association, but people like Theodore Roosevelt were simultaneously writing popular tracts about the need for effete American men to retain some contact with the hard, wild edges of the world. Roosevelt himself had cured his own American nervousness by pursuing what he called "the strenuous life" on a rough Dakota ranch during parts of the 1880s and '90s. And the 1890s in fact ushered in a period of egregious American imperialism: when Roosevelt first spoke of "the strenuous life," he was actually urging America to bolster its military presence overseas. In other words, Roosevelt had moved deftly from his celebration of Indian fighters and buffalo hunters to his recruitment of virile "Rough Riders" in the effort to win the Spanish-American War, claim new territories for the United States, and redeem the white race, which was threatened by new waves of dark-skinned immigrants. His ideas about the "winning of the west," bolstered by scientific racism and social Darwinism and by Turner's positivistic lament over the passing of the American frontier, had translated into a new, hubristic foreign policy.[98]

Clarence King did not dispute the value of frontier experiences, but he drew very different lessons from them. To him, they meant uncertainty, broad-mindedness, careful negotiation, imaginative interpretation, and, most of all, humility. They left him befuddled and broken—but also inspired.

King's ultimate attitude toward frontiers, and the distinction between his and Roosevelt's primitivism, become clearest in the context of the United States' foreign relations in the 1890s. The breakdown King suffered in 1893 seems not to have had a lasting impact, for by early the next year he had joined Henry Adams in the Caribbean—and was eagerly interviewing Cuban revolutionaries. And within another couple of years, his trip through the Cuban highlands had inspired him to embrace the Cuba Libre movement and pen passionate condemnations of both Spanish and American attitudes toward this tiny island nation where "a people living in Paradise, with every gift of nature to ensure human content and cherish social joy, have been stung and tormented into flinging their lives into a vortex of war."[99] King's two articles for *Forum*, published in 1895 and 1896, read like a direct tribute to Humboldt's "Political Essay on the Island of Cuba." (King may even have remembered Humboldt's expressions of outrage in the 1850s, when the U.S. government considered annexing Cuba and a translation of Humboldt's "Essay" was published without the sections condemning slavery.) Adapting his geological imagination to the study of history—"most great wrongs have their tap-roots deep in the past"—King painted a gory picture of the first catastrophic meeting between the Spanish conquistadores and the island's Siboney Indians, who, "according to all early writers . . . , were amiable, innocent, hospitable, and graceful." And from that point on, King argued, the brutal exploitation of the natives had gone hand in hand with the brutal exploitation of the isle's natural resources, resulting in the horrible irony, noted also by Humboldt, of a country "rich beyond description" whose population, with the exception of a few plantation owners, had been reduced to absolute poverty.[100]

"Coarse greed," King wrote, "underlay the enslaving of both Indians and Africans, and the oppression born of that greed, and practiced on peoples whom it was safe to maltreat, became . . . ingrained in the class that governed Cuba." Now, Spain was refusing to relinquish its grasp on the island, and the United States saw Cuba as little more than a wilderness overrun with "nigger bandits," where American investments in the sugar industry were under constant threat. But King exhorted his compatriots to celebrate the blows struck by the rebels not only against the Spanish troops but also against American investors, and, symbolically, against colonial exploitation and white supremacy: "Fires lighted up the Caribbean shore,—they were everywhere,—till two whole provinces were burning together and weaving their smoke into one great black pall that hung over Cuba for weeks, veiling the stars and quenching the sun to a murky ball of rayless red. Sixty million dollars of cane became drifting ashes and blackened sky!"[101]

King's cynicism and despair resulted in a general silence over the last five years of his life. His family and friends rarely laid eyes on him, and his correspondence gradually petered out. Again and again, he would fall desperately ill with malaria or pneumonia or, finally, tuberculosis; remove himself to a warmer climate; wallow in doubt and self-pity; and then rush off excitedly to the next mining job in some other remote location, certain that this time he would strike gold. King's overriding sensation, as on the top of Mount Shasta, was isolation—but no longer could he call it "uplifted."[102] When he did write to his friends, it was generally to inform them of his suffering. John Hay, after receiving a particularly bleak missive, immediately expressed his response to Henry Adams: "There you have it in the face! The best and brightest man of his generation, who with talents immeasurably beyond any of his contemporaries, with industry that has often sickened me to witness it, with everything in his favor but blind luck, hounded by disaster from his cradle, with none of the joy of life to which he was entitled, dying at last, with nameless suffering, alone and uncared-for in a California tavern. *Ça vous amuse, la vie?*"[103]

Part of the suffering King revealed to Hay and Adams stemmed directly from the intense feelings he harbored for these two friends—feelings that had to remain as secret as his love for Ada. Whether or not sexual love bound the three men together, their hearts were enmeshed in a web of emotional dependence. "Why can you not drop in here at the comfortable Bristol and make me glad for a day or two?" King had asked Hay while they were both in England in 1882. "If you knew how lonesome I am and how much good it will do me, you would think twice before you gave me the go by. . . . Come here and get a *Blouse and Knickerbocker*; we will retire to a lonely moor together and put them on and dress the part for a half hour together, where none shall make us afraid." Echoing the language he used with Jim Gardiner in his younger days, King frequently told Hay how difficult it was to be away from him, how he yearned for a time when they could "turn our souls inside out together." And he almost always associated his connection to Hay with the freeing experience of getting lost in nature, whether on English moors or in the pastures of New Hampshire, where the Five of Hearts wanted to buy property together: "I shall think of you incessantly, as I wander over the fields, and drink deep of the pleasures of hope of our living there side by side many a summer day."[104]

As things turned out, King died in circumstances parallel to those of J. N. Reynolds's passing. The lonely, aging explorer, suffering from ailments acquired in the field, fled to a healthier climate on medical advice, seeking rest and clean air, while his nation suffered the pains of political and economic expansion. King wheezed for the last time on Christmas Eve in 1901, in Phoenix, Arizona—generally alone, though cared for by a local doctor. Once it became clear that he was slipping away, he wrote to Ada and confessed his true identity, telling her to get in touch with Jim Gardiner, who would handle his affairs and make sure she and the children remained comfortable.[105] Perhaps the confession gave him some relief. Regardless, King seems to have found a bit of amusement in life during his final weeks. In one last letter to Hay, referring to his tubercular affliction, King mused: "That the microbes do not respect an old and tired constitution shows them to be no better than the Supreme Court."[106]

Besides his sense of humor, King had also retained his imagination. As a friend noted, "his imagination was his dominant, at moments his dominating, quality."[107] Even in his last days, he dreamed of a Humboldtian expedition through a liberated Cuba, during which he might survey all the island's bounteous natural resources. His environmental imagination did not tend instinctively toward wilderness preservation; the Yosemite Valleys of the world were inspiring, but incidental. Rather, like both Humboldt and Reynolds, he usually dreamed about how humanity might better adapt to nature, might develop natural resources more carefully and more democratically. The key was to avoid the first-come, first-served, winner-take-all mentality that had defined the California gold rush of 1849, and the Klondike rush of fifty years later, whose glittering fields and streams King had investigated in the summer of 1900. In Cuba, he wished to see resource exploitation resulting in the "development of vast popular wealth," without any kind of social exploitation, nor any huge capitalist projects that strove to bring wide swaths of nature under human control.[108] King's views of human ambition were firmly grounded, for they had been shaped by glaciers and volcanoes. Throughout his career, King exposed himself on the frontier, and what had become imprinted most vividly on his imagination was nature's unwillingness to be controlled. He understood the environment's cosmic interrelatedness, its ability to instill a sense of connection and peace; he also understood its sheer, catastrophic power.

Yreka: Just North of Mount Shasta

Clarence King had been looking northward ever since he first encountered Mount Shasta, which, with its "ridge and gorge, glacier and snow-field, all cold and still under the icy brightness of the moon, produced a scene of Arctic terribleness such as I had never imagined."[1] He longed to see a country defined by the kind of ice and snow that, in moderate climes, one could find only at the top of the highest peaks. His discovery of active glaciers on the north side of Shasta in 1870 was a step in the right direction. But it paled in comparison to his Klondike trip of 1900: as a friend explained, King "returned from Alaska simply bubbling over with pleasure and stories of his trip, which he declared to have been the most interesting he had ever taken."[2] It was torture for him to retreat to the deserts of Southern California and Arizona.

Back in 1863, during his first season in the field, he had set off northward immediately after climbing Shasta, and then, after a few miles, turned back to stare fondly at his favorite volcano—without, of course, realizing that it was capped by living glaciers: "From this hill we had a view out between the pines over the Yreka plain, saw the beautiful autumnal desert colors, the volcanic cones like little mole hills clustered along the base of grand old Shasta. Adieu, grand mountain, with thy refined severed lines, thy lava flows and gorges, the deep curved furrows, the spotty snow, the dark forest struggling up into arctic climates." King wanted more of those arctic climes, even after he found the glaciers, but, until 1900, he didn't actually get much farther north than the small town of Yreka, where in 1863 he stayed at a rough-and-tumble inn filled with drunks and gamblers.[3]

I imagine him wondering: who named this dirty little outpost?

If you're one of the eccentrics who study American geography, you know—as King knew—that our place-names are relentlessly multicultural. Northern California provides countless examples. Scanning the map near Humboldt County, with all its Prussian pride, you'll find French Gulch, Spanish Ranch, and the Russian River. You can also see Scots influence: note McKinleyville, McCloud, Montague, Dunsmuir. Not too far from Yreka (pronounced Why-REEK-a), you'll also find *Eu*reka—just below Humboldt Bay, in fact. "Eureka" comes from the Greek verb "heuriskein," "to find"; Archimedes called out "Heureka!" (the first-person singular, past tense) when he discovered how to measure the purity of gold.

Check the gold-bearing regions, and you'll find numerous traces of Alta California, the old Spanish colony: the Las Plumas River feathers its way through the snowy Sierra Nevada right above El Dorado County. But you'll also see Emigrant Gap, which Americans started pouring through in the 1840s.

The place-names adopted by white Anglo American settlers erased innumerable native geographies, just as the U.S. Army wiped the continent clean of many actual tribes. Yet Indians persist, and so do their toponyms. You can hike through Hoopa Valley, swim in the Klamath River, visit Siskiyou and Modoc counties. Some people even climb Mount Shasta, the one feature of this landscape distinguishable from up to a hundred miles away. "Go where you will" in northern California, wrote John Muir in 1877, and "there stands the colossal cone of Shasta, clad in perpetual snow, the one grand landmark that never sets."4

For all of the volcano's visibility, its denomination remains obscure. Peter Skene Ogden, the infamous fur trader of the Hudson's Bay Company, who sketched the Humboldt River twenty years before John Charles Frémont, noted in his diary on February 14, 1827, that he had "named the mountain 'Sastise.' "5 A little more than a decade later, when the Wilkes Expedition passed through northern California, its philologist, Horatio Hale, attempted to do ethnographic work among the local natives, who he said were called the Saste or Shasty people, or simply the Shasties. Hale's rather vague scribblings suggest that this was probably a name employed by the Indians themselves, but it could also have been a derogatory name for them used by a neighboring tribe. It may also have been the name of a tribal chief. Hale, who had a hard time finding any willing subjects, claimed that these Indians led "a wandering, savage life, and subsist on game and fruits. They are dreaded by the traders, who expect to be attacked in passing through their country."6

Frémont saw Shasta's sunbathed southern side in 1846: "The snowy *Peak of Shastl* bore directly north, showing out high above the other mountains." Frémont's 1848 map listed the mountain as "Tshastl."7 There is a Russian word, "tshastal," which means "white" or "pure." And check your French, too: "chaste," as we know from the English derivative, has a similar connotation of blinding purity. All bets are off.

Go a few miles north, and you hit Yreka. Then look back, on a clear day, and, if you know what to look for, you can still see the blue-white glint of glacial ice on the volcano's north summit.

The eminently reasonable Hubert Howe Bancroft, in the sixth volume of his *History of California*, claimed that Yreka was "a corruption of Wyeka, whiteness, the Indian term for the adjacent snow-crowned Shasta."[8] But *which* Indian term? There were several tribes here, several languages and dialects. Did one group call the mountain Shasta, another Wyeka? And if the term suggested whiteness, could the Indians have also used it to refer to white-skinned Europeans? While Hale jotted down the name "Shasty," were the local natives chuckling at the ignorance of this strange paleface? In any case, Bancroft was probably right to call it a corruption. History is all about corruption.

Maybe an Indian chief, or his tribe, named this place hundreds of years ago. But we'll never know; the violence has been done. Besides, this is California, where everything is constantly remade. Yreka could just as easily have been named for the first time in the 1820s, by an explorer like Ogden. Maybe someone discovered gold here and named the town in a rush of excitement— "Yreka!"—but didn't much care for orthography. Or maybe the namer was actually an orthographic *expert,* a lover of palindromes and fresh bread who dreamed of opening the Yreka Bakery. ■

PART FOUR

NORTH

George Wallace Melville and John Muir
in Extremis

CHAPTER 8

In the Lena Delta

Arctic Tragedy and American Imperialism

Where is the proper Herbarium—the true cabinet of shells—and Museum of skeletons—but in the meadow where the flower bloomed—by the sea side where the tide cast up the fish—and on the hills and in the valleys where the beast laid down its life—and the skeleton of the traveller reposes on the grass.

—Henry David Thoreau, *Journals* (1842–44)

I know the true history of the Expedition will never be written.

—George W. Melville, letter to Emma De Long (August 10, 1883)

Clarence King's "Eureka!" moment on Shasta's north slope was all the more satisfying—certainly it was more exciting than his unearthing of the cephalopoda in Hell's Hollow back in 1864—because even Humboldt had not expected live glaciers to exist there. Humboldt was never an expert on ice, though: he had always been more interested in fire. When he wrote about the Shasta region, it was simply to point out that "the Shasty, or Tshashtl Mountains, contain basaltic lavas, obsidian, of which the natives make arrowheads." And when he cast his gaze farther north, he focused not on bergs, moraines, and crevasses, but on peaks that hid flames within their snowcapped crests—peaks like Mount St. Helens, "still burning, and always smoking from the summit crater, a volcano of very beautiful, regular, conical form, and covered with perpetual snow. There was a great eruption on the 23d of November, 1842; which, according to Frémont, covered everything to a great distance round with ashes and pumice."[1] To Humboldt, the cold, upper latitudes were most attractive for their paradoxical harboring of some of the earth's most fiery landforms. Approaching the end of his life, he kept coming back to volcanoes as potential keys to unlocking the internal secrets of the globe. Climate, of course, was crucial; but how did the outer air and the inner earth interact? The Ice Age was a powerful concept, but perhaps volcanoes trumped glaciers, since a forceful enough eruption could potentially block out the sun long enough to *cause* an ice age. One ultimately had to look at molten rock in order to understand terrestrial catastrophe.

In the final volume of *Cosmos*, however, Humboldt noted that there were no known volcanoes above the Arctic Circle. The South Pole, then, was in some ways a more magnetic draw for scientific explorers: after all, in 1841, just after the Wilkes Expedition confirmed the existence of Antarctica, Sir James Ross had seen actual flames flickering over the cone of Mount Erebus, at 77° south.[2] Yet Humboldt relished the mystery of the North, and celebrated its surprises: "Even when snow and ice have covered the ground," he wrote, of the boreal zone, "the inner life of vegetation, like Promethean fire, is never extinguished on our planet."[3] Besides, there was plenty of volcanic activity to investigate on the way to the Arctic. The planet's inner fire burned especially bright, for instance, in the Bering Strait: a nice, warm ocean current swept ships northward, and "the range of the Aleutian Isles, stretching over 960 geographical miles, seems to contain above thirty-four volcanoes, the greater part of them active in modern historical times. Thus we see here . . . a strip of the whole floor of the ocean between two great continents in a constant state of formative and destructive activity. How many islands in the course of centuries . . . may there not be near becoming visible above the surface of the ocean, and how many more which, after having long appeared, have sunk either wholly or partially unobserved!"[4]

The scientific justification for the *Jeannette* expedition, therefore, seemed perfectly reasonable to its more serious and idealistic members—as did the captain's decision to drive the ship directly into the polar ice pack. As George Wallace Melville later explained, "the *Kuro-Shiwo* (the black current of Japan) runs . . . northward to Behring Strait, where it . . . passes into the Arctic Ocean, streaming up into the northeast, and was lately regarded as one of the thermometric gateways to the Pole." Yet "no polar expedition had ever set out by way of Behring Strait." The explorers figured the current might carry them pretty far through the pack, but, regardless, they already would have gotten to investigate the Alaskan coast and islands, and it seemed "fair to presume that, if caught in the ice north of Herald Island, the ship would drift on the coast of Wrangel Land, or to the northeast toward Prince Patrick Land"—that is, toward their polar goal.[5] Most of the *Jeannette*'s crewmen had only a passing interest in the wonders of vulcanism, but a number of them shared a fascination with the seemingly blank landscape of the North, and they yearned to understand how human beings might relate to such an extreme environment. So, in the spring of 1879, twenty years after Humboldt's death, they

32 Officers of the *Jeannette* (souvenir, 1879).

made their final preparations in San Francisco, installing new boilers and a steam winch, taking on huge loads of food and coal, and carefully stowing a manual generator that powered fifteen brand-new arc lamps, the whole apparatus having been custom-designed by Thomas Edison himself to brighten the long Arctic winter.[6]

Thousands of locals gathered on the bay to watch the *Jeannette* sail out the Golden Gate in the fading late-afternoon sun, as the fog spread and the soldiers at the Presidio offered an eleven-gun salute. To some onlookers, the whole undertaking may have seemed like little more than a publicity stunt engineered by the infamous owner of the *New York Herald*, James Gordon Bennett, the expedition's financial sponsor—and they would not have been completely wrong. A year earlier, John L. O'Sullivan, coiner of the phrase "Manifest Destiny," had told Bennett that he was "delighted to hear of your intention of 'going for' the North Pole. And I feel sure your man will either hang out our flag on that famous piece of timber or else get down into Symmes's Hole."[7] Various publications and institutions, however, hailed the voyage as a "geographical adventure" and a great scientific opportunity: "It may determine laws of meteorology, hydrography, astronomy and gravitation, reveal ocean currents, develop new fisheries, discover lands and peoples hitherto unknown, and by extending the world's knowledge of such fundamental principles of earth-life as magnetism and electricity, and the various collateral branches of atmospheric science, solve great problems important to humanity."[8] The spirit of J. N. Reynolds seemed to have risen again. But the intervening years of Civil War and Depression and even tremors in geological theory had taught that scientific "progress" tended to be accompanied by catastrophes.

———

Five years later, in the fall of 1884, Eskimos along the southwestern coast of Greenland discovered a cap, a pair of oilskin trousers, and a wooden box frozen in the ice. Once in the hands of the local Danish governor, who took considerable interest in Arctic exploration, these scraps were quickly identified as having belonged to the crew of the *Jeannette*. A document from the wooden box signed by the late George Washington De Long, the ship's commanding officer, as well as two name tags sewn onto the clothing, were dead giveaways. What at first puzzled Governor Carl Lytzen was the fact that the *Jeannette* was known to have sunk in the vicinity of the New Siberian Islands—more than four thousand miles away. Within a few weeks, though, his puzzlement fading, Lytzen published an article proposing a new theory of Arctic drift.[9]

This chance confirmation of the Arctic ice pack's clockwise rotation around the North Pole became the primary scientific contribution of the *Jeannette* Expedition.[10] Ice had surrounded the ship after just two months, slowly squeezing the pitch and oakum from its seams. The *Jeannette*'s frame actually managed to withstand the pressure of the ice pack for almost two years, but the only discoveries made by the crew were due entirely to the northwestward drift of the ice past a few formerly unknown islands. Commander De Long stubbornly named the isles and claimed them as the property of the United States. Later, the navy would honor him by referring to the newly charted group as the De Long Islands. But even this geographic legacy just seemed to reinscribe the expedition's failure. Here, the maps declared, lie the worthless clumps of rock past which Commander De Long helplessly drifted.[11]

The geographic breakthroughs associated with the drift of the *Jeannette*'s relics from Siberia to Greenland, though slightly more useful, did little to ameliorate this sad appraisal of De Long's efforts. In an increasingly imperial age, science was supposed to help humanity assert its dominion over the natural world. De Long's contribution did the opposite, reinforcing the insignificance of human striving in the face of nature's overwhelming power. If

33 Detail of a "Map of the North Polar Region (Arctic Ocean), from the latest information, 1885." The sinking of the *Jeannette* is marked as "De Long July 1881"; the actual date was June 12, 1881. Note the Bering Strait (spelled "Behring" on the map) at the bottom left and the Lena Delta at the top right.

the *Jeannette* represented the ship of state, America's only Manifest Destiny seemed to be inexorable drift.

Twenty men had died, apparently for naught. The public greeted the thirteen survivors, saved by the Tungus and Yakut people of Siberia, with sympathy, but the navy almost immediately sequestered them for an official investigation, and subsequently the House Committee on Naval Affairs demanded that they testify under oath as to what had caused this terrible tragedy.[12] Incompetence, dishonesty, pettiness, and naïveté seem to have plagued the expedition from the start. It turned out, for instance, that De Long's navigation officer and second in command, John Danenhower, had lied about his health in order to secure his assignment. But the syphilitic infection he had concealed wound up erupting after only three months in the ice, rendering Danenhower unable to perform his duties for the duration of the ordeal. Meanwhile, the man hired as the expedition's meteorologist and photographer had brought the wrong darkroom chemicals on board. So, although the survivors managed to salvage De Long's journal and logbook, they came back with no photographic record of the northern latitudes, and precious few scientific papers. Indeed, the reputation of the *Jeannette*'s officers was perhaps worst among men of science. Had the *Jeannette*'s men truly believed that the earth's internal warmth might melt them a passage through the Arctic Ocean? On June 3, 1880, De Long wrote in his journal: "I pronounce a thermometric gateway to the North Pole a delusion and a snare."[13] It is no surprise that, in the end, officials of the navy and the House of Representatives did all they could to send the whole affair, with all its dark undercurrents, drifting back out to sea.[14]

To this day, Americans know little about the Arctic journey of George Washington De Long and his thirty-two crewmates.[15] Of course, America has almost always had a problem acknowledging mistakes. Yet the tragedy of the *Jeannette* was front-page news for several months in the United States, and, at the time, the mission's failure sparked much public debate about the significance of exploration in American culture.[16] For a while, Americans took seriously the limits of human potential, the arguments against expansion. A few of the returned explorers expressed a newfound respect for nature and the indigenous peoples who knew how to adapt to the most extreme ecosystems. Lieutenant Danenhower, who had escaped public disgrace thanks to the navy's cover-up, went on speaking tours in the mid-1880s insisting that exploring expeditions were pure hubris: "I unhesitatingly record myself as opposed to further exploration of the central polar basin. . . . It is time to call a halt."[17] Yet it took the experience of exploring to bring him to such a conclusion. Survivors often came back changed, chastened. In the case of the *Jeannette* expedition, as America drifted out of Reconstruction toward the patriotic fervor of the Spanish-American War, the humbled survivors served as draganchors, reminders of their country's vulnerabilities.

Unfortunately, most of those survivors quickly faded into obscurity as the nineteenth century came to a close. Danenhower, unable to overcome his private torment, shot himself through the temple in 1887. Raymond Newcomb, the ship's naturalist, forced to abandon all but three of the countless specimens he had collected, wound up in an asylum—for significantly longer than Clarence King.[18] Emma De Long published her late husband's "ship and ice journals," but her main goal was to emphasize the commander's accomplishments, so she edited out most of his doubts and frustrations. Only one man from the *Jeannette*'s crew went on to a career of some prominence and remained strong enough to contribute to ongoing debates over exploration and expansion: George Wallace Melville.[19] But the catastrophic power of the Arctic

ice would be forever engraved on his flexible mind, his determined face, his
frostbitten hands.

———

Melville's graven features survive today thanks largely to the painter Thomas
Eakins. The country's foremost portraitist, Eakins sought out prominent,
quirky subjects, individuals skilled with both their head and hands, "the best
of his times"—scientists, surgeons, musicians, champion athletes.[20] If Eakins
himself did little exploring outside his native Philadelphia, he nevertheless
craved the company of explorers, men and women whose committed intelligence took them to literal and figurative frontiers, whose cutting-edge discoveries defined the modern age. Professor Henry Rowland, for instance, author
of "A Plea for Pure Science," attracted the artist's attention by founding the
field of spectroscopy: the first person to make a photographic map of the solar
spectrum, Rowland had actually seen and recorded something that few people could even envision. Eakins painted him in 1897.[21]

As much as ambitious discovery, though, as much as knowledge of the
ever-expanding world, Eakins valued introspection and *self*-knowledge. The
modern age, he felt, took its toll; constant exploration entailed constant sacrifice. His heroes were not conquerors but survivors, keenly aware of the
passage of time, of their mortality, of their isolation. They were the leaders
of a newly emergent professional society, but they understood that their
expertise depended on narrow specialization, that their very proficiency in
their field just widened the chasms that separate all human beings from
one another. Frontiers can be lonely places. The vast majority of Eakins's
subjects, despite their many achievements, seem scarred, weathered by
their experiences, burdened by thought, by their very existence. Estranged
in a mass society, they cannot meet the viewer's eye, but stare intently into a
void.[22]

The void into which George Melville stares, in Eakins's 1905 portrait, is
probably the blinding, reflective white of the relentless Arctic ice pack—
which Melville had experienced not only on the *Jeannette* but also aboard the
Tigress in 1873 and as a member of the Greely Relief Expedition in 1884.
Melville was the embodiment of Eakins's ideal. Indeed, it would have been
shocking had Eakins not painted his fellow Philadelphian and near-exact contemporary. The rear admiral sat for Eakins twice, at the painter's request, and
both canvases display Eakins's Romantic opinion of Melville's heroism. The
first time around, in 1904, Eakins also inscribed his admiration on the back
of his painting, using bold Roman capitals to list the admiral's most important accomplishments. But Eakins preferred the second (1905) version, displaying it much more frequently and prominently, and it is this direct, frontal

34   *Rear-Admiral George W. Melville*, 1905. Thomas Eakins.

image that truly seems to capture both Melville's inner strength and his turmoil.²³

Melville's head and hands anchor the portrait: one's eye drifts between them, following the lines of brass buttons on his uniform. Just as in Eakins's portrait of Professor Rowland, these two areas of exposed flesh are by far the brightest parts of the canvas, standing out against the black background and the navy blue of Melville's coat, and highlighting both his mental ingenuity and his manual dexterity. Of course, Eakins manifestly appreciated Melville's character as a military leader: his stark pate, his firm lips, his sturdy posture, and his clenched hands all point to the rear admiral's imposing physicality and the force of his resolve.²⁴ But Eakins's celebration of Melville seems to go far beyond any classically heroic characteristics. The intense light on his brow, the extreme elongation of the head, and the determined focus of the eyes make Melville seem less a man of action than a man of ideas, a truly modern egghead. After all, Melville was not only an accomplished writer— his 1885 narrative of the *Jeannette* expedition, *In the Lena Delta*, was deemed by one reviewer "more vivid, effective, and exciting" than any of its several competitors[25]—but also a scientist, committed to the abstract, antiutilitarian goal of reaching the North Pole simply in order to expand humanity's knowledge of the earth. Moreover, Eakins's emphasis on Melville's rugged, active hands identifies this military man as a brother to the artist: here was no blunt, brutal commander, but a trained professional, a technical expert, renowned for his craftsmanship and creativity in design, which radically transformed the world-class United States Navy just in time for the Spanish-American War. Aboard the *Jeannette*, after watching him repair pumps, boilers, and engines and invent several useful contraptions on the spot, Commander De Long noted that Melville could probably "make an engine out of a few barrel hoops."[26] Melville's journals are filled with diagrams and trigonometric calculations, with musings about pistons, conducers, and fly wheels, strokes and revolutions.[27] Eakins's portrait of the engineer, then, is in many ways an early-twentieth-century version of John Singleton Copley's *Paul Revere*, an image of a master artist and staunch ideologue, whose revolutionary intellectual labor directly served his country.

Yet all is not well with Rear Admiral George W. Melville. Especially in contrast to Eakins's 1904 portrait, this one seems stark, somber. The scale has been reduced—forty inches by twenty-seven inches hardly make for heroic proportions—and Eakins has removed Melville's elaborate epaulets and his two American medals, leaving only the Russian Order of St. Stanislaus, presented to him by the czar.[28] Indeed, in 1905 Eakins gave Melville a mysterious, darker side—a black current.[29]

The admiral seems at odds with himself, literally divided by a vertical shadow line. To the right of his long nose, his face is cast into a solemn gloom; farther down, the shadow causes the star and stripes on his left sleeve to fade away almost entirely. And, following the shadow line down his torso, one's eye notes an unexpected, subtle curve in the right-hand column of brass buttons, which undercuts the upright rigidity of the admiral's posture. In addition, the bright light shining from the upper left of the portrait casts into sharp relief the dark circles around Melville's eyes, suggesting some kind of anxiety in his character, a secret torment. A hint of martyrdom resides in the cross he wears on his breast, a chilly reminder of his sufferings in Siberia. This literal cross, in turn, suggests a larger formal one, made by the two vertical lines of the brass buttons and the sides of Melville's head in conjunction with the horizontal line of his shoulders, highlighted by gold bands. At the center of this cross is Melville's wispy beard, tinged with a disconcerting ice blue, evoking both the admiral's frightful experiences in the Arctic and perhaps the menace of Bluebeard as well. The same cirrus-cloud wispiness characterizes Melville's long shock of hair, which could be seen as a leonine mane,[30] but which could also be interpreted as strangely incongruous, almost feminizing: wouldn't it have been more manly, more in step with naval discipline, to have closely cropped locks? Wouldn't Melville have wanted the bold knob of his brow to define the shape of his head? Military styles were different back then (elaborate mustachios were all the rage), but not radically different: a leader like Theodore Roosevelt would never have allowed himself to be seen with long hair. From this perspective, Melville's hands, too, take on a darker meaning. Despite their obvious strength and dependability, they appear battered, permanently chapped by the cold, and Eakins has foregrounded a deep scar just above the left pinkie finger. Moreover, there is obvious tension in the straining of his fingers, the clasping of his hands below his belt, as if in defense of his genitals. Romantically heroic in some respects, in others Eakins's portrait seems to undermine Melville's manhood.

The Arctic had of course humbled many a man before Melville. Though Shasta's temperate-zone glaciers were a revelation in 1870, the iceworks of the Far North had been attracting the attention of American explorers (and whalers) since the late 1840s, when Britain's infamous expedition led by Sir John Franklin, searching for a Northwest Passage, had disappeared somewhere in the vicinity of Melville Bay (no relation), on Greenland's west coast. Most famously, Humboldt's friend Dr. Elisha Kent Kane had led an expedition to look for Franklin in 1853, though he succeeded only in losing three men of his own, and several of his party's survivors came back crippled from frostbite. Kane's narrative of his failure turned out to be phenomenally successful, selling sixty-five thousand copies in its first year; sadly, the author died before that year was up, of exhaustion and rheumatic fever, at the age of thirty-seven. The young doctor's legacy was ensured by his vivid writing, though, depicting his careful ethnographic study of some previously unknown Greenland natives and his discovery of a particularly interesting sixty-mile stretch of coastline, "a plastic, moving, semi-solid mass, obliterating life, swallowing rocks and islands, and ploughing its way with irresistible march through the crust of an investing sea"—a mass Kane dubbed the Great Glacier of Humboldt.[31]

The United States had essentially abandoned South Seas exploration since the return of Wilkes in 1842, and the nation's focus was certainly concentrated on the West at midcentury, but every few years saw another expedition tackle the Arctic regions, for a raft of different reasons, with both private and public funding.[32] Sometimes the goal was parallel to that of most contemporaneous western missions: to reconnoiter the land's potential wealth and feel out the attitudes of local natives, in advance of possible investments and settlement schemes. This characteristic frontier approach was deployed in Alaska, for instance, even before William H. Seward bought the territory for the United States in 1867. One of the early classics of Alaska exploration was not a narrative of tragic confrontation with sublime glaciers but the bluntly titled *Alaska and Its Resources*, by William Healey Dall, who had been part of a surveying team sent out by Western Union at the end of the Civil War to investigate the possibility of laying a telegraph line all the way to Siberia. (Dall would also, a decade later, oversee a study for the Coast and Geodetic Survey which ultimately proved that the Black Current was not even as warm as the Gulf Stream—but which was not completed, alas, until just after the *Jeannette* set sail from San Francisco.) The great Russian-American Telegraph venture never quite caught on, but the research conducted by Dall and his men led directly to the founding of the Alaska Commercial Company, whose virtual monopoly on the fish and fur trade would ravage the territory's natural resource base and disrupt countless native communities within a couple of decades.[33]

In light of the unrelenting exploitation of the West, though, the pace of Alaskan development seemed, one might say, glacial, and Seward's purchase was roundly criticized by many of his governmental colleagues: they failed to see why the nation needed an extra "icebox."[34] So most promoters of northern exploration—both businessmen and congressmen—actually sought to differentiate their endeavors from commerce-oriented surveys, playing up idealistic, humanitarian goals like scientific progress, rescue, and simple heroism in the face of nature's most challenging environments. By the 1840s, even those searching for the Northwest Passage realized that any such feasible trade route would be passable only in certain seasons, and perhaps only in particularly warm years. Many Arctic expeditions of the 1850s, '60s, and '70s, then, while certainly tinged with hubris and the ambitiousness of American expansion, attracted eccentric, idealistic explorers in the mold of J. N. Reynolds, explorers who, in turn, came to be understood in American culture not simply as conquering heroes but as symbols of humanity's difficult, contradictory relationship to the cosmos.

Dr. Kane and his two most famous counterparts, Dr. Isaac Israel Hayes (the surgeon aboard Kane's expedition) and Charles Francis Hall, all believed in the open polar sea. Indeed, Captain John Cleves Symmes and his theories enjoyed considerable cultural currency until the very end of the nineteenth century. Joking about the prospects for polar exploration in the spring of 1887, the *New York Times* suggested that an "adventurous gentleman may confidently be expected to sit astride of an iceberg fishing for gamy walrus in Symme's Hole."[35] Another Humboldtian scientist, the oceanographer Matthew Fontaine Maury, known as the world's foremost expert on winds and currents in the 1850s, continually insisted on the possibility of unfrozen chan-nels at the highest latitudes; it was his endless comparative charts that pro-vided the scientific impetus for Kane, Hayes, and Hall to make their generally disastrous voyages, between 1850 and 1871.[36] Most of their expeditions were partially framed as further attempts to find survivors or relics from the lost Franklin expedition—until 1878, there were dozens of these searches, most inspired by the untiring efforts of Lady Jane Franklin, Sir John's widow[37]—but each man also hoped to sail across the open sea to the North Pole, while map-ping new coastlines, learning the ways of native peoples, conducting research in meteorology and magnetism, acquiring evidence for Louis Agassiz's theory of the Ice Age, and marveling at glaciers.[38]

Dr. Hayes, having returned from Kane's expedition with three fewer toes and a reputation as a deserter, nevertheless spent the next several years at-tracting financial and scientific backing for another attempt at Ultima Thule. His highly popular but privately funded venture, aboard a small schooner he named the *United States*, sailed in 1860 amid great public excitement and ex-pectation. The ship was barely noticed, though, when it cruised back into Boston Harbor the following year: the *United States*, unlike the majority of Arctic barks, had remained intact, but the actual United States had just bro-ken up. With his plans for a return trip blocked by the Civil War, Hayes ulti-mately contributed little to Arctic lore; in fact, much of his geographical data turned out to be wrong, since his expedition's most capable scientist had fallen through the ice and died, leaving the calculations to Hayes himself. But Hayes at least had the wisdom to learn from his experiences with Kane that the natives knew better than even the most clever and fearless white explorers how to find food, stay warm, and travel over ice—how to adapt to the Arctic environment. While many earlier leaders had gone to great lengths to hold themselves aloof from any natives they encountered—Franklin lugged with him all the trappings of high British culture, from china tea sets to the finest thick-handled silverware to countless white cashmere gloves[39]—Hayes made a point of hiring Greenlandic hunters and dog drivers and allowing them to formulate the expedition's survival strategies.[40] The natives had essentially come to the rescue of the Kane expedition, after all: as one seaman put it, "When we lived as Esquimaux, we immediately recovered and enjoyed our usual health."[41] Hayes was a classic midcentury racist, believing that he was lowering himself to a kind of barbarism by living like an Eskimo, but his ges-ture of acknowledging the natives' superiority in certain realms nevertheless marked a key watershed in Arctic exploration. From now on, though the Indi-ans of the West became bloodthirsty villains in most popular depictions (they were, after all, competing with white settlers for resources), northern natives were often seen as high-minded heroes.[42]

To Charles Francis Hall, native cultures offered not only the coping techniques that might allow an expedition to reach the pole, but also the oral-history traditions that might finally reveal the fate of Franklin and his men. More thoroughly than almost any other Arctic explorer, Hall overcame the scientific race theory of his day and conscientiously, respectfully, "went native," hoping to discover new points of view.[43] When he found evidence that some Eskimos may have refused to help Franklin's lost crew, he grew somewhat disillusioned, but in general his conversion to the native perspective, developed during seven years of roughing it in the Canadian Arctic, over the course of two 1860s expeditions, suggests the dawning of a new kind of cultural relativism, made possible only by extended contact with these seemingly alien peoples. "My opinion is," he explained, on the subject of gastronomy, "that the Esquimaux practice of eating their food *raw* is a good one. . . . To one *educated* otherwise, as we whites are . . . feasting on uncooked meats is highly repulsive; but *eating meats raw or cooked is entirely a matter of education*." When Hall beheld "the natives *actually feasting on the raw flesh of the whale*," he found that suddenly the meat began to seem as "*white and delicious* as the breast of a Thanksgiving turkey."[44] In the barren North, Hall appealed to his fellow Americans by invoking their nation's founding myth of cross-cultural cooperation and celebration—and this time, thankfully, the white invaders were not attempting to dispossess the very people who had shown them how to survive.

Perhaps inspired by the example of his fellow Ohioan J. N. Reynolds, Hall had taught himself geography and astronomy and garnered public support for his first two expeditions by giving lectures and meeting with potential sponsors, such as Henry Grinnell, who as both whaling mogul and president of the American Geographical and Statistical Society had financed Dr. Kane's voyage. Then, upon his return from the Canadian Arctic during the summer of 1869, just after the completion of the transcontinental railroad and just before the celebration of the Humboldt Centennial, Hall took his agenda to Washington, where his congressional supporters from Ohio managed to push through a bill granting him fifty thousand dollars for a trip to the pole. Shortly before he departed again, in a speech before Grinnell's society, Hall explained that his journeys northward did not require the "bold heart" ascribed to him by his admirers, that his connection to this particular environment had reached a spiritual level: "The Arctic is my home. I love it dearly; its storms, its winds, its glaciers, its icebergs; and when I am there among them it seems as if I were in an earthly heaven or a heavenly earth."45 The *Polaris* sailed in July 1871. Grinnell conducted a farewell ceremony in which he offered the explorer a flag from the Wilkes Expedition, and the whole nation seemed to reunite in support of this rough-edged visionary.

Hall promptly guided his ship to a new "Farthest North," and then just as promptly became the voyage's first and only casualty. He died, under bizarre and suspicious circumstances, from what seems to have been an overdose of arsenic—which he may have administered himself, in the form of arsenious acid, a common remedy for headaches, gout, and syphilis, or which he may have ingested with his coffee, poisoned by someone who perhaps felt that the commander had gone a little bit *too* native. All of his shipmates did manage to survive, but, typically, the *Polaris* did not (it got trapped in something that was definitively not an open polar sea), and nineteen members of the crew, separated from the fourteen others during a storm, had to endure six months of living on an ice floe. Fortunately, among the unlucky nineteen were four adult Eskimos, who built igloos and did enough seal hunting to keep everyone alive while they drifted the two-thousand-odd miles down to Labrador.46

Commander George Washington De Long drifted into the Arctic theater in 1879 with much less experience living in the ice than Charles Francis Hall, and much less of a commitment to native ways. And his expedition aboard the *Jeannette* wound up being a much more spectacular and tragic failure than any other since that of Sir John Franklin himself. It's little solace, but at least De Long's final voyage helped spur the evolution of a powerful new turn in the thinking of his friend George Melville—who had first seen the Arctic six years earlier as an assistant engineer aboard the *Tigress*, the ship sent out to search for Hall's lost *Polaris* expedition.

Both Melville and De Long were serious about scientific discovery, but they also seem to have boarded the *Jeannette* with some fairly typical Rooseveltian attitudes regarding their superiority as Euro-American white men, and with dreams of conquering nature. The *Jeannette* expedition was a classic Gilded Age venture, and these two men were brash, blustery risk takers in the capitalist mold. There were still Humboldtian mists in the scientific atmosphere, still a lingering idealism about understanding the shape and inner workings of the globe, still a sense of the universe's humbling power. But the struggle for resources had commenced. As Clarence King realized, becoming director of the newly consolidated USGS just three months before the *Jeannette* left San Francisco Bay, exploration had to be closely linked to development. While a North Pole expedition may not have been as pragmatic an undertaking as a survey of Nevada, it still had important implications for America's sense of its own status and wealth. Were the whale and seal populations stable? Might there be undiscovered minerals on the Alaskan coast or Siberian islands? Could the United States bring the native populations of the Far North under control without suffering the kind of humiliation meted out to George Armstrong Custer in the summer of 1876, when the nation was supposed to be celebrating its centennial?

King's survey work could at least help to rationalize the use of public lands and resources, could help take power away from capitalists. But would anyone benefit from the *Jeannette* expedition besides James Gordon Bennett? The crew were mostly navy men, and the navy helped to outfit the ship, but the government in general seemed to realize, by the end of the 1870s, that the political and scientific benefits of Arctic exploration were perhaps not worth the risks. Kane had somehow managed to be a sublime, noble failure. Romanticism was fading, though, and the Hayes and Hall expeditions added up to little more than bad publicity. Maybe publicity ought to be left to experts like Bennett, who had clearly succeeded Henry Grinnell as America's leading sponsor of exploring parties (De Long had asked Grinnell for funds before approaching Bennett). It was Bennett, after all, who had sent Henry Morton Stanley into Darkest Africa, and Bennett who had profited the most, through sales of his sensationalist *New York Herald*, when Stanley, in 1871, supposedly posed that mythic question, familiar even to children of the twenty-first century: "Dr. Livingstone, I presume?" Exploration for exploration's sake had been a hard sell even in J. N. Reynolds's day. If men died now, in 1879, then someone would have to answer for this bizarre, quixotic quest. (While the *Jeannette* was stuck in the ice and needed more power to run its pumps, Melville actually constructed a windmill on the bridge, at which the frustrated De Long may well have tilted.) In any case, both Congress and the navy preferred to let Bennett run this particular show. Exploration had generally become less of a scientific and existential search for connection. Now, it was more often an investment, or a sport.[47]

As Thomas Eakins may have known from his direct acquaintance with Melville, the rear admiral had felt the need to defend himself ever since his return from the *Jeannette* expedition. The inquiries conducted by his navy superiors and the congressional committee were trying ordeals. Eventually, both the naval investigators and their colleagues in the House of Representatives decided that it was in the government's best interests to protect the honor of all involved by papering over any potential scandals. Indeed, Congress rightfully recognized the heroism of the *Jeannette*'s survivors and gave them pecuniary as well as symbolic rewards. Melville received a gold medal. Yet the press had already spread persistent, troubling rumors that the tragedy of the *Jeannette* had not been utterly unpreventable—that, in particular, as leader of the survivors, Melville should have done more to find and help the party of crewmates from whom his own party had become separated. John Danenhower had suggested as much in his 1882 *Narrative of the "Jeannette,"* and the journalist John P. Jackson, who interviewed Danenhower in Irkutsk, also cast aspersions on the expedition's engineer.⁴⁸ It seems that Melville published his own narrative in 1885 largely as an apologia.

By the time Melville's book came out, the American public was thoroughly familiar with the bare-bones story of the *Jeannette*. No one dared fault De Long for deciding to plow ahead into the ice pack in the late summer of 1879. "It is useless to send a ship to the Arctic for the purposes of discovery," declared the report of the congressional committee, "unless it is put in the ice."⁴⁹ But De Long did not anticipate being stuck there for more than twenty-one months. And for the first year, as Melville noted, "the ship was frozen in with a list to starboard at an angle of ten or twelve degrees, rendering motion on the decks or sleeping in the berths very uncomfortable." In 1880, the summer's warmth caused the ice to shift and eased the angle somewhat, but no leads offered the ship access to open water. Food supplies ran low, and quarters began to feel even more cramped. Melville suggested in his narrative just how difficult it would be for his readers to imagine "the hideous results of forty dogs and thirty-three men living in one spot" over such a long period of time.⁵⁰

Their isolation was complete; no one knew their fate. Moreover, they felt utterly useless. De Long's journal entries pulsate with his frustration at his lack of accomplishments: "we and our narratives [will be] thrown into this world's dreary wastebasket and recalled . . . only to be vilified and ridiculed." All his men could do each day was play cards, measure the daily drift of the ship, take a little exercise on the ice, and sip lime juice to prevent scurvy. "The steady strain on one's mind is fearful," De Long wrote.⁵¹ Like J. N. Reynolds stuck in a dead calm, De Long thought of this period of inaction as "a clear waste of life . . . , an existence without present tangible results, a mechanical supplying the system with food, heat, and clothing, in order to keep the human engine running."⁵²

There were sporadic reprieves. Melville had a rich, booming voice, and he liked to sing and joke; many of the survivors would refer to his determined cheer as having been a saving grace aboard the wedged ship. Indeed, his sense of hope was perhaps his best Humboldtian quality. The whole crew, following a tried-and-true tradition of polar exploration, also worked to dispel the constant darkness of winter by holding special holiday celebrations, which featured music, poetry, jigs, skits, drag shows, and lots of brandy. Better than the unlucky men of most stranded expeditions, these rugged sailors actually managed to battle their loneliness, at least on occasion, by building some semblance of fraternal community.[53]

Both De Long and Melville even noted moments of glowing beauty and joy while they were caught in the ice. Once, after a particularly harrowing escape from the "grinding masses" of the pack, Melville noted that "with light hearts the men dispersed themselves upon the ice, climbing the slopes of the marble-like basin, leaping from block to block, clambering up pinnacles and tumbling down with laughter, calling each other's attention to the marvelous shapes."[54] But Arctic scenes were almost always more confusing than pleasing to the eye. On the evening of February 17, 1881, De Long recorded in the ship's log "a mass of Auroral light, in shape resembling a Cornucopia, at ESE horizon." After about eighteen months of glacial entrapment, thirty-three men subsisting on the barest rations gazed up at what seemed to be a horn of plenty in the sky. If that hadn't been enough to convince them that they were hallucinating from hunger, they thought they saw land three months later, and De Long allowed himself a little shiver of exhilaration: "It appears to be an island. . . . As no such land is laid down upon any charts in our possession, belief that we have made a discovery is permissible." Of course, the men had little faith in the reliability of their senses by this time, but De Long insisted that he could see "apparent rocky cliffs, with a snow-covered slope extending back to the westward . . . and culminating in a conical mass like a volcano top." Could they have found the first volcano above the Arctic Circle?

De Long never got to find out. Less than three weeks later, there was suddenly "great activity in the ice": it was "grinding and crashing and piling up" all around the *Jeannette*. At first, the commander hoped his ship might finally free itself and be launched into open water—but the leads he thought he saw may well have been mirages. On June 11, 1881, with the sound of heaving, cracking wood filling his ears, De Long ordered his men to abandon ship. At 4:00 a.m. the next morning, the *Jeannette* disappeared below the ice.[55]

———

Almost happy to be responsible for their own motion again, the men loaded the ship's three boats onto makeshift sleds and turned southward, heading for the edge of the ice pack, from which they would set sail for Siberia. They found the dogs almost entirely useless—Melville referred to their attempts at dogsledding as "a pandemonium of horrors!"—and so had to drag the boats themselves, in harness.[56] Huge ice hummocks, watery cracks, windstorms, and snow blindness made progress unbearably slow. They had started with sixty days' rations and were occasionally able to kill polar bears, walruses, and

35 *The Sinking of the* Jeannette, 1881.

seals for meat, but supplies dwindled quickly, and the northward drift of the ice made Siberia seem that much farther away. Still, all thirty-three men reached the edge of the ice pack together. On Sunday, September 11, having made landfall on one of the New Siberian Islands, the full crew of the *Jeannette* gathered for the last time to read the Articles of War and pray for good weather during their hundred-mile journey across the Laptev Sea, under sail and oar, to the Russian mainland. They had seven days' provisions remaining.

Here the story becomes more mysterious. One of the boats, commanded by Lieutenant Charles Chipp, apparently never made it across the water. The other two got separated in a gale. De Long reached land, but his party became hopelessly lost in the crisscrossing channels of the Lena Delta. Only two of his men survived the march south over thin ice toward the nearest native outposts. Melville, commanding in place of the disabled and bitter Danenhower, successfully landed his whaleboat just east of the delta. Finding one of the Lena's main outlets, rather than the labyrinth De Long's men encountered, Melville's crew actually managed to paddle thirty miles upriver, and on September 19, they encountered three Tungus hunters, who brought them to a nearby settlement and offered them food and water. This was the critical juncture. Perhaps, if Melville had immediately convinced the natives to help him conduct a broad search of the delta, he might have found De Long and his men in time to save them. But he and his party were weak and frostbitten, and the Tungus refused to go anywhere: it was too late in the season to travel safely by boat, too early to take a loaded sled on the newly forming ice. Melville believed his first duty was to his own men, since he did not even know if the other two boats had reached the coast, and an overhasty search expedition might just have added to the final death toll. No one will ever know for sure if De Long's party could have been rescued. Even after almost 1,100 pages of official testimony, the congressional committee found it impossible to determine whether or not Melville had faltered in his duty to his fellow men.[57]

Melville himself, as his "ice journals" and letter books reveal, struggled constantly with the issue of what exactly was within the realm of possibility in the Arctic. And the four hundred pages of his narrative adhered quite closely to his journals and letters in both substance and tone; sometimes he even copied passages verbatim.[58] Melville devoted just four chapters—less than one-fifth of the book—to the first two years of the *Jeannette* expedition. The remainder takes place *In the Lena Delta*, as his title implies, and argues that he did everything in his power to help his lost comrades. "I little thought," Melville later wrote to his colleague Robert Peary, "that when I got home I should be put on the defensive for not going after them sooner."[59] Indeed, his whole crew confirmed that Melville did in fact set out right away to find De Long's party—without the Tungus and Yakut villagers, who would not budge—but that his party had to turn back after just a day, having covered hardly any ground. The wind literally blew the men over, and they continually fell through thin ice into deep snowdrifts. "Our clothes froze fast to our bodies," Melville wrote.[60]

Even once thicker ice made travel possible, Melville's crew and their native guides faced brutal, terrifying, lightless conditions. Food and warmth could never be taken for granted. Melville recorded many missed meals and many meals of carrion, guts, or frozen fish that he could barely swallow. His extremities went through various phases of trauma: "Pulled off the loose toe nails, pulled the dead skin and flesh from toes, heels, and shins, and in a couple of days was all right again. I was absent from Belun 23 days, in the dead of winter, did not see the sun during my absence and not likely to see it until I get to Yakutsk." Often the frostbite in his legs was so bad that, at the end of a long day's journey, with shelter in sight, "it was all I could do to crawl on hands and knees up to the house or hut. I rolled inside, and the natives started a fire, and although it was three days since I had [eaten] anything warm and two days since food had passed my lips I lay for two hours unable to move just thawing out, but when I got some hot tea, and revived a little I [ate] about 2 pounds of boiled venison." In empty huts, Melville left whatever supplies he and his crew could spare—skins, parkas, knives, candles, a sack of sugar—but he harbored little hope that his missing companions would ever stumble upon these caches. The delta seemed a "desert of desolation."[61] When he and his men got separated from the natives who were helping them, they simply didn't know where to go. "There is no beaten track or road in these regions," Melville reminded the readers of his book. "The face of the country changes its appearance every season."[62] Melville's published narrative is primarily about how hard it is to interpret nature, how crucial it is to make humble efforts at connection.

The most movingly humble moment in Melville's story comes in the spring, when he finally finds his friends' bodies—when he reads De Long's last journal entry and the last letter written by John Ambler, the ship's surgeon—when he buries their frozen corpses and erects a simple stone monument in the stark landscape. As the winter had begun to fade, Melville had been wracked with fear and guilt, fully cognizant that his search was subject to a time limit imposed by the sweeping forces of nature. "If I wait until the snow leaves the ground the floods will come and carry *all things* away . . . ," he scribbled in his journal. "People here say that at times when an ice dam breaks up the river a vertical wall and flood of water 10 feet high will press down the river for miles at the speed of a race horse. . . . I have seen drift trees lodged 60 feet above the river."[63] But Melville's determination brought him eventually to one last meeting with those who had perished, brought him some measure of peace in the still, bright Arctic air. He could never again touch or comfort his lost companions, but in his long letter to the secretary of the navy describing the successful completion of his search, he included a poignant, suggestive passage about a recent night, reminiscent of Clarence King's night on Mount Tyndall, when human communion had made all the difference: "Everyone seemed dazed and stupefied, and I feared some of us would perish. . . . How cold it was I don't know, as my last thermometer was broken by my many falls upon the ice. . . . A watch was set to keep the fire going and we huddled around it, and thus our third night without sleep was passed. If Achin [a guide] had not wrapped his seal skin around me and sat alongside of me to keep me warm by the heat of his body I think I should have frozen to death. As it was I steamed and shivered and shook."[64]

To bolster his self-defense, Melville portrayed himself as a man of heroic action, who had endured unbelievable hardships in order finally to find his comrades and give them a decent burial. But his heroism is not the heroism of a conqueror of new lands; it is the heroism of endurance and adaptation, of respect for the forces of nature, of acceptance. "Neither man nor dog could face such a gale," he wrote, again and again; "so we did the only practicable

36 A page from Melville's "Ice Journals," 1882.

thing, and abided its abatement." Above all, Melville pointed out his utter dependence on the natives. Having already described himself frostbitten and crawling, Melville continued his own infantilization by explaining that his Tungus savior, Vasilli Kool Gar, was "as proud and careful of his charge as he might be, in his old age, of a baby."[65] This self-portrayal, too, was strategic, underlining his blamelessness in not being able to save De Long. But it also suggests the transformative power of his experiences, betrays the sobered sensibilities of someone who has understood the limits of human achievement. This is not the rugged individualist explorer overcoming nature, but rather a desperate child, dependent on the love and warmth of his comrades.

To his credit, Melville never tried to cast blame on the Tungus and Yakuts. He even wrote official letters to his navy superiors extolling the natives' conduct and asking that they receive both compensation and commendation. Vasilli, for instance, served as pilot, navigator, and procurer of food during Melville's searches, and Melville thought that in every way his savior had been "efficient"—the highest compliment of an engineer—especially with

37 The cairn Melville erected for his dead companions, 1882
(engraving after a sketch by Melville).

regard to "my health and comfort. His conduct has met with my heartiest approval and I would be very much gratified if some distinction be bestowed
upon him."[66]

Melville's narrative, overall, is respectful toward the natives—though the
admiral was by no means a radical. Invoking the scientific racism of his day,
he made sure to point out the practices of his "dusky neighbors" that he considered uncivilized (many of them seemed "squalid and filthy" to him), and,
anticipating the comforts of home during the last leg of his trip across Russia,
he noted that, "true to the law of progress, everything was materially improv-
ing as we journeyed westward." He was also not above bullying his native
hosts when they were reluctant to help him in exactly the way he desired, and
he often played to his readers' biases in suggesting that his hosts could do
with some Christianizing. Yet Melville admired many native customs; he
even understood the reasons behind their resistance to his demands, and he
seemed perversely pleased when his guides responded to reprimands with
stern pronouncements of their own. At times he could barely communicate

with them, but their lively exchanges, in a combination of Yakut, Russian, English, and a vigorous body language, often led to new, creative strategies in the
search for De Long. Indeed, the bulk of Melville's comments about the Tungus and Yakut peoples serve to praise their amazing ingenuity, their ability to
understand Siberia's wild, forbidding landscape on a level he could scarcely
fathom. "They only lose their way when the snow is swirling in clouds or
columns," he noted, and whenever he asked them for help with directions
while working on his map of the delta, "my compass invariably proved their
calculations to be correct." Over time, his background in engineering helped
him to see the logic behind their strange ways, and he often wound up favoring their methods of adapting to the land over traditional Western techniques.
When it came to making fires in the snow, for instance, he realized that his
matches were no match for the brilliant "Yakut treatment," involving flint,
steel, tiny wood shavings, and the buds of the Arctic willow: "I have watched
this operation a hundred times," he reported, "and never saw it fail." Melville
had not freely chosen to live in the Arctic wilds according to native customs,
the way Hall had in the 1860s, but he eventually found himself echoing many
of Hall's assertions about the natives' wisdom and the unfortunate prejudices
of white men.[67]

Following in Humboldt's footsteps through Siberia, Melville also took opportunities to consider issues of social justice. He got especially interested in
the exiles and political prisoners he met, many of whom struck him as "bright
and intelligent"; these earned several pages in both his journal and his narrative. There were Jews exiled just because they were Jews, and young students
banished for life just for having taken part in demonstrations and protest
marches. "To see the Best Brains of a country sent out here to die for political
reasons," Melville wrote, to see them become nihilists and dream only of flee-
ing over the Laptev Sea and the very ice in which he himself had so recently
been entrapped—this was almost enough to make him yearn for home, for
the very first time. He was a guest in Russia, "and a continuing recipient of its
succor and hospitality," so he could "not honorably abet the exiles in their
plans for escape; yet as a Republican I am free to say that all my sympathies
were with them,—the oppressed for speech's sake."[68]

His own and the exiles' suffering led Melville to refer to Siberia as "an Arctic desert," but ultimately he did not condemn the region as a wasteland. Indeed, he came home with a deep appreciation for the Siberian landscape.
Often this appreciation sprang from the humbling force of the elements in
boreal climes: "Here we stood lost in the contemplation of the wild tumult
and rout before us," Melville wrote, of a particular patch of crumpled ice. But
even overwhelming, surreal landscapes possessed a "wondrous beauty": "the

weird ride over hill and mountain, skirting ravine and precipice; the breaks along and across the numerous water-courses . . .—with this quick succession of scenery wild and strange was I kept constantly awake and charmed." Nature in the northern latitudes—on terra firma, anyway—impressed Melville with its simple, direct power. "It was severely cold—ah! ferociously so . . . ," he declared, recalling an inspirational evening, "but the soft, clear, moonlight was gorgeous and glorious. . . . Above us on either side, the gigantic peaks lifted their hoary heads far into the blue vault of heaven; silent, frigid, and white. Ah! what grandeur! I rejoiced that it was night, and so cold and still." Melville may have considered himself primarily an engineer, devoted to the practical applications of science, but it was by no means unusual for him to break into these poetic musings on the Siberian sublime, to pause, turn his head, and extol "the wonderful wilderness of white rolling endlessly around us."[69]

Like Clarence King, Melville tended to invoke not pragmatic explorers like Zebulon Pike or Lewis and Clark, surveying America's potential resources, but aesthetes like John Ruskin.[70] A lifelong military man, the admiral nevertheless seemed to scorn conquest; instead, he favored the sense of his own smallness conveyed by awe-inspiring landscapes. Cold and stillness, whiteness and endlessness—these were signifiers of death, and this voyage had been fatal, tragic, but at least Melville had come face-to-face with the infinite cosmos. Plagued by his own doubts and the horrific loss of so many comrades, resentful that his superiors had questioned his actions, he seemed to take solace in the purity and genuineness of his experiences. He had understood something about himself and the world, something inaccessible to those who stay forever in cities. And he embraced humility. Seeming to mock American presumption, he noted that his "critics" were "10,000 miles or more distant from the scene of action."[71] His best defense was the implication, which drifts just below the surface of his entire narrative, that even the brightest, best-equipped white Americans could never understand the Arctic landscape well enough to master it.

———

What, then, was George Melville doing in the Arctic?

Significantly, Melville never wound up sharing Danenhower's cynicism about sending men northward: he supported polar exploration for the rest of his life. At the end of his 1885 narrative, he tacked on a brief section entitled "A Proposed Method for Reaching the North Pole," and he suggested himself as the man best qualified to test this method. Yet he also acknowledged his hope that his own success in reaching the pole "may prevent other fools from going there."72 His experiences had not shattered his ambition, but they had tempered it.

As Eakins well recognized, Melville's defensiveness encompassed much more than his response to suspicious government officials. At issue was his fundamental utility as a citizen. If Arctic exploration held no value, then he had wasted several prime years, and his comrades had died in vain. Despite all his doubts and regrets about American ambition, he *had* to believe in exploring expeditions. But what had he accomplished amid those unimaginably awe-inspiring landscapes? Had he really earned his government pay, the tax dollars of his fellow citizens? As Eakins got to know Melville—they apparently went to several prizefights together73—he perhaps understood that they shared a parallel anxiety about their role in society. Eakins, thanks to his father's fortune, never even had to earn his living. Like explorers, artists occasionally translated their skills into splashes of national glory, but overall they seemed to contribute little to the common good in a rapidly modernizing, industrializing country. Making maps and making pictures hardly entailed real production. What was the point of all their professional training?

Both Melville and Eakins could at least fall back on their craftsmanship, their manual dexterity, their agency as artisanal workers: both men at least *made* things. Though Melville became famous as an Arctic explorer, at the end of his life he embraced his identity as an engineer, even helping to establish a graduate program in naval engineering. He prayed that society would at least recognize the concrete contributions made by people (like himself) who could build engines. "When some future Macaulay describes the condition of the United States at the beginning of the twentieth century," he proclaimed, in a 1910 address before the American Society of Mechanical Engineers, "and attempts to award the credit for the existing comforts and conveniences, the major part must be given to the profession of engineering." Yet he worried that his brethren were becoming mere technicians, "content to do the work and then fade into the background." To ensure their utility as Americans, they ought to step forward into the political arena, "take an active part in all public questions, great or small, where their knowledge and experience will enable them to contribute to the common good."74

Melville had embraced professionalization, but he had also noticed that even the most practical, utilitarian professions tended to remove people from the public sphere.75 One contribution engineers certainly could offer to the public, though, was their wisdom about how best to utilize the productions of nature. Melville's appreciation of beautiful landscapes, his belief in contact with the physical world, and his respect for nature's limitations, provided him with the cornerstones of an ecological worldview. Certainly, he seemed to bow to Humboldt's legacy of international cooperation and global research when, in Siberia, he sang the praises of a Russian navy scientist "who was then at Yakutsk preparing to set out for the mouth of the Lena River, to establish a meteorological station and make a survey of the Delta."[76] More than cosmical connections, though, Melville, in step with his times, emphasized efficiency. Engineers knew how to maximize agricultural yields, how to draw power from rivers, how to run engines with less fuel. Rooseveltian conservation immediately ran into problems—like the ones raised by John Muir, who didn't want to see dams in his favorite wild gorges. But efficiency was better than inefficiency, and it can still be argued that hydropower projects, despite inevitable drawbacks, may well serve the health of particular ecosystems far better than many other energy-harnessing schemes. In any case, as Melville told his fellow engineers in 1910, "our own Society and others which have taken part in the movement for conservation of our natural resources have set a good example."[77] Perhaps, then, his clasped hands in Eakins's 1905 portrait ultimately suggest his capacity for self-control. In this light, the admiral's grip calls to mind a line from Wallace Stegner: "The best thing we have learned from nearly five hundred years of contact with the American wilderness is restraint, the willingness to hold our hand."[78] This is not as rich or as complicated a worldview as that held by earlier, more committed Humboldtians like J. N. Reynolds and Clarence King—but it is something, at least, to hold onto.

Both Melville and Eakins also tried to take comfort in physicality. To counterbalance their cerebral inclinations, they embraced vigorous activity, upheld direct confrontation with wilderness as a tonic for the overcivilized lifestyle led by people of their class. They had a decided antimodernist streak, despite their obvious modernism.[79] In other words, Melville shared more with Theodore Roosevelt than just a belief in conservation (and the receipt of an honorary degree from Columbia—something both men also shared with J. N. Reynolds).[80] Melville's three trips to the Arctic obviously helped bolster his manhood, and the spirited, active prose of *In the Lena Delta* spread his reputation for vitality and robustness. The novelist Wilkie Collins went so far as to list Melville's narrative among the top ten adventure stories ever written.[81] Yet what most stands out about the book are its abundant episodes of immobilization, when storms trap the admiral inside tiny huts; when his swollen, frozen legs refuse to move; when he can do nothing but work on his maps and write in his journal. Eakins, meanwhile, an avid sportsman, frequently painted hunting and rowing scenes in his early career, shifting to boxing and wrestling matches later on. Even in these overtly active canvases, though, an air of reflection prevails.[82] Eakins's scullers seem completely absorbed in thought, and the surface of the water shines back at us like a blank page waiting to be filled in.[83]

The truth was that both Melville and Eakins were intellectuals, head-men more than hands-men, and they knew it. And what drew Eakins to make such a telling portrait of the admiral was his realization, whether conscious or subconscious, that Melville's pursuit of the North Pole—his defining ambition— was a perfect metaphor for the pursuit of knowledge, the defining ambition of an entire class of professionalizing Americans. Both men hoped that the knowledge they and their colleagues were acquiring—at the nation's swiftly multiplying graduate and professional schools, for instance—would turn out to be useful. They weren't entirely sure it would be, though, and they wanted to hedge their bets by arguing that utility wasn't necessarily the most important thing in life. Melville, in particular, felt obliged in almost every one of his writings to address—or explicitly refuse to address—the usefulness of polar exploration and the pure science of geography.

In introducing his "Proposed Method for Reaching the North Pole," for instance, Melville immediately—and defensively—stated his intention to describe his plan "without entering into the question of the utility of Polar exploration."[84] When Danenhower dared to condemn Arctic exploration in Melville's presence at a symposium in October 1885, the admiral grew even more testy. He, too, had grave doubts about America's ambitions. But he interpreted Danenhower's comments as endorsing only those expeditions that would prove commercially profitable, and he could not abide such a conservative, utilitarian outlook. "If men must die," Melville proclaimed, responding to the lieutenant and seeming to invoke J. N. Reynolds, "why not in honorable pursuit of knowledge? Far be it . . . that our ideas of manhood should be dwarfed to the size of a golden dollar. Woe, woe, to America when the young blood of our nation has no sacrifice to make for science."[85] Melville struck a similar chord in a 1911 preface, just a year before his death. When considering the grim price of Arctic expeditions, he wrote, "we are impelled to ask 'To what purpose?' I may answer now as I have a hundred times before:—to every purpose that is noble, for the benefit of mankind,—that all may have knowledge though at the price of trial and suffering on the part of the investigator of earth's phenomena. To the same purpose that Galileo endured the punishments of Rome for the acquiring of knowledge, to the same purpose that scientists and thinkers in every age have endured hardship that they might know the truth." Melville assumed that knowledge and truth would result in "power, wealth, and happiness."[86] But these were just incidental effects; no matter what the result and whether or not the public approved, the important thing was to seek a deeper understanding of humanity's place on the planet—even if that understanding would always remain partially obscured by mists. The North Pole was Melville's Cotopaxi and Chimborazo, his Symmes's Hole and Mocha Dick, his Shasta and Whitney—his abyss.

The *Jeannette* expedition actually marked the dawning of a new Arctic craze: there was much public debate about the risks involved in glacier hopping, but as frontier settlement proceeded apace in the 1880s and '90s, Americans tended to embrace expansion and ambition, and the North provided a welcome challenge. Aspiring explorers no longer sought training, as Clarence King had, in the far West, but rather lined up for opportunities to tackle the pole. Melville, though, was one of the last Arctic explorers from the United States to make a genuine case for the scientific value of what they were doing. The government did send Adolphus Greely northward in 1881 to set up a meteorological station, in what was supposed to be a straightforward joint venture with the Russians, based on Humboldt's old cooperative model; and there were certainly other official scientific expeditions. But as arrogant, ambitious characters like Robert Peary and Frederick Cook got more involved, the whole enterprise came to seem like more of a stunt, a game, a race. The new generation focused primarily on proving America's heroic manliness and the supremacy of its whiteness, in this whitest of all landscapes.

———

Melville was a hard, complicated man, deeply divided within himself. In a profile of his "many-sided career" for *Cassier's Magazine*, William Ledyard Cathcart described his life as having been "marked by lofty patriotism, by heroic endeavor, by strenuous toil, and by more of sorrow than is its share."[87] The admiral did many different things very well and received commendations from his government throughout his long period of service. Even during the congressional investigation of the *Jeannette* expedition, the lawyer seeking to incriminate Melville backed off for a moment to praise the engineer's testimony as "beautiful and vivid": "in that wonderful oration which he delivered here, under oath, from half past 10 in the morning until 6 at night, I have never witnessed, in my experience, more physical and mental endurance."[88] Yet Melville must have chafed at the skeptical treatment he received from his superiors after returning from his Arctic ordeal. His patriotism never lagged, but neither did he fully trust his government after the *Jeannette* expedition. In his work, Melville managed to unite his head and hands and have a powerful impact on the public sphere in turn-of-the-century America. But his heart, as Eakins's 1905 portrait perhaps suggests, remained with his buried comrades in remotest Russia.

Another reason for Melville's internal divisions was perhaps his awareness of his country's internal divisions. His first naval service had come in 1861, during the Civil War, and his memories of that conflict often haunted him. Indeed, the deep scar on his left hand that Eakins emphasized was actually the result of an ax wound he received in 1864 off the coast of Brazil, while helping to capture a Confederate cruiser that had been disrupting American shipping operations throughout the Atlantic.[89] When he spoke at ceremonies for Memorial Day—then called Decoration Day—he of course invoked the frozen corpses of his shipmates from the *Jeannette* expedition, but he also recalled the men he had seen die in the tropics during the Civil War.[90] And in later years, Melville became acutely aware of the class divisions in America, as his professional friends gained more and more expertise but also withdrew from the public realm, focusing exclusively on the "engrossing and exacting" work of their particular specializations.[91] Like J. N. Reynolds, he wanted colleagues who were engaged rather than engrossed.

Yet another key to Melville's dark side was his tortured private life, which erupted in scandal immediately upon his return from the *Jeannette* expedition. Unlike Humboldt, Reynolds, and King, Melville was a conventional family man, though he was rarely at home. By the time he set sail for the Arctic in the *Jeannette*, he had three daughters, but his wife of fifteen years had started drinking heavily in order to cope with his long, uncertain absences, and the couple's relations had deteriorated. When the news came from Siberia in December 1881 that Melville had survived the disaster, his wife, Henrietta, strangely went to the press with the claim that her husband had expected the expedition to fail from the very beginning. The *New York Times* of March 15, 1882, published a copy of a letter allegedly from George to Henrietta, his last communication from San Francisco before setting sail in June 1879. The engineer's missive revealed his conviction that the construction of the steamer "would never do." "Some of us may weather it . . . ," Melville had supposedly written, "but I am sure the *Jeannette* will never come back." Mrs. Melville retracted the letter the next week, but the damage to the admiral's manhood had been done. When Melville finally arrived home in September 1882, he immediately confronted his wife about her behavior, and, finding her drunk, stormed out of the house. The very next day, he forcibly committed her to the Pennsylvania State Asylum. Though some acquaintances did testify that Henrietta had become mentally ill, wheeling an empty baby carriage around her suburban neighborhood and describing fantastic visions to anyone in earshot, others accused Admiral Melville of abandoning his family.[92] Perhaps it was this sordid episode in the admiral's past that made Eakins think of Bluebeard while he was painting his friend. The legendary villain, after all, killed six of his wives for supposedly betraying his trust.

The blackest current in Melville's character, though, was surely his choppy relationship with his own government. When he returned from the *Jeannette* expedition in 1882, his admirers in the Senate immediately introduced a bill offering him their thanks and advancing him thirty numbers in the ranks of the navy. But Congress didn't pass the bill until 1890, at which point they reduced the promotion to fifteen numbers and struck out the clause that would have thanked Melville on behalf of the people of the United States of America.[93] In the meantime—on August 9, 1887—President Grover Cleveland had appointed him over forty-four senior officers to be the navy's engineer-in-chief.[94] Congress's ambivalence, however, must have rankled so proud a man: he was already dealing with his survivor's guilt and his own self-doubt about whether he might have saved De Long and his men. And though he was fully cleared by the end of the official inquiry into the *Jeannette* affair, his reputation had been sullied. The opposing lawyer, despite admiring Melville's eloquence and endurance, had gone on record claiming his certainty that the admiral had not done his utmost to save Commander De Long's party.[95]

Throughout his career, Melville's superiors had lavished him with praise, both for his technical expertise and for his honorable conduct and good fellowship. De Long himself wrote in his journal that "Melville is more and more a treasure every day. He is not only without superior as an engineer, but he is bright and cheerful to an extraordinary extent . . . , a tower of strength in himself."[96] Yet, between 1882 and 1884, his job was on the line, with his character being questioned by his wife, the press, officers of the navy, and several members of Congress. Melville was never again as "bright and cheerful" as he had been on the *Jeannette.* Indeed, he sometimes showered his bitterness upon the U.S. government. "We Americans are fond of claiming that we have the greatest country and the most free and best government in the world," he once wrote.[97] But he had his doubts. The admiral relished his opportunity to ship aboard the *Thetis* on the 1884 mission to relieve the party of Arctic explorers led by Lieutenant Adolphus Greely, but he would never forgive Congress and the navy for failing to authorize the relief expedition that he had proposed in 1883, which might have saved the nineteen men (of Greely's party of twenty-five) who perished in the frozen isles above Greenland.[98] When Melville published *In the Lena Delta* in 1885, he included a newly penned section about his experiences aboard the *Thetis,* explicitly condemning America's abandonment of the Greely expedition. Greely had been waiting "at the point of safety where our government had promised to deposit supplies and have a vessel awaiting to carry him and his band away." Melville could scarcely believe "how strangely, if not criminally, the government's

efforts were thwarted by carelessness, incompetency, or inexperience."99
Greely's abandonment seemed to mirror his own.

As the United States entered the Spanish-American War and began to embrace expansionism, Melville's distrust of his government increased. It was
here that he broke with the masculine bluster of Theodore Roosevelt and
aligned himself with the anti-imperialists, who were so often ridiculed as
"womanish" by the hawks and boosters. Melville's stance was unusual in the
military, but it was not altogether uncommon among men of his age and class
along the eastern seaboard—men who did not necessarily care much about
the rights of Filipinos, but who defined both their manhood and their patriotism in terms of sacrifice and limitations rather than conquest and expansion.
Indeed, they formed a generation of Civil War veterans who had paid dearly
for their nation's expansion westward and who were now trying to rein in the
young bucks, who seemed to lack a sense of tragedy.100 Needless to say,
Melville was not a radical anti-imperialist. His own innovations in enginery
had helped Americans improve their mastery over nature and had ushered
the navy into the modern era: his designs directly contributed to a vast military buildup. Before war broke out, Melville had even supported the possible
annexation of Hawaii, citing the islands' strategic position, the acquisition of
other Pacific islands by potentially threatening European powers, and petitions filed by the residents themselves in favor of U.S. intervention.101 Yet he
also decried the "feverish hunger for island territory" that seemed to be running rampant among the Western powers. As a naval adviser, he felt he had to
take a defensive approach, remaining suspicious not only of maneuvers by
Germany, France, and Britain to extend their reach in the South Pacific, but
also of the "seemingly inevitable Europeanizing of the long littoral of China"
and of Japan's expansion to Formosa. He did not think it his country's place,
though, to indulge in "the land hunger of mere territorial aggrandizement."
Moreover, perhaps thinking back to his respect for the natives of Siberia, he
could not abide the arrogance of states that sought to conquer weaker peoples.
He did not believe in any master race. Echoing Humboldt, Melville lamented
the fate of "Ancient Peru—peaceful, rich, unwarlike," which was sacrificed to
"Pizarro and his adventurers, soldiers less of Spain than of the lust of gold.
The Inca fell, and the land was stripped of its fatal wealth; its people were en-
slaved, and in slaughter, torture, and rapine a noble civilization perished."102
The admiral—just like Clarence King—supported U.S. intervention in Cuba,
as long as the goal was Cuban independence, but he balked at the McKinley
administration's greed for Spain's territories in the Pacific. In 1905, the year
of Eakins's second portrait, Melville published a pamphlet called *Abandon the
Philippines*.103 He knew from his experience in the Arctic that it did not pay to

overstep certain natural limits. His country, he felt, had become recklessly bold, irresponsible, imperious. For all his enthusiasm in planting the American flag on one of the barren islands discovered by the *Jeannette*'s crew, Melville realized that America's expansive ambition might be self-destructive.

When Eakins had Melville strip himself of his American medals and don the cross of St. Stanislaus for his 1905 portrait, perhaps the two friends discussed the history behind the medal. Perhaps they both knew that Stanislaus had, as a prominent eleventh-century bishop, joined in a revolution to overthrow his country's cruel, overbearing king. The rebellion was successful, but the new ruler's hold on power was tenuous, and he did little to protect his bishop. Stanislaus's martyrdom was secured when, seven years later, the old king stormed back into the land and murdered the bishop in full public view.[104] Perhaps, even as Melville clung to his patriotism in the last decade of his life, he came to identify with the rebellious Stanislaus. Perhaps Eakins respected his determination to remain true to his principles, to his experience, in the face of his nation's growing hubris. Perhaps the two men both realized that, as intellectuals, as professionals, they would always feel like strangers in their own country.

CHAPTER 9

The Cruise of the Corwin

Nature, Natives, Nation

I could only occasionally perceive his trail in the moss, and yet he did not appear to look down nor hesitated an instant, but led us out exactly to his canoe. This surprised me, for without a compass, or the sight or noise of the river to guide us, we could have retraced our steps but a short distance, with a great deal of pains and very slowly, using a laborious circumspection. But it was evident that he could go back through the forest wherever he had been during the day.

—Henry David Thoreau, writing about Joe Polis, his Penobscot guide on his third trip to Maine, "The Allegash and East Branch" (1857)

Arctic explorations are exciting much interest among the natives here. Last evening the shamans called up the spirits supposed to be familiar with polar matters. The latter informed them that not only was the Jeannette forever lost in the ice of the Far North with all her crew, but also that the Corwin would never more be seen after leaving St. Michael this time, information which caused our interpreter to leave us, nor have we as yet been able to procure another in his place.

—John Muir, letter from St. Michael, Alaska, to the San Francisco Evening Bulletin (1881)

On May 4, 1881, another exploring vessel left San Francisco Bay, heading north: the U.S. steamer Thomas Corwin, named for the old Ohio politician and friend of J. N. Reynolds. Aboard ship as the captain's cabinmate, resident naturalist, and correspondent for San Francisco's Evening Bulletin, was forty-three-year-old John Muir, about to spend his third consecutive summer studying Alaskan glaciers. The Corwin's primary mission was to search for some lost whalers—and for any potential survivors of the Jeannette expedition.

38 John Muir as a young man, 1860s? Photographer unknown.

Muir had been obsessed with northern ice since March of 1871, when
Clarence King published his news about the existence of active glaciers on
Mount Shasta. That October, Muir found his own living ice fields in the
Merced group of the Sierras, and over the next four years he claimed to discover a total of sixty-five glaciers in the same general region, always "on the
north sides of the loftiest peaks, sheltered beneath broad, frosty shadows."[1]
King and Muir do not seem to have spent much time together,[2] though they
maintained an antagonistic contact through their publications: they used
books and articles not only to criticize each other's climbing routes but also to
attack each other's arguments about glaciology. While Muir emphasized the
significance of glaciers in forming such impressive topographical features as
Yosemite Valley, King proposed explanations involving other catastrophic geological forces as well, and he expressed a rather cutting skepticism about how
Muir distinguished between glacial "activity" and dormancy: "Motion alone is
no proof of a true glacier," he remarked in *Systematic Geology*. Indeed, "in the
dry season of 1864–'65," King had "examined many of the regions described

by Mr. Muir in the Sierra Nevada, and in not a few cases his so-called glaciers had entirely melted away. The absurdity of applying the word 'glacier' to a snow-mass which appears and reappears from year to year will be sufficiently evident."3

King's scorn, though couched in scientific terms, probably arose as much from emotion as from reason, since Muir the independent climber had often mocked his governmental counterpart in print for having failed at various mountaineering exploits. From Muir's perspective, King seemed to have exaggerated the difficulties of climbing not only Mount Tyndall but also Mounts Ritter and Whitney, and when King found it impossible to get down into the Great Tuolumne Canyon, Muir was quick to express his certainty "that it may be entered at more than fifty different points along the walls by mountaineers of ordinary nerve and skill."4 But the two men actually had a great deal in common—including their admiration of Humboldt—as perhaps became most clear in 1875, when Muir published an article in *Harper's New Monthly Magazine* called "Living Glaciers of California," in which he at least acknowledged King's discovery of Shasta's Whitney Glacier.5 Most important, Muir's article revealed his keen interest in reconstructing geological history and in exploring ways for humanity to relate to the catastrophic power of blinding sheets and rivers of ice.

From "a wide, shadowy amphitheatre, comprehended by the bases of Red and Black mountains," Muir had "set out to trace the ancient ice current back to its farthest recesses, filled with that inexpressible joy experienced by every explorer in nature's untrodden wilds." As he gained higher altitudes, he seemed to be traveling simultaneously to higher latitudes—"the scenery became more rigidly arctic"—and also deeper into the past, to a time defined primarily by the groaning and grinding of ice. At last he climbed deep enough into some characteristic glacial crevasses—"marginal, transversal, and the jagged-edged *Bergschrund*"—to feel immersed in a "weird ice world," into which a "thin subdued light pulsed and shimmered with indescribable loveliness," while far below him echoed "strange solemn murmurs from currents that were feeling their way among veins and fissures" in the ice. Here, Muir became overwhelmed not with joy but with a classic case of explorer's ambivalence, of confusion over whether he was meant to be in such a place, whose undeniable beauty nevertheless seemed utterly stark, blank, inhuman. A living glacier destroys almost all traces of life. And yet those shiny, mirroring surfaces sometimes have a spellbinding power. "Ice creations of this kind," Muir explained, "are perfectly enchanting, notwithstanding one feels so entirely out of place in their pure fountain beauty. I was soon uncomfortably cold in my shirt sleeves, and the leaning wall of the *Schrund* seemed ready to engulf me. Yet it was hard to leave the delicious music of the water, and still more the intense loveliness of the light."[6]

It was an intense experience atop Mount Shasta, though, under extreme conditions, that ultimately drove Muir farther north, to discover how people could actually adapt to Arctic climes and Arctic climbs. On the last day of April in 1875, well before the customary climbing season had begun, Muir and his friend Jerome Fay left the village of Strawberry Valley to examine the "Geological and Botanical Characteristics of Shasta" on behalf of the U.S. Coast Survey—and to make a definitive measurement of the volcano by getting barometric observations from the summit while a survey scientist took periodic readings at the base.[7] Like Humboldt on Chimborazo, Muir explained the peak's vegetation in terms of climate zones, and just as Humboldt gleefully described a butterfly gracing Chimborazo's high flanks, Muir took great pleasure in noting that "a vigorous bumble-bee zigzagged around our heads, filling the air with a summery hay-field drone, as if wholly unconscious of the fact that the nearest honey flower was a mile beneath him."[8] The rest of Muir's account of this particular climb, however, takes the two carefree naturalists from the Alpine zone of vegetation to a realm that could only be called Arctic. Muir and Fay did in fact make it to the top, but before they could start down, a storm burst upon them, and within minutes the trail they had made on the ascent no longer existed: "the cliffs were covered with a remarkable net-work of hail rills that poured and rolled adown the gray and red lava slopes like cascades of rock-beaten water." Then the temperature dropped more than twenty degrees, "hail gave place to snow, and darkness came on like night."

When the same thing happened to Humboldt and his climbing party on Chimborazo, they were able to descend quickly enough to avoid the worst of the snowfall, but Muir and Fay, "flanked by steep ice slopes on one side, and by shattered precipices on the other," had no choice but to seek out a small "patch of volcanic climate" and lie down in some gravel and mud near a "group of hissing fumaroles," where the earth's internal heat came sputtering out in "scalding gas jets." Now Muir was recapitulating Reynolds's experience on the peak of Antuco, and Humboldt's on Tenerife, suffering "the pains of a Scandinavian hell, at once frozen and burned." Fay "was not in talking condition," but Muir desperately worked to maintain both his sense of hope and his sense of humor, looking ahead to the morrow (May Day!), holding onto hallucinations of "dry resiny pine logs suitable for camp fires," and staring up at the strangely distinct constellations in open parts of the sky. The two men lay in the same spot for seventeen hours without moving, and life seemed to them "a mere fire, that now smolders, now brightens, showing how easily it may be quenched." Having first made an impressive wilderness conquest, they were then penetrated by the brute forces of nature: "The night wind rushed in wild uproar across the shattered cliffs, piercing us through and through, and causing violent convulsive shivering, while those portions of our bodies in contact with the hot lava were being broiled." Yet they came out of it, in the late morning of May 1, not only alive but mostly unharmed. Muir's feet were actually terribly frostbitten, and they would hurt for the rest of his life, but at the time they hardly seemed worth mentioning.[9] Muir gloried in the "reserve of power after great exhaustion" that he thought was granted to all people willing to expose themselves to the elements: he and Fay had found themselves with "a kind of second life only available in emergencies like this."

The experience had been harrowing, but the rewards of self-exposure were immense. "How fresh and sunful and new-born our beautiful world appeared!" Muir cried, on reaching Strawberry Valley. He suspected, though, that he might be hard-pressed to keep extending these adventures in the gentle Sierras, that such mild latitudes might not be able to provide him with the requisite sublimity.

His tale of immersion in fire and ice on Shasta became an article in the fall of 1877. Less than two years later, Muir was scrambling up glaciers in Alaska. He had finally reached the actual Arctic zone.

Muir's trips to Alaska, like most of his wilderness excursions, were in part escapes from an unhealthy civilization to a pure, restorative nature, where all is as it should be. "Climb the mountains and get their good tidings," Muir would counsel in 1901. "Nature's peace will flow into you as the sunshine into the trees. The winds will blow their freshness into you, and the storms their energy, while cares will drop off like autumn leaves."[10] The Muir of 1901 was most interested in the peace, sunshine, and trees, for he had become nature's white knight, its publicist, its official appreciator. In his younger days, though, as suggested by the exhilaration he derived from the treeless, lightless, and far-from-peaceful regions of Mount Shasta, he embraced nature's winds and storms, and these were among the most salient features of the Alaskan environment. At times, his inclination to climb tall, slender trees during near-hurricanes, or explore cliff walls in the midst of earthquakes, or slide down glaciers after sunset, could come across as masculine posturing or even dangerous naïveté: nature, after all, does not care how much of an admirer you are, and will not hesitate to kill you. While scholars have tended to portray

Muir as embodying Zen submissiveness, as playing the mountain monk to Clarence King's proud conqueror, he may actually have done more rhetorical damage to nature than King ever did, because of his tendency to minimize the experience of real distress in the outdoors.[11] Yet his confrontations with environmental challenges seemed more honest in Alaska. The perilous, storm-filled summers of 1879, '80, and '81 produced some of Muir's richest, most complicated responses to the natural world.

Alaska inspired conflicts in Muir that Humboldt (and Reynolds, King, and Melville) would have recognized instantly. Nature, here, was fantastical, awe-inspiring, breathtakingly magnificent; it was also terrifying, confusing, over-whelming, fatal. In almost all of Muir's letters and journal entries during these three trips, ambivalence dominates. His very first week on a northbound ship led him to remark on the equal balance between "the trouble and the en-joyments of the voyage." Upon reaching Alaska, Muir wrote a newspaper arti-cle emphasizing how "tranquil" and unexpectedly "cloudless" the atmosphere was; it seemed to have the healing power of a "universal poultice." Yet he also described the days as being "intensely calm, gray and brooding in tone. . . . The air has an Indian-summerish haze along the horizon, and the same kind of brooding stillness." Clearly, it was a calm before the storm. "Indian sum-mer," after all, still had a potent double meaning in the nineteenth century. Taking on a particular resonance in the context of American frontier life, the term referred to the brief season just after the autumn harvest when the weather often warmed up again for a few days—and when, in colonial times, local Indians launched their final attacks of the year. In any case, a few days later, when Muir at last saw Alaska's glaciers stretching out above long, forested valleys, he couldn't decide exactly how he felt about them, or even how they compared to the ones he'd seen in California: "How strange seem these un-tamed solitudes of the wild free bosom of the Alaska woods. Nevertheless they are found necessarily and eternally familiar."[12] Muir was feeling his way, getting a little lost, playing with paradoxes—exploring.

He soon became enthralled by the dazzling ice fields and auroras, by the very complexity of the northern environment and his response to it. But, de-cades later, he would divide these three summers into two very different types of experiences: in 1879 and '80, he was essentially a wandering tourist, free to make his own plans, to switch from steamers to canoes to walking shoes whenever he pleased, to tackle glaciers on a whim; in 1881, though, he was at-tached to an official rescue mission, and he had specific responsibilities as the expedition's naturalist. At the very end of his life, then, when writing the book *Travels in Alaska* (the last thing he ever worked on), Muir decided to include only episodes from his first two trips (plus a few scenes from an 1890 excursion). His summer aboard the *Corwin* seemed to him too much like work, and its context too somber, to include in a book of northern romps.

Indeed, since *Travels in Alaska* was composed as yet another of his propaganda pieces on behalf of American wilderness areas, Muir even wound up cutting out the most intriguing aspects of his responses to the North in those first two summers.[13] As many students of Muir's writing have commented, he was his own worst editor.[14] Most often, the narrative of *Travels in Alaska* drowned out the specifics of landscape in simple exclamations of Muir's general love for nature: "I ran down the flowery slopes exhilarated, thanking God for the gift of this great day . . . , while every feature of the peak and its traveled boulders seemed to know what I had been about and the depth of my joy, as if they could read faces."[15] This particular Alaskan environment was transformed by Muir's voice and gaze into little more than a confirming echo and mirror of his exploits.[16]

Today, people who read only *Travels in Alaska* get a whitewashed version of Muir's perspective on the North. *The Cruise of the* Corwin, largely forgotten in Muir's oeuvre,[17] is a much darker book. Thankfully, Muir never edited his journal from the summer of 1881, nor did he retouch the articles he wrote that summer for the *Evening Bulletin*. So when his literary executor, William Frederic Badè, put these materials together in book form in 1917, three years after Muir's death, he did us, and Muir himself, a great service. It is possible that Muir would have objected to the ways in which parts of the tome seem to undermine his efforts, as an old man, to celebrate the natural world as utterly uplifting, but he could not have objected to the freshness with which it brings us his frank, middle-aged confrontation with an extreme environment.[18] His 1881 trip was perhaps the bleakest and most complicated of all his adventures in the Arctic regions, and the writing one finds in *The Cruise of the* Corwin is that much more surprising, more nuanced—more Humboldtian.

———

The *Thomas Corwin* sailed on May 4, 1881, but it was not until late June that her crew first got an extended look at the sun: "I have never before seen so cloudy a month," Muir wrote to the *Evening Bulletin*, "weather so strangely bewildering and depressing."[19] Pondering the fate of the *Jeannette*'s crew, Muir realized that the Arctic winds and storms he was coming to love could also bring death, that wilderness could bewilder as easily as it could inspire. There existed no accurate maps of these lonely places, and few clear trails in the ice and snow. This was a frontier environment, full of hidden icebergs and crevasses. Of course, the danger and novelty of these explorations held a deep appeal for Muir. He had missed out on the western frontier: by 1867, when he arrived in California, the state was already widely settled, the Yosemite Valley already preserved by earlier explorers and wilderness advocates like Clarence King.[20] In Alaska and Siberia, he experienced the thrill of the pioneer. No one else—no nonnatives, anyway—had ever studied these places. Yet it was no easy thing "to read the ice record,"[21] or even to survive in the "mysterious polar world."[22] Muir made it clear that the native peoples knew this land better than anyone else, but even they became vulnerable during lean years. One of the most gruesome scenes Muir witnessed in 1881 was on St. Lawrence Island, in the Bering Strait, where just one bad winter had killed about a thousand people—two-thirds of the isle's population. "In seven of the villages not a single soul was left alive," Muir wrote to the *Evening Bulletin*, and it seemed particularly tragic to see the "ghastly and desolate" ruins of decimated native settlements in the midst of such beautiful boreal scenery.[23]

Death may well have been haunting Muir that summer, for, in the months leading up to the trip, his health had been steadily failing.[24] At midlife, he seemed suddenly an old man: he was coughing his lungs out and having trouble eating, his characteristic ruggedness draining from his face with each passing week, as his weight dropped to just over one hundred pounds. Indeed, he almost decided not to go to Alaska on the *Corwin*, for even if he wound up having the strength for the trip, he was needed on his California fruit ranch, and his constant flights to the wilderness were beginning to weigh on his conscience. In 1879 he had left California immediately after agreeing to get married and stayed away for half a year; in 1880 he'd taken off just three months after the wedding, with his wife Louie pregnant and suffering terribly from morning sickness. Now he was a father, and forty-three years old: it was time to settle down. Muir chafed against societal expectations and financial obligations, but he loved his wife and daughter. Wishing to prove that he was man enough to do whatever was necessary for their well-being, Muir had labored through the fall and winter, and he was proud of his accomplishments—the rich grape harvest, the pruning and plowing and replanting. Ultimately, though, this life seemed to remind him too much of his strict childhood on a Wisconsin farm. Or maybe domesticity in any form would have felt constraining. In any case, Louie saw what was happening and convinced him that he needed a change. Her parents were there to help with little Annie Wanda and the summer's chores. She wanted him back for the fall harvest, and for a winter of getting to know their daughter, but she recognized that both his health and his ability to write depended on a certain amount of outdoor wandering. The glaciers beckoned.

Muir, too, knew that wilderness was always recuperative for him, in at least some respects, and *The Cruise of the* Corwin does acknowledge the power of the Alaskan environment to drive away the concerns of civilization. "All seems sure and true and righteous as to the trip I am making," he wrote to Louie from San Francisco, "and I feel sure that I shall return to you better than I go—better in every way."[25] In fact, the previous two summers had not fully prepared either Muir or his wife for the extent of the brooding constraint he would feel in 1881. But this very sensation led Muir to think about the limits of human endeavor and the relationship between humanity and nature with a new sense of urgency. The mission undertaken by the *Corwin* focused all of its crew on the question of survival in the Arctic, and everyone quickly realized that the expedition's success would depend on the expertise of the local indigenous peoples. While Muir and the other professionals might be able to navigate the coastlines and read the patterns in the glaciers, only the natives could take them deep enough into this icy wilderness to conduct a proper search—only they could handle the dogsleds—only they knew how to stay warm enough and hunt enough game. In *Travels in Alaska*, there is a chapter devoted entirely to Indians, and Muir shows hints of respect for them, but the natives appear mostly as exotic curiosities or elements of the environment, and Muir thinks of their best quality as being a willingness to convert to Christianity. In *The Cruise of the* Corwin, though, the natives actually bring nature to life, through their skills as guides, builders, makers of clothes, storytellers, educators. Ultimately, thanks to Muir's Arctic journey of 1881, it was the Aleuts and Inuits, the Tlingits and Chukchis, who taught him his most important lessons about living with nature.[26]

In most of Muir's later writings, the natives simply disappear. All people disappear, in fact, except for Muir himself and his ideal readers, men and women of acute sensibilities who are capable of appreciating the pure beauty of a theoretically untouched wilderness. The John Muir who has come down to us as a founder of modern environmentalism is simply a defender of nature. When the Sierra Club recently republished his books in a paperback series called The John Muir Library, it was attempting to bolster the preservationist movement, offering contemporary readers Muir's genteel, Victorian justification for the setting aside of large tracts of "virgin" land where, as the 1964 Wilderness Act states, "man is a visitor and does not remain."[27] The Muir of the Sierra Club rarely talked about *living in nature*, about remaking our overall relationship with the land; human beings, for him, were merely tourists in the wilderness. Pure nature, in other words, was defined as separate from all that was human.

In opposing nature and humanity—or culture, or civilization—Muir was tapping into fairly conventional turn-of-the-century sentiments about the oppressiveness of American urbanization and industrialization. Romanticism was alive and well, and Muir invoked its century-old tropes—just as the Sierra Club today invokes Muir's century-old tropes—in order to convince any potential patrons that wilderness held the key to human health and well-being. In American cities, he wrote, at the beginning of his celebration of *Our National Parks* (1901), are "thousands of tired, nerve-shaken, over-civilized people"; in parks, these same people can be found "jumping from rock to rock, feeling the life of them, learning the songs of them, panting in whole-souled exercise, and rejoicing in deep, long-drawn breaths of pure wildness."[28] Of course, the wildness, while representing everything that was not Our National Cities, was nevertheless thoroughly domesticated: "The snake danger is so slight it is hardly worth mentioning. Bears are a peaceable people. . . . As to Indians, most of them are dead or civilized."[29] South of Alaska, Muir seemed to embrace scientific racism. The effete, sophisticated urbanites to whom he addressed his writings could find something different in the woods, but they did not have to question their basic assumptions and prejudices there. Muir was not interested in changing the nature of cities, or criticizing factory farms, or abolishing social hierarchies—mostly because he believed that human beings were fundamentally sinful and that society would inevitably tend toward "gross heathenism."[30] Instead, he simply offered a redemptive escape to small, primitivist enclaves, where there were no city mobs, no savages, where you could renew yourself in the bosom of your family: "going to the mountains is like going home."[31]

The stated goal of the Sierra Club was "to enlist the support and cooperation of the people and government in preserving the forests and other natural features of the Sierra Nevada mountains."[32] So Muir was after converts, and almost all of his post-1892 writings tended toward propaganda.[33] Because he wanted his readers to appreciate nature enough to preserve it, he depicted it in purely positive terms. The complex realism of his earlier Arctic articles no longer served his purposes. Now, dark, life-threatening experiences suddenly became epiphanies; foreboding landscapes became calmingly beautiful. Climbing Mount Ritter for the first time, Muir "was suddenly brought to a dead stop, with arms outspread, clinging close to the face of the rock, unable to move hand or foot either up or down. My doom appeared fixed. I *must* fall." But this description occurs in *The Mountains of California*, Muir's first full-length book, which officially launched his career as a wilderness publicist. It appeared in 1894, just two years after Muir founded the Sierra Club (and twenty years after he scaled Mount Ritter), so we know everything will turn out for the best. "I became nerve-shaken for the first time since setting foot on the mountains, and my mind seemed to fill with a stifling smoke. But this terrible eclipse lasted only a moment, when life blazed forth again with preternatural clearness. . . . Had I been borne aloft upon wings, my deliverance could not have been more complete." Moreover, the "yawning chasms and gullies" and the "savagely hacked and torn" mountains suddenly appeared "truly glorious" and "noble": once you know how to look at them, "their far-reaching harmonies become manifest." To Clarence King, this would have seemed a denial of experience and of nature's devastating power, an unwillingness to admit the blackness of the abyss. To Muir, though, even "the darkest scriptures of the mountains are illumined with bright passages of love that never fail to make themselves felt."[34]

Muir's religious language is not unintentional. "How glorious a conversion," he wrote, describing his feelings soon after he first arrived in the Sierras,[35] and his works abound with conversionary episodes. He had what William James would call the "will to believe": he insisted on seeing only the good, the confirming, the invigorating, in nature.[36] Ever pragmatic, Muir figured that if enough of his readers also had such experiences in the Sierras, then they might help him save the wilderness spots that he deemed most sacred. It was this attitude, too, that made him particularly scornful of the infidel John Ruskin, who filled book 5 of *Modern Painters* with what Muir abbreviated as "Mtn gloom and mtn evil and mtn devil & the unwholesomeness of mtn beauty."[37] Clearly, John Muir and Clarence King had much to argue about.

In contrasting nature and humanity and favoring nature, Muir did mount a moderate critique of Victorian civilization, and in pressing for wilderness preservation he certainly attacked his society's tendency toward thoughtless development and the overexploitation of natural resources. Americans might not have preserved any wilderness at all if they hadn't perceived it as glorious and comforting. In this sense, Muir truly did inaugurate a key element of modern environmentalism, by providing Americans with a convincing rationale for making sure that at least some parcels of land would always remain undeveloped—comfortable homes for animals and plants as well as fonts of inspiration for human beings.

A much more radical legacy than preservationism, though, is latent in Muir's midlife devotion to Humboldt. In his later books he could be hopelessly anthropocentric—as when he envisioned the landscape nodding back at him in approval of his exhilaration—and even utilitarian (nature is made to ease your cares). But when he walked through the southeastern United States in 1867 and, especially, when he explored the Arctic between 1879 and 1881, his approach was significantly more ecological. Indeed, given Muir's genuine respect for nonhuman entities and intact ecosystems, some scholars and activists have claimed him as a founding father of the so-called Deep Ecology movement.[38] Defending alligators in the journal he kept during his thousand-mile walk to the gulf, Muir railed against human prejudices: "How narrow we selfish, conceited creatures are in our sympathies!"[39] He even defended rattlesnakes and poison oak and mosquitoes as being good in and of themselves, because they were part of the environment as a whole, "however noxious and insignificant" they might seem to us.[40] "Nature's object in making animals and plants might possibly be first of all the happiness of each one of them, not the creation of all for the happiness of one," he wrote. "Why should man value himself as more than a small part of the one great unit of creation?" Further, in a direct echo of his mentor, Muir asserted that nature's defining characteristic was its combination of "essential unity with boundless variety," and that "the cosmos . . . would be incomplete without man; but it would also be incomplete without the smallest transmicroscopic creature that dwells beyond our conceitful eyes and knowledge."[41] Here is a much more substantive challenge to the spirit of his age. In constantly decrying human blindness and self-centeredness, Muir was directly attacking the complacent social Darwinist view that white Americans were the fittest of earth's creatures, the ones whom all the rest were meant to serve. Indeed, Muir felt that the "dogma taught by the present civilization" prevented people from achieving "a right understanding of the relations which culture sustains to wildness."[42]

In 1881, Muir attempted to relate to the wild by following Humboldt's lead in studying meteorology, climatology, glaciology, and botany. He tied together weather patterns and animal behavior, envisioned the "colossal ice-flood grinding" of Agassiz's glacial era, and gleefully compiled unexpectedly long lists, organized climatologically, of plant species that he found blooming in the boreal wastes. Plants, he now believed, were far better colonizers than human beings: "many species show but little climatic repression"—just as Humboldt had suggested—"and during the long summer days grow tall enough to wave in the wind, and unfold flowers in as rich profusion and as highly colored as may be found in regions lying a thousand miles further south."[43] And, following in the footsteps of J. N. Reynolds, Muir had a new species—the daisy-like *Erigeron muirii*—named after him by the Harvard botanist Asa Gray, to whom he sent his *Corwin* specimens. (There is also a Muir Glacier in Alaska, not to mention Mount Muir, the Muir Inlet, and Muir

Point.)[44] Muir's final reports on the geography of boreal plants and "The Glaciation of the Arctic and Sub-Arctic Regions" were the two closest things to scientific monographs he would ever publish. But, as the *Corwin* expedition bounced back and forth across the Bering Strait, as Muir puzzled out the ecological relationships between frozen water and the earth's inner warmth, between the "dense, spongy plush of lichens" and the "fissured, frost-cracked limestone" below the soil, his most important field of Humboldtian research became anthropology.[45] If you study the Humboldt volumes that survive today in Muir's personal library, if you consider his pencil underlinings, his exclamation points, his quirky, makeshift indexes, you'll find that, next to passages about trees and plants and natural scenery in general, Muir spent the most time on Humboldt's musings about Indians.[46]

The close engagement Muir developed with native peoples in Alaska and Siberia, during the *Corwin*'s search, led him eventually to make powerful connections between social injustices and the domination of nature. Moreover, his awareness of ecological limitations in these extreme environments made him question American hubris to an even greater extent than in his musings about the inherent value of nonhuman species: in the North, he challenged not only his compatriots' general arrogance but their specific assumptions about expansion and empire. And in examining native cultures, Muir made observations about the potential for human beings to live in harmony with nature, to adapt to the land, even in the houses and villages they construct, in the jobs they do, in their normal, day-to-day lives. Here, wilderness represented not just an edenic escape, a separate peace, a temporary relief from the pressures of society; to some, Muir realized, wilderness could be a permanent home.

———

The comparison between foreign and home cultures is a classic trope of travel writing, and late-nineteenth-century nature writers like Muir—and Clarence King and George Melville—easily adapted it to celebrate the Wilderness they were visiting over the Civilization from which they came. Muir could have been reading from one of King's essays as he wrote in his journal about feeling his "hard, money-gaining, material thoughts loosen and sink off and out of sight," as his ship sailed "through the infinite beauty enchanted."[47] Even classic tropes became extremely complicated in *The Cruise of the* Corwin, though. All of Muir's usual criticisms of human society do surface here—he scorns the effete quality of citified life, the dishonesty of white sailors and traders, their narrowness and wastefulness and violence—but he also critiques certain elements of civilization that seem almost above reproach in

39   Sketches of natives, 1881. John Muir.

most Victorian-era writings penned by enlightened, liberal intellectuals. Muir himself was on the *Corwin* expedition as a scientist, yet some of his most cynical comments are reserved for the futility of science in the face of this inscrutable landscape. "Knowledge concerning this mysterious country," Muir noted, came only in "fragments," and those were "of so vague and foggy a character"—just like the weather—as to be useless. Civilized science did not bring light to these impenetrable regions but rather proved itself in every way inferior to the knowledge and practices developed by the local "savages." "Seamen with charts and compass" were helpless in a thick fog, while the uneducated natives never seemed to have any trouble navigating. Time after time, Muir juxtaposed not just civilization and wilderness but civilization and the people who *inhabited* the wilderness. Because there were figures in this landscape, and because this type of wilderness was not classically picturesque and inspiring, but confusing and dreary, Muir's boomeranging criticisms of his society took on an added layer of complexity and sophistication.[48]

Particularly appalling to Muir was his society's narrowness of vision. Like Humboldt in South America, Muir felt that he had discovered a new world in the Arctic, one that required new ways of seeing and understanding nature. But his fellow travelers and scientists could not break free from their traditional assumptions and prejudices. "In trying to account for the observed similarity," wrote Muir, exasperated, "between the peoples of the opposite shores of Asia and America, and the faunas and floras, scientists have long been combating a difficulty that does not exist save in their own minds." They knew that the most sophisticated white navigators had trouble making the crossing between Siberia and Alaska in steamships, so they simply assumed that the lowly natives could never accomplish such a feat in their flimsy canoes. Indeed, it was primarily this assumption, and not any geological evidence, that had led theorists to speculate about the land bridge across the Bering Strait. But Muir knew the simple truth. "As to-day, so from time immemorial canoes have crossed for trade or mere pleasure, steering by the swell of the sea when out of sight of land." Again and again, in both his journals and his newspaper articles, Muir described his ship getting lost, the navigator's charts and calculations never adequate to the task of getting from one safe harbor to the next. Muir's solution was to throw away the map and practice a kind of empathy, to put himself in the fur-lined moccasins of the locals and do as they did, which in this case was to study birds. The gulls, auks, and murres that lived on the rocky coasts and forayed out into the ocean for food would always lead you right where you needed to go. "To persons acquainted with their habits it is not difficult to determine whether their flight is directed homewards or away from home." The key was to open your mind, to be alive to new possibilities, to overcome the apathy and complacency of civilization. As an explorer-scientist, Muir might have celebrated the accomplishments that resulted from the great tradition of Western curiosity, but instead he recognized that the northern natives far surpassed his own people in engagement with the world. "Though savage and sensual, they are by no means dull or apathetic like the sensual savages of civilization . . . , for these Eskimos, without newspapers or telegraphs, know all that is going on within hundreds of miles";[49] indeed, they possessed "more sound sense and natural reason than are found among the so-called enlightened and religious of our own race."[50]

American narrowness and disrespect for native peoples also translated into a particularly brutal form of imperialism, Muir realized. The Eskimos "probably were better off before they were possessed of a single civilized blessing—so many are the evils accompanying them!" In the North, the exploitation of nature went hand in hand with the exploitation of the natives. Muir discovered at St. Paul, in the Fur Seal Islands, that a law barring all non-natives from the pursuit of certain furs had "induced some fifteen white men to marry Indians for the privilege of taking sea-otter." Meanwhile, the Western Fur and Trading Company and Alaska Commercial Company hired natives "as butchers, to kill and flay the hundred thousand seals that they take annually here and at the neighboring island of St. George." The Aleut hunters received forty cents per skin, each of which was worth about fifteen dollars in the London market, where they were all sent. Of course, if the seal and sea otter populations started to dwindle because of the companies' industrial-style approach, it was the local natives who would suffer.[51]

White traders also brought the natives repeating rifles, Muir noted, "which tempt them to destroy large amounts of game which they do not need. The reindeer has in this manner been well-nigh exterminated within the last few years." The Indians themselves were hardly blameless, but Muir thought that the introduction of an inappropriate technology was at the root of this disaster. After all, the natives were merely proceeding according to the dictates of their religion: they felt obliged "to kill every animal that comes within reach, without a thought of future scarcity, fearing, as some say, that, should they refuse to kill as opportunity offers . . . , then the deer-spirit would be offended at the refusal of his gifts and would not send any deer when they are in want." In the end, Muir believed that many of the natives were "better behaved than white men, not half so greedy, shameless, or dishonest." Those white invaders who gave the natives whiskey and rum, he thought, were guilty of murder. And "even where alcohol is left out of the count," Muir wrote to the *Evening Bulletin*, "the few articles of food, clothing, guns, etc., furnished by the traders, exert a degrading influence, making them less self-reliant." Muir was sure that "unless some aid be extended by our government which claims these people, in a few years at most every soul of them will have vanished from the face of the earth."[52]

Of course, Muir here was buying into the classic American myth of "the disappearing Indian."[53] Especially since he rarely wrote about Indians again after *The Cruise of the* Corwin, he can be seen as depicting the Eskimos as already on the point of vanishing simply so as to avoid accountability with regard to their fate. For all his admiration of the northern natives, he never did propose much in the way of policies for protecting them against white encroachment. Yet his suggestion of the possibility of government assistance does at least recognize a kind of official responsibility. Much worse off in his eyes than the Siberian Chukchis, the American Eskimos were the latest in a long line of imperial subjects used and discarded by agents of U.S. industry and government in the ongoing conquest of this continent. American society was not just dull and worn out, Muir argued, but systematically oppressive and destructive. So much for the home of the brave.

While the Indians maintained a fairly mutualistic relationship with the land, the whites came to the North only temporarily, and then only to kill, for profit and pleasure. Natives used walruses for food, clothing, and canoe construction. When American schooners reached the northern waters, though, "these magnificent animals are killed oftentimes for their tusks alone, like buffaloes for their tongues, ostriches for their feathers, or for mere sport and exercise. In nothing does man, with his grand notions of heaven and charity, show forth his innate, low-bred, wild animalism more clearly than in his treatment of his brother beasts. From the shepherd with his lambs to the red-handed hunter, it is the same; no recognition of rights—only murder." Invoking animal rights, Muir again seemed to embrace biocentrism, but his northern biocentrism had a social edge, because he was criticizing *only* his fellow whites. In a powerful reversal of cultural imagery, redness, the classic indicator of Indians' racial difference and inferiority, here reverts to its biblical symbolism, marking the white hunters as guilty, bloody savages. The native hunters actually seemed admirable to Muir. Their distinctly nonindustrial approach arose from a deep respect for the natural systems on which they depended. "The Eskimos hunt and kill [polar bears] for food, going out to meet them on the ice with spears and dogs." But when the crew of the *Corwin* went after "the master existences of these frozen regions," "the whole affair was as safe and easy a butchery as shooting cows in a barnyard from the roof of the barn. . . . Of all the animals man is at once the worst and the best." Overall, to Muir, the Eskimos were "a mixed, jolly multitude" who easily "bid defiance to the most extreme cold."54

Muir did not even stop here, though, at Rousseau's idealized vision of Indians in a timeless state of noble savagery.55 He very carefully observed how American culture was affecting Eskimo culture, and though he sometimes despaired of the future, he occasionally came out with a defiant defense of the fullness and resiliency of the civilization built by the northern natives—a kind of defense almost unheard-of in conventional Alaska travel accounts.56 They had manufactures, trade, education, arts, religion, ingenuity—and they did not disrupt their environment. Perhaps, in the end, they would resist white encroachment and prove that they had created the more enduring culture:57

The extent of the dealings of these people, usually regarded as savages, is truly surprising. And that they can keep warm and make a living on this bleak, fog-smothered, storm-beaten rock, and have time to beget, feed, and train children, and give them a good Eskimo education; that they teach them to shoot the bow, to make and throw the bird spears, to make and use those marvelous kayaks, to kill seals, bears, and walrus, to hunt the whale, capture the different kinds of fishes, manufacture different sorts of leather, dress skins and make them into clothing, besides teaching them to carry on trade, to make fire by rubbing two pieces of wood together, and to build the strange houses—that they can do all this, and still have time to be sociable, to dance, sing, gossip, and discuss ghosts, spirits, and all the nerve-racking marvels of the shaman world, shows how truly wild, and brave, and capable a people these island Eskimos are.

———

Muir especially appreciated the natives' ability to come to grips with such a challenging landscape. In the Sierras, in our national parks, people knew instinctively what to do, knew why they had come: the beauty was usually manifest, the dangers scarce. But the Arctic posed direct threats. Muir's constant refrains, in *The Cruise of the* Corwin, are invocations of the unpredictability of the ice and the blinding power of the weather. On a typical day, Muir awoke to "a cold, bleak, stormy morning, with a close, sweeping fall of snow, that encumbered the deck and ropes and nearly blinded one when compelled to look windward." The crew were frequently "baffled by darkness and close-packed ice." Every few days, Muir described having to turn back and change plans, and mentioned the ever-present possibility that the ship would get stuck in the Arctic ice pack and be immobilized for the entire winter. His readers thought of ice as solid and stationary, but Muir brought it alive for them: "the jagged, tumbling ice blocks that formed the edge of the pack" quite often "struck the ship and made her tremble in every joint."[58] Yet Muir also found inspiration here: this, too, was nature. Thanks in part to the example of the native inhabitants, both human and nonhuman, who persisted here with a down-to-earth sense of cheer and calm, Muir was able to make a full, realistic acknowledgment of the dark side of this wild world, and then look further and see its beauty as well. Because he was not simply trying to convince his readers to come visit an ideal wilderness paradise, as in his later works, he left us, in *The Cruise of the* Corwin, a much more balanced portrait of nature. Walking near Cape Thompson, he took pains to point out contrasts in the land. "Where hills of this rock have steep slopes," he explained, ". . . they bleach white and present a remarkably desolate aspect in the distance. . . . These barren slopes, however, alternate with remarkably fertile valleys, where flowers of fifty or more species bloom in rich profusion, making masses of white, purple, and blue."[59]

When Muir stressed the beauty of the Arctic regions in his letters back to the *Evening Bulletin*, or, for that matter, the appealing qualities of the natives, his main point seemed to be to upset the prejudices of his readers. The actual subjects of such passages almost don't seem to matter; like J. N. Reynolds, Muir sought primarily to make people think about nature and culture (their own and others') in new ways. Watching an infant girl sleep soundly through a snowstorm, and perhaps thinking of his own newborn daughter, Muir exclaimed: "No happier baby could be found in warm parlors, where loving attendants anticipate every want and the looms of the world afford their best in the way of soft fabrics." Repeatedly, he sought to adjust his own and his readers' notions of home: "When the dogs got upon the ice, their native heath, they rolled and raced about in exuberant sport. The rough pack was home sweet home to them, though a more forbidding combination of sky, rough water, ice, and driving snow could hardly be imagined by the sunny south." Approaching a village on St. Lawrence Island, Muir noted that "the storm-beaten row of huts seemed inexpressibly dreary through the drift. Nevertheless, out of them came a crowd of jolly, well-fed people."[60] Though he never explicitly suggested that his readers move to the North and follow the ways of the Indians, the very structure of his observations reveals that he was attempting to rethink how human beings fit into their environments, how one might live happily even outside of temperate climates and without the amenities of modern civilization. The northern frontier represented an opportunity to achieve a full-fledged, honest communion with nature.

On balance, though, Muir's vision of the North, as expressed in *The Cruise of the* Corwin, is a fairly dreary one. His trip seems to have been a chastening experience; though the *Corwin* remained intact, the search for the missing explorers and whalemen yielded no survivors. Every now and then, the theme of the crew's morbid task reentered Muir's newspaper articles, darkening them considerably, as when he described some of the officers "erecting a cairn" on Wrangell Land. And though the message they left under the rocks said "All well on board," Muir wrote home about the bleakness that defined their experience on shore, about the threats posed by this "severely solitary" land, about his discovery of a "much scoured and abraded" fragment of a boat mast.[61]

Here in the harsh, forbidding North, the assumptions of white men from southern climes simply didn't apply. Even classically picturesque events became distorted. In his journal, Muir noted "a weird red sunset; land miraged into most grotesque forms." Extremes of light and dark sent his senses spinning: "The general effect of this confusing interblending of the hours of day and night, of the quick succession of howling gales that we encountered, and of dull black clouds dragging their ragged, drooping edges over the waves, was very depressing, and when, at length, we found ourselves free beneath a broad, high sky full of exhilarating light, we seemed to have emerged from some gloomy, icy cave. How garish and blinding the light seemed to us then!"[62] Muir often wrote in the language of seeming and appearances, because he was never sure what was happening. And the long traditions of Western art and science, of keen observation and understanding, did not offer much help. Muir wrote to his wife that he had "tried an hour ago to make a sketch of the mountains along the shore for you, to be sent with this letter, but my fingers got too cold to hold the pencil, and the snow filled my eyes, and so dimmed the outlines of the rocks that I could not trace them."[63] No language, no skill, no human artifice, could make sense of the terrain; interpretation itself, our most basic adaptation, was impossible. We are left with only blank pages, reflecting the blankness of the snow and ice. This land simply would not be mastered.

Compare Jack London's popular Arctic works, which were almost contemporaneous with Muir's: here, the dark, forbidding aspects of the North became explicit challenges to "exploit" the land—this was London's own word, repeated in several different tales of adventure. The call of the wild brought men to conquer the elements in search of gold, oil, fish, and wood.[64] Muir's boreal bleakness also posed a challenge, but it was the challenge of living with a respect for the ways of nature and a recognition of the limits to human endeavor and expansion.

———

Clarence King and George Melville were more outspoken than Muir about the hubris of American imperialism, but Muir ultimately offered a fuller picture of how people could eschew such hubris and live in a way that acknowledged environmental restrictions—because he offered a fuller picture of native cultures. And perhaps the most compelling thing about northern natives, for Muir, was their humility. He suspected that they did not find the Arctic environment so overwhelming in part because they simply never *expected* to master it. The white, Western, scientific, imperialist tradition dictated that explorers and pioneers adapt the land to their needs. In the North, Muir recognized an alternative tradition: the natives followed the example of Muir's beloved animals and plants, and adapted themselves to the land. If moss and lichens could survive "upon a stratum of solid ice," then so could people.[65] At the same time, Muir did not simply romanticize the natives as "ecological

Indians": they are not two-dimensional heroes who fade into nature but rather complex individuals seeking out specific, sophisticated ways of utilizing natural resources.[66]

While daily life in most parts of America was already at least one step removed from nature,[67] all of native culture in the North depended directly and visibly on elements of local ecosystems. The Indians knew, for example, that the hair of the reindeer was "very heavy, with fine wool at the bottom, thus making a warm covering sufficient to enable the animal to resist the keenest frosts of the Arctic winter without any shelter beyond the lee side of a rock or a hill." So, in Siberia, the Chukchis became herders, and now the reindeer "supply every want of their owners . . . —food, warm clothing, coverings for their tents, bedding, rapid transportation, and, to some extent, fuel"—"thus rendering these bleak and intensely cold regions inhabitable." Indeed, all the Arctic animals and plants seemed to play a role in native adaptation. Muir found the typical Siberian hut, "each having a private polog [nest] of deerskins," to be "the snuggest storm nest imaginable." From the outside, it seemed a "cold, squalid shell," but once within you discovered the "furry sanctum." Each had a few "luxurious bedrooms, whose sides, ceiling, and floor were made of fur; they were lighted by means of a pan of whale-oil with a bit of moss for a wick." Muir thought the Eskimos on the American side "show good taste and ingenuity in the manufacture of pipes, weapons, knickknacks of a domestic kind, utensils, ornaments, boats, etc.," simply because they used local, natural materials. The Eskimos always looked warm in their "superb fur clothes"; Muir could not understand why some would choose to dress themselves in "shabby foreign articles." Perhaps most important, the natives "make everything durable." Most American manufactures seemed to Muir "flimsy and useless" when compared to the Chukchis' reindeer trousers, or sealskin shoes, or their parkas "made of the breast skins of ducks," or one of their sacks, which, "made from the intestine of the sea-lion, while exceedingly light, is waterproof."[68]

Muir was particularly impressed by the natives' skill in making and handling boats and sleds. Perhaps because he himself was traveling mostly on a ship, whose primary mission was to locate three other ships, all of which had probably been crushed in the Arctic ice pack, Muir found himself making constant comparisons between the whites' and the natives' experiences on the water and ice. The differences went far beyond the natives' superior navigation techniques. Part of the edge the locals had, beyond knowledge of the ecosystem, was simply their attitude. They often crossed the Bering Strait in canoes for "mere pleasure," and they took to their sleds, no matter how inaccessible and jagged the ice pack seemed, with "everyday commonplace confidence," always going "merrily on, up one side of a tilted block or slab and down the other with a sudden pitch and plunge, swishing round sideways on squinted cakes, and through pools of water and sludge in blue, craggy hollows, on and on, this way and that, with never a halt."[69]

The white sailors and explorers, in contrast, were usually on tight schedules, and they tended to cruise not for pleasure but for profit or glory. Bent on a particular goal, they often ignored signs of danger. While the natives slipped and slid along the contours of the land and ice, the whites just barreled straight ahead. The whalers were already wreaking environmental havoc.[70] The region above the Bering Strait was a "comparatively new hunting ground," as Muir pointed out. "Nevertheless it is being rapidly exhausted." To Muir, the whole operation seemed foolhardy: "It is not long since the first whale-ship passed Bering Strait, and yet no less than forty-seven have been crushed hereabouts, or pushed ashore, or embayed and swept northward to nobody knows where." Most often, the pack simply "closes upon them and crushes them as between huge crunching jaws." The native sailors, of course, rarely lost even their largest vessels. As for Muir's fellow white sailors, the only time they really earned his respect was when the ship's rudder broke, and the crew had to forgo their previous plans and adapt to a brand-new situation, acting like the natives in that they could use only locally available materials for their repairs. Fortunately, the men responded bravely and stoically: "The whole was brought into complete working order in a few hours, nearly everybody rendering service, notwithstanding the blinding storm and peril, as if jury-rudder making under just these circumstances were an everyday employment." Muir's most heartfelt admiration, though, was reserved for the natives' ability to build canoes that were perfectly adapted to their environment:[71]

> The gale has been loosening and driving out past the vessel, without doing us any harm, large masses of the ice. . . . One large piece drifted close past the steamer and immediately in front of a large skin canoe capable of carrying thirty men. The canoe, which was tied to the stern of the ship, we thought was doomed to be carried away. . . . Greatly to our surprise, however, when the berg, rough and craggy, ten or twelve feet high, struck her bow, she climbed up over the top of it, and, dipping on the other side, glided down with a graceful, launching swoop into the water, like a living thing, wholly uninjured. The sealskin buffer, fixed in front and inflated like a bladder, no doubt greatly facilitated her rise. She was tied by a line of walrus hide.

Muir often expressed his regard for the natives' adaptability by comparing them and the things they made to local animal species. Although Muir was not free of the racism that so dominated his culture, in this case he was specifically overturning assumptions about the natives' "savagery," since he considered animals more humane than most civilized humans. His descriptions of natives as natural were parallels to his descriptions of elements of nature as embodying certain praiseworthy human qualities. An Arctic seal, for instance, with "humanlike eyes," seemed "most wonderful" to Muir, in that "it could live happily enough to grow fat and keep full of warm red blood with water at 32 degrees F. for its pasture field, and wet sludge for its bed." A "remarkable Chukchi orator," then, might be described as having "wolfish eyes" not to suggest his wild viciousness but rather his vigor in attacking particular topics: he was perfectly suited to his cultural role. Muir associated this man even with the power of water: "he poured forth his noisy eloquence late and early, like a perennial mountain spring," and, when interrupted, "he quickly recovered and got underway again, like a wave withdrawing on a shelving shore, only to advance and break again with gathered force." Chukchis in their canoes glided by like "gulls, auks, eider ducks"; the King Island natives built huts on steep slopes that seemed "accessible only to murres" and other seabirds. One Chukchi settlement was so well integrated into "the face of a hill that the entire village makes scarcely more show at a distance of a few hundred yards than a group of marmot burrows."[72] To Muir, the natives of the North, in their conscientious merging with nature, seemed far more fit for survival than the invaders from the South. Perhaps Americans, back home in the temperate zone, might take a lesson from their northern cousins as they made decisions about land use; perhaps they could become more adaptable.

———

One of Muir's most important personal discoveries during the *Corwin* expedition was his recognition of a sense of kinship between white Americans and Indians. He went out of his way, in his earliest letters and journal entries, to humanize the peoples on both sides of the Bering Strait. Casting his gaze at some American Eskimos, he saw "a response in their eyes which made you feel that they are your very brothers." In Siberia, witnessing the separation of a woman and her husband, who was joining the *Corwin* expedition, Muir noted: "One touch of nature makes all the world kin, and here were many touches among the wild Chukchis."[73] This perspective, fairly progressive for 1881, is remarkably similar to the one expressed in the ethnographic essays Franz Boas wrote after his sojourn with the Eskimos on Baffin Island in 1883–84.[74] "I had seen," Boas summed up, ". . . that the Eskimo is a man as

40 Sketch of a native village, Diomede Island, 1881. John Muir.

we are; that his feelings, his virtues, and his shortcomings are based in human nature, like ours." For Boas, these observations led directly to the conviction that "the idea of a 'cultured' individual is merely relative," and, indeed, working out of New York's Columbia University, Boas became the twentieth century's most important theorist of cross-cultural understanding. Or perhaps one ought to call his doctrine a kind of "cosmopolitanism," since he often claimed that much of his intellectual inspiration derived from his fellow Prussian, Alexander von Humboldt.[75]

Muir was not quite so radical. He still at times fell back on the basic assimilationist program of his contemporaries: he argued vehemently that the American government owed the Eskimos something, but this something sometimes turned out to be "civilization."[76] While he defended the northern natives' culture, he never offered a real vision of pluralism. His relationship to Indians, then, was mostly appropriative: he admired their deep connection to nature and wanted to bring that home to his own people as a model. Still, this in itself entailed an important step forward in American attitudes toward natives, during an era dominated by raciology, polygenetic theory, assimilation campaigns, and, indeed, outright warfare against Indians. And Muir's attempt at reshaping American interactions with the land, at integrating nature and culture, at bringing wilderness home to everyday life, was revolutionary. Before diving into a less radical preservationism, Muir—like Humboldt, Reynolds, King, and Melville—realistically acknowledged that every human activity required the use of natural resources, and then argued for use that was careful, efficient, and perhaps even beautiful, based on the kind of environmental understanding frequently possessed by native peoples. "I often wonder what man will do with the mountains," Muir once wrote in his journal, "that is, with their utilizable, destructible garments. . . . Will human destructions, like those of Nature—fire and flood and earthquake—work out a higher good, a finer beauty? Will a better civilization come in accord with obvious nature, and all this wild beauty be set to human poetry and song? Another universal outpouring of lava, or the coming of a glacial period could scarce wipe out the flowers and shrubs more effectually than do the sheep. And what then is coming? What is the human part of the mountains' destiny?"77

Unfortunately, Muir rarely followed up on his initial Humboldtian efforts in the Arctic. After 1881, he hardly ever wrote about Indians again. Indeed, in later years, when editing his earlier works for publication, he even expunged passages about natives.78 Muir did not lose his awareness of Indians, though, or simply dismiss all of them as racially inferior. Some of the volumes in his Humboldt collection are editions from the first decade of the twentieth century—he seems to have bought several new tomes shortly before his own trip to South America in 1911—and Muir's markings in them clearly reveal a continuing fascination both with indigenous cultures and with the injustices suffered by natives throughout the Americas. A simple quantitative analysis of the notes he took in volume 2 of the 1907 edition of the *Personal Narrative*, for instance, suggests that he paid more attention to Humboldt's valorization of Indians than to any other subject in the book (descriptions of trees were a distant second).79 But neither South American Indians nor those in Alaska were competing with the Sierra Club for pieces of California: they did not insinuate themselves into local parks and spoil the pristine quality of Muir's favorite wilderness scenes. When it came to California Indians, Muir either tried to ignore them or wound up referring to them as "dirt-specks in the landscape."80 Both personal resentment and political expediency, then, led him to push Indians out of the picture, to shape his philosophies, after 1881, in ways that would cater to broad cultural assumptions about native inferiority. If his message was going to emphasize humanity as the great disturber of nature's harmonies, then he couldn't afford to draw fine distinctions about different human ways of acting on nature; the safest approach would be to keep people away from the wilderness, except when he or someone he trusted was leading them on hiking trips. His ultimate vision of America as a nation would be defined by nature rather than natives.

Almost as soon as Muir returned from his voyage aboard the *Corwin*, he threw himself into the type of preservationist battles that would define the rest of his career in the nascent environmental movement. With a few friends from the California Academy of Sciences, he spent the late fall of 1881—after the grape harvest—drawing up two bills for submission to Congress: one that would enlarge Yosemite State Park, and another that would set aside new preserves in the southern Sierras.[81] Muir had now entered a new phase of life: unlike most explorers, he actually did settle down. His family, and the affairs of the fruit ranch, took almost all of his time and energy. He essentially stopped writing in the 1880s. Nor would there be any more trips northward to the frontier until the end of the decade; his only wilderness rambles would be brief visits to nearby peaks, in the Sierras or the Coast Range.[82] When he thought about nature, in other words, he thought about escaping his routines, his responsibilities, and being a tourist in an exhilarating, but comforting, parklike setting. Civilization and wilderness had fallen back into their separate categories.

Perhaps, as Muir got older and gave more of his time to the task of supporting his family, he felt even more resentment toward the burdens of civilization—much as he loved Louie and Annie (whom he always called by her middle name, Wanda) and Helen, his second daughter, born in January 1886. Perhaps he also grew to accept the shape of white American civilization as inevitable. He seemed to retreat to the position that, if we were going to have such an effete, meaningless, money-driven culture in our cities and even our rural areas, we ought at least to preserve a few parcels of slightly more wild land. In any case, after the two 1881 bills failed to get out of committee, and after Muir had put in a solid seven years of labor directed solely toward achieving financial stability, when he finally got back to writing, he clearly had lost his edge. His contributions to *Picturesque California*, a volume he edited in 1888–89, read like brochures advertising West Coast scenery to prospective settlers. Meanwhile, his life as a fruit rancher was leaving him increasingly bitter and conservative, and unable to think creatively about new ways of relating to people or the land. He was ashamed to be manipulating the environment purely for the sake of making money, but not ashamed enough to stop exploiting his day laborers, whom he referred to in an 1889 letter as a "horde of oriental heathen."[83] By the time Muir founded the Sierra Club in 1892, turning full-time to political activism on behalf of wilderness areas, he no longer seemed to be thinking about unity, or diversity, or the concept of cosmos; almost all the radicalism had seeped out of his agenda. As a result, his main legacy became tied to the assumption that the only positive relationship a human being could have with the natural world was as a visitor. Muir's rich social and ecological visions had been reduced to "the cause of saving samples of God's best mountain handiwork."[84]

John Muir's near-religious philosophy, based on the preservation of natural beauty, and George Melville's instrumental philosophy, based on conservation, efficiency, and the recognition of natural limits, represented the two withering branches of Humboldtian thought, and they would knock against each other repeatedly in the windstorm of twentieth-century industrial development. In the first major clash of these two approaches, over the Hetch Hetchy dam, between 1906 and 1913, conservationism won a clear victory: it seemed more obviously compatible with economic growth and the general ethos of capitalism. Yet preservationism would become the dominant force in U.S. environmental politics for most of the rest of the century. Muir and the Sierra Club had lost Hetch Hetchy Valley, but they had gained a national following and a reputation for hard-nosed political maneuvering. When Muir died in 1914, he was essentially canonized, and his love of wilderness inspired countless new activists. His friend C. Hart Merriam, the former chief of the U.S. Biological Survey, started a memorial essay in Muir's honor by asserting that the founder of the Sierra Club "was doubtless more widely known and more generally loved than any other Californian."[85] Certainly, the 1916 passage of the National Park Service Act provided fairly immediate evidence of Muir's impact.

Merriam was also close enough to Muir, though, and enough of a Humboldtian himself, to recognize that Muir had sacrificed an important part of his vision when he embraced the role of wilderness propagandist. Like many explorers, Muir had vast reserves of optimism in public, but in private often fell prey to a lonely melancholy. He threw himself into politics, fighting for something he believed in with all his soul; yet he would have preferred to throw himself into nature. In truth, as Merriam noted, "Muir abhorred politics, and once, when speaking of a man whom he regarded as having fallen from grace, remarked, 'This playing of politics saps the very foundations of righteousness.'"[86] One has to wonder, of course, whether Muir meant to direct this comment at himself as well, whether he felt that he, too, had made an excessive number of compromises, had been drained of his moral vigor. In any case, it is difficult to argue with his conclusion that his political battles were absolutely necessary. Like George Melville, Muir had come of age at a time when Humboldtian intellectuals constituted a thriving, interconnected, interdisciplinary community, and he grew old during a time of rapid modernization and radical discontinuity. The United States had perhaps never seen such shock waves before, in the form of electricity and mass immigration and professionalization and automobiles and moving pictures and air pollution and urban ghettos. The compromise Muir made in his environmental ethic, then, his shift from a fluid cosmopolitanism to a nativist lococentrism, may ultimately suggest the widespread sense of desperation created by America's breakneck drive to develop its resources. Clarence King, in his prime, could make a living in the field, could create a wilderness culture that thrived for more than a decade; for John Muir, in the early twentieth century, the frontier was closed.

Home: The Harriman Expedition

In the late spring of 1899, the last Humboldtian exploring expedition launched from the United States set sail for the Bering Strait. Aboard the steamship *George W. Elder*, out of Seattle, were enough famous scientists and artists to fill most of the faculty positions at a small liberal arts college. Among the all-star crew were John Muir; the nature writer John Burroughs; Clarence King's field mentor, William Henry Brewer; King's ornithologist, Robert Ridgway; one of King's division chiefs in the USGS, the geologist Grove Karl Gilbert; the artist Frederick Dellenbaugh, a veteran of John Wesley Powell's famous expedition down the Colorado River; George Bird Grinnell, renowned Indian advocate and editor of *Forest and Stream*; the young bird artist Louis Agassiz Fuertes; William H. Dall, the early Alaska explorer and author of *Alaska and Its Resources* (1870); and the avant-garde photographer Edward S. Curtis. The majority of these men had been recruited during meetings at Washington's Cosmos Club, by the head of the expedition's scientific corps, the J. N. Reynolds of his day—Muir's soon-to-be close friend C. Hart Merriam.[1]

Back in 1875, in his report on Mount Shasta for the U.S. Coast Survey, Muir had divided the great volcano into three distinct ecological zones: the chaparral, the fir, and the Alpine. Twenty-three years later, in 1898, Merriam, overseeing the U.S. Biological Survey, followed in Muir's footsteps—and Clarence King's, and William Henry Brewer's, and John Charles Frémont's, and James Dwight Dana's—and attempted a new analysis of Shasta, using what he had started calling in 1890 his "Life Zone" theory. Merriam's official report, *Results of a Biological Survey of Mount Shasta, California*, retained Muir's Alpine zone, but replaced the older naturalist's two lower regions with four distinct ones that he called (from lowest to highest) the Upper Sonoran, the Transition, the Canadian, and the Hudsonian.[2] The direct comparison of altitude to latitude thus became clear—as did Merriam's direct debt to Humboldt.

Merriam's theory of geographical distribution became an important guiding principle in American ecology in the twentieth century.[3] His pioneering 1889 study of the San Francisco Mountain region of Arizona—he had told his superiors that a great deal could be learned about the relative significance of

temperature and humidity by surveying high peaks surrounded by deserts—
is still cited today. But in 1935, a few years before his death, Merriam was quite
open about the actual source of his theory:[4]

> My father, reaching up to the Humboldt shelf of his library, gave me the vol-
> ume entitled "Views of Nature"—a great work, containing a most revealing
> discussion of the distribution of animals and plants. I was deeply impressed
> by Humboldt's account of animal and plant life in the lofty Andes, particu-
> larly on the great Chimborazo, where the various species are grouped one
> above another in successive belts or zones according to differences of tem-
> perature and humidity. Thus, while still a boy I had become impressed by
> the *zonal distribution of breeding birds,* and in the single state of New York had
> recognized three faunal areas. These, as I soon learned, were already known
> to naturalists, as the *Canadian, Alleghenian,* and *Carolinian* faunas, estab-
> lished by the elder Agassiz in 1854. . . . In other words, the subject of the geo-
> graphical distribution of animals and plants—the very substance and
> essence of a biological survey—had taken form in my mind at an early date.

Merriam, who lived until 1942, was perhaps the last great Humboldtian
scientist bred in America. Franz Boas has proved more important to American cultural history, but he shared Humboldt's nationality as well as his
worldview. And, interestingly, Merriam's later work actually became more
and more Boasian: he spent the last twenty-five years of his life doing ethnological studies of California Indians, in the general vicinity of Mount Shasta.
He even wrote an article in 1926 on the "source of the name Shasta."[5] While
Muir steered clear of natives once he started advocating for the preservation of
California's wilderness areas, Merriam followed the example of his old governmental colleague John Wesley Powell and studied non-Western ways of
interacting with the environment, meanwhile collecting hundreds of photographs and compiling several books' worth of complex vocabularies. Today,
there is a Merriam-Powell Center for Environmental Research located not far
from San Francisco Mountain in Arizona.

Merriam's most entertaining project, however, was the so-called Harriman
Alaska Expedition of June and July 1899. As John Muir put it: "Nearly all my
life I wandered and studied alone. On the *Elder,* I found not only the fields I
liked best to study, but a hotel, a club, and a home, together with a floating
University in which I enjoyed the instruction and companionship of a lot of
the best fellows imaginable, culled and arranged like a well balanced bouquet,
or like a band of glaciers flowing smoothly together, each in its own channel,
or perhaps at times like a lot of round boulders merrily swirling and chafing
against each other in a glacier pothole."[6]

The short cruise up Alaska's coast, as Muir suggested, was far from avant-garde. More explicitly programmed than even the *Corwin* expedition, this one at times bordered on ecotourism: the scientists enthusiastically scrambled ashore at certain scheduled stops, but they generally had to be back on board shortly after the ship's whistle blew a few hours later. There was neither the time nor the space for them to become truly disoriented, to leave aside the trappings of their elite civilization, the comforts of home. Indeed, the cruise's comforts were sumptuous, even, one might say, *extravagant*—in the sense of capitalist luxuriance rather than the Thoreauvian sense of extreme vaga-bondism. The expedition was sponsored by the railroad magnate Edward H. Harriman, who had decided rather whimsically to turn his bear-hunting vacation, prescribed by the family doctor to relieve the businessman's nervous tension, into a philanthropic publicity stunt. There would actually be some significant scientific results from the cruise, because Harriman demanded them: while the United States expanded its political borders in the Spanish-American War, he would single-handedly see to it that the frontiers of American knowledge were also stretched outward. In a sense, Harriman played the czar to Merriam's Humboldt: the scientists were free, even encouraged, to pursue whatever arcane studies they might have time for, though what counted most were the prospects for mineral investments and recommendations for new railroad routes. To Harriman, thinking globally meant envisioning track being laid all around the world. As in Humboldt's Siberian expedition, then, the politics generally dropped out of nature's political economy: this was a crew of mostly older, established scientists, flexing their expert muscles during the day and then banqueting at night, not much interested in rocking the boat except in raucous revelry (provided by Harriman's Committee on Music and Entertainment). Harriman himself set a very particular tone, especially toward the end of the cruise. The hunting hadn't worked out terribly well, and, besides, vacations can get a little tiresome in the second month. Called by Merriam to the shore side of the ship so as not to miss the grand beauty of the renowned (if misnamed) Fairweather Range, the tycoon retorted, "I don't give a damn if I never see any more scenery!"[7]

Still, Harriman had sponsored an exploring expedition, and, for all his committees and efficiencies and ironclad schedules, he presumably knew that, in the field, anything can happen. A July Fourth celebration aboard the *Elder* featured cannon blasts and patriotic hymns and John Philip Sousa on the graphophone, but it also brought out the rebel in the young poet Charles Keeler, who could not abide the invasion of the Philippines: "Ye who have failed to rule a wilderness," he declaimed, "Now preach of liberty in tropic seas, / Forsooth our sway the Orient hordes shall bless, / While politicians fat-ten at their ease; / O Lord, must our dear sons be slain, such men to please?"[8] Keeler had been John Muir's roommate for more than a month at this point, so perhaps he had taken to heart the older man's grumblings about the crew's wanton killing of animals and about the general bluster and scriptedness of the expedition. Certainly, Muir was not shy about expressing his yearning "to break away from the steamboat and its splendid company, get a dugout canoe and a crew of Indians and . . . poke into the nooks and crannies of the mountains and glaciers which we could not reach with the steamer."[9] George Bird Grinnell also broke with the Harriman script on occasion: though he shared the general view that all Indians would be better off Christian and civilized, and he did not make a peep in protest when the scientists plundered a recently abandoned native village for artifacts and souvenirs, he at least published an attack on the rampant exploitation of natives by capitalists in the mining and canning industries. "White men uncontrolled and uncontrollable," he wrote, "already swarm over the Alaska coast and are overwhelming the Eskimo. . . . In a very short time they will ruin and disperse the wholesome, hearty, merry people whom we saw at Port Clarence and Plover Bay."[10]

The Harriman Alaska Expedition ultimately produced thirteen thick volumes of scientific writing, including a distinctly Humboldtian essay by Merriam on the volcanic Bogoslof Islands.[11] Most of the work was highly specialized, though: the twenty-five scientists had rarely collaborated in any fundamental ways, and their writing seemed directed exclusively at fellow experts rather than ordinary folk who were actively engaged with nature in their daily lives. The expedition's publications covered a diverse range of subjects, but even Merriam's careful supervision of the tomes could not provide any unity.

Merriam's greatest success was perhaps his selection of young Edward S. Curtis as the expedition's official photographer. Curtis was not part of the Cosmos Club crowd. Unknown outside the Pacific Northwest, he had introduced himself to Merriam in 1897 under unusual circumstances: Merriam, George Bird Grinnell, and their climbing partner Gifford Pinchot, who would later become Theodore Roosevelt's chief forester and the nation's chief conservationist, had gotten stuck on the slopes of Mount Rainier, and Curtis, who happened to be photographing the peak that day, rescued them. In a way, with his homages to the power and mystery of glaciers and his stark documentation of native life in the North, Curtis also rescued the Harriman expedition. His pictures continued the tradition launched by Timothy O'Sullivan on the King survey—the tradition of combining scientific explanation with aesthetic experimentation, of confronting harsh landscapes and humanity's smallness, of registering ambivalence about frontier development. In a photograph like "House and Hearth—Plover Bay, Siberia," with its blanched whalebones in

41 "House and Hearth—Plover Bay, Siberia," 1899. Edward S. Curtis.

the foreground and ghostly mountains in the background, Curtis captured both the elegant adaptability of the natives and the fragility of their culture. There is a sense of home, here, but also a sense of the cosmos.[12]

This balance of the local and the universal, and of economy and ecology, was a rare achievement on the Harriman expedition. Few of the scientists tried to make real connections with the people and landscapes of the North. Curtis would go back again and again in the coming years, in a spirit of exploration, but most of his shipmates were done with Alaska and Siberia, and most of his countrymen were done with old-fashioned exploring expeditions. A decade later (or thereabouts), Peary and Cook would both make it to the North Pole (or thereabouts)—but their expeditions were essentially performances of manliness.[13] It no longer made sense in the United States to send out parties hoping to come to some kind of holistic understanding of other environments and cultures. Since there seemed to be so little left to discover in the world, since the frontier had closed, all eyes were fixed on the home front. As Curtis realized, though, to understand your home, you occasionally have to make yourself a stranger. ■

CHAPTER 10

The Grounding of
American Environmentalism

What was the meaning of that South-Sea Exploring Expedition, with all its parade and expense, but an indirect recognition of the fact that there are continents and seas in the moral world to which every man is an isthmus or an inlet, yet unexplored by him, but that it is easier to sail many thousand miles through cold and storm and cannibals, in a government ship, with five hundred men and boys to assist one, than it is to explore the private sea, the Atlantic and Pacific Ocean of one's being alone. . . . It is not worth the while to go round the world to count the cats in Zanzibar. Yet do this even till you can do better, and you may perhaps find some "Symmes' Hole" by which to get at the inside at last.
   —Henry David Thoreau, *Walden* (1854)

Psychiatrists, politicians, tyrants are forever assuring us that the wandering life is an aberrant form of behaviour; a neurosis; a form of unfulfilled sexual longing; a sickness which, in the interests of civilisation, must be suppressed.

  Nazi propagandists claimed that gipsies and Jews—peoples with wandering in their genes—could find no place in a stable Reich.

  Yet, in the East, they still preserve the once universal concept: that wandering re-establishes the original harmony which once existed between man and the universe.
   —Bruce Chatwin, *The Songlines* (1987)

C. Hart Merriam and Edward Curtis lived well into the twentieth century, and so did Humboldt's legacy. In Latin America, perhaps because Humboldt was indelibly linked to Simón Bolívar, he actually still has the status of a celebrity. His memory is more muted in Europe, but there scholars in many fields at least know who he was and why he mattered. In the United States, meanwhile, the Humboldt current has experienced a fate similar to that of the Colorado River: blocked by numerous dams, drained of much of its power, losing more and more volume to evaporation in a rapidly warming climate, it still flows, but there are times when it can be hard to find.

The decisive blow to Humboldtianism in the United States was likely the rise of a rabid anti-German sentiment during World War I: there could be no Prussian heroes after 1917. But, as the careers of John Muir, George Melville, and even Clarence King suggest, Humboldtian approaches to the world were already in decline by the end of the nineteenth century. Part of the problem arose directly from the professionalization of science. Even in the day of J. N. Reynolds, it was difficult to earn a living as a synthesizer of all the different subdisciplines; to succeed, you had to specialize. Moreover, as science became disconnected from fieldwork, fewer people actually had the opportunity to see and experience entire ecosystems operating as organic units. And, by the time of the Spanish-American War, Humboldt began to seem absurdly naive in the realm of social theory, as the United States took up the white man's burden and set out to impose its model of masculine development on the rest of the world. In an era of rapid growth and acute fragmentation, of assembly lines and riots and cubist art, the very concept of unity began to seem outdated. With the American frontier seemingly closed, with flags planted at both poles (by the second decade of the century), with technology ascendant, people interested in exploring were inclined to seek out models who were flashier, more modern, than Alexander von Humboldt.

Despite the decline of scientific exploration, though, a number of Americans in the twentieth century managed to mount Humboldtian-style expeditions—even if, like Thoreau, they did much of their traveling in or near their own hometown. The key is perhaps to integrate the spirit of journeying into one's everyday life. "Rise free from care before the dawn," Thoreau wrote, "and seek adventures." Indeed, as the literary critic Stephen Greenblatt has commented, there is a scholarly tradition dating back to Herodotus suggesting that all real "knowledge depends upon travel, upon a refusal to respect boundaries, upon a restless drive toward the margins."[1] Fieldwork will always be crucial to the Humboldtian creed, but it can be metaphorical fieldwork: more important than the ability to visit exotic locales is the effort simply to expose oneself to a variety of environments and cultures and then to take up the role of mediator, interpreter. There are marginalized people and places in every city and town (and marginalized subjects in every library). What matters, ultimately, is receptivity, subjection, the genuine attempt to connect with differentness.[2]

Merriam and Curtis, then, with their volumes documenting Native American traditions, their countless vocabularies, their boxes of photographs, placed themselves squarely in the twentieth-century version of the Humboldt current. Indeed, much twentieth-century anthropology could be considered Humboldtian: the students who took to the field after studying with Franz Boas—people like Ruth Benedict and Margaret Mead—not only replaced Samuel George Morton's polygenism with a new cultural relativism but also embraced the fundamental concept of unity in diversity. The Boasian school, in turn, influenced any number of radical race theorists, such as Carey McWilliams, author of *Brothers under the Skin* (1943). McWilliams's understanding and appreciation of pluralism, developed while he traveled through America's agricultural areas and interviewed migrant farmworkers, could even be said to have a deep connection to his awareness of nature. In two of his earliest books, *Factories in the Field* (1939) and *Ill Fares the Land* (1942), he argued that the exploitation of natural resources, especially in California, had always gone hand in hand with the exploitation of ethnic minorities. Then, continuing to echo Humboldt's nineteenth-century formulations, McWilliams went on to suggest that the embrace of a unified multiculturalism, certainly a good in its own right, might also help to restore a less industrial, more "organic relation" between humanity and our surrounding environment, itself a model of diversity.[3]

Twentieth-century writers who defined themselves explicitly as students and defenders of nature also did some wading in Humboldtian eddies. At the very beginning of the century, Mary Austin wrote a long ode in prose called *The Land of Little Rain* (1903), which did for the California desert what Humboldt had done for the oppressively sultry tropics and what Muir had done for the frigid Arctic. "In Death Valley, reputed the very core of desolation, are nearly two hundred identified species," Austin explained, hoping to shock her readers; indeed, "the desert floras shame us with their cheerful adaptations to the seasonal limitations." Moreover, as she herself took up life in the desert, thus doing Muir one better, she strove to put the principle of adaptation into practice by following the lead of the local Paiutes. Austin "[learned] how Indians live off a land upon which more sophisticated races would starve, and how the land itself instructed them." Then, in later years, she turned to a much more thorough study of Native American cultures and tried to bolster native art and poetry against the tide of assimilationism, becoming "a fierce and untiring opponent of the colossal stupidities, the mean and cruel injustices, of our Indian Bureau."[4]

California, with its ecosystems ranging from Death Valley to the glaciers of

Mount Shasta, actually became something of a hotbed of Humboldtianism. C. Hart Merriam's anthropological efforts were sustained in part by two Boas students based at Berkeley, Alfred Kroeber and Robert Lowie, and in the 1920s and '30s they joined forces with a younger professor named Carl Sauer, who was determined to keep Humboldt's tradition alive in the discipline of geography.⁵ Sauer, a pioneering and influential cultural geographer, became an expert on Latin America and examined broad interactions between peoples and landscapes. While he kept a first edition of Humboldt's 1805 *Essay on the Geography of Plants* in his office, Sauer preferred studying his mentor's theories in the actual contact zone.⁶ At a time when almost all scientists were eschewing fieldwork in favor of laboratory experiments, he continually argued that geographers ought to adopt the Boasian model of periodic immersion in foreign cultures, and he himself especially enjoyed following in Humboldt's footsteps through Mexico.⁷ Ultimately, his body of work tracked rich, living traditions of plant and animal lore, agricultural innovation, and the shaping of nature for human habitation. But, in keeping with Humboldt's politics, Sauer also condemned the social dynamics that resulted in the *misshaping* of nature: one fairly typical article he wrote is called "Destructive Exploitation in Modern Colonial Expansion."⁸

Sauer's interest in human beings' power to mar and abuse the environment, especially through commerce and colonial relationships, was heightened considerably by his familiarity with the work of the cultural critic Lewis Mumford, one of whose scholarly projects in the 1930s was the restoration to public awareness of the nineteenth-century Humboldtian George Perkins Marsh, author of *Man and Nature: or, Physical Geography as Modified by Human Action* (1864). Today, Marsh is recognized as a founder of conservation, but not necessarily as a disciple of Humboldt. Environmentalists cite him for having labeled humanity as being "everywhere a disturbing agent," but they generally fail to consider his recognition that certain social structures were more disturbing than others.⁹ Mumford, though, a tireless explorer in various fields of knowledge, tried to honor Marsh's and Humboldt's political edge by connecting them to other radical theorists, such as the anarchist geographers Peter Kropotkin and Elisée Reclus and Mumford's own mentor Patrick Geddes. Drawing on the writings of all these Humboldtians, Mumford, over a career that lasted until 1990, developed a blueprint for a society in which people could enjoy "the vitalities and creativities of a self-sustaining environment and a stimulating and balanced communal life."¹⁰ Replacing the megalopolis/suburbs/countryside model of development would be a network of interconnected, midsized "garden cities," each of which would combine urban patterns with agriculture and even some wilderness areas.[11] Perhaps most important, Mumford insisted that the only way of sustaining his ideal society was by abolishing hierarchies and fostering an intensely participatory democracy. He even discredited property rights as dividing human beings from each other and from nature. "Enjoy the land," as Thoreau said, "but own it not." Assaulting social Darwinism by citing Kropotkin's finding that "mutual aid is as much a law of animal life as mutual struggle," Mumford sought a shift "from capitalism and fascism to co-operation and basic communism" and suggested that "political life, instead of being the monopoly of remote specialists, must become [a] constant . . . process in daily living."[12]

When politics exploded into daily life in the 1960s and environmentalism finally became a genuine social movement, Murray Bookchin pieced together Mumford's theories and started writing treatises on what he called social ecology. Though Bookchin tended not to cite Humboldt by name—the reference would not have meant anything to most people, anyway—he did explicitly take as his fundamental law the "principle of unity in diversity."[13] Not nearly as peripatetic as people like Boas and Sauer—for much of his life, he has rarely left Vermont—Bookchin has nevertheless become perhaps the United States' foremost Humboldtian cosmopolite, arguing that, "from an ecological viewpoint, balance and harmony in nature, in society, and, by inference, in behavior, are achieved not by mechanical standardization but by its opposite, organic differentiation."[14] Modern ecologists have cast the concepts of nature's balance and harmony into doubt, but Bookchin hedges his bets by focusing on how diversity promotes stability in the *social* realm, on the ways in which multicultural, egalitarian, communal living situations could remake our very perceptions of nature. The monolithic tendencies of capitalism were what first led to environmental abuses: "it was not until organic community relations . . . dissolved into market relationships that the planet itself was reduced to a resource for exploitation. . . . Owing to its inherently competitive nature, bourgeois society not only pits humans against each other, it also pits the mass of humanity against the natural world. Just as men are converted into commodities, so every aspect of nature is converted into a commodity, a resource to be manufactured and merchandised wantonly."[15] In Europe, where class is often a central political issue and socialist-oriented candidates sometimes come to power, Bookchin's social ecology has had a significant influence on the Green Party's platform.[16] In the United States, though, few environmentalists have even heard of Bookchin. Here, it remains difficult to question basic capitalist structures, and environmental thought remains focused on the protection of particular pieces of ground, which are treated as elite, even sacred landscapes. By celebrating only faraway "wilderness" areas allegedly full of biodiversity, we tacitly endorse the industrial system that has fractured our society and devastated the environments where the majority of Americans live.[17] We seem to have forgotten that all places are connected and, ultimately, equally valuable, that life depends on all the mutually dependent features of the cosmos.

———

John Muir is the patron saint of American environmentalism. Through all the permutations of environmental thought in the twentieth century, the one consistently resonant voice has been his, boldly insisting that our failure to save at least a few special tracts of wilderness would constitute a betrayal of both the planet and our own heritage. Muir, with his long, messy beard, his genuine love of the outdoors, and his general curmudgeonliness, is an appealing rebel. And it makes sense to think of him as a radical of sorts, for America was dominated by utilitarianism at the turn of the century, and he was preaching aesthetics. Americans are truly blessed to have so many beautiful, wild places within their borders, not just for the sake of recreation but for the sake of our cultural self-awareness. No "wilderness" is pure, and of course the "wild" is accessible in our own backyards and alleyways, but it is easier to understand what "civilization" means when you can at least point to places where its imprint seems fainter—where you might have the chance to experience utterly unfamiliar, and uncomfortable, environments. My own feeling is that such places can be deeply inspiring: by, paradoxically, connecting with them and also leaving them mostly alone, we acknowledge the world's separate peace, its usefulness primarily to itself. Yet, at the same time, the American wilderness ethic tends to distract us from the ways in which humanity and nature are always implicated in each other. Muir's legacy can be powerfully constraining, not to mention undemocratic.

Of course, there are other important elements of American environmental thought—including Humboldtian ecology—and other heroes and heroines. A 1993 study called *The Environmentalist's Bookshelf* suggested that the two most significant books in the launching of the twentieth-century environmental movement were Aldo Leopold's *A Sand County Almanac* (1949), in some senses an update of *Walden*, and Rachel Carson's call to arms, *Silent Spring* (1962).[18] Both Leopold and Carson were, fundamentally, ecologists, and they succeeded spectacularly in raising awareness of the weblike interconnectedness of the natural world. But because their ecology had almost no social edge to it, their legacies have become hopelessly entangled with American lococentrism. Leopold, an enthusiastic backpacker, is remembered primarily as being "among the earliest government officials to see things John Muir's way and support setting aside vast expanses of land as wilderness."[19]

Carson, meanwhile, did shift attention to the issue of environmental toxins, by pointing out how many pesticides reside in "the common salad bowl." But her strategy of evoking a small American town where the birds no longer sang in the springtime suggested to many readers that the most important contribution they could make to the environmental movement was simply to become defenders of their own homes.[20] Many schoolchildren began to learn how chemicals got passed through the food chain, began to join community protests against incinerators that spewed poisons into their local air and water supply—but they rarely learned about how pesticides got sprayed onto farmworkers as well as crops, or how the majority of toxic waste dumps wound up in African American towns. Unfortunately, even the environmental justice movement, which takes social ecology very seriously and may well represent the best hope of American environmentalism, has been hampered by its provincialism, its not-in-my-backyard (NIMBY) ethos. While it has mounted powerful, admirable defenses of communities threatened by various toxins, it has failed to generate a broad attack on the social forces that allow those toxins to be created in the first place.[21]

There can be a kind of selfishness, then, to the American emphasis on place, whether one is defending a favorite far-off wilderness or one's own hometown. Indeed, many historians locate the rise of American environmentalism as a real movement in the immediate post–World War II era, in the context of a general rise in expectations regarding the quality of life—including environmental amenities.[22] Large groups of people first started coming together in support of nature not because of pesticides but because of threats to a federally protected park. Preservationism and conservationism had chugged along through the first half of the century, operating only at elite levels of society, with efficiency experts touting tree farms and dam projects and parkways, and wealthy mountain climbers founding organizations like the Wilderness Society. By the late 1940s, though, conservation had come to seem like merely another branch of industry, while the preservationists, working largely through Aldo Leopold's Wilderness Society and John Muir's Sierra Club, captured Americans' attention and cornered the market on environmental thought, thanks to their protests against a proposed dam in Colorado's Dinosaur National Monument. It was Hetch Hetchy all over again, but this time the wilderness types won—though in the deal-making process they sacrificed the Colorado River's Glen Canyon to the Bureau of Reclamation. The bureau quickly erected a monumental dam there, which inundated a huge tract of land now known as the Lake Powell Recreation Area, named in honor of the Humboldtian explorer John Wesley Powell, who probably would have objected to the whole damn debate. Before the lake was created, though, the Sierra Club sent Eliot Porter to see Glen Canyon in its "natural" state and take some final, regretful photographs, which were published in a book called *The Place No One Knew*.[23] Many people still read this book, and the environmental movement is still stuck in a rut of nostalgic yearning for beautiful scenery.

At the heart of today's American environmentalism, in the early twenty-first century, is a new place no one knows: the Arctic National Wildlife Refuge. For the entire span of George W. Bush's presidency, this seems to have been the number one issue not only for the Sierra Club and the Wilderness Society but also for such groups as the Natural Resources Defense Council, the Audubon Society, and Environmental Defense. I have received numerous letters and e-mails myself asking for donations to help prevent Bush from drilling for oil in this pristine wilderness area, where the precious Porcupine herd of caribou travel every year to have their calves. Rarely do the mailings mention the two native groups who live on the edge of the refuge, the Iñupiat and the Gwich'in, who vehemently disagree with each other about what kind of industrial development ought to occur in their homeland. Rarely do we hear much about how we could lower our consumption of oil—in order to make drilling unnecessary—or how we could invest in alternative sources of energy. Indeed, the major environmental organizations seemed powerless in the 1990s to stop the oil and auto lobbies from prolonging the SUV craze, which, given the current progress of global warming, may have constituted the gravest environmental policy disaster in recent memory. Taking their cue from John Muir, American environmentalists seem to have decided that the most expedient strategy, politically, is to garner support for the preservation of certain wilderness jewels, no matter what other issues they might have to ignore. They seem afraid to attack our ravenous consumer culture, and they never even seem to notice the problem of urban environments.

Our major environmental organizations, focusing narrowly on regulatory politics, have made decent progress in preserving tracts of land and cleaning up the water and air. But the environmental movement has been essentially stalled since 1974 (the assertion that it is, in fact, "dead"—as a couple of Greens put it in 2004—was long overdue). In the last thirty years, each small step forward has tended merely to deflect ecological (and social) damage somewhere else, because American environmentalists have never embraced a holistic approach to activism. It's impossible to disentangle all the potential reasons for this state of affairs. Certainly, though, it is significant that environmentalism in the United States is so deeply rooted in traditional conceptions of property, and that, unlike feminism or the civil rights or labor movements, it did not arise in response to injustice or oppression. Indeed, American Greens, in prizing hard-to-get wilderness experiences, have consciously chosen elitism over egalitarianism. The environmental movement has *never* had much of a social edge, in other words—at least, not since the nineteenth century. And therein may lie a central problem. Perhaps the previous century's soaring ecological activism has been essentially grounded for the last hundred years—recent histories have been titled *Losing Ground* and *The Fading of the Greens*—because for all that time environmental thought has been disconnected from Humboldtian exploration.[24]

When I looked through the collections made 150 years ago by the Boston Society of Natural History, I found that the vast majority of its books were in fact travel narratives. Ecology would have been unthinkable outside the context of nineteenth-century exploration. Of course, that context has disappeared forever: we can't simply relaunch those expeditions. But perhaps the ideas developed by nineteenth-century explorers can themselves travel, can slip beyond the bounds of the original milieu in which they first came to light.

Humboldtianism inevitably constituted a social vision of nature. It started internally, in the emotional, intellectual, and spiritual realms: humanity was linked to the world through perception, and connectedness brought comfort and sometimes even joy. But then the Humboldtian explorer always had to make a full acknowledgment of the external, the Other, that which he could never fully see or understand. Still, though, he remained engaged: knowing the world was out of reach, he nevertheless reached out to it, in the hope of experiencing unity in diversity.

Most environmental histories suggest that widespread awareness of ecological damage or resource scarcity had to be minimal in the nineteenth century, because the world appeared limitless back then.[25] Yet it seems to me that our forebears may actually have been much more aware of how they shaped nature and used resources than we are today, for their survival and development were so obviously dependent on ecological processes. In Humboldt's day, you would notice if the water level sank in your local lake, while in the twenty-first century it's quite common for nature lovers to drive SUVs into the backcountry without ever thinking about the real impact of their needlessly inefficient vehicles. It's relatively easy to be Green, today, at least superficially, because so many of us are so far removed from environmental realities.[26] Two centuries ago, though, to explore and interrogate nature was to consider the very foundations of society.

In the 1820s and '30s, J. N. Reynolds upbraided his fellow Americans for not contributing enough to the international effort to come to grips with the planet. He had some fairly outrageous theories about the cosmos; he also had some timeless ones. "You used to take an interest in Exploring Expeditions," he wrote to his old friend Thomas Corwin in 1850. "If you have not lost all taste for the Fine Arts, Science, and the humanities in general—just request [to see] . . . a recent report sent me by Major Chapman . . . of the successful voyage of a keel boat up the Rio Grande."[27] Even late in life, then, Reynolds remained convinced that basic human wisdom could be gained only through journeys. But he also enjoyed travel for its own sake—or, perversely, because travel sometimes seemed to be unsettling, seemed to make advances in the arts and sciences virtually impossible. To keep moving was in a sense to reclaim certain impulses that might be stifled in the regulatory regime of the city. Indeed, the centers of settled civilization could get boring, especially in capitalist societies, where every activity was supposed to lead efficiently toward the acquisition of wealth. True individualism lay on the frontier.

Clarence King, in particular, appreciated this aspect of exploration. His wandering represented a type of irresponsibility, but also a type of radical rebellion. He wanted to be a Don Quixote, eschewing financial gain in order to follow his heart: "Were it not for my family cares I would gladly wander on anywhere a mule or a canoe could bear us, and search till death for the Garden of Eden or the Fountain of Eternal Wit or any other thing we were sure not to find."[28] Indeed, his friends remembered King as perpetually in motion, the consummate "travel-dressed cosmopolitan," and this quality stretched even into the realm of metaphor: "He never expected discussion to lead to anything. Sometimes indeed he would not permit it to. He enjoyed 'travel, not arriving.' I fancy he thought that things capable of settlement had been settled long since."[29] Ever embracing paradoxes, King acknowledged the hardship of the road and direct confrontations with nature, the sting of a cold night, but he also believed that only the external world could ease his internal pain. "To me the camp nights are of a charm greater than daylight," he wrote to John Hay toward the end of his life. "I shall never lose the sharp individual memories of many heavens full of cold stars which have kept me awake even after a hard day's work and which haunt me still. This summer I have spent in presence of my noble and charming mountain landscape, and though always alone I have suffered but little with impatience or ennui. I seem to have settled down into a patient calm condition in which fate may kick me as it will and I do not cry out."[30]

Ultimately, though, King found the catastrophic environment overwhelming. His explorations of nature seem to have furnished him with a sense of tragedy—which was hard for him to bear, but which, it should be said, represents a significant human accomplishment. Tragedy is an acknowledgment of higher powers, an embrace of humility. For King, it was simply reality: "The human organism has rarely been subjected to a severer test than the study of scientific problems, nor is there a truer hero than an investigator who never loses heart in a life-long grapple with the powers of the universe. It requires courage of the highest order to stand for years face to face with one of the enigmas of nature; to interrogate patiently, and hear no answers, to try all known methods and weapons of attack, and yet see the lips of the sphinx compressed in stony immobility; to invoke the uttermost powers of imagination; to fuse the very soul in the fire of effort, and still press the listening ear against the wall of silence. It is easier to die in the breach."[31]

George Melville also cut something of a tragic figure in the annals of Humboldtian exploration. Twenty of his companions did in fact die in the breach, and he was haunted by their memory. By the late 1890s, people started noticing an air of distraction and melancholy in his general demeanor. Living in Washington and serving as chief engineer of the navy, Melville would take a daily walk in the afternoon, past "a famous art school. As, day after day, there went slowly by, on the shady side, the powerful figure—hat in hand and lost in thought, with drooped head and white hair floating—the girl students opposite noted it all with artistic eyes and the silent officer, all unaware, was made to serve as model for their studies of King Lear."[32] Despite his Shakespearean qualities, though, and despite his sometimes mechanistic, instrumental way of seeing nature, he always stood firm in his beliefs about exploration as a form of art, as something to do purely for its own sake, as something that helped keep us human. Indeed, one reason so many Washington politicians hoped essentially to write the *Jeannette* disaster out of our history books was that the whole expedition, in retrospect, seemed so pointless—but, to Melville, that was the point. In some ways an utterly pragmatic expert on the efficient harnessing of nature's forces for human needs, he nevertheless counseled that, every now and then, human beings ought to put their ships into the ice, and flow with nature's currents.

John Muir, toward the end of his days, had to go far from home for his ebbs and flows, for the rush of real contact, the paradoxical combination of gentle undulations and electric shocks that he liked to experience in the natural world. His favorite California valleys had become political battlegrounds. So his 1911 trip to South America, like his early voyages to the Arctic, was both an escape and a deeper immersion in life.[33] It was not the same kind of adventure he would have had if he had gone in 1868, just after his walk to the Gulf of Mexico, when he wrote, "I am sometimes lonely but I dispel my sore thoughts—with the hope of seeing the snowcapped Andes and the flowers of the Equator."[34] At age seventy-three, he was not able to clamber up mountains the way he once could. But he nevertheless spent several days in the cold Chilean Andes searching for groves of the so-called monkey spider tree, the rare *Araucaria imbricata*, with its wonderfully overlapping, intertwining branches. His family and friends had worried that he would risk his life on just such a quest, and Muir acknowledged that there were difficulties involved. Ultimately, though, travel always proved worthwhile. Rereading Humboldt's *Personal Narrative* before his departure for the southern hemisphere, Muir made some vigorous pencil markings next to the second sentence of the introduction, in which Humboldt noted the "enjoyments" he claimed to have experienced in his travels—"which have amply compensated for the privations inseparable from a laborious and often agitated life."[35] Muir is known today for his placedness, for his commitment to particular sacred locations in California, but that commitment would have been impossible without the balancing force of his wanderlust, without the difficult but ultimately enlightening journeys that took him away from everything he had previously known.

———

"Get lost."

"Take a walk."

These are typically contemptuous utterances, attempts to get rid of people. I'd rather take them as mantras, mandates, periodic reminders to wander into the world and make connections. If you can't walk, or don't like to walk, then explore whatever edges and boundaries might be accessible to you—in your reading, your cooking, your conversations. Unsettle yourself. Turn off your lights and stare at the cold stars; allow yourself to be both haunted and enchanted. Get lost in someone else's world.

In England, today, people of all classes still talk about groups like the Association for the Protection of Ancient Footpaths, founded in 1824, because such organizations still sponsor "mass trespasses," in which large numbers of citizens tread Ancient Footpaths through both public and private lands, thus asserting their right of way, their claim to connections across lines of ownership. A body of laws, built up over the last two centuries, now protects that claim, as long as the paths are walked at least once a year: the trespassers, then, are actually legal aliens who bind the country together through their activism. Environmental groups in the United States continue to endorse a kind of elitist ordering of our "wild" and "tame" places, a rigid drawing of boundaries that sometimes succeeds at keeping capitalist development out of certain areas but that also, in what is becoming a terrible irony, concentrates recreational activities—and their attendant pollutants—in our favorite parks. The

British approach, born in a more socialistic context, in angry response to the exclusive aristocratic privilege of strolling in private gardens, explicitly attacks the very concept of boundaries. Americans go on vacation, passing from civilization to wilderness; Britons are much more likely to go on daily rambles.[36]

In the Australian culture of landscape there exists what Bruce Chatwin called a "labyrinth of invisible pathways," travel routes "known to Europeans as 'Dreaming-tracks' or 'Songlines'; to the Aboriginals as the 'Footprints of the Ancestors' or the 'Way of the Law.'" Ancient beings laid them down, attaching to each a particular totemic animal and particular tones, notes, melodies. From then on, visions guided people across these paths so that they could communicate with other tribes over vast, unreadable deserts. You sang as you walked, hoping to attract just the right creature, who might show you the way to your next landmark. The tune remained the same, though the words might pass through dozens of different languages as new peoples were encountered. And the words, as Chatwin put it, eventually expressed an "endless accumulation of detail," until the journey produced a cosmos of sorts, "in which the structures of kinship reach out to all living men, to all his fellow creatures, and to the rivers, the rocks and the trees."[37]

How many different ways might there be of moving through a landscape? Of living in nature? Chatwin considered the contrast of two peoples making their home in central Australia: "The Aranda, living in a country of sage waterholes and plentiful game, were arch-conservatives whose ceremonies were unchangeable, initiations brutal, and whose penalty for sacrilege was death. They looked on themselves as a 'pure' race, and rarely thought of leaving their land. The Western Desert People, on the other hand, were as open-minded as the Aranda were closed. They borrowed songs and dances freely, loving their land no less and yet forever on the move."[38] The point is not to romanticize nomadic life, nor to subvert legitimate claims to land rights, but merely to explore the cosmopolitan possibilities of something like a diaspora model for modern society. Approached with a sense of stewardship instead of a sense of ownership, the local can contain the global. To embrace exile is both to cling to one's individuality and distant commitments and also to seek connections wherever one happens to be.[39]

"Conceive a space that is filled with moving." Gertrude Stein was probably thinking of the buzz and bustle of New York City when she offered this description of the United States, but she also managed to express the general perception that America's history is a particularly rootless one.[40] Indeed, many pioneers rampaged across the country, pulling up stakes at the first sign of crowding, and the cultural myths of abundant land and absolute freedom reinforced each other to foster a general ethos of individualistic itinerancy. Strong communities did of course exist on the frontier, but many of America's most potent cultural icons—from Daniel Boone and Davy Crockett to Jack Kerouac and James Dean—have stood for a kind of transience bordering on escapism.[41] It makes sense, then, that many writers, especially within the environmental tradition, have taken great pains to try to counter American dislocation with exhortations to get back to the land, settle down, develop local associations, cultivate our gardens, and defend sacred pieces of wilderness.[42] But we may already have overcompensated. As the cultural geographer Yi-Fu Tuan has remarked, "in the United States, which once acclaimed mobility and space, 'place' now seems the favored word, and in the world generally, cultural particularism and ethnic heritage carry greater resonance— arouse more positive feelings—than cosmopolitanism and universalism." Tuan does not mean to defend or reestablish a simple Enlightenment-style universalism, or even to celebrate travel simply as a broadening and tolerance-promoting activity. But, in his book *Cosmos and Hearth*, he does admit that his primary goal is, as it were, to speak a word for Cosmos—since it is this concept that "now needs intellectual restitution."[43]

To me, the history of Humboldtian exploration suggests that there is a need for balance between cosmos and hearth. It is clearly important to cultivate connections to particular pieces of land and to particular heritages and traditions. But it is also important to get lost in the world, or at least to pay attention to those people who lose themselves, who have long desires, who know deep sadness—for without their example, we would never understand what it meant to be at home. Thoreau assuredly achieved a powerful intimacy with the world of Walden Pond, yet he also lost himself, almost every day, on long, sauntering rambles. Rootedness and deracination go hand in hand. The philosopher Stanley Cavell has commented that "the achievement of the human requires not inhabitation and settlement but abandonment, leaving."[44] Connections are too easily taken for granted, in other words, unless they are occasionally broken; we cannot forget to acknowledge the reality of tragedy. Of course, our ability to get lost does depend to some extent on having access to a diverse set of landscapes, so wilderness preservation will always have a place, so to speak, in the environmental movement. "Our village life would stagnate," Thoreau wrote, "if it were not for the unexplored forests and meadows which surround it."[45] It may be even more important, though, to cultivate a sense of constant exploration, of potential connection with all people and places, no matter how alien or familiar.

A Humboldtian social ecology would have to be as fluid and inclusive as the world itself. Many people have no land to go back to, no garden to cultivate, no park to hike in. We can't simply start living the way Thoreau did at Walden Pond. We can't even begin to reconnect with nature until we address the inequalities that have plagued our society for centuries—the skewed distribution of resources, the racism, the sexism. We need a new chain of connection.

What we can do, immediately, is remind ourselves that we are interacting with nature not just when we seek out a wilderness area but every time we flip a light switch and take advantage of the electrical current running through the modern, industrialized world. We can remember that each person's consumption of natural resources leaves less of nature for others. Perhaps most important, we can remember, with Thoreau, that there have always been different ways of living, and that "it is never too late to give up our prejudices." Cultural values, like ocean currents, can shift; it is still possible to remake society as if everyone mattered. If it were not acceptable to pollute anyone's community, then we would have to stop polluting. And then maybe a significant number of us could get excited about exploring new ways of preventing pollution and saving resources—about solar power and farmers' markets and light-rail networks and city parks and leisure activities that allow us to slow down, make contact, simplify, slough off our "desperate haste to succeed," our "desperate enterprises."[46]

It's ultimately a good sign, I think, that American culture is being swept by a tidal wave of natural-disaster movies and reality shows set in the wilderness. A little antimodernism, the celebration of real experiences in nature, can be quite healthy when it comes with an acknowledgment of the environment's power over us. But our response to the threats of industrialism must not be to speak of shutting our borders and protecting our own lands, especially since we import so many of the resources we consume and export so much of the waste we produce. The world is irreversibly cosmopolitan. Pollution crosses borders; global warming is, well, global. To think internationally about natural forces, then, in the Humboldtian fashion, is now both a radical and a critical act. Every nation, every community, will be remade every day, as the flow of people and goods follows the flow of wind and water.[47]

Preservationism and lococentrism treat the world as static, finished. But even if our aim is to be custodians of the land, even if we take some sort of maintenance as our highest priority, we can only maintain some sort of flux. We can't simply "save" environments. We have to interact with nature, have to use natural resources, have to reshape the land.[48] I think the key is to do these tasks delicately, to approach them as experiences, with the right balance of art and science, in a spirit of awe and humility. We should be grateful, as Thoreau pointed out, to the people who walk the land, the explorers, the folks who go fishing: "Early in the morning, while all things are crisp with frost, men come with fishing reels and slender lunch, and let down their fine lines through the snowy field to take pickerel and perch; wild men, who instinctively follow other fashions and trust other authorities than their townsmen, and by their goings and comings stitch towns together in parts where else they would be ripped."[49] Hard work, love, and the explorer's sense of stubborn hope can help make our environmental interactions enriching both to the world and to our own lives. We'll never complete our journeys of enrichment, but there could be countless moments of beauty and communion along the way.

Humboldt on Chimborazo

God keep me from ever completing anything.
—Ishmael, in *Moby-Dick*

"We got quite far up," Humboldt wrote in his journal, "further than I had dared to hope."[1]

We found quite a curious trail, a bare knife-edge. The path was only 5 or 6 inches wide, sometimes a mere 2 inches. To the left the slope was of a horrifying steepness and covered with ice-encrusted snow. To the right there was not a mote of snow, but the slope was strewn with masses of rocks. We had to decide whether it would be better to break our limbs by falling on these rocks, which extended downward for a good 1,000 to 1,300 feet, or if we'd prefer to slide down the snow into an abyss that seemed to stretch even further down. The latter option seemed the more frightful. . . . So, the whole time, we tried to lean our bodies to the right. . . .

The slope was so steep that we had to cling to the side of the mountain with both our feet and our hands. We had injured all our extremities; all of us were bleeding; the stones had sharp corners. You couldn't tell where to plant your foot; the rocks kept shifting in their bed of fine sand. . . .

We found that we still had enough strength to keep going, although we could barely feel our feet from the cold, the snow having melted through the miserable boots that they make in this part of the world. The air temperature was 2.9° C. . . . We ascended still higher; the trail became slightly more gentle, but the cold increased with each step. Also, our breathing was terribly constrained, and what was even more uncomfortable was that everyone felt a kind of malaise, a need to vomit. A local farmer (Chagra de San Juan) who accompanied us quite cheerfully, a very robust man, assured us that his stomach had never before been so upset as it was right then. Along with all this we were bleeding from our gums and lips. The whites of our eyes were injected with blood. . . . We all felt light-headed, and suffered from a constant vertigo, which was quite dangerous, given our situation. All these symptoms of asthenia doubtlessly resulted from the lack of oxygen received by our blood. . . . We climbed for just one more half-hour. It was so cloudy that we could no longer see the top of the mountain. The trail of stones continued upward, though, which gave us a glimmer of hope that we might actually achieve the summit. But a huge crevasse put an end to our attempt. It was at least 600 feet deep and maybe 65 feet across. . . . We were only 1,300 feet from the top. The air temperature was now −1.6° C. . . . We could not have taken the cold much longer, in part just because we had lived in the tropics for the last three years. . . . Our hair, our beards, our eyebrows were covered with ice.

It was June 23, 1802, near Quito, Ecuador. Humboldt, with three companions, had just experienced Mount Chimborazo, then thought to be the highest peak in the Americas, at just under 21,000 feet. This was both Humboldt's proudest and his most humble moment.

Using Laplace's barometric formula, Humboldt calculated that his party had attained an altitude of 19,734 feet above sea level—a new record. Humboldt's native guides had abandoned the attempt almost 3,000 feet lower, insisting that the whole group would perish for lack of air. So Humboldt, and his dashing young friend Carlos Montúfar, and the botanist Aimé Bonpland, and the farmer Chagra de San Juan, could all claim a significant accomplishment. The "terrain" had been "magnificent."[2] Bloodied, frozen, exhausted, nauseous, out of breath, they were nevertheless inspired and exhilarated. They had tested their bodies against the cosmos.

Humboldt was genuinely disappointed a few years later when the physicist Joseph Gay-Lussac reached a greater altitude in a balloon, and he even felt some pangs in the 1820s and '30s when British mountaineers routinely surpassed his climbing record in the mighty Himalayas. He also occasionally expressed frustration that he had not achieved the actual summit of Chimborazo. Despite the dense fog, which would have prevented any panoramic view, reaching the top of such a mountain would have had a powerful symbolic value: there is something deeply satisfying about having no further to go. And yet attaining a summit is also a bit like dying.

Albert Camus summed up one of his core philosophies by suggesting that Sisyphus was happy[3]—that to live fully was to burn with desire to roll your boulder up the hill, was to take pleasure in the actual rolling, in human initiative, all the while realizing that what matters most is the hill itself, and that the hill will always prove more powerful than you. Humboldt, thinking back on his encounters with the cosmos, his contact with fire and sky and rock and ice, swirling currents and gaseous fumes, seemed to come to this same conclusion—seemed, in fact, to weave it into his science.

"Experimental sciences," he wrote in *Cosmos*, "based on the observation of the external world, cannot aspire to completeness; the nature of things, and the imperfection of our organs, are alike opposed to it. . . . No generation of men will ever have cause to boast of having comprehended the total aggregation of phenomena."[4] Indeed, Humboldt had noted in the first volume of his *Personal Narrative* that he came down from studying the volcano of Tenerife, in the Canary Islands, feeling that the different aspects of nature now seemed "still more isolated, more variable, more obscure" than they had before.[5] He strove to unify his views, knowing all the while that they would keep fracturing. And almost every Humboldtian explorer-scientist of the nineteenth century had a similarly keen awareness of humanity's limitations in comprehending the cosmos. Think of J. N. Reynolds returning snow-blind from the volcano of Antuco, daydreaming in New York of blank icebergs and white whales. These men spent their careers gazing into an impenetrable abyss.

That act of gazing, though, that direct engagement, made for a delightfully vertiginous journey along the edge of life itself. The explorer, Humboldt suggested near the beginning of *Cosmos*, eventually learns to take comfort in his own insignificance and impermanence: "in the midst of the universal fluctuation of phenomena and vital forces—in that inextricable net-work of organisms by turns developed and destroyed—each step that we make in the more intimate knowledge of nature leads us to the entrance of new labyrinths; but the excitement produced by a presentiment of discovery, the vague intuition of the mysteries to be unfolded, and the multiplicity of the paths before us, all tend to stimulate the exercise of thought in every stage of knowledge. . . . As men contemplate the riches of nature, and see the mass of observations incessantly increasing before them, they become impressed with the intimate conviction that the surface and the interior of the earth, the depths of the ocean, and the regions of the air will still, when thousands and thousands of years have passed away, open to the scientific observer untrodden paths of discovery."[6]

Humboldt's awareness of the limits of human perception even led him to the protoenvironmentalist insight that we can never know the full ecological impact of our resource use, since all natural forces are so intimately interdependent—which, in turn, dictates that we tread on the earth as lightly as possible. And Humboldt also noted that every act of scientific observation actually changed the environment being observed, and thus always opened up new questions. By volume 2 of *Cosmos*, he was absolutely adamant about recognizing humanity's smallness and nearsightedness: "Weak minds complacently believe that in their own age humanity has reached the culminating point of intellectual progress, forgetting that by the internal connection existing among all natural phenomena, in proportion as we advance, the field to be traversed acquires additional extension, and that it is bounded by a horizon which incessantly recedes before the eyes of the inquirer."[7]

It was inevitable, then, that the final outpouring of Humboldt's thought would break off in midstream—that volume 5 of his great exploration of the *Cosmos* would be left unfinished. What book isn't?

In most of this book, I've tried to rein in my effusive Romantic side—or at least balance it with doses of rationality and realism. But be warned: I'm about to undo the bridle.

———

Christine Evans accompanied me on every step of this journey, and I am eternally grateful for her companionship—and her love. She also put in a great deal of hard work on various aspects of this project and put up with all my obsessions and compulsions, in a spirit of bemused goodwill. She's the strongest person I know.

Samuel Evans Sachs came along while this manuscript was still an unfinished dissertation and I was a seemingly unemployable grad student. He immediately made it all worthwhile. Taking him on various expeditions has been my greatest pleasure in the past couple of years. Christine and Sam also provided excellent motivation to finish the writing. They are my cosmos as well as my hearth. And now it seems we'll be joined by another: "if you want to go explore, the number you should have is four."

For the last fifteen years, Lou Masur has called me up, read me quotations, and kept me sane. I could say that he's the reason I became a historian, but that would suggest only a small fraction of my indebtedness to him. It's notable that our bond has transcended one of the most heated sports rivalries in the world. It's also notable that Lou essentially outlined this book for me at the end of my first year of grad school. He's put up with my Clarence King obsession ever since that fateful moment when he agreed to be my college adviser, despite being warned by a departmental administrator that I was just coming back from a year off, and that I was something of a "special case." I'm eternally grateful for his patience and his friendship.

In the American Studies Program at Yale, Johnny Faragher oversaw the dissertation that became this book with his characteristic vigor. He proved an invaluable guide during my intellectual explorations. As I told many of the students who signed up for his lecture class on the U.S. West, I can't think of anyone in the country with whom I'd rather study frontiers. In addition, Johnny is an expert builder of intellectual community: for six years, the westerners' lunch he organized was a highlight of my week.

I am also thrilled to count myself among the countless Yale grad students in History and American Studies who found a historical voice while taking John Demos's celebrated seminar on the writing of history. John is not only a passionate and devoted teacher and a phenomenal reader, writer, and editor, but also one of the warmest presences in academe. When I think of grad school, I think of having breakfast with John and of meandering conversations that eventually covered all the most important subjects—life, death, reading, writing, teaching, children, and of course the Red Sox. As I left New England in the summer of 2004 to penetrate the Yankee wilderness of upstate New York—John and I actually said good-bye at Fenway Park—I would have scoffed at the notion that when we saw each other again the following spring the Sox would have become world champions. Go figure. Or, as the *Boston Globe* put it: YES!!!

Robert Johnston introduced me to the profession of history and became one of my closest friends. Seven years later, I could not survive academia without him. I continually draw inspiration from his curiosity, intellectual breadth, kindness, joy, thoughtfulness, radicalism, and hope. He gave me a new way of thinking about the history of politics and the politics of history. He also turns a damn good double play and hits to the opposite field.

Both John and Robert, bless them and their right arms, made page-by-page notations for me as I was turning this manuscript into a book. Their deep engagement with my work was one of the best gifts I have ever received.

It was an exhilarating experience to sit in the same room with Bryan Wolf and talk about art and literature. But he is not only a creative thinker: he also happens to be the kind of generous spirit any student would be lucky to encounter. I never believed those stories about teachers giving students the perfect book at the perfect moment, but Bryan proved it could happen when he urged me to read Raymond Williams's *Culture and Society*.

I was incredibly lucky to join the American Studies community at the same time as Victorine Shepard. Vicki became the center of all good feelings in the Hall of Graduate Studies as soon as she moved in. Her office was a second home to me, and it was there and at Royal India, on her lunch breaks, that I had some of my most honest and significant conversations during my six years of grad school. I knew everything was going to be OK when, in my second semester, Vicki had me bottle-feeding a brand-new kitten.

I am also fortunate to have had a group of amazing friends among the Yale grad students—people who knew not only how to sustain an intellectual community but also how to retain their sense of humor. Christian McMillen was the perfect griping and drinking companion; the only problem with him is that his investigative brilliance makes everyone else's research seem thin and shoddy. Finger-popping Sandy Zipp knows creative nonfiction, and there's not much that's better than talking with him about it over lunch at the New York Public Library. Now we just have to find the Ping-Pong tables at the YMCA, and we can actually pretend to be Alfred Kazin and Richard Hofstadter. Bob Morrissey would be a smashing success in any endeavor he cared to try, but I'm grateful he chose history and shared some of his gusto and his wisdom—not to mention his home brew—with me.

Roxanne Willis helped make me feel at home in New Haven from the moment I first visited, and she is one of the strongest voices I have ever heard in the field of humanistic environmental studies. Catherine Whalen has the gift of kindness, and she has patiently been trying to teach me to understand material culture for the last seven years. Brian Herrera, my pop-culture phone-a-friend, taught me new ways to think about movies. Adriane Smith became Sam's Aunt Cookie as soon as he was born; to me, she went from cherished friend and colleague in New Haven to lifeline and anchor in Ithaca. Kip Kosek is a man of few words—but they're choice, and refreshingly dry. Kat Charron always put things in perspective; she also led me to the poetry of Mary Oliver. Rob Campbell was only a shadowy presence at the beginning of my time at Yale, but by the end I had come to depend on his boundless energy and his deep insight into the writing of history. Rob also offered me rides and hikes in California, not to mention a place to sleep and access to his wonderful book collection. Paul Sabin, along with his dog Shadow and his son Eli, was an ideal intellectual companion on regular walks up East Rock; forgoing his company was one of the hardest things about moving to the West Rock neighborhood. Lila Corwin Berman, besides being a softball stalwart and thrower of delightful Chanukah parties, was a model of efficiency for all of us.

In addition, my time as a grad student was powerfully enriched by the presence of Mark Brilliant, Brenda Carter, Carrie Lane Chet, Guy Chet, Caitlin Crowell, Andrew Friedman, Scott Gac, Michael Kral, Dan Lanpher, Roger Levine, Elaine Lewinnek, Kari Main, Karen Marrero, Michelle Nickerson, Amy Reading, Neeraja Sankaran, Scott Saul, Josh Tyrangiel, Wendy Warren, and Owen Williams.

My writing, in particular, was influenced and improved by many of my colleagues, especially within the friendly confines of the Writing History working group; Christian, Sandy, Bob, Roxanne, Catherine, Adriane, Lila, and Scott were all particularly generous with their time and constructive criticism. Three other exceptional editors deserve special mention. Mark Krasovic, a mensch among men, made critical comments that somehow always seemed *exactly right* (when he wasn't razzing me about his Nets' dominance of my Celtics). Barry Muchnick, perhaps the most creatively energetic person I know, paid me what I take to be the ultimate compliment (one I'm definitely not worthy of): he brought the first half of my dissertation with him on vacation. His perceptive reading of those five chapters gave me exactly the kind of juice I needed to reach the end. And Adam Arenson reinspired me to proceed word by word. I am also grateful to him for keeping Writing History alive once I moved to Ithaca.

In the realm of writing outside the academy, I was incredibly fortunate to meet Rebecca Solnit and Jenny Price during my work on this project. I already admired their books, but both also provided huge doses of creative and intellectual nutriment in person and over e-mail, and Rebecca recommended me to the Mesa Refuge, for which I am eternally grateful. My writing retreat at Mesa, in the company of two other inspiring writers, Sarah Silbert and Betsey Herbert, was brief but breathtaking. It's something every struggling key-tapper should experience. Rebecca also introduced me to Paul Slovak, who wound up becoming the editor of this book.

In the environmental community, I had the benefit of an "ideal reader" who was also an old friend: Lisa Hymas gave me some invaluable responses to an early version of chapter 1, and her commentary echoed in my mind through the writing of all the subsequent chapters. Then, at probably the last possible moment in the writing process, Lynn Dixon came through for me with a priceless reference. Who could have guessed that Humboldt was quoted in the latest Jimmy Buffett novel?

Back inside the academy, I learned a huge amount from Jon Butler, Nancy Cott, Wai Chee Dimock, Glenda Gilmore, Paul Gilroy, Gil Joseph, Ben Kiernan, Jim Scott, and Stuart Schwartz—whether in the classroom, at colloquia, or on the softball field. Norma Thompson and Maria Menocal generously offered me an office at the Whitney Humanities Center in my final year of grad school and welcomed me into the Whitney's community of scholars. Jay Gitlin made the Lamar Center a truly collegial and congenial place. Jean-Christophe Agnew embodied intellectual and departmental leadership, and the commentary he wrote for me as an official dissertation reader was an absolute marvel of acute, penetrating, generous criticism. Jean-Christophe was also a constant unofficial adviser at Friday-morning basketball games; now if only I could figure out how to stop his hook shot. Jules Prown taught me a new way of seeing in my very first semester. Alan Trachtenberg was the reason I first got interested in Yale's American Studies program, and it was a pleasure to study with him. George Miles is the best archivist a researcher could ever hope to meet. I admired Steve Pitti and Jonathan Holloway from the moment I met them, and have been trying to learn from them ever since; they're excellent teachers, and I only wish I'd gotten to know them better. Mary Lui has been a rock the last few years, and I will always treasure our trip to Changbaishan, with Robert Johnston and Jace Weaver and the Yale-China Association. Mike Quinn was an absolute joy to be around during his year at the Lamar Center, and I'm grateful to him for his extensive help with my work on Mormon history. It's an embarrassment to the profession that Mike doesn't have a full-time teaching job. Jennifer Brinley not only helped keep me afloat through her work in the financial aid office but also pointed me to the museum at the base of Mount Shasta.

I owe a special debt to the students of the 2002 "Wilderness" seminar. It's a real blessing when a class clicks like this one did. I have no doubt that these folks taught me more than I taught them. Huge thanks, then, to Chris Bogia, Talley Burns, Erin Coughlin, Christian Dietrich, Adam Domby, Florian Freitag, Louise Levy, Greg Pasquali, Sam Ryerson, Amanda Seaton, Brady Weeks, Liba Wenig-Rubenstein, and Aaron Young.

I was also the beneficiary of intellectual support from various professors who were sometimes near and sometimes far. My hearty thanks go out to Larry Buell, Ray Craib, Bill Cronon, and Jim Goodman. Larry showed me the path into the woods. Ray gave me incredibly helpful comments on my prospectus at a crucial time. Bill offered genuine support and also genuine challenges to my approach. And Jim took me aside at Ben Masur's bar mitzvah and invited me to be a part of the first AHA History Slam, which changed my life. Jim, I owe you big-time. I am also grateful to six extraordinary Humboldt scholars: Andreas Daum, Kent Mathewson, Miguel Ángel Puig-Samper, Ingo Schwarz, Laura Dassow Walls, and Petra Werner. All six have been magnanimous and supportive.

At a conference in October 2004, I met another Humboldt aficionado and historian of exploration to whom I have quickly become deeply indebted. Michael Robinson read and critiqued this entire manuscript with incredible thoroughness and acuity and as a result saved me from many blunders. I hardly know him, but he seems to me a model scholar, marked by integrity, nuanced thinking, and generosity.

In addition, I feel compelled to mention the names of a few authors who were particularly inspiring to me as I worked on this project: Rebecca Solnit's books stayed with me all the way through, and I also reached again and again for works by Christopher Lasch, Yi-Fu Tuan, and Raymond Williams. Much of this project would have been impossible without the pioneering research of William H. Goetzmann and Richard G. Woodbridge III. Back in college, Donald Worster's book *Nature's Economy* gave me a glimpse of what I might want to do when I grew up. And for all of my adult life I have been trying to learn from the writerly examples of Joseph Conrad, Guy Davenport, Carey McWilliams, Wallace Stegner, and E. B. White.

More than any other writer, though, Henry David Thoreau haunts this manuscript. I trace my connection with Thoreau to my eleventh-grade English teacher, David Outerbridge, who essentially taught me to write. He also helped me get over sixteen years of near-debilitating shyness. David is quite simply the best teacher I have ever encountered, and he has remained a true friend. Through countless hours of conversation in the Newton North bookroom, and countless pages of extra commentary on my essays, he gave me the greatest gift I could have hoped for: he showed me that he cared about what I thought and who I was. I try my best to follow his example with my own students, but I'm not sure I'll ever be able to express the depth of my gratitude to him, much less match his kindness and perceptiveness.

Several institutions have provided me with generous financial assistance (or a roof) along the way. I am grateful to Yale University; the Jacob K. Javits Fellowship Program of the Department of Education; the Andrew W. Mellon Foundation; the Huntington Library; the Beinecke Library; the Mrs. Giles Whiting Foundation; the Mesa Refuge; the Howard R. Lamar Center for the Study of Frontiers and Borders; and the Cornell University History Department.

My thanks also go out to the various institutions where I was permitted to do research, and to the many archivists, librarians, and staff members who made it possible for me to see materials. I got to know a few of them, but many more remained anonymous. It was my great pleasure to work with Theresa Salazar and Bill Brown at the Bancroft Library; Rob Cox and Roy Goodman at the American Philosophical Society; Pilar Castro at the Archivo Histórico Nacional in Madrid; Sarah Hartwell and Joshua Shaw at the Dartmouth College Special Collections; Peter Blodgett, Anne Mar, Amy Meyers, and Mona Naureldin at the Huntington Library; the amazingly generous Carolyn Kirdahy at Boston's Museum of Science; Jocelyn Wilk of the University Archives and Columbiana Library at Columbia University; Abe Parrish and Fred Musto at the Yale University Maps Collection; and of course George Miles at Yale's Beinecke Library. I am also grateful for the time I was able to spend in the John Muir Center at the Holt-Atherton Special Collections of the University of the Pacific; the Rare Book Room at the New York Public Library; the Special Collections at Princeton University's Firestone Library; the National Archives in Washington, D.C., and College Park, Maryland; the Library of Congress; the Boston Public Library; the Houghton Library at Harvard University; the John Hay Library at Brown University; the Massachusetts Historical Society in Boston; and the Hoover Institution at Stanford University.

One of the pleasures of travel research is the opportunity it affords to visit friends and family. Numerous people made my work that much more rewarding and comfortable by providing beds, coffee, wine, ice cream, and good company. Many thanks to all of you. I could not have managed my stint at the Huntington without the unstinting generosity of Bonnie Kaufman, "The Wizard of Bras." Carole Kaufman, meanwhile, gave me the best haircuts of my life. Peter Mancall had me over to watch football, and Kathleen Donegan broke out some Grande Chartreuse to finish off what had already been a spectacular dinner. Rodger Roundy provided art, music, burritos, gelato, and general hilarity. Naomi Meyer always showed me new restaurants and let me hang out with her cats. Tania Simoncelli drove me up to Point Reyes, hiked with me over hill and dune, and taught me a thing or two about table tennis. Carl Lee and Dorothea Braemer hosted me in both Philadelphia (where they joined enthusiastically in a search for a Humboldt statue) and Berlin. Ben Liebman offered me an incredible apartment in Manhattan while I did some work at Columbia and Princeton, and then offered me a ticket to a certain playoff game that he and I will have nightmares about for the rest of our lives. Tom Iurino came east and helped me find Humboldt in Central Park—and then met me in Kathmandu. He also told me about Theodore Winthrop. Midori Evans met me for walks in Massachusetts and Connecticut and peered with me through the window of Edwin Way Teale's writing cabin. Devah Pager welcomed me to Paris, and, wherever she's been in the world, she has been the best friend and confidante a person could ever hope for. Joe Levine and Carrie Schoenbach gave me their senior-citizen discount every time I stayed with them in Maryland. Sasha, Lauren, Ida, and Harriet Cuttler took me biking, and James, Jeannie, Noah, Ruth, and Anna Evans took me hiking. Bobbie and Stuart Hauser offered their beautiful house in Maine as a writing cabin. David and Diana Evans provided tzimmes, puppet shows, and some of the best conversations east of the Mississippi.

Kirk and Midge Evans did not host me on research trips, but they did welcome me into their family and home with incredible openness and generosity. And Bob Evans, warmest of patriarchs, essentially adopted me as a grandson. I will always be grateful for his spirit and for the enthusiasm he showed for this project. I hope, also, that the spirit of two departed grandparents, Sheldon and Amy Blank, will be evident in these pages to those who knew them.

My sister Debbie has been there for Christine and me so many times and taught us so much that I feel waves of remorse whenever I remember how bratty I was when I was a kid. Debbie and her husband, Ricky Gabor, and their kids, Rebecca, Zachary, and Nathaniel, have been a wonderful part of our lives these last several years, and they were instrumental in facilitating my research into beach conditions on Cape Cod.

A year ago, I was lucky enough to join the faculty at Cornell University, and my colleagues have been incredibly welcoming. Judy Burkhard solved several problems for me before classes had even started, and if I could work as efficiently as Barb Donnell, Maggie Edwards, Jennifer Evangelista, and Katie Kristof, this book would have come out in 1999. Greg Tremblay puts up with my ignorant computer questions, and, I must say, it's a pleasure to work with a high-tech person who knows how to talk to humanists so they can understand. I'm tempted to thank every single Cornell historian by name, because everyone has been so extraordinarily hospitable, but a few people do merit special mention. Sherm Cochran has answered my endless queries with endless patience and kindness; he's always forthcoming yet always discreet. It's been wonderful to be reunited with Ray Craib's invigorating intellect. And I'm thankful for Nick Salvatore's warmth and wisdom every time I meet up with him.

I'm also extremely grateful for the various forms of support offered by Ed Baptist, Holly Case, Derek Chang, Oren Falk, Maria Cristina Garcia, Sandra Greene, Peter Holquist, Michael Jones-Correa, Michael Kammen, Vic Koschmann, Fred Logevall, Tamara Loos, Larry Moore, Mary Beth Norton, Jon Parmenter, Dick Polenberg, and Eric Tagliocozzo.

And then there's Beer and History. If not for this group of Ithaca's rowdiest historians, I might never have spent an evening outside my home or office this past year—so I raise my glass high to Derek Chang, Jeff Cowie, Adriane Smith, Michael Smith, Michael Trotti, and Rob Vanderlan. I'm especially grateful to Adriane for making the initial invitation to join. But the whole Chapter House gang has been extraordinarily welcoming, and our monthly dissections of each other's writing have made me feel truly at home in Ithaca. Thanks to one of those sessions last fall, I received several pints' worth of invaluable advice about how to trim chapter 7 of this book.

The main work I did on the manuscript while based in Ithaca was related to images, and in that department I am indebted to the mastery of Devorah Cohen, an independent researcher who went far beyond the call of duty. At the Cornell Library's Map Department, Susan Greaves committed a huge amount of time and effort to assisting my pursuit of appropriate maps and my bumbling attempts to scan them—and her kindness ultimately saved me from a looming data-storage disaster. I am also grateful to the many kind people who granted me permission to use various images and who made reproductions for me. I don't know all of their names, but they include the incredibly well-organized Lisa Hinzman at the Wisconsin Historical Society; Peter Huestis at the National Gallery of Art; Sylvia Inwood at the Detroit Institute of Arts; Theresa Heinz Kerry; Janea M. Milburn at the Naval Historical

Foundation; Susan Snyder at the Bancroft Library; Michael Wurtz at the Holt-Atherton Special Collections of the University of the Pacific; and Jeff Zilm at the Dallas Museum of Art. My thanks also to the Metropolitan Museum of Art and to Art Resource of New York. The generous policies of Yale University's Beinecke Library and the Dartmouth College Library allowed me to use their reproductions without additional fees.

As this manuscript begins its transformation into a published book, I find myself remembering the AHA History Slam of January 2000 and feeling waves of gratitude to Eric Chinski for first noticing my work there. More recently, Zoe Pagnamenta has been a wonderful advocate and has guided me through this process with effectiveness and grace. At Viking, I've been lucky to have the skilled support of Barbara Campo and Mike Brennan. And I could not have asked for a better editor than Paul Slovak. His enthusiasm and understanding have bowled me over from the beginning, and, all along, he's offered challenging, productive suggestions while remaining keenly sensitive to my concerns.

Finally: my parents, Miriam and Murray Sachs, are the ones who got me into this mess in the first place. I can never thank them enough. And I can't help but think of this book, if it has any merits at all, as owing its existence to thirty-five years of their incredible nurturing and cultivation.

———

There's much—and there are many people—I've either forgotten or had to leave out. You can't study history without becoming acutely aware of mistakes and omissions. But I want to acknowledge one last debt of gratitude: to the subjects of this book. I have tried to be as true to their lives and characters as possible—but how could I not have mangled them? Meanwhile, though, these explorers have managed to temper my curmudgeonly cynicism with a sense of determined hopefulness, and for that I am profoundly thankful. Humboldt, in particular, will be an inspiration to me for the rest of my life. As I struggled to find a vision and a voice, his basic human decency gave me hope, and his books helped me finally articulate my desire to blend scholarship and creativity, analysis and narrative, argumentation and suggestion, scientific precision and artistic intuition.

"Dear Friend," Clarence King once wrote, "men are such undemonstrative creatures that I do not know if I ever said in words how greatly I value our friendship. If I have not, no matter, you have felt my meaning." I hope you have.

August 2005
Ithaca, New York

— CHRONOLOGY —

1769 Humboldt is born in Berlin. Napoleon is born in Corsica.
1790 Humboldt travels through Europe with Georg Forster, naturalist of Captain Cook's second expedition (1772–75).
1796 Humboldt's mother dies.
1798 Humboldt meets botanist Aimé Bonpland in Paris.
1799 J. N. Reynolds is born to a Quaker family in Pennsylvania.
1799 Humboldt launches his expedition to the Americas.
   June 5, 1799 Humboldt departs La Coruña, Spain, aboard the corvette *Pizarro*.
   June 21, 1799 Climbs the volcano of Tenerife, in the Canary Islands.
   July 16, 1799 Lands at Cumaná, Venezuela.
   February and March 1800 Crosses the high desert plains, or llanos.
   March 30, 1800 Launches exploration of the Orinoco and its tributaries. Covers 2,250 kilometers by boat over seventy-five days.
   November 1800–March 1801 First trip to Cuba.
   April–August 1801 Explorations of the regions around Cartagena and Bogotá.
   September 1801–January 1802 Travels overland, crossing the Andes, from Bogotá to Quito, Ecuador.
   June 23, 1802 Attempt on Mount Chimborazo.
   September 2, 1802 Arrives in Lima, Peru.
   January 3, 1803 Sails for Acapulco.
   February 1803–January 1804 Mexico.
   January–April 1804 Second trip to Cuba.
   June 1804 Meets with Jefferson in Washington shortly after Lewis and Clark start up the Missouri; returns to Europe in July.
August 1804 Humboldt arrives back in France. Settles in Paris, where he bases his operations until 1827.
1805 Humboldt prints *Essay on the Geography of Plants*.
1808 J. N. Reynolds, orphaned, moves to the Ohio frontier with his stepfather, Job Jeffries.
1808 Humboldt publishes *Views of Nature*.
1808 Humboldt completes *Political Essay on the Kingdom of New Spain*.
1810 Humboldt publishes *Views of the Cordilleras and Monuments of the Indigenous Peoples of the Americas*.
1814 Humboldt publishes the first volume of *Personal Narrative of a Journey to the Equinoctial Regions of the New Continent*. The second and third volumes follow in 1819 and 1825.
1818 John Cleves Symmes announces his proposed expedition to the Arctic and then on to the center of the earth, to prove his "Theory of Concentric Spheres and Polar Voids," with Humboldt as his "protector."
1820 A man calling himself Captain Adam Seaborn publishes *Symzonia*.
1825–26 Symmes and J. N. Reynolds go on a lecture tour to the major cities of the East Coast.
August 1826 Reynolds meets Secretary of the Navy Samuel Southard in Washington.
1826–29 Reynolds meets with politicians and scientists and lobbies for a South Sea Expedition.
1827 Reynolds publishes *Remarks on a Review of Symmes' Theory*.
April 1827 Humboldt leaves Paris, resettling in Berlin a month later.

November 1827 Humboldt launches his "Cosmos" lecture series at the University of Berlin (now called Humboldt University).

June 1828 Reynolds is hired as a special agent of the Navy Department to research the sealing and whaling industries.

January 1829 The House of Representatives passes the bill launching Reynolds's expedition.

March 1829 Andrew Jackson takes office as president and cancels the expedition.

April 1829 Humboldt leaves Berlin on his expedition to Siberia and the Chinese border.

September 14, 1829 Humboldt turns sixty.

October 1829 At the age of thirty, Reynolds completes the arrangements for his privately sponsored expedition and sails for the South Sea on the brig *Annawan*.

November 1829 Humboldt asks Czar Nicholas in St. Petersburg to establish geomagnetic and meteorological observation stations throughout Russia.

December 1829 Humboldt arrives in Berlin, having traveled about fifteen thousand kilometers since April.

January–April 1830 Reynolds skirts the 60th Parallel, exploring Antarctic islands.

April–July 1830 Based at Valparaiso, Chile, Reynolds makes several inland excursions through northern Chile and southern Peru.

May 1830 President Jackson approves the Indian Removal Act.

July 1830–April 1831 Reynolds explores Araucania, climbing volcanoes and meeting Indians, with his friend John Frampton Watson.

December 1830 Ralph Waldo Emerson starts reading Humboldt's *Personal Narrative*.

1831 Humboldt publishes *Fragments of Asiatic Geology and Climatology*.

October 1832 Reynolds, again based in Valparaiso after more inland expeditions, signs on as private secretary to Commodore John Downes of the frigate *Potomac*.

December 1832–May 1834 Reynolds sails to Boston aboard the *Potomac* after exploring the Galápagos Islands.

July–October 1834 Reynolds donates his specimen collections from South America and the Antarctic to the Boston Society of Natural History.

April 1835 Wilhelm von Humboldt dies.

Spring 1835 Reynolds publishes *Voyage of the Potomac*.

October 1835 Reynolds receives an honorary degree from Columbia.

1836–39 Humboldt publishes *Critical Examination of the History of the Geography of the New Continent*.

April 1836 Reynolds addresses Congress on the subject of a new South Sea expedition.

May 1836 President Jackson authorizes the United States Exploring Expedition.

May 1837 Banks start failing and the Panic of 1837 begins, launching an unprecedented economic depression throughout the country. The price of western lands collapses.

June 1837 Reynolds publishes the first of his letters attacking the conduct of Navy Secretary Mahlon Dickerson, who has managed to delay the sailing of the expedition for more than a year.

January 1838 War Secretary Joel Poinsett takes over primary responsibility for the expedition.

April 20, 1838 Lieutenant Charles Wilkes, who distrusts Reynolds and all non-navy personnel involved in the expedition, is appointed commander.

April 21, 1838 John Muir is born in Dunbar, Scotland.

August 1838 The Wilkes Expedition sails, without Reynolds.

1839–44 Reynolds works as a lawyer and lectures extensively in New York as an advocate for the working classes and a supporter of Whig politicians, especially Henry Clay.

May 1839 Reynolds publishes "Mocha-Dick."

June 1839 Reynolds publishes "A Leaf from an Unpublished Manuscript" and "A Leaf from an Unpublished Journal."

1841 Reynolds publishes *Pacific and Indian Oceans; or The South Sea Surveying Expedition: Its Inception, Progress, and Objects*.

January 10, 1841 George Wallace Melville is born in New York City.

January 6, 1842 Clarence Rivers King is born in Newport, Rhode Island.

1843 Humboldt publishes *Central Asia*.

December 1843 Reynolds publishes "Rough Notes of Rough Adventures."

1845 Humboldt publishes volume 1 of *Cosmos*. Volumes 2 through 5 follow between 1847 and 1862.

1845 John Charles Frémont launches his Third Expedition, to explore the Great Basin.

1846–48 The United States fights its war with Mexico.

1847 Lieutenant James W. Abert laments the destruction of the prairies and the oppression of Indians.

1848 Revolutions in Prussia and France.

June 1848 Clarence King's father, James King, dies near Formosa while engaged in the China trade.

1851 Herman Melville publishes *Moby-Dick*.

1851 The Mariposa Battalion attacks the Indians of Yosemite Valley.

August 1851 Yale Professor Benjamin Silliman visits Humboldt in Berlin.

1853 Elisha Kent Kane leads an Arctic expedition in search of Sir John Franklin.

August 1858 J. N. Reynolds dies in Canada.

1859 Charles Darwin publishes *On the Origin of Species*.

April 1859 Frederic Church displays his painting *The Heart of the Andes* in New York.

May 6, 1859 Humboldt dies in Berlin at age eighty-nine.

September 1860 King enrolls at Yale's Sheffield Scientific School; studies with Professors George Brush and James Dwight Dana.

June 1861 Theodore Winthrop is buried at Yale.

Fall 1862 Having finished his degree at Yale, King travels to Boston to hear Louis Agassiz lecture on glaciology.

January–March 1863 King lives in New York, affiliating himself with the Association for the Advancement of Truth in Art.

April–September 1863 King crosses the prairies with his bosom companion, Jim Gardiner, arrives in California, meets William H. Brewer, and begins work with the California Geological Survey.

September 1863 King climbs Mount Shasta for the first time.

1864 George Perkins Marsh publishes *Man and Nature; or, Physical Geography as Modified by Human Action*.

January 1864 King finds a telltale fossil, the cephalopoda, in Hell's Hollow, thus dating the gold-bearing regions of California.

July 1864 King makes his first attempt to climb Mount Whitney but is stopped just below the summit by a sheer granite cliff.

Summer–Fall 1864 Congress grants Yosemite Valley to the state of California to be preserved "for public use, resort and recreation," and King and Gardiner do the park's first official survey.

November 1864 The Third Colorado Regiment commits the Sand Creek Massacre.

Summer 1866 King returns to Yosemite with the photographer Carleton Watkins.

October 1866 The Union Pacific Railroad crosses the hundredth meridian.

March 1867 King launches his Geological Exploration of the 40th Parallel and hires the photographer Timothy H. O'Sullivan. Fieldwork will continue until 1873.

September 1867 John Muir launches his thousand-mile walk to the Gulf of Mexico.

September 14, 1869 Americans throughout the United States celebrate the Humboldt Centennial.

September 1870 King climbs Shasta again and discovers active glaciers.

1871 Charles Francis Hall achieves a new "Farthest North" on the *Polaris*.

1871 Darwin publishes *The Descent of Man*.

June 1871 King makes another attempt on Whitney, and believes he has succeeded.

August 1871 King meets Henry Adams in Wyoming.

October 1871 Muir finds his first glacier in the Sierras.

February 1872 King publishes *Mountaineering in the Sierra Nevada*.

March 1872 Yellowstone becomes the first national park.

November 1872 King exposes the Great Diamond Hoax.

1873 George Wallace Melville sails as an assistant engineer aboard the *Tigress*, sent north in search of Hall and the *Polaris*.

August 1873 W. A. Goodyear announces to the California Academy of Sciences that King actually climbed Sheep Rock in 1871, having mistaken it for Mount Whitney.

September 1873 King climbs the real Mount Whitney.

December 1873 King completes his fieldwork in the West.

Spring 1875 Muir gets caught in a snowstorm atop Mount Shasta; the weather forces him to lie in the same spot for seventeen hours.

1876 The Centennial Exposition in Philadelphia displays survey photographs by Timothy H. O'Sullivan.

June 1877 King delivers the commencement address at Yale's Sheffield Scientific School: "Catastrophism and the Evolution of Environment."

1878 King publishes *Systematic Geology*.

December 1878 John Wesley Powell founds the Cosmos Club in Washington, D.C.

1879–81 King serves as the first director of the U.S. Geological Survey.

Summer 1879 Muir takes his first trip to Alaska.

July 1879 George Washington De Long and George Wallace Melville launch the *Jeannette* expedition to the Arctic. The ship sails out the Golden Gate from San Francisco Bay.

September 1879 The *Jeannette* plows into the Arctic ice pack.

April 1880 Muir marries Louise ("Louie") Strentzel, and then three months later departs again for Alaska, despite the fact that Louie is suffering from morning sickness.

May 1881 Muir joins the crew of the *Corwin* and sails north out of San Francisco to look for survivors of the *Jeannette* expedition.

June 1881 The *Jeannette* finally sinks.

Summer 1881 The *Corwin* cruises through the Bering Strait, searching the Alaskan and Siberian coasts; Muir consults with natives and takes careful botanical notes.

Autumn 1881 Muir works on two wilderness preservation bills.

September 11, 1881 The *Jeannette* expedition's three boats attempt to cross the Laptev Sea from one of the New Siberian Islands to the Russian mainland. One boat is lost in a gale.

September 19, 1881 Melville's crew is saved by three Tungus hunters.

March 1882 Melville finds the body of his commander, George Washington De Long.

September 1882 Melville commits his wife, Henrietta, to the Pennsylvania State Asylum.

1882–83 Congress and the navy both conduct investigations into the *Jeannette* disaster.

1882–83 King travels in Europe, collecting art and mining investments.

1883–84 Franz Boas does his fieldwork on Baffin Island.

1884 Melville goes back to the Arctic, as a member of the Greely Relief Expedition.

1885 Melville publishes *In the Lena Delta*.

August 1887 President Grover Cleveland promotes Melville above forty-four senior officers to serve as chief engineer of the navy.

September 1888 King secretly marries Ada Copeland.

1892 Muir founds the Sierra Club.

July 1893 Frederick Jackson Turner lectures on "The Significance of the Frontier in American History" while the nation endures the worst financial panic it has ever seen.

October 1893 King gets arrested in Central Park and committed to the Bloomingdale Asylum.

1894 Muir publishes his first major book, *The Mountains of California*.

September 1895 King publishes "Shall Cuba Be Free?"

1896 Franz Boas begins teaching at Columbia.

1898 The United States goes to war with Spain.

Summer 1899 The Harriman expedition sails up and down the Alaskan coastline.

1900–1930 Edward S. Curtis travels throughout North America photographing Native Americans.

Summer 1900 King visits the Klondike.

1901 Muir publishes *Our National Parks.*
December 24, 1901 Clarence King dies alone in Phoenix, Arizona.
1903 Mary Austin publishes *The Land of Little Rain.*
1904 Melville receives the Russian Order of St. Stanislaus.
1904 and 1905 Thomas Eakins paints two portraits of Melville, now a rear admiral.
1905 Melville publishes *Abandon the Philippines.*
1906–13 Muir fights, ultimately unsuccessfully, to save the Hetch Hetchy Valley.
April 1909 Robert Peary (probably) reaches the North Pole.
1911 Muir finally follows in Humboldt's footsteps to South America.
1912 George Wallace Melville dies in Philadelphia.
1914 John Muir dies in California.
1915 Muir's literary executor, William Frederic Badè, publishes *Travels in Alaska.*
1916 Congress passes the National Park Service Act.
1917 Badè publishes *The Cruise of the* Corwin.
1923 Carl Sauer joins the Geography Department at Berkeley.
1942 C. Hart Merriam, perhaps the last great Humboldtian naturalist, dies in California, after spending the final twenty-five years of his life studying the native peoples of the Mount Shasta region.
1943 Carey McWilliams publishes *Brothers under the Skin* (one year after *Ill Fares the Land*).
1949 Aldo Leopold's *A Sand County Almanac* is published (posthumously).
1955 Lewis Mumford organizes a conference entitled "Man's Role in Changing the Face of the Earth" to honor the career of nineteenth-century Humboldtian George Perkins Marsh.
1956 The Sierra Club and its allies manage to block the proposed dam in Dinosaur National Monument.
1962 Rachel Carson publishes *Silent Spring.*
1964 Murray Bookchin begins publishing his tracts on social ecology.
1970 The first Earth Day celebration is held in the United States.
1990s The SUV revolution sends the average fuel efficiency of American cars spiraling downward after nearly two decades of improvement.
2000 Robert Redford begins sending letters and e-mails to concerned citizens asking them to help save the Arctic National Wildlife Refuge from oil companies.
September 2004 Michael Shellenberger and Ted Nordhaus release their pamphlet *The Death of Environmentalism.*
October 2004 Diverse Humboldtians unite in New York to celebrate the two-hundredth anniversary of Humboldt's visit to the United States.

For a slightly more extensive set of notes, see Aaron Jacob Sachs, "The Humboldt Current: Avant-Garde Exploration and Environmental Thought in 19th-Century America" (Ph.D. diss., Yale University, 2004).

Prologue: Humboldt in America: 1804–2004

1. Humboldt to Delambre (of the National Institute in Paris), November 25, 1802, Lima, in *Lettres Américaines d'Alexandre de Humboldt*, ed. E. T. Hamy (Paris: Librairie Orientale et Américaine, 1904), p. 147, my translation.

2. Humboldt's letter is quoted in Helmut de Terra, *Humboldt: The Life and Times of Alexander von Humboldt, 1769–1859* (New York: Knopf, 1955), p. 173. Also see Felix M. Wasserman, "Six Unpublished Letters of Alexander von Humboldt to Thomas Jefferson," *Germanic Review* 29 (October 1954), pp. 191–200, and de Terra, "Alexander von Humboldt's Correspondence with Jefferson, Madison, and Gallatin," *Proceedings of the American Philosophical Society* 103 (December 1959), pp. 783–806. Much of the Humboldt-Jefferson correspondence is also available online, in the digital version of the Library of Congress's Thomas Jefferson Papers, at http://memory. loc.gov/ammem/mtjhtml/mtjhome.html. But the best place to go for Humboldt's correspondence with any American is the recently published volume edited by Ingo Schwarz, *Alexander von Humboldt und die Vereinigten Staaten von America: Briefwechsel* (Berlin: Akademie Verlag, 2004), which has commentary in German but includes the letters in their original language.

3. Humboldt, Scientific Diary, "Voyage de la Havane à Philadelphia, 1804," American Philosophical Society, Humboldt Collection, BH88.42, my translation.

My account of the storm comes from Humboldt's multivolume personal diary, much of which is published in *Reise auf dem Río Magdalena, durch die Anden und Mexico*, vol. 1, ed. Margot Faak (Berlin: Akademie Verlag, 1986). This volume has commentary in German but prints the journal entries in their original language—usually French, sometimes German, occasionally Spanish. For the storm, see pp. 396–99.

For the number of specimen cases Humboldt had with him, see his letter to Gottlob Christian Kunth of August 3, 1804, in *Lettres Américaines*, pp. 170–71. On the other hand, he says that he actually had forty-two cases with him at the end of his trip in his *Personal Narrative of Travels to the Equinoctial Regions of America, 1799–1804*, trans. Thomasina Ross. (London: Henry G. Bohn, 1852–53; orig. pub. in French between 1814 and 1825), 1: xii.

4. As published in Faak, p. 397, my translation (from the original French).

5. Among the relatively recent studies of Humboldt, besides de Terra's, are Douglas Botting, *Humboldt and the Cosmos* (London: Michael Joseph, 1973); Anne M. Macpherson, "The Human Geography of Alexander von Humboldt" (Ph.D. diss., University of California, Berkeley, 1972); and Hanno Beck, *Alexander von Humboldt*, 2 vols. (Wiesbaden: F. Steiner, 1959 and 1961)—this is one of the more authoritative works on Humboldt, but it has never been translated from the German. One of the best recent essays on Humboldt appears in the middle of a book on Thoreau: see Laura Dassow Walls, *Seeing New Worlds: Henry David Thoreau and Nineteenth-Century Natural Science* (Madison: University of Wisconsin Press, 1995), pp. 76–166.

6. Humboldt, while visiting Madrid in 1799 and attempting to get permission to explore the Spanish colonies in America, wrote a "Memoria" to the king of Spain as well as a summary of his career up until that point. In both letters, he made important and revealing comments about his approach to science and exploration. Both documents, written in French and dated March 11,

1799, can be found at the Archivo Histórico Nacional in Madrid, Sección Estado, legajo 4709. These quotes are my translations.

The expert on Humboldt's time in Madrid is Miguel Ángel Puig-Samper, who just discovered these documents a few years ago, and for whose help I am grateful. He has reprinted these documents in an article, "Humboldt, un Pruisano en la Corte del Rey Carlos IV," *Revista de Indias* 59 (Mayo–Agosto 1999), quotes on pp. 337 and 354.

7. On Humboldt's visit to the States, see H. R. Friis, "Alexander von Humboldts Besuch in den Vereinigten Staaten von Amerika," in *Alexander von Humboldt: Studien zu Seiner Universalen Geisteshaltung*, ed. Joachim H. Schultze (Berlin: De Gruyter, 1959), pp. 142–95. A condensed version of Friis's article was translated into English and printed in Francis Coleman Rosenberger, ed., *Records of the Columbia Historical Society of Washington, D.C., 1960–1962* (Washington, D.C.: Columbia Historical Society, 1963), pp. 1–35. There is some uncertainty as to the exact dates of Humboldt's U.S. sojourn, but it is clear that he arrived no earlier than May 20 and departed no later than July 9. Other relevant sources include Helmut de Terra, "Studies of the Documentation of Alexander von Humboldt" (two articles with the same title), *Proceedings of the American Philosophical Society* 102 (February and December 1958), pp. 136–41 and 560–89; and de Terra, "Motives and Consequences of Alexander von Humboldt's Visit to the United States (1804)," *Proceedings of the American Philosophical Society* 104 (June 1960), pp. 314–16. I have used May 24 as Humboldt's arrival date because while going through the collection at the American Philosophical Society I found an official U.S. Customs statement showing that Humboldt's bags had been inspected in Philadelphia on that date.

8. Quoted in Silvio A. Bedini, *Jefferson: Statesman of Science* (New York: Macmillan, 1990), p. 354.

9. Gallatin to his wife, Hannah Nicholson Gallatin, June 6, 1804, American Philosophical Society, Collection of Materials Relating to Humboldt, BH88.1.

10. Peale quoted in Botting, p. 171.

11. Quoted in de Terra, *Humboldt*, p. 173.

12. In 1851, the eminent Yale scientist Benjamin Silliman, passing through Berlin, reported that Humboldt spoke of having visited Jefferson at Monticello in 1804—but there is no other documentary evidence suggesting that such a visit took place, and it seems likely that either Humboldt or Silliman made a mistake. Humboldt certainly did visit Jefferson at the president's house in Washington, and he also made a stagecoach trip to Mount Vernon, George Washington's country estate. The confusion has also been compounded by a mistranslation of one of Humboldt's letters to Jefferson by Helmut de Terra, who then took the liberty of imagining a Monticello scene for his biography of Humboldt—a scene that has been copied in many subsequent studies of Humboldt. In general, German scholars have realized that "the conversations between Jefferson and Humboldt beneath the trees of Monticello are a legend," in large part thanks to research presented by H. R. Friis (see note 7), who determined that Humboldt almost certainly didn't have enough time to travel to Monticello and back, and that, in any case, there was not even any evidence that Jefferson himself had spent time at Monticello during the weeks of Humboldt's tenure in the United States. See Ingo Schwarz, "Alexander von Humboldt's Visit to Washington and Philadelphia, His Friendship with Jefferson, and His Fascination with the United States," in "Alexander von Humboldt's Natural History Legacy and Its Relevance for Today," special issue no. 1, *Northeastern Naturalist* (2001), p. 49; and Benjamin Silliman, *A Visit to Europe in 1851* (New York: George P. Putnam, 1853), 2:319. Humboldt's letter to Jefferson is dated June 12, 1809, Paris, and can be found in the Thomas Jefferson Papers at the Library of Congress (see note 2). De Terra's mistranslation is in his article, "Alexander von Humboldt's Correspondence," p. 790. Most of his translations are in fact reliable.

13. See Donald Jackson's important study, *Thomas Jefferson and the Stony Mountains: Exploring the West from Monticello* (Urbana: University of Illinois Press, 1981).

14. Humboldt lent his main map of Mexico *(Carte générale du Royaume de la Nouvelle Espagne)* to Jefferson and Gallatin to be copied (officially) for the government's records; a letter from Humboldt to Gallatin on June 20, 1804, served as a reminder that he needed to get it back before his (imminent) departure. But Pike made his copy surreptitiously and then utilized it a few years

later (without attribution) when he published his map of "The Internal Part of Louisiana" (1810), supposedly based on his own explorations in 1805–7. See de Terra, "Alexander von Humboldt's Correspondence," p. 802 (Humboldt to Gallatin) and p. 792 (Humboldt to Jefferson, December 20, 1811, noting Pike's plagiarism). On Pike, also see Jackson, pp. 104 and 273–74. Both the official copy and Pike's bastardized one went on to exert a great influence on subsequent exploration of the American Southwest.

15. Emerson is quoted in Louis Agassiz, *Address Delivered on the Centennial Anniversary of the Birth of Alexander von Humboldt; Under the Auspices of the Boston Society of Natural History; with an account of the evening reception* (Boston: Boston Society of Natural History, 1869), pp. 71–72, in the "account of the evening reception" section. Emerson also said of Humboldt that "a university, a whole French Academy, travelled in his shoes."

On Thoreau and Humboldt, see Donald Worster, *Nature's Economy: A History of Ecological Ideas* (New York: Cambridge University Press, 1977), p. 65; and Walls, pp. 76–166.

16. Humboldt was particularly interested in painting. Volume 2 of his magnum opus, *Cosmos*, contains a long section on landscape painting that had a significant impact on art both in Europe (in part through John Ruskin, an avid reader of Humboldt's works) and in America. For the United States, a good source is Barbara Novak, *Nature and Culture: American Landscape and Painting, 1825–1875* (New York: Oxford University Press, 1980), pp. 66–77 and 138–39. And see Humboldt, *Cosmos: A Sketch of a Physical Description of the Universe*, trans. E. C. Otté (London: Henry G. Bohn, 1849), 2:440–57.

17. Albert Gallatin, *A Synopsis of the Indian Tribes within the United States* (Cambridge, Mass.: American Antiquarian Society, 1836). Humboldt and Gallatin cemented their relationship while Gallatin served as an envoy in Paris, from 1816 to 1823. See de Terra, "Alexander von Humboldt's Correspondence," pp. 800–801.

18. On Humboldt and Whitman, see David S. Reynolds, *Walt Whitman's America* (New York: Knopf, 1995), pp. 244–46.

In Whitman, see the late poem "Kosmos" in *Leaves of Grass* (New York: New American Library, 1980), p. 310. "Kosmos" is from "Autumn Rivulets," originally added to *Leaves of Grass* in 1881. In *Song of Myself* (1855), the poet proclaims himself "Walt Whitman, a kosmos," p. 67. Whitman may have chosen this spelling in order to refer both to the ancientness of the concept and to its modern coinage by Humboldt (*Kosmos* was the original, German title of Humboldt's book, though it was rendered as *Cosmos* in English editions).

I've generally used the translation of *Cosmos* by E. C. Otté, in five volumes, the second volume of which is cited in note 16 (London: Henry G. Bohn, 1849, 1851, 1852, and 1870; note that the very first edition of volume 1 of this translation actually came out in 1848). But there are several editions in English, published both in Europe and in America, including a slightly earlier translation by Elizabeth and Edward Sabine (volume 1 of which came out in London in 1846). The original German edition, in five volumes, with an index of more than a thousand pages, was published between 1845 and 1861. Note that the British editions were just as common as the American in the United States. The first two volumes of Otté's translation were also published in paperback in 1997 by the Johns Hopkins University Press (Baltimore). These excellent, accessible volumes are reprints of an American edition of 1858 by Harper and Brothers and come with introductions by two accomplished Humboldt scholars, Nicolaas A. Rupke (vol. 1) and Michael Dettelbach (vol. 2).

19. One of the cards (about five inches long and two inches wide) on which the letter was printed can be found in the Rare Books Department of the Boston Public Library, "Alexander von Humboldt on the Fugitive Slave Law"; the letter, to the New York–based abolitionist John Matthews, is dated October 12, 1858.

20. Nevertheless, as David McCullough lamented in a 1973 essay, "it is doubtful that one educated American in ten today could say who exactly Humboldt was or what he did." And the situation has hardly improved. See McCullough, "Journey to the Top of the World," in *Brave Companions: Portraits in History* (New York: Touchstone, 1992), pp. 3–19, quote on p. 5. This essay was first published as "The Man Who Rediscovered America," *Audubon* 75 (September 1973), pp. 50–63.

21. Jefferson to Humboldt, June 9, 1804; letter reprinted in de Terra, "Alexander von

Humboldt's Correspondence," p. 789. Humboldt also answered Jefferson's queries in writing, in the form of a nineteen-page report, which Jefferson then incorporated into a compilation volume of information about what he called "Louisiana-Texas." This volume, like Humboldt's map, had a major impact on Jefferson's further plans to explore the Southwest. See Bedini, pp. 355–56.

22. Jefferson's instructions to Lewis are reprinted in full in Jackson, pp. 139–44. For a thorough discussion of various myths of the West, see Henry Nash Smith's *Virgin Land: The American West as Symbol and Myth* (1950; repr., Cambridge, Mass.: Harvard University Press, 1978).

23. For a much darker reading of Jefferson's intentions with regard to the Lewis and Clark expedition, see Mark Spence, "Let's Play Lewis and Clark: Strange Visions of Nature and History at the Bicentennial," in *Lewis and Clark: Legacies, Memories, and New Perspectives*, ed. Kris Fresonke and Mark Spence (Berkeley: University of California Press, 2004), pp. 219–38.

24. Humboldt, *Essai Politique sur le Royaume de la Nouvelle Espagne* [Political Essay on the Kingdom of New Spain], tome II (Paris: F. Schoell, 1811; orig. pub. 1808), p. 353, my translation. Humboldt referred to this particular spot, just to the southeast of Havana, as "one of the most delightful locations in the New World," with such "rich and fertile" soil that it certainly should have been used to produce "harvests that contribute to the nourishment of humanity" rather than luxury trade goods, mere "objects of commerce."

25. Humboldt, *Personal Narrative*, 3:272; and letter to William Thornton of June 20, 1804, Philadelphia, in Ulrike Moheit, ed., *Alexander von Humboldt: Briefe aus Amerika, 1799–1804* (Berlin: Akademie Verlag, 1993), pp. 299–300, my translation. While it is impossible to know exactly what Humboldt and Jefferson said to each other, it seems highly likely that Jefferson's stated interest in "red" and "black" people would have spurred Humboldt's usual rebukes with regard to Native Americans and slaves, in addition to his usual cautions about the scientific understanding of natural resources.

26. Quoted in Philip S. Foner, "Alexander von Humboldt on Slavery in America," *Science and Society* 47 (Fall 1983), p. 335. In 1854, frustrated that the United States still had not abolished slavery and furious about the passage of the Fugitive Slave Law, Humboldt had written that the United States "presents to my mind the sad spectacle of liberty reduced to a mere mechanism in the element of utility. . . . Hence indifference on the subject of slavery." Humboldt, letter of July 31, 1854, in *Letters of Alexander von Humboldt to Varnhagen von Ense, from 1827 to 1858*, translated from the second German edition by Friedrich Kapp (New York: Rudd and Carleton, 1860), p. 305. Also see Foner, ed., *Alexander von Humboldt über die Sklaverei in den USA* (Berlin: Humboldt-Universität zu Berlin, 1981), a book printed in both German and English that preceded Foner's article and includes more material—for instance, some long excerpts from letters written by Humboldt to prominent American abolitionists, several of which were published in American newspapers in the 1850s—and E. R. Brann, *The Political Ideas of Alexander von Humboldt: A Brief Preliminary Study* (Madison, Wis.: Littel Printing, 1954), pp. 10–17.

27. Humboldt to Jefferson, June 12, 1809, reprinted in de Terra, "Alexander von Humboldt's Correspondence," p. 790. The reference is to Humboldt's *Essai Politique sur le Royaume de la Nouvelle Espagne* (see note 24). Also see Schwarz, *Alexander von Humboldt und die Vereinigten Staaten von America*, pp. 112–14.

28. Humboldt, *Personal Narrative*, 3:271.

29. From the *Philadelphia Public Ledger*, July 8, 1853, quoted in Thomas Bailey, *A Diplomatic History of the American People*, 9th ed. (Englewood Cliffs, N.J.: Prentice-Hall, 1974), p. 285. Many thanks to Paul Sabin for leading me to this quotation.

Chapter 1: "The Chain of Connection"

1. For Frémont's explanation of the names, see his *Geographical Memoir upon Upper California* (1848; repr., San Francisco: Book Club of California, 1964), p. 9.

2. Frémont, *Report of the Exploring Expedition to the Rocky Mountains in the year 1842, and to Oregon and north California in the years 1843–'44* (Washington, D.C.: Blair and Rives, 1845).

3. Frémont, *Geographical Memoir*, p. 12.

4. Ibid., p. 9. When Frémont refers to Humboldt as the *"Nestor of scientific travellers,"* he is

quoting directly from a fellow admirer, the surveyor Joseph N. Nicollet, who served as Frémont's practical mentor in the field. See J. N. Nicollet, *Report intended to illustrate a map of the hydrographical basin of the Upper Mississippi River* (Washington, D.C.: Blair and Rives, 1843), p. 95. On Humboldt and Frémont, see William H. Goetzmann, *New Lands, New Men: America and the Second Great Age of Discovery* (1986; repr., Austin: Texas State Historical Association, 1995), pp. 168–73 and 181–84.

5. *New York Times*, September 15, 1869, p. 1.

6. Andreas Daum has documented a kind of Humboldt craze in American popular culture in the mid-to-late nineteenth century; see, for instance, his article "Celebrating Humanism in St. Louis: The Origins of the Humboldt Statue in Tower Grove Park, 1859–1878," *Gateway Heritage* 15 (Fall 1994), pp. 48–58. I am grateful to Professor Daum for sharing his work with me, and for organizing the conference "Alexander von Humboldt and North America," held at the German Historical Institute in Washington, D.C., on June 3–5, 2004.

7. Agassiz's speech was also published separately: Agassiz, *Address Delivered on the Centennial Anniversary of the Birth of Alexander von Humboldt; Under the Auspices of the Boston Society of Natural History; with an account of the evening reception* (Boston: Boston Society of Natural History, 1869).

8. See, for instance, Donald Worster, *Nature's Economy: A History of Ecological Ideas* (New York: Cambridge University Press, 1977), p. 192, and Anna Bramwell, *Ecology in the Twentieth Century: A History* (New Haven: Yale University Press, 1989), pp. 39–63.

9. Humboldt, *Cosmos: A Sketch of a Physical Description of the Universe* (London: Henry G. Bohn, 1849), 1:36.

Note that Humboldt himself was a fanatical empiricist, classifying about sixty thousand plant species during his American expedition (this figure is cited in Helmut de Terra, *Humboldt: The Life and Times of Alexander von Humboldt, 1769–1859* [New York: Knopf, 1955], p. 375)—and scholars have on occasion mistakenly identified him as merely another encyclopedist. But Humboldt also criticized list making: "The depiction of the whole picture of nature requires more, however, than detailed research into the individual organisms and cannot be considered complete if it merely lists every form of life, every phenomenon and natural process. The tendency to reduce everything observed and collected to countless fragments must be overcome and the orderly thinker must try to avoid the danger of empirical profusion" (quoted in Adolf Meyer-Abich, *Alexander von Humboldt, 1769/1969* [Bonn: Inter Nationes, 1969—in English], p. 93).

10. Humboldt, *Personal Narrative of Travels to the Equinoctial Regions of America, 1799–1804* (London: Henry G. Bohn, 1852–53), 1:x–xi. Note that in the text I sometimes follow the Penguin Classics edition and render this title as *Personal Narrative of a Journey to the Equinoctial Regions of the New Continent* (London: Penguin, 1995).

11. Humboldt, *Cosmos*, "Introduction," 1:1; Wilhelm quoted in Ottmar Ette, *Literature on the Move* (Amsterdam: Rodopi, 2003), p. 125.

12. See Mary Louise Pratt, *Imperial Eyes: Travel Writing and Transculturation* (London: Routledge, 1992), p. 140. The phrase "intellectual annexation" actually appears in Pratt's book as a quotation from the Italian scholar Antonello Gerbi, author of *The Dispute of the New World: The History of a Polemic*, trans. Jeremy Moyle (Pittsburgh: University of Pittsburgh Press, 1983; originally published in Italian in 1955). Pratt, a pioneering scholar in the field of postcolonialism, has influenced every scholar attempting to write about travel and exploration in the past decade; her book has achieved canonical status. Certainly, Pratt and her scholarly compatriots have enriched the study of travel by focusing on "Others"—nonwhite explorers, female tourists, mixed-blood guides. Justifiably and effectively, they have struggled, in the words of James Clifford, "to free the related term 'travel' from a history of European, literary, male, bourgeois, scientific, heroic, recreational meanings and practices." But in so doing they have obscured any potentially useful, even radical, ideas that might have been developed by European, literary, male, bourgeois scientists. See Clifford, *Routes: Travel and Translation in the Late Twentieth Century* (Cambridge, Mass.: Harvard University Press, 1997), pp. 17–46, esp. p. 35 on Humboldt.

I have tried to respond to Pratt and reevaluate elements of postcolonialism in Aaron Sachs, "The Ultimate 'Other': Post-Colonialism and Alexander von Humboldt's Ecological Relationship with Nature," theme issue on the environment, *History and Theory* 42 (December 2003), pp. 111–35. A healthier postcolonialism, in my opinion, would embrace Humboldt as—dare I say it?—a founding father.

13. Humboldt, *Political Essay on the Kingdom of New Spain* (New York: Knopf, 1972; orig. pub. in French in 1811), p. 34; I've used the John Black translation, edited (and abridged) by Mary Maples Dunn. Of course, in keeping with the widely circulating "black legend," it was always easiest to criticize Spanish colonialism as opposed to British or French.

14. See especially volume 3 of Humboldt, *Personal Narrative*, pp. 153–284, esp. 228–84. On its own, this long section, sometimes referred to as the "Political Essay on the Island of Cuba," provides sufficient evidence to refute Pratt's claim that Humboldt's writings covered up the human disasters of the conquest.

15. A very good atlas might list some fifty place names around the world that honor Humboldt. For a partial listing, see de Terra, *Humboldt*, pp. 377–78, and Ingo Schwarz, "Das Album der Humboldt-Lokalitäten in der Neuen Welt," *Magazin für Amerikanistik* 20 (no. 2, 1996), pp. 64–66, and another article of the same name in the next issue of the same magazine (no. 3, 1996), pp. 52–54.

16. See *Don Juan*, canto 4, verse 112 (1821), p. 217 in the version edited by T. G. Steffan, E. Steffan, and W. W. Pratt for Yale University Press (London: 1982): "Humboldt, 'the first of travellers,' but not / The last, if late accounts be accurate, / Invented, by some name I have forgot, / As well as the sublime discovery's date, / An airy instrument, with which he sought / To ascertain the atmospheric state, / By measuring the intensity of blue. / Oh Lady Daphne, let me measure you!"

17. The poem was published as a broadsheet and entitled "HUMBOLDT'S BIRTHDAY: BONAPARTE, AUG. 15TH, 1769—HUMBOLDT, SEPT. 14TH, 1769" (copy consulted at the H. E. Huntington Library, Rare Book Division). A copy of the poem is also included at the end of Agassiz, *Address Delivered on the Centennial Anniversary of the Birth of Alexander von Humboldt* (Holmes read the poem at the Humboldt Centennial celebration in Boston). Holmes described and compared the two men thus: "Master and Servant of the sons of earth. / Which wears the garland that shall never fade, / Sweet with fair memories that can never die? . . . / Bring the white blossoms of the waning year, / Heap with full hands the peaceful conqueror's shrine / Whose bloodless triumphs cost no sufferer's tear! / Hero of knowledge, be our tribute thine!"

The *Encyclopedia Britannica* has claimed, in many of its editions, that Humboldt "with the exception of Napoleon . . . was the most famous man in Europe." Quoted in E. R. Brann, *Alexander von Humboldt, Patron of Science* (Madison, Wis.: Littel Printing, 1954), p. 5. (See, for instance, *Encyclopedia Britannica*, 11th ed. [New York: Cambridge University Press, 1910], 13:874.)

18. Darwin asserted that his "whole course of life is due to having read and re-read" Humboldt's work, and, in a note to Humboldt himself, he thanked his mentor for contacting him: "That the author of those passages in the Personal Narrative which I have read over and over again and have copied out that they might ever be present in my mind should have so honoured me, is a gratification of a kind which can but seldom happen to anyone." The first quote appears in Worster, p. 132; the second is from Darwin's letter of November 1, 1839, American Philosophical Society, Humboldt microfilm collection, F#870, reel 2.

19. Bancroft, journal entry of May 28, 1821, describing a dinner at Benjamin Constant's house, printed in M. A. DeWolfe Howe, *The Life and Letters of George Bancroft* (New York: Charles Scribner's Sons, 1908), 1:103. After a later meeting, Bancroft noted that Humboldt "detests slavery": 2:78.

20. The graphologist is cited in de Terra, *Humboldt*, p. 234 (the words quoted here are de Terra's); Humboldt's own words are from a letter to Henry Wheaton, June 10, 1837, American Philosophical Society, Humboldt Collection, BH88.38. This document is probably a translation made by the recipient (an American).

21. Quoted in Pierre Gascar, *Humboldt L'Explorateur* (Paris: Gallimard, 1985), p. 38, my translation.

22. Humboldt, "Mes Confessions, à lire et à me renvoyer un jour" [My Confessions, to read and send back to me one day], a long autobiographical letter Humboldt wrote to Auguste Pictet, a Swiss colleague, in 1806, for use in advertising a future publication; the quote is my translation

of the French. The entire document is printed in Albert Leitzmann, ed., *Georg und Therese Forster und die Brüder Humboldt* (Bonn: Verlag Ludwig Röhrscheid, 1936), pp. 197–212, quote on p. 198. It can also be found in E. T. Hamy, ed., *Lettres Américaines d'Alexandre de Humboldt (1798–1807)* (Paris: Librairie Orientale et Américaine, 1904), pp. 236–44.

23. Leitzmann, p. 210, my translation.

24. Quoted in de Terra, *Humboldt*, p. 68.

25. "Memoria" of March 11, 1799, Archivo Histórico Nacional, Madrid, Sección Estado, legajo 4709. These quotes are my translations.

Also see Miguel Ángel Puig-Samper, "Humboldt, un Pruisano en la Corte del Rey Carlos IV," *Revista de Indias* 59 (Mayo–Agosto 1999), quotes on pp. 337 and 354.

26. A new book retraces Humboldt's equinoctial expedition and does admirable work in helping to restore Humboldt in historical memory: Gerard Helferich, *Humboldt's Cosmos: Alexander von Humboldt and the Latin American Journey That Changed the Way We See the World* (New York: Gotham Books, 2004).

27. Quoted in Tzvetan Todorov, *The Conquest of America: The Question of the Other* (New York: HarperPerennial, 1992), p. 250. On the deep significance of Humboldt's "moral unrest," see Ette, pp. 111–13.

28. Letter of March 8, 1853 (in French; my translation), American Philosophical Society, Elisha Kent Kane Papers, BK132.

29. Quoted in Dale L. Morgan, *The Humboldt: Highroad of the West* (New York: Farrar and Rinehart, 1943), p. 6.

30. Goetzmann, *New Lands, New Men*, p. 421.

31. The key here is the word "engagement," as Paul Carter has convincingly argued in his nuanced study, *The Road to Botany Bay: An Exploration of Landscape and History* (New York: Knopf, 1987); see esp. pp. 25 and 18. In this and his later book, *The Lie of the Land* (London: Faber and Faber, 1996), Carter asserts the need to distinguish between such different (and differingly colonialist) modes of landscape experience as discovery, exploration, surveying, and settlement. Better than anyone else, Carter has captured the ambivalence of explorers and their flexible, "dynamic" way of understanding what they observe—an approach not only encouraged by their intellectual backgrounds but also, quite often, necessitated by the physical conditions of exploring, by fog and sweat and insects and waterfalls.

32. Goetzmann, *Exploration and Empire: The Explorer and the Scientist in the Winning of the American West* (New York: W. W. Norton, 1966), quote on p. xi.

33. Unfortunately, the "current objectives" approach still dominates the field of exploration studies. In a 1999 essay called "Where We Are and How We Got There: Surveying the Record of Exploration Studies," the geographer John L. Allen sums up the current state of the historiography by quoting Goetzmann's framework from 1966. See Allen in *Surveying the Record: North American Scientific Exploration to 1930*, ed. Edward C. Carter II (Philadelphia: American Philosophical Society, 1999), pp. 3–18. The Goetzmann quote is on p. 18. Allen is also the editor of an important three-volume scholarly compilation of current research on exploration: *North American Exploration* (Lincoln: University of Nebraska Press, 1997).

34. See Louis P. Masur, *1831: Year of Eclipse* (New York: Hill and Wang, 2001), pp. 115–28, quote on p. 118.

35. J. W. Abert, *Western America in 1846–1847: The Original Travel Diary of Lieutenant J.W. Abert, who mapped New Mexico for the United States Army*, ed. John Galvin (San Francisco: John Howell Books, 1966), p. 90. Abert's father, Colonel John J. Abert, was the head of the Corps of Topographical Engineers. Under the corps and its core of Humboldtians, as David A. White has noted, "The Humboldtean pattern of scientific investigations for government surveying was set." See David A. White, ed., *News of the Plains and Rockies, 1803–1865: Original Narratives of Overland Travel and Adventure Selected from the Wagner-Camp and Becker Bibliography of Western Americana* (Spokane: Arthur H. Clark, 1998), 5:17.

Note that Abert's use of the phrase "Great Being," in particular, reflects the respect he felt for the worldview of certain native groups. Note also that Humboldt sometimes used the phrase "Great Spirit" in the same vein—in discussing Indians with the painter George Catlin, for instance: see his letter to Catlin of September 12, 1855, in Thomas Donaldson, *The George Catlin Indian Gallery in the U.S. National Museum* (Washington, D.C.: United States National Museum, 1886), p. 758.

36. Raymond Williams, *Culture and Society: 1780–1950* (1958; repr., New York: Columbia University Press, 1983), p. xi. Jonathan Bate has made a similar argument with regard to Wordsworth in his compelling study, *Romantic Ecology: Wordsworth and the Environmental Tradition* (New York: Routledge, 1991), and this strand of literary analysis has been productively pursued in several other works, including Bate, *The Song of the Earth* (Cambridge, Mass.: Harvard University Press, 2000); Karl Kroeber, *Ecological Literary Criticism: Romantic Imagining and the Biology of Mind* (New York: Columbia University Press, 1994); Eric Wilson, *Romantic Turbulence: Chaos, Ecology, and American Space* (Houndmills, U.K.: Macmillan, 2000); and Laura Dassow Walls, *Seeing New Worlds: Henry David Thoreau and Nineteenth-Century Natural Science* (Madison: University of Wisconsin Press, 1995). An earlier collection of essays that develop Williams's insights is U. C. Knoepflmacher and G. B. Tennyson, eds., *Nature and the Victorian Imagination* (Berkeley: University of California Press, 1977); and also see Lee Clark Mitchell, *Witnesses to a Vanishing America: The Nineteenth-Century Response* (Princeton: Princeton University Press, 1981).

Competing points of view can be found in Renato Rosaldo, "Imperialist Nostalgia," in *Culture and Truth: The Remaking of Social Analysis* (Boston: Beacon Press, 1989), pp. 68–87, and Richard Slotkin, *Regeneration through Violence: The Mythology of the American Frontier, 1600–1860* (1973; repr., New York: HarperPerennial, 1996), p. 563.

37. As cultural theorist Fredric Jameson has commented, it may be more useful to think of past radicals "as failures—that is, as actors and agents constrained by their own ideological limits and those of their moment of history—than as triumphant examples and models in some hagiographic or celebratory sense." Jameson, *Postmodernism, or, The Cultural Logic of Late Capitalism* (Durham, N.C.: Duke University Press, 1991), p. 209.

38. The handbill, now in the National Archives, is reproduced in Herman J. Viola and Carolyn Margolis, eds., *The U.S. Exploring Expedition, 1838–1842* (Washington, D.C.: Smithsonian Institution Press, 1985), p. 10. On Symmes, see Elmore Symmes, "John Cleves Symmes, The Theorist," an article in three installments in the *Southern Bivouac*, n.s., 2 (February, March, and April 1887), pp. 555–66, 621–31, and 682–93, and Paul Collins, "Symmes Hole," in *Banvard's Folly: Thirteen Tales of People Who Didn't Change the World* (New York: Picador, 2001), pp. 54–70.

39. As a young adult in Ohio, he was consistently called Jeremiah; in later records, he is almost always referred to as James. "John" appears to be simply a mistake. (Reynolds did have a contemporary named John Reynolds who was a congressman from Illinois.)

Convincing proof that he was called Jeremiah as a young adult is given in Robert F. Almy, "J. N. Reynolds: A Brief Biography with Particular Reference to Poe and Symmes," *Colophon*, n.s., 2 (Winter 1937), p. 228. But the bulk of the existing documentation suggests that his official name was James Neilson Reynolds—and thus that virtually every published treatment of his life and career in the last hundred years has been mistaken in referring to him as Jeremiah. Important evidence of his real name is found in his obituary, the reporting of his honorary degree at Columbia (see note 43), and in nineteenth-century accounts of his career, as well as numerous newspaper articles from the 1820s, '30s, and '40s—see, for instance, *National Intelligencer*, May 31, 1834, p. 3, listing James N. Reynolds among the crew of the just-returned *Potomac* (see chapter 5). Also see A. E. Carrell [sic], "The First American Exploring Expedition," *Harper's New Monthly Magazine* 44 (December 1871), pp. 60–64 (the author's named should be spelled Carroll).

Among the mistaken references to the explorer "John" Reynolds are John Wells Peck, "Symmes' Theory," *Ohio Archaeological and Historical Publications* 18 (1909), pp. 41–42; R. S. Garnett, "Moby-Dick and Mocha-Dick: A Literary Find," *Blackwood's Magazine* 226 (December 1929), pp. 841–58; and Edwin Swift Balch, *Antarctica* (Philadelphia: Allen, Lane, and Scott, 1902), p. 92.

40. I have read dozens of his letters and publications; every single one is signed "J.N. Reynolds." See Reynolds, *Remarks on a Review of Symmes' Theory, which appeared in the American Quarterly Review* (Washington, D.C.: Gales and Seaton, 1827). Reynolds seems to have earned his

high status as an exploration lobbyist by 1834; see Richard G. Woodbridge III, "J.N. Reynolds: Father of American Exploration," *Princeton University Library Chronicle* 45 (Winter 1984), p. 108.

41. Reynolds, ibid., p. 53.

42. On Reynolds's expedition and his career in general, see Richard G. Woodbridge III, "Condensed Biography of J.N. Reynolds—American," typescript (Princeton, 1995), copy consulted at the New York Public Library, and especially Woodbridge, "J.N. Reynolds—American," typescript, Princeton University Library, Special Collections; Goetzmann, *New Lands, New Men*, pp. 244–46, 258–64, 270–73; William E. Lenz, *The Poetics of the Antarctic: A Study in Nineteenth-Century American Cultural Perceptions* (New York: Garland Publishing, 1995), pp. 45–50; Kenneth J. Bertrand, *Americans in Antarctica, 1775–1948* (New York: American Geographical Society, 1971), pp. 144–97; Almy, pp. 227–45; Aubrey Starke, "Poe's Friend Reynolds," *American Literature* 11 (May 1939), pp. 152–59; William Coyle, ed., *Ohio Authors and Their Books* (Cleveland: World Publishing, 1962), pp. 523–25; and Henry Howe, "The Romantic History of Jeremiah N. Reynolds," in *Historical Collections of Ohio in Two Volumes: An Encyclopedia of the State* (Columbus: Henry Howe and Son, 1889), 1:430–32. (Note that this last source seems to contain a few mistakes, including the suggestion that Reynolds was made some sort of native chief during his South Seas expedition. Reynolds himself wrote extensively about his experiences with native peoples, and he never mentioned any chieftainship.)

43. Quote is from Reynolds, introduction to *Pacific and Indian Oceans; or The South Sea Surveying and Exploring Expedition: Its Inception, Progress, and Objects* (New York: Harper and Brothers, 1841), no pagination.

On the 1835 Columbia commencement, see the *Columbia University Annual Commencement Scrapbook, 1830–1849*, in the Columbiana Archives, Columbia University. The scrapbook also has a clipping from the *New York Evening Post* of October 8, 1835. Also see *Minutes of the Trustees of Columbia College*, vol. 3, pt. 2 (May 6, 1828, to December 4, 1837), p. 1618 (typescript in the Columbiana Archives).

Note also that Reynolds wrote a letter to Columbia President William A. Duer to thank him for the honor: "In any future labours, I trust that this literary distinction will prove such an incentive to higher efforts, as in some measure to justify the propriety of your favour." The letter, dated October 13, 1835, New York, is in the College Papers, Special Manuscript Division, Columbia University.

44. Reynolds, *Voyage of the United States Frigate* Potomac, *under the Command of Commodore John Downes, during the Circumnavigation of the Globe, in the Years 1831, 1832, 1833, and 1834* (New York: Harper and Brothers, 1835); on the four editions, see Starke, p. 153.

45. Reynolds, *Address on the Subject of a Surveying and Exploring Expedition to the Pacific Ocean and South Seas. Delivered in the Hall of Representatives on the Evening of April 3, 1836* (New York: Harper and Brothers, 1836), p. 72. Newspaper accounts and government documents show that the date was actually April 2.

46. Poe's review appeared in the January 1837 edition of the *Southern Literary Messenger*. I consulted a reprint in *The Complete Works of Edgar Allan Poe*, ed. James A. Harrison (New York: AMS Press, 1965), 9:306–14. On Poe's plagiarism, see Robert Lee Rhea, "Some Observations on Poe's Origins," *University of Texas Bulletin, Studies in English* 10 (July 8, 1930), pp. 135–46; J. O. Bailey, "Sources for Poe's *Arthur Gordon Pym*, 'Hans Pfaal,' and Other Pieces," *PMLA* 57 (June 1942), pp. 513–35; Keith Huntress, "Another Source for Poe's *Narrative of Arthur Gordon Pym*," *American Literature* 16 (March 1944), pp. 19–25; and Randel Helms, "Another Source for Poe's *Arthur Gordon Pym*," *American Literature* 41 (January 1970), pp. 572–75.

47. J. N. Reynolds, Esq., "Mocha-Dick; or, The White Whale of the Pacific: A Leaf from a Manuscript Journal," *Knickerbocker* 13 (May 1839), pp. 377–92.

48. On O'Sullivan, see chapter 6, and also Alan Trachtenberg, *Reading American Photographs: Images as History, Mathew Brady to Walker Evans* (New York: Hill and Wang, 1989), pp. 99–102 and 119–63, and Weston J. Naef and James N. Wood, *Era of Exploration: The Rise of Landscape Photography in the American West, 1860–1885* (Buffalo: Albright-Knox Art Gallery and the Metropolitan Museum of Art, 1975), pp. 125–66.

49. The most comprehensive work on King is Thurman Wilkins, *Clarence King: A Biography*, rev. and enlarged ed. (Albuquerque: University of New Mexico Press, 1988).

50. On the art scene in New York, see Roger Stein, *John Ruskin and Aesthetic Thought in America, 1840–1900* (Cambridge, Mass.: Harvard University Press, 1967), and David Dickason, *The Daring Young Men: The Story of the American Pre-Raphaelites* (Bloomington: Indiana University Press, 1953).

51. Humboldt, *Cosmos*, 2:456–57.

52. King Papers, Huntington Library, D-23, Notebooks, Scientific: "Journal of a trip in the Northern Sierras and Grass Valley," September 20, 1863, p. 26, and A-2, #4, Notebooks, Personal: "Scientific Notes, Private," 1863.

53. King, "The Biographers of Lincoln," *Century Magazine* 32 (October 1886), p. 868.

54. Lorin Blodget, *Climatology of the United States* (Philadelphia: J. B. Lippincott, 1857). The quote is from Humboldt's 1829 "Address before the Imperial Academy of St. Petersburg."

55. This global reach is the aspect of Humboldt's approach most commonly referred to by historians of science as "Humboldtian science"—a phrase coined by Susan Cannon in her 1978 book, *Science in Culture: The Early Victorian Period* (New York: Dawson and Science History Publications, 1978); see pp. 73–110. Humboldt famously made himself a sort of clearinghouse for any data collected by traveling scientists on their expeditions: he wanted statistics not only on temperature but also on pressure, terrestrial magnetism, elevation, the chemical composition of the atmosphere, humidity, electrical tension, general meteorological phenomena, tides, even rock formations. See also Felix Driver, *Geography Militant: Cultures of Exploration and Empire* (Oxford: Blackwell, 2001), esp. pp. 1–37, and Michael Dettelbach, "Humboldtian Science," in *Cultures of Natural History*, ed. N. Jardine, J. A. Secord, and E. C. Spary (Cambridge: Cambridge University Press, 1996), pp. 287–304.

56. On Petermann, see Leonard F. Guttridge, *Icebound: The Jeannette Expedition's Quest for the North Pole* (Annapolis: Naval Institute Press, 1986), pp. 24–41. My references are to this edition, but note that there is a newer, more accessible edition published as a trade paperback by Berkley Books (New York) in December 2001.

57. Blodget himself sorted through the reports of countless exploring expeditions in compiling the data for his *Climatology*. J. N. Reynolds also consulted almost every existing account of Arctic travel in assessing Captain Symmes's claim that there might be warmer climes at the higher altitudes.

58. Guttridge's book, *Icebound*, is the best recent history of De Long's expedition. Other sources focusing exclusively on this voyage include A. A. Hoehling, *The Jeannette Expedition: An Ill-Fated Journey to the Arctic* (New York: Abelard-Schuman, 1967), and a fictionalized account by Commander Edward Ellsberg, *Hell on Ice: The Saga of the Jeannette* (New York: Dodd, Mead, 1938).

59. Melville refers to Humboldt in his *Views of Commodore George W. Melville, Chief Engineer of the Navy, as to the Strategic and Commercial Value of the Nicaraguan Canal, the Future Control of the Pacific Ocean, the Strategic Value of Hawaii, and its Annexation to the United States* (Washington, D.C.: Government Printing Office, 1898), p. 29.

60. Melville, *In the Lena Delta* (London: Longmans, Green, 1885), pp. 265 and 146.

61. Humboldt, *Cosmos*, 1:3.

62. See Samuel P. Hays, *Conservation and the Gospel of Efficiency: The Progressive Conservation Movement, 1890–1920* (Cambridge, Mass.: Harvard University Press, 1959).

63. Melville, *In the Lena Delta*, p. 101.

64. Letter of John Muir to Mrs. Jeanne Carr, September 13, 1865, Trout's Mills, near Meaford, in Bonnie Johanna Gisel, ed., *Kindred and Related Spirits: The Letters of John Muir and Jeanne C. Carr* (Salt Lake City: University of Utah Press, 2001), p. 29; also quoted in Stephen Fox, *John Muir and His Legacy: The American Conservation Movement* (Boston: Little, Brown, 1981), p. 47.

65. Quoted by William Frederic Badè, Muir's literary executor, in his introduction to Muir's posthumously published travel diary, *A Thousand-Mile Walk to the Gulf* (1916; repr., San Francisco: Sierra Club Books, 1991), p. xix.

66. Ibid., p. 56.

67. Ibid., p. 79.

68. Muir, *Our National Parks* (1901; repr., Boston: Houghton Mifflin, 1916), p. 4.

69. Muir, *The Cruise of the Corwin* (1917; repr., San Francisco: Sierra Club Books, 1993), p. 103.

70. See, for instance, Thurman Wilkins, *John Muir: Apostle of Nature* (Norman: University of Oklahoma Press, 1995), pp. 223–42, and Roderick Nash, *Wilderness and the American Mind*, 4th ed. (1967; repr., New Haven: Yale University Press, 2001), pp. 161–81.

71. This was a catchphrase used commonly in the period, but most often by Theodore Roosevelt's chief forester, Gifford Pinchot, a conservationist who generally opposed Muir's preservationism.

72. Quoted in Wilkins, *John Muir*, p. 238.

73. See Douglas Botting, *Humboldt and the Cosmos* (London: Michael Joseph, 1973), pp. 13–14 and 199–201.

74. Quoted ibid., p. 44.

75. See the wonderful collection of essays by Evan S. Connell, *A Long Desire* (New York: Holt, Rinehart, and Winston, 1979), and Eric Leed, *Shores of Discovery: How Expeditionaries Have Constructed the World* (New York: Basic Books, 1995).

76. Estwick Evans, *A Pedestrious Tour, of Four Thousand Miles, through the Western States and Territories, during the Winter and Spring of 1818* (Concord, N.H.: Joseph C. Spear, 1819), p. 6. Evans also notes that "the season of snows was preferred, that I might experience the pleasure of suffering, and the novelty of danger." Note that George H. Daniels, more than most historians, has recognized "the ambivalent justifications underlying American exploration in the nineteenth century"; see his *Science in American Society: A Social History* (New York: Knopf, 1971), p. 178, as well as William H. Goetzmann, "A 'Capacity for Wonder': The Meanings of Exploration," in *North American Exploration*, ed. Allen, 3:521–45, which is more nuanced than some of Goetzmann's earlier work; and Mitchell *(Witnesses to a Vanishing America)*, who writes at great length about "the ambivalence felt even among those who participated in the nation's triumphant conquest of the wilderness" (p. xiv). An excellent recent study of the contradictions of American expansionism, as expressed in the literature of the day, is Stephanie LeMenager, *Manifest and Other Destinies: Territorial Fictions of the Nineteenth-Century United States* (Lincoln: University of Nebraska Press, 2004).

77. There is an emerging literature that examines "in-between" figures as the keys to cross-cultural understanding; see, for instance, Clifford, *Routes;* Frances Karttunen, *Between Worlds: Interpreters, Guides, and Survivors* (New Brunswick, N.J.: Rutgers University Press, 1994); Paul Gilroy, *The Black Atlantic: Modernity and Double Consciousness* (Cambridge, Mass.: Harvard University Press, 1993); and Stephen Greenblatt, *Marvelous Possessions: The Wonder of the New World* (Chicago: University of Chicago Press, 1991), esp. pp. 119–51. One of the most interesting angles on such cultural mediators comes from Lewis Hyde, who sees them as trickster figures. See his broad and original study, *Trickster Makes This World: Mischief, Myth, and Art* (New York: North Point Press, 1998).

78. See Paul Zweig, *The Adventurer: The Fate of Adventure in the Western World* (1974; repr., Pleasantville, N.Y.: Akadine Press, 1999), pp. 83 and 96.

79. See especially the famous passage in which Thoreau walks through the burned, barren landscape of Mount Ktaadn in *The Maine Woods* (1864; repr., New York: W. W. Norton, 1950; version arranged with notes by Dudley C. Lunt), p. 278.

80. Humboldt, *Views of Nature: Or Contemplations on the Sublime Phenomena of Creation; with Scientific Illustrations* (London: Henry G. Bohn, 1850), p. x. I've used the translation by E. C. Otté and Henry G. Bohn from the German of the third edition, published in Stuttgart and Tübingen in 1849. *Views of Nature* was originally written in German and published in 1808; note that in another widespread English translation of this work, originally called *Ansichten der Natur*, the title is rendered as *Aspects of Nature*, translated by Mrs. Sabine (Philadelphia: Lea and Blanchard, 1849).

81. To explore this tension, see David Lowenthal, *The Past is a Foreign Country* (Cambridge: Cambridge University Press, 1985), and Richard Grove, *Green Imperialism: Colonial Expansion, Tropical Island Edens and the Origins of Environmentalism, 1600–1860* (New York: Cambridge University Press, 1995), esp. pp. 1–15.

82. On the Nevada Test Site, see Rebecca Solnit's moving and pathbreaking book *Savage Dreams: A Journey into the Landscape Wars of the American West* (1994; repr., Berkeley: University of California Press, 1999).

83. As the historian Robert Johnston recently commented, "Our most pressing responsibility as citizen-historians . . . [is] continuously opening up the past in order to explore the hope embedded in history." See his article "Beyond 'The West,'" *Rethinking History* 2 (Summer 1998), p. 264. And, on the political culture of the early twenty-first century, see Lewis Lapham, *Gag Rule: On the Suppression of Dissent and the Stifling of Democracy* (New York: Penguin, 2004).

84. Parkman quoted in David Levin's introduction to the Penguin Classics edition of Parkman's *The Oregon Trail* (New York: Penguin, 1985), p. 8.

85. National Archives, Record Group 57, Geological Survey Records, Copies of letters and reports sent to the Chief of Engineers, March 28, 1867–January 18, 1879: letter of December 18, 1867.

86. Richard A. Bartlett, in *Great Surveys of the American West* (Norman: University of Oklahoma Press, 1962), p. 169, says that the lightning strike took place in "a mountain range west of the Reese River, almost in the heart of central Nevada," which certainly means it couldn't have been the Carson Range in his opinion. Note also that King was accompanied by the topographer F. A. Clark, one soldier, and an Indian they called "Son," who scouted for them and was chief of the Humboldt Paiutes.

Michael S. Durham says explicitly that it was the Stillwater Range in *Desert Between the Mountains: Mormons, Miners, Padres, Mountain Men, and the Opening of the Great Basin* (Norman: University of Oklahoma Press, 1997), p. 247.

Stephen Trimble says that it was the Carson Range in *The Sagebrush Ocean: A Natural History of the Great Basin* (Reno: University of Nevada Press, 1989), p. 219.

87. "Then it seemed to me the Sierra should be called not the Nevada, or snowy range, but the range of light . . . the most divinely beautiful of all the mountain chains I have ever seen." The quote is printed on a transparent page that fits over a full-color photo of a sunset above the Carson Range, taken from the valley below. The Carson Range is actually just east of the main Sierra Chain.

88. I revisited the National Archives to reread the letters King wrote describing his movements in 1867 (Record Group 57) and compared them to letters I found at the Huntington Library written by William Whitman Bailey, a member of King's survey who was confined to his bed in Unionville, Nevada, at the time of King's excursion to Job's Peak. See, for instance, Bailey's letter to his uncle William M. Bailey of September 20, 1867 (two days before the lightning strike), Manuscripts Collection, HM39967. Unionville is closer to the Stillwaters than to the Carsons.

89. See Ulrich-Dieter Oppitz, "Der Name der Brüder Humboldt in aller Welt," in *Alexander von Humboldt: Werk und Weltgeltung*, ed. Heinrich Pfeiffer (Munich: R. Piper, 1969), p. 290. Other possibilities debated at the territory's constitutional convention in 1863 were "Washoe" and "Esmeralda."

90. Stanley W. Paher, *Nevada Ghost Towns and Mining Camps* (Berkeley: Howell-North Books, 1970), pp. 130–33; also see Helen S. Carlson, *Nevada Place Names: A Geographical Dictionary* (Reno: University of Nevada Press, 1974), pp. 138–39.

91. Mark Twain, *Roughing It* (1872; repr., Berkeley: University of California Press, 1993), p. 184.

92. See, for instance, Johannes Fabian's intriguing study, *Out of Our Minds: Reason and Madness in the Exploration of Central Africa* (Berkeley: University of California Press, 2000), esp. pp. 3–11.

Excursion: Exile: Napoleon's France

1. Humboldt to Jefferson, June 27, 1804, reprinted in Helmut de Terra, "Alexander von Humboldt's Correspondence with Jefferson, Madison, and Gallatin," *Proceedings of the American Philosophical Society* 103 (December 1959), p. 789.

2. On Humboldt and Napoleon in Paris, see Douglas Botting, *Humboldt and the Cosmos*

(London: Michael Joseph, 1973), pp. 177–201; Helmut de Terra, *Humboldt: The Life and Times of Alexander von Humboldt, 1769–1859* (New York: Knopf, 1955), pp. 190–231; and Wolfgang-Hagen Hein, ed., *Alexander von Humboldt: Life and Work* (Ingelheim am Rhein: C. H. Boehringer Sohn, 1987), p. 16. Humboldt himself, describing his meeting with Napoleon, later wrote that the emperor "seemed full of hatred towards me" (quoted in Botting, p. 179). Humboldt's sister-in-law, Caroline von Humboldt, is quoted in Botting, p. 177.

Chapter 2: *Personal Narrative of a Journey:* Radical Romanticism

1. Humboldt, *Personal Narrative of Travels to the Equinoctial Regions of America, 1799–1804* (London: Henry G. Bohn, 1852–53; orig. pub. in French in 1814), 1:79.

2. Volcano quotes are from Humboldt, *Fragmens de Géologie et de Climatologie Asiatiques* (Paris: A. Pihan Delaforest, 1831), pp. 1–4, my translations; last quote of the paragraph is Humboldt, quoted in William H. Goetzmann, *New Lands, New Men: America and the Second Great Age of Discovery* (1986; repr., Austin: Texas State Historical Association, 1995), p. 59.

For Humboldt's further musings on volcanoes (including the erupted-fish episode), see *Cosmos: A Sketch of a Physical Description of the Universe* (London: Henry G. Bohn, 1849; orig. pub. in German in 1845), 1:221–46, esp. pp. 230–31; and *Views of Nature: Or Contemplations on the Sublime Phenomena of Creation; with Scientific Illustrations* (London: Henry G. Bohn, 1850; orig. pub. in German in 1808), pp. 353–75.

Note: the modern romanizations for "Pe-schan" and "Ho-tscheu" are "Ba Shan" and "He Zhu."

3. Letter to his brother Wilhelm of November 25, 1802, in E. T. Hamy, ed., *Lettres Américaines d'Alexandre de Humboldt (1798–1807)* (Paris: Librairie Orientale et Américaine, 1904), p. 132, my translation. Humboldt was sometimes more emotional in his letters, just because they were less public than his books. But his correspondents were also the people he thought of as the main audience for his publications, so there is generally a close correlation between the rhetorical tropes of his published works and the language of letters he wrote while in the field.

4. Humboldt, *Cosmos.* 3:1.

5. See, for instance, Marjorie Hope Nicolson, *Mountain Gloom and Mountain Glory: The Development of the Aesthetics of the Infinite* (1959; repr., Seattle: University of Washington Press, 1997); and Barbara Novak, *Nature and Culture: American Landscape and Painting, 1825–1875* (New York: Oxford University Press, 1980), esp. pp. 34–44.

6. Letter of November 25, 1802, my translation; see note 3.

7. Humboldt, *Personal Narrative,* 1:xxi.

8. Humboldt, "Mes Confessions, à lire et à me renvoyer un jour" [My Confessions, to read and send back to me one day], in *Lettres Américaines,* pp. 236–44.

9. Humboldt's agnosticism is clear from his correspondence with his friend Varnhagen von Ense; see his *Letters of Alexander von Humboldt to Varnhagen von Ense, from 1827 to 1858,* translated from the second German edition by Friedrich Kapp (New York: Rudd and Carleton, 1860), pp. 182–83, 194, 239, 271, 339, 385, and 397.

10. Humboldt, *Essai sur la Géographie des Plantes* (Paris: F. Schoell, 1807), with its accompanying plate, published separately and entitled *Géographie des Plantes Équinoxiales.* This volume bears an original *publication* date of 1807, but it was actually first *printed* in 1805.

11. Quoted in Helmut de Terra, *Humboldt: The Life and Times of Alexander von Humboldt, 1769–1859* (New York: Knopf, 1955), p. 207—for a slightly different translation, see Humboldt, *Views of Nature,* p. 173; Humboldt, *Aspects of Nature* (Philadelphia: Lea and Blanchard, 1849), p. 170—for a slightly different translation, see *Views,* p. 154; and *Views,* p. ix.

12. Humboldt, *Political Essay on the Kingdom of New Spain* (New York: Knopf, 1972; orig. pub. in French in 1811, though the volume was completed in 1808). I've used the John Black translation, edited (and abridged) by Mary Maples Dunn.

13. *Vues des Cordillères et Monumens des Peuples Indigènes de l'Amérique,* 2 vols. (Paris: Chez L. Bourgeois-Maze, 1840; orig. pub. 1810). I've made a literal translation of the title, above, but when this book was actually translated and published in English it came out as *Researches, Concerning the Institutions and Monuments of the Ancient Inhabitants of America, with Descriptions and*

Views of Some of the Most Striking Scenes in the Cordilleras, also in two volumes, translated by Helen Maria Williams (London: Longman, Hurst, 1814). Both versions were common in the United States; the translation I consulted, at the New York Public Library, was owned by the historian George Bancroft, who, as noted in chapter 1, met Humboldt in Paris in 1821.

14. On the joint tradition of nature and travel writing, see, for instance, Peter A. Fritzell, *Nature Writing and America: Essays upon a Cultural Type* (Ames: Iowa State University Press, 1990); Norman Foerster, *Nature in American Literature* (New York: Macmillan, 1923); and Rebecca Solnit, *Wanderlust: A History of Walking* (New York: Viking, 2000).

15. Humboldt, *Personal Narrative*, 1:67, 65, 69–70, 86, and 69.

16. Ibid., p. 82; Humboldt, *Cosmos*, 1:3.

17. Humboldt, *Personal Narrative*, 1:131 and 143; on seeing, see John Berger, *Ways of Seeing* (New York: Penguin, 1972), and Jonathan Crary, *Techniques of the Observer: On Vision and Modernity in the Nineteenth Century* (Cambridge, Mass.: MIT Press, 1990).

18. Humboldt, *Personal Narrative*, 1:87.

19. Humboldt, *Cosmos*, 1:19.

20. Humboldt, *Personal Narrative*, 1:1 and 80.

21. Ibid., pp. 19 and xx. Humboldt's long tangent on ocean currents is on pp. 15–26.

22. The book's original French title was *Relation Historique du Voyage aux Régions Équinoxiales du Nouveau Continent*; quotes are from ibid., pp. 96 and 104–5, my emphasis. A more obvious choice for a French title would have involved the word "*récit*" or "*histoire*" or even "*narration*"; this was a relatively rare usage, and therefore probably represents a significant decision. My thanks to Murray Sachs, a scholar of nineteenth-century French literature, for confirming this suspicion.

Note that Ottmar Ette sees Humboldt's jumpy, eclectic approach to writing as suggesting the fundamental conditions of modernity: see Ette, *Literature on the Move* (Amsterdam: Rodopi, 2003), pp. 11–27.

23. Emerson is quoted in Louis Agassiz, *Address Delivered on the Centennial Anniversary of the Birth of Alexander von Humboldt; Under the Auspices of the Boston Society of Natural History; with an account of the evening reception* (Boston: Boston Society of Natural History, 1869), p. 72, in the "account of the evening reception" section.

24. On the cyanometer, see Humboldt, *Personal Narrative*, 1:84–85 and 408; on the Golden Gate, see John Charles Frémont, *Geographical Memoir upon Upper California* (1848; repr., San Francisco: Book Club of California, 1964), p. 29.

25. Letter to the physicist Antoine-François Fourcroy, January 25, 1800, La Guayra, in *Lettres Américaines*, p. 62, my translation. The parallelism of strata is discussed with greater precision in Humboldt, *Personal Narrative*, 3:367–76.

Also see *Personal Narrative*, 1:11: "The most curious geological phenomena are often repeated at immense distances on the surface of continents; and naturalists who have examined different parts of the globe, are struck with the extreme resemblance observed in the rents on coasts, in the sinuosities of the vallies, in the aspect of the mountains, and in their distribution by groups. The accidental concurrence of the same causes must have everywhere produced the same effects; and amidst the variety of nature, an analogy of structure and form is observed in the arrangement of inanimate nature, as well as in the internal organization of plants and animals."

26. Today, change is seen as one of nature's central characteristics. Yet modern chaos theory has demonstrated that beneath almost every manifestation of disorder lurks some sort of pattern or equilibrium. As Daniel Botkin, author of *Discordant Harmonies: A New Ecology for the Twenty-First Century* (New York: Oxford University Press, 1990), has noted, "certain rates of change are natural, desirable, and acceptable, while others are not" (p. 12).

27. Anthony Pagden, *European Encounters with the New World* (London: Yale University Press, 1993), quote on p. 36, but also see pp. 24–38, 104–15, 166–69, and 183–88; and Mary Louise Pratt, *Imperial Eyes: Travel Writing and Transculturation* (London: Routledge, 1992), p. 124.

My response to Pratt is Aaron Sachs, "The Ultimate 'Other': Post-Colonialism and Alexander von Humboldt's Ecological Relationship with Nature," theme issue on the environment, *History and Theory* 42 (December 2003), pp. 111–35.

On the colonial use of science, see, for instance, Bernard S. Cohn, *Colonialism and Its Forms of*

Knowledge: The British in India (Princeton: Princeton University Press, 1996); Benedict Anderson, "Census, Map, Museum," in *Imagined Communities: Reflections on the Origins and Spread of Nationalism*, rev. ed. (New York: Verso, 1991), pp. 163–85; and James Clifford, *Routes: Travel and Translation in the Late Twentieth Century* (Cambridge, Mass.: Harvard University Press, 1997), esp. pp. 197–219.

28. Humboldt, *Personal Narrative*, 1:126 and 134.

29. Humboldt, *Cosmos*, 1:2, my emphasis. On ecosystems, which Humboldt usually referred to as geographical "regions," he wrote: "the character of different regions of the earth may depend upon a combination of . . . the outline of mountains and hills, the physiognomy of plants and animals, the azure of the sky, the forms of the clouds, and the transparency of the atmosphere"; quoted in Malcolm Nicolson's "Historical Introduction" to the Penguin Classics abridgement of Humboldt's *Personal Narrative of a Journey to the Equinoctial Regions of the New Continent* (London: Penguin, 1995), p. xiii.

On biodiversity, Humboldt wrote of the need to respect "the universal profusion of life," and he constantly insisted that any real understanding of the cosmos depended on opportunities "to contemplate nature in all her variety"; see *Views of Nature*, p. 210, and *Personal Narrative*, 1:2. He also noted that, in the tropics, "a single trunk displays a greater variety of vegetable forms than are contained within an extensive space of ground in our countries" (*Personal Narrative*, 2:256).

30. Humboldt, *Cosmos*, 1:63; and Humboldt, *Examen Critique de l'Histoire de la Géographie du Nouveau Continent* [Critical Examination of the History of the Geography of the New Continent], (Paris: Librairie de Gide, 1837), 3:154, my translation (this work has never been translated into English to my knowledge; it originally came out in five French volumes, published between 1836 and 1839).

31. See Humboldt, *Personal Narrative*, 1:81 and 114–21; Humboldt, *Essai sur la Géographie des Plantes;* Malcolm Nicolson, "Alexander von Humboldt, Humboldtian Science and the Origins of the Study of Vegetation," *History of Science* 25 (June 1987), pp. 167–94; and Klaus Dobat, "Alexander von Humboldt as a Botanist," in *Alexander von Humboldt: Life and Work*, ed. Wolfgang-Hagen Hein (Ingelheim am Rhein: C. H. Boehringer Sohn, 1987), pp. 167–93: the English edition of this book, which I consulted, is translated from the German original (Ingelheim am Rhein: C. H. Boehringer Sohn, 1985) by John Cumming and edited by Peter Newmark and is extremely useful in making some relatively recent German scholarship on Humboldt available to a much wider audience. On "occult forces," see Humboldt, *Views of Nature*, p. x.

On graphic representations, note this comment from a review of *Cosmos* in "Alexander von Humboldt and his Cosmos," *Methodist Quarterly Review*, 4th ser., 12 (July 1860), pp. 422–23: "[Humboldt's] contributions to geography itself were equally great with the change he wrought in the mode of its presentation by giving those graphic and pictorial illustrations of the chief features of land and sea, and of generalized facts, and of complex physical phenomena, by which means the pupil obtains a far more ready, correct, and lasting view of the same than by the old method." Also see Anne Marie Claire Godlewska, "From Enlightenment Vision to Modern Science? Humboldt's Visual Thinking," in *Geography and Enlightenment*, ed. David N. Livingstone and Charles W. J. Withers (Chicago: University of Chicago Press, 1999), pp. 236–75.

32. Humboldt quoted in de Terra, p. 44; second quote is from "Notice sur la vie littéraire de Mr. de Humbold" [Memo on the literary life of Mr. Humboldt], one of the documents Humboldt presented to the king of Spain in 1799, written in French and dated March 11, 1799, my translation, in the Archivo Histórico Nacional in Madrid, Sección Estado, legajo 4709; also see Miguel Ángel Puig-Samper, "Humboldt, un Pruisano en la Corte del Rey Carlos IV," *Revista de Indias* 59 (Mayo–Agosto 1999), p. 353.

33. Georg Forster, *Ansichten vom Niederrhein, von Brabant, Flandern, Holland, England, und Frankreich* (1792; repr., Leipzig: F. A. Brodhaus, 1868); this book has never been translated into English to my knowledge. Also see Douglas Botting, *Humboldt and the Cosmos* (London: Michael Joseph, 1973), pp. 16–21.

34. Humboldt, *Cosmos*, in the Sabine translation (London: Longman, Brown, Green, and Longmans, 1848), 2:70.

35. The influence of Kant and Goethe on Humboldt's intellectual development, not to

mention that of Schelling and Schiller, has been thoroughly studied; that of Herder is particularly significant, in my view, because of Herder's cosmopolitanism and especially his interest in Eastern religions. I also think the influence of French thinkers has been underestimated. Key texts here might include Jean-Jacques Rousseau, *A Discourse on Inequality*, translated by Maurice Cranston (London: Penguin, 1984; orig. pub. 1755); Bernardin de St. Pierre, "Voyage à l'île de France" (orig. pub. 1773), in *Ile de France: Voyage et Controverses* (Mauritius: AlmA, 1996); and Denis Diderot, "Supplement to Bougainville's *Voyage*," in *Rameau's Nephew and Other Works*, trans. Jacques Barzun and Ralph H. Bowen (Indianapolis: Bobbs-Merrill, 1964; orig. pub. 1796). For explorations of Humboldt's intellectual background, see Margarita Bowen's crucial book, *Empiricism and Geographical Thought: From Francis Bacon to Alexander von Humboldt* (Cambridge: Cambridge University Press, 1981), esp. pp. 210–22; Alexander Gode-von Aesch, *Natural Science in German Romanticism* (New York: Columbia University Press, 1941); Andrew Cunningham and Nicholas Jardine, eds., *Romanticism and the Sciences* (Cambridge: Cambridge University Press, 1990); Richard Grove, *Green Imperialism: Colonial Expansion, Tropical Island Edens and the Origins of Environmentalism, 1600–1860* (New York: Cambridge University Press, 1995), pp. 364–74; and Richard Hartshorne, "The Concept of Geography as a Science of Space, from Kant and Humboldt to Hettner," *Annals of the Association of American Geographers* 48 (1958), pp. 97–108.

36. Quoted in Botting, p. 33; on this period in Humboldt's life, see Botting, pp. 21–49, and de Terra, pp. 45–60. The same principle applied to the writing up of one's travels; Humboldt commented that "the narratives of distant travels, too long occupied in the mere recital of hazardous adventures, can only be made a source of instruction, where the traveller is acquainted with the condition of the science he would enlarge, and is guided by reason in his researches" (*Cosmos*, 1:32).

37. Forster quoted in Christopher McIntosh, *The Rose Cross and the Age of Reason: Eighteenth-Century Rosicrucianism in Central Europe and Its Relationship to the Enlightenment* (Leiden: E. J. Brill, 1992), p. 88.

38. Forster, *A Voyage Round the World* (1777; repr., Honolulu: University of Hawaii Press, 2000), p. 330. Diderot expressed a similar perspective, in the voice of the Tahitian chief ("Supplement," p. 188): "When one of our lads carried off some of the miserable trinkets with which your ship is loaded, what an uproar you made, and what revenge you took! And at that very moment you were plotting, in the depths of your hearts, to steal a whole country!"

39. Humboldt, *Personal Narrative*, 1:134, 131, 136, 139, and 137.

40. Ibid., pp. 128–29.

41. Ibid., p. 140.

42. Humboldt, in *Lateinamerike am Vorabend der Unabhärigigkeitsrevolution: eine Anthologie von Impressionen und Urteilen, aus seinen Reisetagesbüchern*, ed. Margot Faak (Berlin: Akademie-Verlag, 1982), pp. 63–64. This is a scholarly compilation of excerpts from Humboldt's travel journals, showing his attitude toward such things as colonialism, Indians, slavery, corruption, and Christian missions. All the commentary is in German, but the journal entries are given in their original language, usually French, as is the case with these quotations (my translation). In this same passage, dated January 4–February 17, 1803, Guayaquil, [Ecuador], Humboldt goes on to say: "Nowhere should a European be so ashamed of his nationality as in the island colonies, whether French, English, Danish, or Spanish. To argue about which Nation treats Negroes with more humanity is to mock the very word humanity, is to ask whether it's nicer to be disemboweled or skinned alive" (p. 64).

43. Humboldt, *Personal Narrative*, 1:121 and 57; also see 2:214, where Humboldt mentions "the poor labourers in the east of Europe, whom the barbarism of our feudal institutions has held in the rudest state."

44. Ibid., 1:144–47.

45. Humboldt, letter of February 3, 1800, to the Baron von Forell, in *Lettres Américaines*, p. 65, my translation. Michael A. Bryson, in his recent book, *Visions of the Land: Science, Literature, and the American Environment from the Era of Exploration to the Age of Ecology* (Charlottesville: University Press of Virginia, 2002), expresses the prevailing view of the gender dynamics involved in

nineteenth-century exploration when he writes that almost every "scientist-explorer" portrayed himself as a "hero, a distinctly masculine figure who strives to 'conquer' a feminine nature" (p. 5; also see p. xiii, where Bryson notes that the "unabashedly" male explorer is also wholly "rational" and that nature, in this model, is always "passive").

46. Humboldt, *Personal Narrative*, 2:199 and 319; on El Dorado, see 2:181, 378–79, and 400, and 3:25–65, as well as Evan S. Connell, *A Long Desire* (New York: Holt, Rinehart, and Winston, 1979), pp. 162–95.

47. Humboldt, *Personal Narrative*, 2:151 and 153.

48. Ibid., 1:216 and 220–22, and 2:162–63, 50, 47–49, 494–504, 184–94, and 4.

49. Ibid., 2:155; letter of February 21, 1801, to Karl Ludwig Wildenow, *Lettres Américaines*, p. 112, my translation.

50. On curare, see Humboldt, *Personal Narrative*, 2:438–48.

51. Ibid., p. 273.

52. Keats quoted in W. Jackson Bate, *John Keats* (Cambridge, Mass.: Harvard University Press, 1963), p. 249.

53. Humboldt, *Personal Narrative*, 1:255 and 2:4.

54. Ibid., 1:273–74. Note also 2:253: "In the physical, as in the moral world, the contrast of effects, the comparison of what is powerful and menacing with what is soft and peaceful, is a never-failing source of our pleasures and our emotions."

55. This was uncharted territory. The expedition proved that there can be rivers flowing across watersheds (armchair cartographers in Europe had placed a major mountain range between the Orinoco and Amazon basins simply because geographic theory said that such a range had to be there). See Botting, pp. 102–41, esp. pp. 125–29, referring to the French geographer Philippe Buache.

56. As Mary Louise Pratt has noted, "Humboldt is steadfastly revered and revived in South American official culture precisely for his unconditional, intrinsic valorization of the region." See Pratt, p. 141.

57. Humboldt, *Personal Narrative*, 2:519.

58. Botting, p. 283.

59. Quoted in Bowen, p. 251; quoted in de Terra, p. 208.

60. The first letter is quoted in de Terra, p. 31. Also see another letter quoted on p. 62: "A man should early accustom himself to stand alone. Isolation has much in its favor. One learns thereby to search inwardly and to gain self-respect without being dependent on the opinions of others." The second letter, to Wilhelm, written on October 17, 1800, is in *Lettres Américaines*, p. 87, my translation.

61. On Bonpland, see, for instance, Humboldt's letter of October 17, 1800, to Wilhelm von Humboldt, in *Lettres Américaines*, pp. 88–89; on Montúfar, Haeften, and the other friend, Wilhelm Gabriel Wegener, see de Terra, pp. 147–48, 198, 62–68, and 26–29, quotes on pp. 64, 66, and 27. The richest sources for exploring this side of Humboldt are, clearly, his early letters, which have been published in their original languages (mostly German, some French) in Ilse Jahn and Fritz G. Lang, eds., *Die Jugendbriefe Alexander von Humboldts: 1787–1799* (Berlin: Akademie-Verlag, 1973).

62. E. Anthony Rotundo has argued quite convincingly that overtly romantic letters between youthful members of the educated classes were fairly routine in America and Europe in the late eighteenth and early nineteenth centuries—but that such expressions of passion generally ceased once the men reached their twenties and started to fulfill their adult obligations. See his study *American Manhood: Transformations in Masculinity from the Revolution to the Modern Era* (New York: Basic Books, 1993). So it makes sense, as Leila Rupp suggests, that "if not all men in fact forgot their male friends in the interests of marriage and manhood . . . , the persistence of such attachments began to shade into more questionable behavior." Rupp also notes that, while it is incredibly difficult to interpret the evidence of sexual behavior in times past, "it would be a mistake . . . to assume that the social acceptance of romantic friendship means that sexual acts never occurred between romantic friends." See Leila J. Rupp, *A Desired Past: A Short History of Same-Sex Love in America* (Chicago: University of Chicago Press, 1999), pp. 48–50.

63. The painter Carl von Steuben, for instance. See de Terra, pp. 236–39, and Botting, pp. 197–99. For the full range of artists with whom Humboldt consorted, see Halina Nelken, *Alexander von Humboldt: His Portraits and Their Artists; A Documentary Iconography* (Berlin: D. Reimer, 1980).

64. Quoted in de Terra, p. 227.

65. Humboldt, *Personal Narrative*, 2:68 and 152.

66. Ibid., p. 362; on the same page, Humboldt also notes: "The nations of the Upper Orinoco, the Atabapo, and the Inirida, like the ancient Germans and the Persians, have no other worship than that of the powers of nature." See also 1:296, on certain Indians' "tendency to the worship of nature and her powers. This worship . . . recognises no other sacred places than grottoes, valleys, and woods."

67. Ibid., 1:240 and 2:377.

68. Ibid., 2:248–49, 1:300, and 2:346. Humboldt also defends natives in his introduction to an exploration narrative written by one of his countrymen, referring to them as "Indios bravos" and explaining: "A melancholy experience has taught us that, in almost all climates, the vicinity of European or North American settlers has always tended to the destruction of the uncivilized races." See Humboldt in Baldwin Möllhausen, *Diary of a Journey from the Mississippi to the Coasts of the Pacific* (London: Longman, Brown, Green, Longmans, and Roberts, 1858), 1:xii–xiv.

69. Humboldt, *Personal Narrative*, 1:299, 2:412, and 1:294. Humboldt, then, was a monogenist, opposed to the polygenists who believed that different human "races" were actually different species, and of course that all dark-skinned species were inherently inferior. See Richard H. Popkin, "The Philosophical Bases of Modern Racism," in *Philosophy and the Civilizing Arts*, ed. Craig Walton and John P. Anton (Athens: Ohio University Press, 1974), pp. 126–65.

70. Humboldt, in letters written on November 24, 1800, and December 23, 1800, to Jean-Baptiste-Joseph Delambre and D. Guevara Vasconcellos, in *Lettres Américaines*, pp. 92 and 105, my translations; *Personal Narrative*, 1:295, 293, and 124.

On the complexity of pre-Columbian civilizations, see also Humboldt's *Political Essay*, pp. 48–49 and 53–70.

In his journal, Humboldt also remarked: "One sees here that the Incas still had active architecture projects when the Spanish arrived, and with them barbarity and indifference toward the arts." Humboldt, in *Lateinamerike am Vorabend der Unabhärigigkeitsrevolution*, ed. Faak, p. 75, my translation.

71. Humboldt, letter of November 25, 1802, to Wilhelm von Humboldt, in *Lettres Américaines*, pp. 135–36; *Researches, Concerning the Institutions and Monuments of the Ancient Inhabitants of America*, 1:20; *Personal Narrative*, 1:328.

Humboldt's appreciation of native languages clearly flouts what J. M. Blaut, in a classic work of postcolonial geography, has called "the colonizer's model of the world": "Closely connected to this theory [of the 'primitive mind'], " Blaut writes, "was the notion that there are 'primitive languages,' languages incapable of expressing higher theoretical and abstract thought. This old notion (which had been used in one form by William [Wilhelm] von Humboldt) was joined to the proposition that people cannot think beyond the limitations of their natural language, and so a primitive language entails a primitive mind." See Blaut, *The Colonizer's Model of the World: Geographical Diffusionism and Eurocentric History* (New York: Guilford Press, 1993), p. 97.

72. Humboldt, *Personal Narrative*, 1:200–201 and 293–94. This wording calls to mind a radical quote from the French novelist Bernardin de St. Pierre, whom Humboldt admired greatly: "I do not know if coffee and sugar are essential to the happiness of Europe, but I know well that these two products have determined the unhappiness of two other regions of the world. We have depopulated America so as to have land to plant them; we have depopulated Africa so as to have people to cultivate them." See Bernardin de St. Pierre, p. 88, my translation. On St. Pierre, see Humboldt, *Cosmos*, 2:433.

73. Humboldt, *Personal Narrative*, 1:202; Humboldt, *Political Essay*, p. 240.

74. Humboldt, *Political Essay*, p. 185. On the *Political Essay on the Island of Cuba*, see Philip S.

Foner, ed., *Alexander von Humboldt über die Sklaverei in den USA* (Berlin: Humboldt-Universität zu Berlin, 1981), pp. 15–37. In 1856, an American named John Thrasher published a highly edited and abridged version of this essay, omitting Humboldt's condemnations of slavery, in order to provide geographic and demographic information to a pro-slavery contingent in the United States interested in annexing Cuba. Humboldt was furious, and responded with public letters thrashing Thrasher and reiterating his hatred of slavery and even his low opinion of the American "slave States," whose treatment of blacks he referred to as "inhuman and atrocious."

Note also that Humboldt had read the so-called Ostend Manifesto of 1854, which revealed the U.S. *government's* desire (under President Franklin Pierce) to obtain Cuba, by force if necessary, so as to expand slavery; Humboldt called it the "most outrageous political document ever published" and also characterized it as "savage." Quoted in Ingo Schwarz, "Alexander von Humboldt—Socio-political Views of the Americas," in *Ansichten Amerikas: Neuere Studien zu Alexander von Humboldt*, ed. Ottmar Ette and Walther L. Bernecker (Frankfurt am Main: Vervuert Verlag, 2001), p. 113; and Foner, p. 11.

75. See Pratt, pp. 136 and 146–47.

76. Humboldt, *Political Essay*, p. 145. Also see de Terra, p. 307, and Jason Wilson's introduction to the Penguin Classics abridgement of Humboldt's *Personal Narrative*, p. liv.

77. The land issues are elaborated in the text. On violence against native peoples, see, for instance, the concluding chapter of the *Political Essay*, in which Humboldt condemns "the petty warfare carried on incessantly by the troops stationed in the presidios with the wandering Indians" (p. 235). On what Humboldt calls "the barbarous law of the *mita*," see p. 42. On general work conditions in Mexico, see the chapter entitled "State of Manufactures and Commerce" (pp. 185–220). Here, Humboldt details "the unhealthiness of the situation and the bad treatment to which the workmen are exposed" (p. 189). Further: "Every workshop resembles a dark prison," and "all are unmercifully flogged if they commit the smallest trespass" (ibid.). We could easily be in present-day maquiladoras.

78. Ibid., pp. 141–42, 145, and 183.

79. Humboldt, *Personal Narrative*, 3:264.

80. Quoted in Foner, p. 24; Humboldt, *Cosmos*, 1:368, my emphasis.

Even in his earliest comparative studies, Humboldt asserted the unity of the human species: "When we shall have more completely studied [humankind] . . . , the Caucasian, Mongul, American, Malay, and Negro races, will appear less insulated, and we shall acknowledge, in this great family of the human race, one single organic type, modified by circumstances which perhaps will ever remain unknown." Humboldt, *Researches, Concerning the Institutions and Monuments of the Ancient Inhabitants of America*, 1:14–15.

81. Humboldt, *Personal Narrative*, 1:175–76 and 3:272 and 263, and see pp. 228–84.

82. Ibid., 3:271.

83. Ibid., 1:90–91.

84. See Bowen, p. 250, and Wilson, p. xliv.

Chapter 3: *Cosmos:* Unification Ecology

1. Quoted in Robert D. Richardson, *Emerson: The Mind on Fire* (Berkeley: University of California Press, 1995), p. 135.

2. Ibid., p. 101.

3. Ibid., p. 121.

4. Emerson, *The Journals and Miscellaneous Notebooks of Ralph Waldo Emerson*, ed. Alfred P. Ferguson (Cambridge, Mass.: Harvard University Press, 1964), 4:16.

5. Humboldt, in Helmut de Terra, *Humboldt: The Life and Times of Alexander von Humboldt, 1769–1859* (New York: Knopf, 1955), p. 277; and in *Cosmos*, 1:51. In this chapter, unless otherwise noted, references will be to the new paperback editions of vols. 1 and 2 of *Cosmos* (Baltimore: Johns Hopkins University Press, 1997), with introductions by Nicolaas A. Rupke (vol. 1) and Michael Dettelbach (vol. 2).

Note that in a speech shortly after Humboldt's death the scientist Arnold Guyot made this comment on the lectures: "All the classes, from the king and the nobleman, to the literary, the scientific, and the simple man of intelligence and education, were represented in the immense audiences which gathered around him, and which no hall was large enough to contain." Guyot's speech is in a book put out by the American Geographical and Statistical Society, *Tribute to the Memory of Humboldt* (New York: H. H. Lloyd, June 15, 1859), p. 138.

Also note that a complete list of the lectures was compiled by one of Humboldt's earliest biographers: see Professor Klencke, "Alexander von Humboldt: A Biographical Monument," in *Lives of the Brothers Humboldt, Alexander and William*, by Klencke and Schleiser, trans. Juliette Bauer (London: Ingram, Cooke, 1852), p. 158.

6. Emerson actually wrote "Kosmos" in the original Greek in his journal. Robert K. Barnhart, in his etymological dictionary of American English, notes that the word "cosmos" seems to have been used once in England in the year 1200 but, after that, "the word disappears from the record until 1848, in a translation of Humboldt's *Kosmos*." (The year should be 1846.) See Barnhart, *The Barnhart Concise Dictionary of Etymology: The Origins of American English Words* (New York: HarperCollins, 1995), p. 165.

7. Humboldt, *Cosmos*, 1:65–70. Henry David Thoreau clearly agreed: "Heaven is under our feet as well as over our heads." See Thoreau, *Walden* (1854; repr., Boston: Beacon Press, 2004), p. 266.

8. Humboldt, *Cosmos*, 1:37.

9. See Rupke's introduction to the 1997 edition of *Cosmos*, 1:xxii–xxix. One admirer in Boston remarked that Humboldt "was not the founder of a religious creed, and yet a 'saviour.'" See Karl Heinzen, *The True Character of Humboldt; Oration Delivered at the German Humboldt Festival, in Boston* (Indianapolis: Association for the Propagation of Radical Principles, 1869), p. 1.

10. Humboldt, *Cosmos*, 1:8, 27, and 55.

11. Humboldt, *Views of Nature: Or Contemplations on the Sublime Phenomena of Creation; with Scientific Illustrations* (London: Henry G. Bohn, 1850; orig. pub. in German 1808), p. x; Humboldt, *Cosmos*, 1:25 and 7.

12. Mary Louise Pratt, *Imperial Eyes: Travel Writing and Transculturation* (London: Routledge, 1992), p. 140.

13. Malcolm Nicolson, "Historical Introduction" to the Penguin Classics abridgement of Humboldt's *Personal Narrative of a Journey to the Equinoctial Regions of the New Continent* (London: Penguin, 1995), p. xxxiii. My textbook was Elton D. Enger and Bradley F. Smith, *Environmental Science: A Study of Interrelationships*, 5th ed. (Dubuque, Iowa: Wm. C. Brown, 1995). On Humboldt and geography, see Margarita Bowen, *Empiricism and Geographical Thought: From Francis Bacon to Alexander von Humboldt* (Cambridge: Cambridge University Press, 1981), and Anne Buttimer, *Geography and the Human Spirit* (Baltimore: Johns Hopkins University Press, 1993). Harvey's article is in *Annals of the Association of American Geographers* 88 (December 1998), pp. 723–30; I am grateful to Kent Mathewson for pointing it out in his paper, "Humboldt and the Development of North American Geography," delivered in Washington, D.C., June 4, 2004, at the conference "Alexander von Humboldt and North America," held at the German Historical Institute.

14. Humboldt, *Personal Narrative of Travels to the Equinoctial Regions of America, 1799–1804* (London: Henry G. Bohn, 1852–53), 2:2, 1, 8, 3, 7, and 9.

15. Ibid., p. 9.

16. Thomas Robert Malthus, *An Essay on Population* (London: J. Johnson, 1798); Humboldt quoted in Mary Maples Dunn, introduction to the *Political Essay on the Kingdom of New Spain* (New York: Knopf, 1972), p. 13; and Humboldt, *Cosmos*, 1:53.

17. Aldo Leopold, *A Sand County Almanac* (1949; repr., New York: Ballantine Books, 1966), p. 190.

18. Humboldt, *Cosmos*, 1:42; Humboldt, *Personal Narrative*, 2:287–88 and 276.

19. Gustav Rose, *Mineralogisch-geognostische Reise nach dem Ural, dem Altai, und dem Kapischen Meere* (Berlin: Verlag der Sanderschen Buchhandlung, 1837), 1:608. My account of this part of Humboldt's Siberian expedition relies primarily on Rose's travel narrative, which does not seem to have been translated into English.

20. Humboldt quoted in de Terra, *Humboldt*, pp. 292–93 and 300.

21. See Rose, pp. 598–608, quotes on p. 602, my translations. Note: the modern romanization for "Tschin-fu" is "Qinfu."

22. Ibid., p. 607, my translation. Note: the modern romanization for *"Sankuetschi"* is *"Sanguo zhi."*

23. See Humboldt, *Personal Narrative,* 2:392.

24. Rose, pp. 607–8.

25. Quoted in Douglas Botting, *Humboldt and the Cosmos* (London: Michael Joseph, 1973), p. 246.

26. Quoted in de Terra, *Humboldt*, p. 290.

27. See, for instance, Rupke's introduction to the 1997 edition of *Cosmos*, 1:xxix–xxxii, quote on p. xxx, and Dettelbach's introduction to the 1997 edition of *Cosmos*, 2:vii–xvi.

28. Quoted in Pratt, p. 141; also see J. Fred Rippy and E. R. Brann, "Alexander von Humboldt and Simón Bolívar," *American Historical Review* 52 (July 1947), pp. 697–703.

29. On Humboldt's politics, see E. R. Brann, *The Political Ideas of Alexander von Humboldt: A Brief Preliminary Study* (Madison, Wis.: Littel Printing, 1954); on Humboldt and Jews and Judaism, see pp. 44–48. Humboldt made a point of supporting the rights of Jews throughout his life and lobbying for their equal treatment not only before the law but in society. His principle of unity was truly meant to include all social groups: "In the last number of the *Journal des Débats* there is a strong and very fine article against the abominable Jew Bill, with which we are threatened, and against which I have already protested. . . . The bill is a violation of all the principles of a wise policy of unity" (Humboldt, *Letters of Alexander von Humboldt to Varnhagen von Ense, from 1827 to 1858*, translated from the second German edition by Friedrich Kapp [New York: Rudd and Carleton, 1860], pp. 120–21).

On the Prussian fugitive slave law, designed explicitly to have the opposite result of the American law, see Brann, pp. 16–17, and Philip S. Foner, "Alexander von Humboldt on Slavery in America," *Science and Society* 47 (Fall 1983), pp. 340–41.

On 1848, see Botting, pp. 267–68.

The quotes are from the published volume of Humboldt's correspondence with Varnhagen von Ense, though the first is actually an excerpt from Varnhagen's diary; see *Letters of Alexander von Humboldt to Varnhagen von Ense*, pp. 246 and 259.

30. Humboldt quoted in de Terra, *Humboldt*, p. 286.

31. Humboldt, *Fragmens de Géologie et de Climatologie Asiatiques* (Paris: A. Pihan Delaforest, 1831), 2:554 and 1:21.

32. Humboldt, *Personal Narrative*, 3:272.

33. Humboldt, *Examen Critique de l'Histoire de la Géographie du Nouveau Continent* (Paris: Gide, 1837), 3:262, 277, and 290–91, my translations (this work has never been published in English to my knowledge).

34. Pratt, p. 130.

35. Humboldt, *Personal Narrative*, 2:3.

36. Ibid., 1:232 and 2:9.

On the commodification of natural resources, see Pratt, pp. 34 and 129–32.

On occasion, Humboldt in fact did go beyond a basic sustainability argument to advocate a profound respect for all natural things regardless of their usefulness to human beings: "The view of nature ought to be grand and free, uninfluenced by motives of . . . relative utility" (*Cosmos*, 1:83).

37. Humboldt, *Cosmos*, 1:53.

38. See, for instance, Dettelbach, p. xi: "Humboldt's ideal was reform introduced by an informed and liberal monarch. . . . Humboldt was quite sincere in his dedication of the *Essai politique sur la Nouvelle Espagne* to Spain's Charles IV, like that of *Cosmos* to Prussia's Friedrich Wilhelm IV."

39. On the diamonds, see de Terra, *Humboldt*, pp. 298–99; for the letter to Jefferson, see de Terra, "Alexander von Humboldt's Correspondence with Jefferson, Madison, and Gallatin," *Proceedings of the American Philosophical Society* 103 (December 1959), p. 790.

40. Quoted in de Terra, *Humboldt*, p. 307.

41. Ibid., p. 306; also see Botting, pp. 252–53.

42. "Baron Humboldt's Cosmos," *Christian Examiner and Religious Miscellany*, 48, 4th series Vol. 13 (January 1850), p. 59.

43. Humboldt, *Cosmos*, 1:178. On Humboldt's massive correspondence, see Botting, p. 278.

44. Humboldt, *Cosmos*, 1:83 and 79; also see Botting, pp. 253–54. On magnetism, see Tim Fulford, Debbie Lee, and Peter J. Kitson, *Literature, Science, and Exploration in the Romantic Era: Bodies of Knowledge* (Cambridge: Cambridge University Press, 2004), pp. 149–75.

45. Humboldt, *Cosmos*, 1:50; Humboldt, *Letters of Alexander von Humboldt to Varnhagen von Ense*, p. 39.

46. Humboldt, *Letters of Alexander von Humboldt to Varnhagen von Ense*, pp. 35–36 and 38.

47. Humboldt, *Cosmos*, 1:79, 343, 352, and 358, and *Views of Nature*, quoted in Ottmar Ette, *Literature on the Move* (Amsterdam: Rodopi, 2003), p. 120.

48. Humboldt's letters quoted in Botting, p. 257, and de Terra, *Humboldt*, p. 315.

49. Humboldt, *Cosmos*, 1:7.

50. "Baron Humboldt's *Kosmos*: A General Survey of the Physical Phenomena of the Universe," *North British Review* 4 (November 1845–February 1846), pp. 202–54, quote on p. 202.

51. Rupke's introduction to the 1997 edition of *Cosmos*, 1:vii. The second volume of *Cosmos* was particularly popular. Right after it came out, in 1847, Humboldt's publisher wrote to him: "Book parcels destined for London and St. Petersburg were torn out of our hands by agents who wanted their orders filled for the bookstores in Vienna and Hamburg. Regular battles were fought over possession of this edition, and bribes offered for priorities" (quoted in de Terra, *Humboldt*, p. 359).

52. Humboldt, *Cosmos*, 1:24–25.

53. On the companion volumes, see Rupke's introduction to the 1997 edition of *Cosmos*, 1:xx. Vintage posters show that Chocolat Poulain (which is still in business today) was publishing chromolithographs with Humboldt's portrait as late as 1900. For a reproduction of one such poster, see the exhibit catalog, *Alexander von Humboldt: Netzwerke des Wissens* (Ostfildern-Ruit, Germany: Hatje Cantz Verlag, 1999), p. 25.

54. Forbes's article, originally published in the *London Quarterly Review*, was reprinted for New York readers in *Eclectic Magazine* 7 (March 1846), pp. 353–75, quote on p. 369.

55. Crosse quoted in Rupke's introduction to the 1997 edition of *Cosmos*, 1:xxiii.

56. "Alexander von Humboldt and his Cosmos," *Methodist Quarterly Review*, 4th ser., 12 (July 1860), p. 428. As another American put it: "There are those who will be horrified by the discovery that Humboldt, who never carried on an open war against belief and religion . . . , was an atheist and a materialist. Their horror, if unfeigned, will only prove that they did not understand the great man, or that they did not meditate sufficiently upon that which he professed." See Heinzen, p. 5.

57. On 1840s America, see, for instance, Arthur M. Schlesinger, *The Age of Jackson* (Boston: Little, Brown, 1945), and Charles Sellers, *The Market Revolution: Jacksonian America, 1815–1846* (New York: Oxford University Press, 1991).

58. James Davenport Welpley, "Humboldt's Cosmos" (published with the Latin epigraph "Mens ingenti scientiarum flumine inundata," translated as "A mind inundated with a great flood of sciences"), *American Review: A Whig Journal of Politics, Literature, Art, and Science* 3 (June 1846), pp. 598–610; and see Daniel Walker Howe, *The Political Culture of American Whigs* (Chicago: University of Chicago Press, 1979), p. 21, my emphasis. For more on the Whigs, also see Michael F. Holt, *The Rise and Fall of the American Whig Party: Jacksonian Politics and the Onset of the Civil War* (New York: Oxford University Press, 1999).

59. Welpley, pp. 598 and 610.

60. Welpley, pp. 604, 598, and 603.

On Whigs and public education, see Howe, p. 22.

61. Welpley, pp. 598 and 602–3.

62. Ibid., p. 610.

Humboldt's influence on the American Renaissance has never been adequately studied, to my knowledge, but an excellent start has been made in Laura Dassow Walls, *Seeing New Worlds:*

Henry David Thoreau and Nineteenth-Century Natural Science (Madison: University of Wisconsin Press, 1995).

63. On Sumner, see Ingo Schwarz, "'Shelter for a Reasonable Freedom' or Cartesian Vortex: Aspects of Alexander von Humboldt's Relation to the United States," *Debate y Perspectivas* 1 (2000), pp. 169–70.

On Symmes, see chapters 1 and 4.

On Irving, see Schwarz, "The Second Discoverer of the New World and the First American Literary Ambassador to the Old World: Alexander von Humboldt and Washington Irving," *Acta Historica Leopoldina* 27 (1997), pp. 89–97.

On Catlin, see Schwarz, "'Shelter for a Reasonable Freedom' or Cartesian Vortex," p. 170; William H. Goetzmann, *New Lands, New Men: America and the Second Great Age of Discovery* (1986; repr., Austin: Texas State Historical Association, 1995), pp. 160–68; and Thomas Donaldson, *The George Catlin Indian Gallery in the U.S. National Museum* (Washington, D.C.: United States National Museum, 1886), pp. 376–77, 676, and 758. Catlin met Humboldt in Paris in 1845, and they exchanged a number of letters. In 1845, Catlin noted that Humboldt "accompanied us quite through the rooms of the Louvre, and took a great deal of interest in the Indians, having seen and dealt with so many in the course of his travels" (Donaldson, p. 676). A decade later, Catlin visited Humboldt in Berlin, and on September 12, 1855, Humboldt wrote to him: "I believe, with you, that the Crows are Toltecs. . . . I believe your discoveries will throw a great deal of light on the important subject of the effect of cataclysms on the distribution of races. . . . Let nothing stop you; you are on a noble mission, and the Great Spirit will protect you" (Donaldson, p. 758).

For more on Indians and American antiquities, see Humboldt, *Vues des Cordillères et Monuments des Peuples Indigènes de l'Amérique*, 2 vols. (Paris: Chez L. Bourgeois-Maze, 1840; orig. pub. 1810), and the 1814 translation by Helen Maria Williams, *Researches, Concerning the Institutions and Monuments of the Ancient Inhabitants of America, with Descriptions and Views of Some of the Most Striking Scenes in the Cordilleras*, 2 vols. (London: Longman, Hurst, 1814). Also see Michael Anthony Wadyko, "Alexander von Humboldt and 19th-Century Ideas on the Origin of American Indians" (Ph.D. diss., West Virginia University, 2000); Robert Silverberg, *Mound Builders of Ancient America: The Archaeology of a Myth* (1968; repr., Athens: Ohio University Press, 1986); Robert Wauchope, *Lost Tribes and Sunken Continents: Myth and Method in the Study of American Indians* (Chicago: University of Chicago Press, 1962); and John C. Greene, *American Science in the Age of Jefferson* (Ames: Iowa State University Press, 1984), pp. 320–408.

There are some fascinating connections between Humboldt's study of American antiquities and the origin story of the Church of Jesus Christ of Latter-day Saints (popularized by Joseph Smith in the Book of Mormon); I cover this in my dissertation, "The Humboldt Current: Avant-Garde Exploration and Environmental Thought in 19th-Century America," Yale University, 2004, pp. 166–73.

64. Humboldt mentions the work of most of these men in various places. Agassiz, Prince Maximilian, Bodmer, and Catlin, for instance, appear in *Cosmos*. For Silliman's link to Humboldt, see Benjamin Silliman, *A Visit to Europe in 1851* (New York: George P. Putnam, 1853), 2:317–22. Agassiz's plea for financial assistance for his museum is in a letter to Humboldt of September 8, 1855 (American Philosophical Society, Humboldt Collection, microfilm, reel 2). Humboldt's influence on Dana and Gray is palpable in their work, and both Americans noted the importance of Humboldt's books to them in letters they sent to the Navy Department in preparation for what became the U.S. Exploring Expedition of 1838–42: see National Archives, Microfilm Publication M75, Letters Received by the Navy Department relating to the Wilkes Expedition, Letters Relating to Preparations for the Exploring Expedition, roll 2, letter of Dana, July 13, 1837, New Haven, and roll 3, letter of Gray, November 7, 1837. Humboldt wrote an introduction to Möllhausen's narrative, *Diary of a Journey from the Mississippi to the Coasts of the Pacific*, 2 vols. (London: Longman, Brown, Green, Longmans, and Roberts, 1858). In a letter of 1849, Humboldt noted his indebtedness to Frémont, Abert, and Bache for their "very valuable hypsometric, astronomic, botanic, and geognostic work" (letter of December 22, 1849, to Dr. J. G. Flügel-Leipsic,

American Philosophical Society, Humboldt Collection, BH88.5). There is also a letter from Humboldt to Bache preserved at the Huntington Library, call no. RH 1565, dated July 10, 1851, Berlin. Maury's extensive debt to Humboldt is noted in Goetzmann, pp. 298–350. Also see: A. Hunter Dupree, *Science in the Federal Government* (Cambridge, Mass.: Harvard University Press, 1957), and William M. Smallwood, *Natural History and the American Mind* (New York: Columbia University Press, 1941), esp. pp. 146–336.

65. J. W. Abert, *Western America in 1846–1847: The Original Travel Diary of Lieutenant J.W. Abert, who mapped New Mexico for the United States Army*, ed. John Galvin (San Francisco: John Howell Books, 1966), pp. 56, 19, 10, and 87.

66. D. D. Barnard, "The War with Mexico," *American Review: A Whig Journal of Politics, Literature, Art, and Science* 3 (June 1846), p. 571; Humboldt quoted in Foner, p. 341. Explanations and appreciations of Humboldt's work in Mexico and of his general relationship with the country can be found in Rayfred Lionel Stevens-Middleton, *La Obra de Alejandro de Humboldt en México: Fundamento de la Geografía Moderna* (México: Sociedad Mexicana de Geografía y Estadística, 1956), and Seminario de Historia de la Filosofía en México, de la Facultad de Filosofía y Letras, *Ensayos Sobre Humboldt* (México: Universidad Nacional Autónoma de México, 1962).

67. Emerson, "Nature," in *Nature/Walking*, ed. John Elder (Boston: Beacon Press, 1991), pp. 37 and 9.

68. Humboldt, *Cosmos*, 2:89 and 81.

69. Ibid., pp. 4, 19, and 98. On the Hudson River School, see, for instance, Barbara Novak, *Nature and Culture: American Landscape and Painting, 1825–1875* (New York: Oxford University Press, 1980).

70. Humboldt, *Cosmos*, 2:93 and 95; and see *Aspects of Nature* (Philadelphia: Lea and Blanchard, 1849); *Views of Nature* (London: Henry G. Bohn, 1850); and *Personal Narrative* (London: Henry G. Bohn, 1852–53).

On Church, see Kevin J. Avery, *Church's Great Picture: "The Heart of the Andes"* (New York: Metropolitan Museum of Art, 1993), pp. 12–22, and David Huntington, "Landscape and Diaries: The South American Trips of F.E. Church," *Brooklyn Museum Annual* 5 (1963–64), pp. 65–99.

On Humboldt's fame, see "Humboldt, Ritter, and the New Geography," *New Englander* 18 (May 1860), p. 282.

71. Thomas Cole, "Essay on American Scenery," in *American Art, 1700–1960: Sources and Documents*, ed. John W. McCoubrey (Englewood Cliffs, N.J.: Prentice-Hall, 1965), p. 109. Also see Perry Miller, "Nature and the National Ego," in *Errand into the Wilderness* (Cambridge, Mass.: Harvard University Press, 1956), pp. 204–16.

72. Walls, *Seeing New Worlds*, pp. 76–166. On Stephens and Taylor, see Larzer Ziff, *Return Passages: Great American Travel Writing, 1780–1910* (New Haven: Yale University Press, 2000), pp. 58–169.

73. See Walls, pp. 3–14 and 121–66. Also note Perry Miller, "Thoreau in the Context of International Romanticism," in *Nature's Nation* (Cambridge, Mass.: Harvard University Press, 1967), pp. 175–83.

74. On Thoreau's relation to gender and sexuality, see, for instance, Michael Warner, "Thoreau's Bottom," *Raritan* 11 (Winter 1992), pp. 53–78, and Laura Dassow Walls, "*Walden* as Feminist Manifesto," *ISLE: Interdisciplinary Studies in Literature and Environment* 1 (Spring 1993), pp. 137–44.

75. Walls, *Seeing New Worlds*, pp. 13 and 143.

76. Thoreau, "Walking" (1862), in Elder, pp. 71 and 110.

77. Humboldt, *Cosmos*, 2:73; on Cole, see, for instance, Ellwood C. Parry III, *The Art of Thomas Cole* (Newark: University of Delaware Press, 1988), esp. pp. 226–59 and 342–61; on Church, see "Mr. Church's New Picture—The Heart of the Andes," *New York Herald*, April 28, 1859, p. 7; on Church and Humboldt, see "Humboldt, Ritter, and the New Geography," p. 292. On Church's reinterpretations of Cole, note Bryan J. Wolf, "A Grammar of the Sublime, or Intertextuality Triumphant in Church, Turner, and Cole," *New Literary History* 16 (Winter 1985), pp. 321–41.

78. See Avery; David Huntington, *The Landscapes of Frederic Edwin Church: Vision of an Ameri-*

can Era (New York: Braziller, 1966); Franklin Kelly, *Frederic Edwin Church and the National Landscape* (Washington, D.C.: Smithsonian Institution, 1988); and Kelly, *Frederic Edwin Church* (Washington, D.C.: National Gallery of Art, 1989).

79. "The Heart of the Andes," *The Crayon* 6 (June 1859), p. 193.

80. "Church's Heart of the Andes," *Harper's Weekly* 3 (May 7, 1859), p. 291; "Mr. Church's New Picture," *New York Times*, April 28, 1859, p. 4.

81. For some of the standard interpretations of *The Heart of the Andes*, see Avery; Huntington, *The Landscapes of Frederic Edwin Church*, pp. 5–10 and 50–53; Kelly, *Frederic Edwin Church*, pp. 55–58; Novak, pp. 37–38 and 71–74; Katherine Emma Manthorne, *Tropical Renaissance: North American Artists Exploring Latin America, 1839–1879* (Washington, D.C.: Smithsonian Institution, 1989), pp. 6–7 and 79–80; Manthorne, "Legible Landscapes: Text and Image in the Expeditionary Art of Frederic Church," in *Surveying the Record: North American Scientific Exploration to 1930*, ed. Edward C. Carter II (Philadelphia: American Philosophical Society, 1999), pp. 133–45; and David C. Miller, *Dark Eden: The Swamp in Nineteenth-Century American Culture* (New York: Cambridge University Press, 1989), pp. 107–17.

82. Humboldt, *Cosmos*, 2:20, 95, and 96.

83. Theodore Winthrop, *A Companion to the "Heart of the Andes"* (New York: Appleton, 1859), pp. 6–7, 22, 14, 39, and 29; note that this pamphlet was reprinted in Winthrop, *Life in the Open Air and Other Papers* (New York: John W. Lovell, 1862), pp. 329–74.

84. Humboldt, *Letters of Alexander von Humboldt to Varnhagen von Ense*, p. 351.

85. Quoted in Botting, p. 268. One of Humboldt's earliest biographers wrote, of his relationship to "the people": "The inhabitants of Berlin and Potsdam all know him personally, and show him as much honor as they show the king. With a slow but firm step, a thoughtful head, rather bent forward, whose features are benevolent with a dignified expression of noble calmness, either looking down, or politely responding to the greetings of the passers-by with kindness, and without pride, in a simple dress, frequently holding a pamphlet in his hand. . . . Wherever he appears he is received by tokens of universal esteem, the passers-by timidly step aside for fear of disturbing him in his thoughts; even the working man looks respectfully after him, and says to his neighbour, 'there goes Humboldt.'" See Klencke, p. 148.

86. Humboldt, *Cosmos*, 2:80.

87. Humboldt's plea and Irving's response to it are in Pierre M. Irving, *Life and Letters of Washington Irving* (New York: G. P. Putnam, 1864), 4:284–85.

88. Richard Henry Stoddard, *The Life, Travels, and Books of Alexander von Humboldt*, with an introduction by Bayard Taylor (New York: Rudd and Carleton, 1859), p. 445; Humboldt, *Cosmos*, 2:236; on Silliman, see Dumas Malone, ed., *Dictionary of American Biography* (New York: Charles Scribner's Sons, 1936), 19:160.

89. On Humboldt's apartment, see Stoddard, pp. 446–56; de Terra, *Humboldt*, pp. 367–74; and Botting, pp. 274–83. Botting also includes the crucial *Humboldt in his Study in Oranienburger Strasse, Berlin*, a grandiose watercolor by Eduard Hildebrandt (1856), pp. 270–71.

90. Silliman, pp. 318–22. (Silliman printed Humboldt's letter in his book.)

91. Taylor quoted in Stoddard, pp. 455–62; Humboldt quoted in Botting, pp. 278 and 277; also see de Terra, *Humboldt*, pp. 363–66.

92. "Humboldt, Ritter, and the New Geography," p. 281.

93. American Geographical and Statistical Society, *Tribute to the Memory of Humboldt* (New York: H. H. Lloyd, June 15, 1859), pp. 150, 117, 123, and 140. There were also speeches given later in the year, on what would have been Humboldt's ninetieth birthday, September 14, 1859. See, for instance, Alfred Stillé, M.D., *Humboldt's Life and Character: An Address before the Linnaean Society of Pennsylvania College* (Philadelphia: Linnaean Association, 1859).

94. Whitman, *Leaves of Grass* (New York: New American Library, 1980), pp. 67 and 310.

95. For examples of speeches given in Richmond, Virginia, and Dubuque, Iowa, see Henry Patrick Aylett, *Address of Henry Patrick Aylett before the German Societies* (Richmond: "Enquirer" Steam Presses, 1869), and Iowa Institute of Science and Arts, *The Opening of the Iowa Institute of Science and Arts* (Dubuque: Iowa Institute of Science and Arts, 1869), esp. pp. 15–49.

96. On Ingersoll, see Malone, ed., *Dictionary of American Biography*, 9:469; all of Ingersoll's quotes come from "HUMBOLDT: The Universe is Governed by Law," in Ingersoll, *The Gods and Other Lectures* (Peoria, Ill.: Religio Philosophical Publishing House, 1874), pp. 95, 103, 115, 112, 116, 107, 99, 115, and the title page. Capitalization in original.

Excursion: Eureka: The Death of Edgar Allan Poe

1. For the narrative of Poe's death, see Kenneth Silverman, *Edgar A. Poe: Mournful and Never-Ending Remembrance* (New York: HarperCollins, 1991), pp. 433–37 and 518–19.

2. Poe, *Eureka*, in *The Complete Works of Edgar Allan Poe*, ed. James A. Harrison (New York: AMS Press, 1965), 16:179–315, quotes on pp. 181 and 183.

3. Quoted in David Dickason, *The Daring Young Men: The Story of the American Pre-Raphaelites* (Bloomington: Indiana University Press, 1953), p. 33.

4. Humboldt, *Cosmos: A Sketch of a Physical Description of the Universe* (London: Henry G. Bohn, 1849), 1:63.

5. Poe, "The Poetic Principle," in *Selected Writings of Edgar Allan Poe: Poems, Tales, Essays, and Reviews*, ed. David Galloway (Harmondsworth, U.K.: Penguin, 1985), p. 505; "Manuscript Found in a Bottle," p. 109.

6. Poe, "A Descent into the Maelstrom," in ibid., p. 236.

7. For a fascinating perspective on Symmes and Poe in the context of a long literary tradition linking Antarctic exploration to an exploration of the self and the unconscious, see Victoria Nelson, "Symmes Hole, or the South Polar Romance," *Raritan* 17 (Fall 1997), pp. 136–66, esp. 148–53.

8. A reprint of Poe's article on Reynolds can be found in Harrison, ed., *Complete Works*, 9:306–14.

9. Poe, *The Narrative of Arthur Gordon Pym*, in *"The Fall of the House of Usher" and Other Tales* (New York: New American Library, 1980), p. 371.

10. Poe, *Marginalia*, in Harrison, ed., *Complete Works*, 16:90 and 88.

11. It is by no means clear that Moran could be trusted: see, for instance, W. T. Bandy, "Dr. Moran and the Poe-Reynolds Myth," in *Myths and Reality: The Mysterious Mr. Poe*, ed. Benjamin Franklin Fisher IV (Baltimore: Edgar Allan Poe Society, 1987), pp. 26–36. Nevertheless, most Poe scholars tend to assume that Poe did in fact cry out for Reynolds. See, for instance, Richard Kopley, ed., *Poe's Pym: Critical Explorations* (Durham, N.C.: Duke University Press, 1992), in which most contributors refer to Reynolds as probably having been one of Poe's close friends.

Chapter 4: "Rough Notes of Rough Adventures": Exploration for Exploration's Sake

1. Captain Benjamin Pendleton's official report of the Palmer-Pendleton Expedition is printed in Edmund Fanning, *Voyages Round the World* (New York: Collins and Hannay, 1833), pp. 478–88. Detailed scholarly accounts of the expedition can be found in Kenneth J. Bertrand, *Americans in Antarctica, 1775–1948* (New York: American Geographical Society, 1971), pp. 144–58; William Stanton, *The Great United States Exploring Expedition of 1838–1842* (Berkeley: University of California Press, 1975), pp. 8–28; and Philip I. Mitterling, *America in the Antarctic to 1840* (Urbana: University of Illinois Press, 1959), pp. 67–100. A nineteenth-century account of the expedition's goals and historical context can be found in A. E. Carrell [*sic*], "The First American Exploring Expedition," *Harper's New Monthly Magazine* 44 (December 1871), pp. 60–64. The name of the author of this article was misspelled; it was actually Anna Ella Carroll, also the author of the popular book *The Star of the West; or, National Men and National Measures* (1856; repr., New York: Miller, Orton, 1857), the 3rd edition of which included a long section about Reynolds's expedition on which Carroll's later article was based (see pp. 13–136). Carroll seems to have been an ardent admirer of Reynolds; she referred to him as "the great oracle of the *future!*" (p. 136).

Annawan was captured by Captain Benjamin Church in late August 1676, at the end of King

Philip's War, in Rehoboth, Massachusetts, where today there is still a monument called "Annawan's Rock" and where Annawan's cooking pot is displayed in a local museum.

Note that "seraph," in addition to its usual angelic denotation, could also refer to a fossil shell in the nineteenth century—an appropriate meaning, given that the first Antarctic fossil was discovered on this expedition.

2. Reynolds, "Rough Notes of Rough Adventures," *Southern Literary Messenger* 9 (December 1843), p. 705; Pendleton in Fanning, p. 479; Reynolds, letter of October 25, 1830, Los Angeles [de Chile], to Samuel Southard, in Samuel L. Southard Papers, Princeton University Library, box 33 (hereafter cited as Southard Papers).

3. Pendleton in Fanning, p. 481; Reynolds, letter of October 25, 1830, Southard Papers, box 33. Little is known about Watson, but Edmund Fanning, the primary sponsor of the expedition, refers to him as a medical doctor and a professor in *Voyages to the South Seas* (New York: William H. Vermilye, 1838), p. 174. (This book should be distinguished from Fanning's *Voyages Round the World*, published five years earlier. There is some overlap between the two volumes, but the later one has an important new section on "The Origins, Authorizations, and Progress of the First American South Sea Exploring Expedition," pp. 152–216.)

4. Reynolds, letter of October 25, 1830, Southard Papers, box 33. Note that much more was learned about Chile and the Araucanians in the late 1840s and early 1850s, when Lieutenant Melville Gilliss, a confirmed Humboldtian, launched his "U.S. Astronomical Expedition in Chili." See William H. Goetzmann, *New Lands, New Men: America and the Second Great Age of Discovery* (1986; repr., Austin: Texas State Historical Association, 1995), pp. 334–37, and Edmond Reuel Smith, *The Araucanians; or, Notes of a Tour among the Indian Tribes of Southern Chili* (New York: Harper and Brothers, 1855).

5. On Reynolds and his career, the best source is an unpublished biography, the typescript of which is in the Special Collections at the Princeton University Library: Richard G. Woodbridge III, "J.N. Reynolds—American."

Other Reynolds sources include Robert F. Almy, "J. N. Reynolds: A Brief Biography with Particular Reference to Poe and Symmes," *Colophon*, n.s., 2 (Winter 1937), pp. 227–45, and Goetzmann, pp. 244–46, 258–64, and 270–73.

6. This account of Reynolds's ascent of the volcano of Antuco, including all quotations, comes from Reynolds, "A Leaf from an Unpublished Manuscript," *Southern Literary Messenger* 5 (June 1839), pp. 408–13.

7. Reynolds, "Rough Notes of Rough Adventures," p. 705.

8. Reynolds, "A Leaf from an Unpublished Manuscript," p. 410.

9. Ibid., pp. 410–13.

10. J. N. Reynolds, *Remarks on a Review of Symmes' Theory, which appeared in the American Quarterly Review* (Washington, D.C.: Gales and Seaton, 1827).

11. On Symmes, see especially Elmore Symmes, "John Cleves Symmes, The Theorist," an article in three installments in the *Southern Bivouac*, n.s., 2 (February, March, and April 1887), pp. 555–66, 621–31, and 682–93. This article includes reproductions of Symmes's original circulars as well as excerpts from his speeches and private papers. It has a smattering of identifiable minor mistakes but remains an important source on the collaboration between Symmes and Reynolds. At the very end of the article, Elmore Symmes takes great pleasure in noting that "the celebrated English astronomer, Halley," as well as Leslie and Kepler, had also thought that the earth might be a hollow sphere, and that "in Humboldt's 'Cosmos' we find the above names enumerated with that of Captain Symmes" (p. 693).

Other sources on Symmes include P. Clark, "The Symmes Theory of the Earth," *Atlantic Monthly* 31 (April 1873), pp. 471–80; E. F. Madden, "Symmes and His Theory," *Harper's New Monthly Magazine* 65 (October 1882), pp. 740–44; John Wells Peck, "Symmes' Theory," *Ohio Archaeological and Historical Publications* 18 (1909), pp. 28–42; and Reginald Horsman, "Captain Symmes's Journey to the Center of the Earth," *Timeline* (September/October 2000), pp. 2–13.

Also see James McBride, *Symmes's Theory of Concentric Spheres* (Cincinnati: Morgan, Lodge, and Fisher, 1826), and Thomas J. Matthews, *Lecture on Symmes' Theory of Concentric Spheres, Read at the Western Museum* (Cincinnati: A. N. Deming, 1824).

Hairs and aerolites from a Symmes quote in Paul Collins, "Symmes Hole," in *Banvard's Folly: Thirteen Tales of People Who Didn't Change the World* (New York: Picador, 2001), p. 57.

12. Captain Adam Seaborn, *Symzonia: Voyage of Discovery* (New York: J. Seymour, 1820). A more recent edition comes with an introduction (Gainesville, Fl.: Scholars' Facsimiles and Reprints, 1965) by the literary scholar J. O. Bailey, who claims that Symmes was probably the real author—though he has little evidence to support this claim, and my own opinion is that the novel was written by some anonymous citizen who had read one of Symmes's circulars.

13. Quoted in Stanton, p. 9. Even Humboldt admitted that "nothing is more marvellous, and nothing is yet known less clearly in a geographical point of view, than the direction, extent, and term of the migrations of birds." See *Personal Narrative of Travels to the Equinoctial Regions of America, 1799–1804* (London: Henry G. Bohn, 1852–53), 2:385.

14. Humboldt, *Cosmos* (Baltimore: Johns Hopkins University Press, 1997), 1:171. The correct spelling of Davy's first name is "Humphry."

15. Humboldt's most widely read study of isotherms was "Des Lignes Isothermes et de la Distribution de la Chaleur sur le Globe" [On Isothermic Lines and the Distribution of Heat on the Planet], *Mémoires de Physique et de Chimie de la Société d'Arcueil* 3 (1817), pp. 462–602.

The Symmes supporter who cited Humboldt is McBride, p. 105.

Belief in the open polar seas lasted well into the nineteenth century: see Edward L. Towle, "The Myth of the Open Polar Sea," *Proceedings of the Tenth International Congress of the History of Science* (Paris, 1965), pp. 1037–41.

Reynolds discusses the possibility of an open sea in the Antarctic in his official report of 1828 to Navy Secretary Samuel Southard, *Pacific Ocean and South Seas*, reprinted in 1835 as H. Doc. 105, 23rd Cong., 2nd sess., p. 26. Of the sealers he interviewed who had been south of 70°, Reynolds noted: "They all agree that . . . the greatest impediment to navigation from ice will be found from 62° to 68° S." Also see his *Remarks on a Review of Symmes' Theory*, p. 40.

Edmund Fanning, in *Voyages Round the World* (1833), pp. 473–74, notes: "The report of all (within the author's knowledge) that have passed beyond the 68th degree is, that above this degree of latitude, the sea was found to be mainly clear of ice, and the climate becoming more mild, with prevailing winds from the southward.

"The report of that persevering navigator, Weddel, who has sailed farther south than any other navigator has been known to do, is this, viz., to the latitude of 74°15' S., he states that at this position the weather was mild as summer, the wind at the time being from the south, while the sea was clear in that quarter, as far as the eye could discern from the masthead."

16. See Elmore Symmes, Collins, and Stanton on this period of Symmes and Reynolds's collaboration. Poem quoted in Elmore Symmes, p. 563.

On America in the 1820s and '30s, see Jean V. Matthews, *Toward a New Society: American Thought and Culture, 1800–1830* (Boston: Twayne Publishers, 1991), and Charles Sellers, *The Market Revolution: Jacksonian America, 1815–1846* (New York: Oxford University Press, 1991).

17. On Glass, see Reynolds, ed., *A Life of George Washington, in Latin Prose: by Francis Glass, A.M., of Ohio* (New York: Harper and Brothers, 1835), pp. ii and vii; and on Reynolds's childhood, see Henry Howe, "The Romantic History of Jeremiah N. Reynolds," in *Historical Collections of Ohio in Two Volumes: An Encyclopedia of the State* (Columbus: Henry Howe and Son, 1889), 1:430.

Reynolds's edited volume of Glass's Latin biography of George Washington is one of the great oddball projects of American literature. (Note the review of it, probably by Edgar Allan Poe, in the *Southern Literary Messenger* 2 [December 1835], pp. 52–54.) Glass had already died by the time Reynolds managed to bring the book into print, but it seems to have been intended as a tribute to the old man's dedication to the life of the mind. Reynolds's introduction to the book is a touching testimonial not only to the worthiness of Francis Glass but to Reynolds's own anxieties and insecurities about living as an intellectual in a rapidly modernizing society. "Such a man," Reynolds wrote, of his former teacher, "is seldom properly appreciated any where, even in the bosom of letters, where many are capable of understanding such gifts; but a new country furnishes few competent judges of high literary acquirement." Glass himself had "often dwelt upon that enlightened age of Greece, when the lecturer at the Academy or Lyceum was a greater man, in public estimation, than the commander of armies" (pp. vi–vii). Glass died impoverished and unknown on the Ohio frontier.

18. Reynolds, *Remarks on a Review of Symmes' Theory*, p. 4. Baudelaire proclaimed Poe to be the greatest genius of America, and his translations made Poe the most popular American writer in France for several decades. Verne was also fascinated with Arctic and Antarctic exploration—as suggested by the titles of several of his novels: *An Antarctic Mystery, The Field of Ice, A Journey to the North Pole, The Sphinx of the Ice Fields*, and *The Adventures of Captain Hatteras*.

The letter of the Harrisburg legislators is printed in "Capt. Symmes' Theory," *Niles' Weekly Register* 29 (February 25, 1826), pp. 427–28.

Wylie is quoted in Mitterling, p. 84.

19. New York *National Advocate*, May 28, 1828, cited in Elmore Symmes, p. 565; *New-York Mirror* 3 (June 17, 1826), p. 375. For more coverage of Reynolds and Symmes in New York, also see the *Mirror* of April 15 (p. 303) and May 27 (pp. 350–51).

20. Elmore Symmes, p. 623.

21. Reynolds, *Remarks on a Review of Symmes' Theory*, pp. 42, 4, and 69; Reynolds, letter of August 3, 1826, Washington City, Southard Papers, box 19; Southard, letter of August 7, quoted in Woodbridge, "J.N. Reynolds: Father of American Exploration," *Princeton University Library Chronicle* 45 (Winter 1984), p. 110.

22. On Lewis and Clark as failures, see, for instance, James P. Ronda, *Finding the West: Explorations with Lewis and Clark* (Albuquerque: University of New Mexico Press, 2001), esp. pp. 117–29. A. Hunter Dupree argues that the expedition's legacy was so inconsequential because the government had not yet established the institutions necessary for processing and analyzing the vast collections Lewis and Clark brought back; see Dupree, *Science in the Federal Government* (Cambridge, Mass.: Harvard University Press, 1957), pp. 26–28.

23. Reynolds, *Remarks on a Review of Symmes' Theory*, p. 53. On American exploration between Lewis and Clark and J. N. Reynolds, see, for instance, Goetzmann, *New Lands, New Men*, pp. 110–58, and Goetzmann, *Exploration and Empire: The Explorer and the Scientist in the Winning of the American West* (New York: W. W. Norton, 1966), pp. 3–145. On the encounter between Long, Symmes, and Reynolds, see Elmore Symmes, pp. 564–65, and William Keating, *Narrative of an Expedition to the Source of the St. Peter's River* (London: Geo. B. Whittaker, 1825), 1:44. Keating, one of the members of Long's party, wrote of Symmes that "he appears conversant with every work of travels, from Hearne's to Humboldt's."

24. Quoted in Stanton, p. 25.

25. Reynolds, *Remarks on a Review of Symmes' Theory*, pp. 69–70; letter of June 11, 1829, New York, Southard Papers, box 31.

On the politics of this era, see Sellers and Harry L. Watson, *Liberty and Power: The Politics of Jacksonian America* (New York: Hill and Wang, 1990).

26. Humboldt, *Personal Narrative*, 1:434.

Cooper's debt to Long is noted in Goetzmann, *New Lands, New Men*, p. 124.

On American literature and culture in the 1820s, see Howard Mumford Jones, *O Strange New World: American Culture: The Formative Years* (New York: Viking, 1964), esp. pp. 312–95; R. W. B. Lewis, *The American Adam: Innocence, Tragedy, and Tradition in the Nineteenth Century* (Chicago: University of Chicago Press, 1955); and Cecelia Tichi, *New World, New Earth: Environmental Reform in American Literature from the Puritans through Whitman* (New Haven: Yale University Press, 1979).

27. Reynolds, *Remarks on a Review of Symmes' Theory*, p. 72; and Reynolds, untitled article in the *New-York Mirror* 3 (May 27, 1826), p. 351.

28. Note that some historians simply assume that he did drop Symmes's theory: see Stanton, p. 16, Goetzmann, *New Lands, New Men*, p. 260, and Collins, p. 65. In fact, Reynolds still did mention the possibility of holes at the poles, and many people who supported his proposed expedition did so *despite* their skepticism about the more extreme aspects of his position. Samuel Southard, for instance, expressed precisely this attitude on two occasions. In 1826, he explained to a friend that "altho' I do not believe his theory, yet I feel anxious that he should be successful in

fitting out his vessels & have every means in his power to render his voyage useful" (quoted in Woodbridge, "J.N. Reynolds: Father of American Exploration," p. 110). And in 1827 he reiterated the point to Reynolds himself: "Altho' no advocate of Mr Symmes' Theory, I feel deep solicitude for the promotion of an enterprise which if successful, cannot fail to add largely to our stock of knowledge" (Southard, letter to Reynolds, July 11, 1827, Southard Papers, box 23).

29. Reynolds, *New-York Mirror* 3 (May 27, 1826), p. 351; Reynolds, *Remarks on a Review of Symmes' Theory*, pp. 74–75.

30. Adams, in Charles Francis Adams, ed., *Memoirs of John Quincy Adams: Comprising Portions of His Diary from 1795 to 1848* (Philadelphia: J. B. Lippincott, 1875), 7:168 (entry for November 4, 1826); Reynolds, letter of November 6, 1827, Boston, Southard Papers, box 23.

31. Reynolds's activities can be tracked during this period through his correspondence with Southard and in various newspaper articles. Evidence of his approach in New York can be found in his article in the *New-York Mirror* 3 (May 27, 1826), p. 351.

32. Reynolds, letter of May 25, 1827, Charleston, Southard Papers, box 23; letter of December 27, 1827, Raleigh, Southard Papers, box 23; letter of December 3, 1828, Southard Papers, box 27. The letter from Washington is in the National Archives, Record Group 45.2.1, General Records of the Office of the Secretary of the Navy, Microfilm Publication M124, Miscellaneous Letters Received by the Secretary of the Navy, 1801–1884, roll 114, vol. 3, letter 88, June 26, 1828. Collected public quotations are included in roll 113, vol. 2, letter 147, dated February 17, 1828, Newton, from Reynolds to Southard, to which was attached an official resolution of the House of Delegates of the State of Maryland.

33. Reynolds, letter of July 15, 1828, New York, Southard Papers, box 27.

On Say, Peale, Mitchill, DeKay, and natural science in general for this period, see Stanton, pp. 19–21; Goetzmann, *New Lands, New Men*, pp. 110–26 and 229–73; and George H. Daniels, *American Science in the Age of Jackson* (New York: Columbia University Press, 1968). Mitchill, by the way, was a very early supporter of Symmes—see Elmore Symmes (who misspells Mitchill's name), p. 563, and Mitterling, p. 73.

34. John Quincy Adams expressed his respect for Reynolds in June: see Charles Francis Adams, ed., *Memoirs of John Quincy Adams*, 8 (pub. 1876):37, entry for June 19, 1828; *National Intelligencer*, June 21, 1828, p. 3.

35. Reynolds, *Pacific Ocean and South Seas*, pp. 1, 3, 18–28. On the history of whaling and sealing, see Lance E. Davis, Robert E. Gallman, and Karen Gleiter, *In Pursuit of Leviathan: Technology, Institutions, Productivity, and Profits in American Whaling, 1816–1906* (Chicago: University of Chicago Press, 1997); Briton Cooper Busch, *The War against the Seals: A History of the North American Seal Fishery* (Kingston, Ont.: McGill-Queens University Press, 1985); and Granville Allen Mawer, *Ahab's Trade: The Saga of South Seas Whaling* (St. Leonard's, N.S.W.: Allen and Unwin, 1999).

36. Reynolds, letter of August 13, 1828, National Archives, RG 45.2.1, roll 115, vol. 2, letter 48; *Pacific Ocean and South Seas*, pp. 3, 4, 26.

37. "South Sea bubble" quoted in Stanton, p. 25; DeKay, letter to Southard, September 24, 1829, quoted in Woodbridge, "J.N. Reynolds—American," chap. 10, p. 11.

38. For an interesting history of the discovery of Palmer's Land, see Captain Edmund Fanning, *Voyages and Discoveries in the South Seas, 1792–1832*, compilation ed. (Salem, Mass.: Marine Research Society, 1924), pp. 306–310; also see Reynolds, *Address on the Subject of a Surveying and Exploring Expedition to the Pacific Ocean and South Seas. Delivered in the Hall of Representatives on the Evening of April 3, 1836* (New York: Harper and Brothers, 1836), pp. 33–34.

On Deception Island, see Reynolds, *Pacific Ocean and South Seas*, pp. 26–27; Reynolds on Fanning, quoted in Stanton, p. 18.

For Fanning's perspective on the expedition, see his *Voyages to the South Seas*, pp. 152–74.

Reynolds, letter to Southard of September 28, 1829, Boston, Southard Papers, box 31.

The *Morning Courier and New York Enquirer* followed the progress of the expedition's preparations throughout September and October; see especially October 17, p. 2, and October 20, p. 1.

39. *New-York Mirror* 7 (September 26, 1829), p. 95.

40. On the concept of *Terra Australis Incognita*, see, for instance, Edwin Swift Balch, *Antarctica* (Philadelphia: Allen, Lane, and Scott, 1902), pp. 11–72.

On Arctic and Antarctic exploration in this period, see, for instance, Goetzmann, *New Lands, New Men*, pp. 30–60, 97–110, and 229–70; Alan Gurney, *Below the Convergence: Voyages toward Antarctica, 1699–1839* (New York: W. W. Norton, 1997); Gurney, *The Race to the White Continent* (New York: W. W. Norton, 2000); and Reynolds, *Address*, pp. 87–96.

41. Reynolds, letter of October 25, 1830, Southard Papers, box 33.

42. Dewey, "Pragmatic America," in *Pragmatism and American Culture*, ed. Gail Kennedy (Boston: D. C. Heath, 1950), p. 59.

43. Humboldt, *Cosmos*, 1:310; Reynolds, letter of October 25, 1830, Southard Papers, box 33; Pendleton, in Fanning, *Voyages Round the World*, p. 487. Pendleton, despite earning Reynolds's scorn, claimed to be a supporter of the expedition's scientific aims, and he certainly put in a good word for Reynolds and his friend John Watson (p. 488): "I regret, extremely regret, that the expedition has thus been cut short, and terminated without realizing the expectations of its friends. . . . I cannot conclude this report without stating, that messrs. Reynolds and Watson, from the perseverance and interest they have manifested throughout this voyage, do deserve the favor and thanks, not only of our government, but our fellow-citizens also."

44. Torrey quoted in Stanton, p. 27; Reynolds, *Remarks on a Review of Symmes' Theory*, p. 4.

45. Reynolds quoted in Stanton, p. 21, and Reynolds, *Pacific Ocean and South Seas*, p. 26.

46. There is a long chapter by Eights summarizing his Antarctic findings in Fanning, *Voyages to the South Seas*, pp. 195–216, from which these quotes are drawn. A good account of Eights's overall scientific significance is given in Bertrand, pp. 146–51. Eights's remarks about erratic boulders, in advance of Darwin, are discussed in Lawrence Martin, "James Eights' Pioneer Observations and Interpretations of Erratics in Antarctic Icebergs," *Bulletin of the Geological Society of America* 60 (January 1949), pp. 177–82; also see Stanton, p. 27. Other sources on Eights and his scientific legacy include John M. Clarke, "The Reincarnation of James Eights, Antarctic Explorer," *Scientific Monthly* 2 (February 1916), pp. 189–202, and W. T. Calman, "James Eights, a Pioneer Antarctic Naturalist," *Proceedings of the Linnean Society of London*, sess. 149, pt. 4 (1937), pp. 171–84. Also see Eights, "Description of a New Animal Belonging to the Arachnides of Latreille; Discovered in the Sea Along the Shores of the New South Shetland Islands," *Boston Journal of Natural History* 1 (Boston: Hilliard, Gray, 1837), pp. 203–6 (communicated to the society on September 17, 1834).

47. Reynolds, "Leaves from an Unpublished Journal," *New-York Mirror* 15 (April 21, 1838), p. 340. This article must of course be distinguished from the previously cited "A Leaf from an Unpublished Manuscript."

For more of Reynolds's observations on his Antarctic travels, also see "Bearding a Sea-Lion in His Den," *Knickerbocker* 13 (June 1839), pp. 524–26, and a longer version of this essay called (here we go again) "A Leaf from an Unpublished Journal," *New-Yorker* 7 (June 29, 1839), pp. 228–29. This second version carries the subtitle "STATEN LAND: Its Location, Discovery, Appearance, General Structure, Vegetables and Minerals—Its Harbors—Entering the Lion's Den—Rookery of the Penguins."

Note that Estwick Evans adopted an attitude similar to Reynolds's with regard to human beings' treatment of animals in the narrative of his 1818 exploration of the American West: "Man . . . is the great devourer. He revels, in pride and in luxury, upon the animal world; and after feasting high, employs himself in the butchery of his own species. . . . The destruction of animal life is necessary to the security, and perhaps to the health of man; but the life and comfort of animals should never be trifled with." See Evans, *A Pedestrious Tour, of Four Thousand Miles, through the Western States and Territories, during the Winter and Spring of 1818* (Concord, N.H.: Joseph C. Spear, 1819), pp. 93–94.

48. Eights, in Fanning, *Voyages to the South Seas*, p. 210.

49. All quotes are from "Leaves from an Unpublished Journal," pp. 340–41.

50. Humboldt, *Aspects of Nature*, trans. Mrs. Sabine (Philadelphia: Lea and Blanchard, 1849), p. 170.

51. Reynolds, "Leaves from an Unpublished Journal." On the Antarctic as a cultural symbol, see William E. Lenz, *The Poetics of the Antarctic: A Study in Nineteenth-Century American Cultural Perceptions* (New York: Garland Publishing, 1995), and Paul Simpson-Housley, *Antarctica: Exploration, Perception, and Metaphor* (London: Routledge, 1992).

52. This course of events is described by Captain Pendleton in Fanning, *Voyages Round the World*, pp. 479–87. The suggestion that Reynolds and Watson were put ashore involuntarily and that the sealers turned to piracy afterward is a myth that started in the nineteenth century (see, for instance, Howe, p. 431) and was carried on in some sources into the twenty-first (see, for instance, Collins, p. 67). The source of the myth seems to be Americus Symmes, *The Symmes Theory of Concentric Spheres* (1878), but this work is extremely rare, and I have not been able to consult it (the copy supposedly owned by the New York Public Library—the only one I know of in the United States—is missing).

53. Reynolds, introduction to *Voyage of the United States Frigate* Potomac (New York: Harper and Brothers, 1835), p. v.

Captain Pendleton, too, suggested that Reynolds had plans to write up the adventures of the Palmer-Pendleton expedition: "A very interesting history of this Araucanian country, its people, & c., from the notes of the scientific gentlemen of the expedition, it is expected will soon be given to the public. They have explored the country from the northern to the southern extremities, from the mouth to the source of several of the principal rivers, as well as visiting volcanoes, and crossing those noted mountains, the Cordilleras, which extend the whole length of the country" (in Fanning, *Voyages Round the World*, p. 485).

Reynolds seems to have laid aside these publishing plans in his rush to launch a new South Sea exploring expedition in the mid-1830s.

On the Araucanians, also see Edmond Reuel Smith.

54. Reynolds, "Rough Notes of Rough Adventures," pp. 706–7.

55. All quotes are from "Rough Notes of Rough Adventures," pp. 705–15. The mention of cider on the South American frontier, given Reynolds's self-identification as a Whig partisan, may well have been a direct reference to the "log-cabin and hard-cider campaign" of 1840, in which Reynolds took part as a vocal supporter of William Henry Harrison. (Whigs also tended to be much more supportive of Indians than Democrats were.) See especially Elizabeth Johns's discussion of William Sidney Mount's painting *Cider Making*, in *American Genre Painting: The Politics of Everyday Life* (New Haven: Yale University Press, 1991), pp. 50–54; and also Edward Pessen, *Jacksonian America: Society, Personality, and Politics* (Homewood, Ill.: Dorsey Press, 1969), pp. 180–274; Watson, pp. 213–30; and Robert Gray Gunderson, *The Log-Cabin Campaign* (Lexington: University of Kentucky Press, 1957).

On frontier life and Indian policy in the 1830s and '40s, see Malcolm J. Rohrbough, *The Trans-Appalachian Frontier: People, Societies, and Institutions, 1775–1850* (New York: Oxford University Press, 1978); Anthony F. C. Wallace, *The Long Bitter Trail: Andrew Jackson and the Indians* (New York: Hill and Wang, 1993); and Robert V. Hine and John Mack Faragher, *The American West: A New Interpretive History* (New Haven: Yale University Press, 2000), pp. 159–218.

Chapter 5: "Mocha-Dick": The Value of Mental Expansion

1. J. N. Reynolds, Esq., "Mocha-Dick; or, The White Whale of the Pacific: A Leaf from a Manuscript Journal," *Knickerbocker* 13 (May 1839), p. 377.

2. On the importance of whale oil to the early Republic, see, for instance, Lance E. Davis, Robert E. Gallman, and Karen Gleiter, *In Pursuit of Leviathan: Technology, Institutions, Productivity, and Profits in American Whaling, 1816–1906* (Chicago: University of Chicago Press, 1997).

3. Reynolds, "Rough Notes of Rough Adventures," *Southern Literary Messenger* 9 (December 1843), p. 705. On the commercialization of literature in this period, see, for instance, Michael T. Gilmore, *American Romanticism and the Marketplace* (Chicago: University of Chicago Press, 1985), and Terence Whalen, *Edgar Allan Poe and the Masses: The Political Economy of Literature in Antebellum America* (Princeton: Princeton University Press, 1999).

4. Washington Irving, "Rip Van Winkle," in *The Sketch-Book of Geoffrey Crayon, Gent.* (1819; repr., New York: G. P. Putnam's Sons, 1888), pp. 56, 55, 69, and 73. On literature's reflection of and response to industrial production, see Gilmore; Whalen; R. Jackson Wilson, *Figures of Speech: American Writers and the Literary Marketplace, from Benjamin Franklin to Emily Dickinson* (New York: Knopf, 1989); and David S. Reynolds, *Beneath the American Renaissance: The Subversive Imagination in the Age of Emerson and Melville* (Cambridge, Mass.: Harvard University Press, 1988).

5. Reynolds, *Voyage of the United States Frigate* Potomac, *under the Command of Commodore John Downes, during the Circumnavigation of the Globe, in the Years 1831, 1832, 1833, and 1834* (New York: Harper and Brothers, 1835). The *New-York Mirror* of Saturday, September 5, 1835 (vol. 13), p. 79, noted that the book "has proceeded to a 4th edition, which is of the rate of more than an edition per month. We are happy to perceive that the opinion we gave of this work on its first publication is thus confirmed by the judgment of the publick."

6. Reynolds, letter of May 25, 1834, Boston, to Samuel Southard, in Samuel L. Southard Papers, Princeton University Library, box 43 (hereafter cited as Southard Papers).

7. Reynolds, letter of August 25, 1834, New York, to Southard, Southard Papers, box 43. On this period of Reynolds's life, see Richard G. Woodbridge III, "J.N. Reynolds—American," chap. 14, typescript, Princeton University Library, Special Collections. On Rhode Island, see the documents attached in Reynolds, *Address on the Subject of a Surveying and Exploring Expedition to the Pacific Ocean and South Seas. Delivered in the Hall of Representatives on the Evening of April 3, 1836* (New York: Harper and Brothers, 1836), pp. 243–50 and 167–70. Also note Anna Ella Carroll, *The Star of the West; or, National Men and National Measures*, 3rd ed. (1856; repr., New York: Miller, Orton, 1857), p. 126; *National Intelligencer*, November 12, 1834, p. 3; and *Register of Debates in Congress*, vol. 11, December 10, 1834 (Washington, D.C.: Gales and Seaton), column 777.

8. Reynolds's address at Tremont House is noted on some of his letters to Southard from this period. Quotes on the Boston Society of Natural History come from Augustus A. Gould, "Notice on the Origin, Progress and Present Condition of the Boston Society of Natural History," *American Quarterly Register* 14 (February 1842), pp. 236 and 239, and from an "Address" by T. T. Bouvé, president of the society, in *Proceedings of the Boston Society of Natural History* 18 (Boston: Boston Society of Natural History, 1877), p. 244. For more on the society, also see volume 1 of the *Boston Journal of Natural History* (Boston: Hilliard, Gray, 1837), pp. 5–14, which includes an "Introduction" and "An Address Delivered before the Boston Society of Natural History, at the opening of their New Hall in Tremont Street, by Rev. F.W.P. Greenwood, Aug. 21, 1833." Also see Max Meisel, *A Bibliography of American Natural History: The Pioneer Century, 1769–1865* (New York: Premier Publishing, 1926), 2:457–517; Sally Gregory Kohlstedt, "The Nineteenth-Century Amateur Tradition: The Case of the Boston Society of Natural History," in *Science and Its Public: The Changing Relationship*, ed. Gerald Holton and William A. Blanpied (Dordrecht: D. Reidel, 1976), pp. 173–90; and William M. Smallwood, *Natural History and the American Mind* (New York: Columbia University Press, 1941), pp. 146–67.

On Thoreau, see Laura Dassow Walls, *Seeing New Worlds: Henry David Thoreau and Nineteenth-Century Natural Science* (Madison: University of Wisconsin Press, 1995), pp. 122–47.

Reynolds's contributions to the society are listed in Boston Society of Natural History, "Donations: Old Accession Book," vol. 1 (handwritten, no page numbers, in a blank ledger book manufactured by Jones and Oakes), under July and October. His membership is noted in Boston Society of Natural History, "Patrons, Honorary, Corresponding and Life Members," as having started on November 5, 1834, and also in Boston Society of Natural History, "Records of the Recording Secretary, 1830–1836," where the secretary also wrote, on October 15, 1834: "Noted— that the thanks of the Society be presented to Mr. Reynolds for the very valuable donations made by him to the Society." The significance of Reynolds's donations is mentioned in Gould, p. 238.

The society eventually became the Boston Museum of Science, and the old collections are now overseen by Carolyn Kirdahy, to whom I am exceedingly grateful and who is perhaps the most welcoming archivist/curator I have ever encountered. In the stacks, in addition to the old records of the society, I found a copy of Reynolds's *Voyage* inscribed by the author to the society, right next to which were two nineteenth-century German volumes about Humboldt's life and career. The early book collections of the society are overwhelmingly dominated by travel narratives, indicating the extent to which members of the society subscribed to the Humboldtian fieldwork model of cutting-edge scientific research.

Unfortunately, Reynolds's fish drawings, and most of the rest of his donations, seem to have been discarded several decades ago.

9. For the full story, see Reynolds, *Voyage*, pp. 17–21 and 88–131. Quallah-Battoo is Kuala Batu in modern spelling.

10. The *Southern Literary Messenger* had just begun publication: see vol. 1 (June 1835), pp. 594–95; also see Reynolds, *Voyage*, p. viii, for his self-consciousness about his additions to the simple chronicle of the mission.

The reviewer for the *Knickerbocker* (vol. 6, July 1835) especially appreciated Reynolds's choice of "minor details, which impart a rare and pleasing air of nature to his descriptions," and asserted that "the important scientific information, of various kinds, which the volume contains, reflects credit upon the industry and research of the author. The whole is valuable, as a national work,— one which will command the attention of the general mind of America" (p. 75).

The quote from Reynolds's dedication is in *Voyage*, p. ii.

Nathaniel Philbrick offers a contrasting view of Reynolds's *Voyage*, citing it as representative of Reynolds's "Jacksonian sense of the United States' imperialist destiny." See Philbrick, *Sea of Glory: America's Voyage of Discovery, The U.S. Exploring Expedition, 1838–1842* (New York: Viking, 2003), pp. 373–74; also see pp. 20–33.

11. Reynolds, letter of May 6, 1830, Valparaiso, Chile, to Southard, in Southard Papers, box 33; Reynolds, *Voyage*, pp. 445, 438, 555–60, 334, 60, and 515; and Reynolds, "Leaves from an Unpublished Journal," *New-York Mirror* 15 (April 21, 1838), p. 341.

12. Reynolds, *Voyage*, pp. 460 and 447; the first quote recalls *Remarks on a Review of Symmes' Theory* (Washington, D.C.: Gales and Seaton, 1827), p. 53, as quoted in chap. 1. The Humboldt quote is from his *Personal Narrative of Travels to the Equinoctial Regions of America, 1799–1804* (London: Henry G. Bohn, 1852–53), 1:57.

13. Reynolds, *Voyage*, pp. 333, 299, and 397. Of another long, slow trip, Reynolds wrote: "During four tedious days, the ship's headway did not average one knot per hour. The heat was oppressive; no variety to relieve the dull monotony; the sick-list was large, and still increasing. The history of one day is a specimen of the rest" (p. 327).

Note that Melville's story "Bartleby the Scrivener," the tale of a clerk whose reduplicative tasks are portrayed as a threat to his free will, did not come out until 1853. But the strains of industrial and industrial-style work had already arisen in American society (in the Northeast) in the 1830s.

14. Reynolds, *Voyage*, pp. 82, 81, 105, 155–56, 46, 477, 417, and 478. Note that Reynolds also bends over backward to insist that the United States should not be equated with the European imperial powers: "*conquest* forms no part of our national policy" (p. 230, Reynolds's emphasis); "we are content with our own extent of territory, and would not accept any portion of another country if it were freely offered us" (p. 385).

15. Ibid., p. 30.

16. Reynolds, *Address*, p. 1; "Reynolds' Lecture on Maritime Discovery," *New-York Mirror* 13 (April 30, 1836), p. 350; "Exploring Expedition," *Niles' Weekly Register* 50 (April 23, 1836), p. 142; and Henry Howe, "The Romantic History of Jeremiah N. Reynolds," in *Historical Collections of Ohio in Two Volumes: An Encyclopedia of the State* (Columbus: Henry Howe and Son, 1889), 1:432. The date of the lecture was definitely Saturday, April 2 (not April 3), as noted in chapter 1.

Sadly, I have been unable to find any images of Reynolds, though it seems that at least two portraits of him were done, one by James Shegogue and one by Jerome Thompson. See Woodbridge, "J.N. Reynolds—American," chap. 27, p. 19; the *New-York Mirror* 14 and 15 (June 3, 1837, and January 6, 1838), p. 391 and p. 224; and the *New-Yorker* 7 (May 4, 1839), p. 109.

17. Reynolds, *Address*, p. 70.

18. Quoted in Howe, p. 432. Note that Edgar Allan Poe was interested enough in Reynolds's personality to try analyzing his signature on two occasions (in 1836 and 1841). See James A.

Harrison, ed., *The Complete Works of Edgar Allan Poe*, vol. 15, *Literati—Autography* (New York: AMS Press, 1965), pp. 159 and 243–44.

19. Reynolds, *Address*, p. 99. On Poe's borrowings, see Robert Lee Rhea, "Some Observations on Poe's Origins," *University of Texas Bulletin, Studies in English* 10 (July 8, 1930), pp. 135–46, and J. O. Bailey, "Sources for Poe's *Arthur Gordon Pym*, 'Hans Pfaal,' and Other Pieces," *PMLA* 57 (June 1942), pp. 513–35.

20. Reynolds, *Address*, pp. 22, 41, 36, 42, 54, 67, 69, 68, 36, and 48.

Melville, *Typee: A Peep at Polynesian Life during a Four Months' Residence in a Valley of the Marquesas* (1846; repr., New York: New American Library, 1964), pp. 145–46.

21. Reynolds, *Address*, pp. 14, 24, 10, 87, and 86.

Just as scholars failed to acknowledge that Reynolds did not in fact stop referring to the abstract, theoretical, Symmesian reasons for launching an exploring expedition in the late 1820s, so too have they failed to notice that Reynolds also continued attacking some of the core economic values of his society while lobbying for another expedition in the 1830s. See, for instance, Whalen, pp. 150–57.

22. Reynolds, *Address*, pp. 86, 24, 99–100, 75–83, and 22; and "Reynolds' Lecture on Maritime Discovery," p. 350.

23. *Register of Debates in Congress*, vol. 12, pt. 3 (Washington, D.C.: Gales and Seaton, 1837), May 9 and May 5, 1836, columns 3565, 3562, 3560, and 3471–72.

24. Committee on Naval Affairs, *Report, with Senate Bill 175*, 24th Cong., 1st sess., 1836, S. Doc. 262, p. 3; Reynolds, *Address*, pp. 103–71, Silliman on pp. 112–15.

Reynolds's own vision for the scientific corps is given in National Archives, Microfilm Publication M75, Letters Received by the Navy Department relating to the Wilkes Expedition, Letters Relating to Preparations for the Exploring Expedition, roll 1, Reynolds to President Jackson, November 16, 1836.

25. On Jackson, see William Stanton, *The Great United States Exploring Expedition of 1838–1842* (Berkeley: University of California Press, 1975), pp. 28–30. On natural science in general, see George H. Daniels, *American Science in the Age of Jackson* (New York: Columbia University Press, 1968), and Charlotte M. Porter, *The Eagle's Nest: Natural History and American Ideas, 1812–1842* (Tuscaloosa: University of Alabama Press, 1986).

Poe's review, "Report of the Committee on Naval Affairs, to whom were referred memorials from sundry citizens of Connecticut interested in whale fishing, praying that an exploring expedition be fitted out to the Pacific Ocean and South Seas. March 21, 1836," appeared in the *Southern Literary Messenger* of August 1836, but I consulted a reprint in Harrison, ed., *Complete Works*, vol. 9, *Literary Criticism—Volume 2*, pp. 84–90, quote from p. 89.

Reynolds, *Address*, p. 73.

26. See Stanton, pp. 33–38, Dickerson quotes on p. 34; Jackson quote from National Archives, Microfilm Publication M75, Letters Received by the Navy Department relating to the Wilkes Expedition, roll 1, letter of Andrew Jackson to Mahlon Dickerson, Washington, July 9, 1836.

27. On Dickerson's dickering, see Stanton, pp. 33–72. On the Panic, see William Charvat, "American Romanticism and the Depression of 1837," *Science and Society* 2 (Winter 1937), pp. 67–82, and Charles McGrane, *The Panic of 1837* (1924; repr., Chicago: University of Chicago Press, 1965).

28. Irving, *Astoria* (1836; repr., Portland, Ore.: Binfords and Mort, 1967), pp. 185–89 (quote on p. 185) and 84–95. Interestingly, Lawrence Buell refers to Irving's 1830s fur-trade trilogy (see also *A Tour on the Prairies*, 1835, and *The Adventures of Captain Bonneville*, 1837) as "the first clear case of commercially successful American literary nonfiction about the environment." Buell, *The Environmental Imagination: Thoreau, Nature Writing, and the Formation of American Culture* (Cambridge, Mass.: Harvard University Press, 1995), p. 399.

29. Poe, *Marginalia*, in Harrison, ed., *Complete Works*, 16:118–19; Poe, *Narrative of Arthur Gordon Pym*, in *"The Fall of the House of Usher" and Other Tales* (New York: New American Library, 1980), pp. 326–66, explosion on pp. 352–53. For scholarly interpretations of the massacre, see John Carlos Rowe, "Poe, Antebellum Slavery, and Modern Criticism," in *Poe's Pym: Critical Explorations*, ed. Richard Kopley (Durham, N.C.: Duke University Press, 1992), pp. 117–38; Scott

Bradfield, *Dreaming Revolution: Transgression in the Development of American Romance* (Iowa City: University of Iowa Press, 1993), pp. 67–82; Leslie Fiedler, *Love and Death in the American Novel*, rev. ed. (New York: Dell, 1966), pp. 396–400; and (for a more highly contextualized, detailed, and nuanced treatment) Whalen, pp. 138–92.

30. Quoted in John T. Irwin, *American Hieroglyphics: The Symbol of the Egyptian Hieroglyphics in the American Renaissance* (New Haven: Yale University Press, 1980), p. 78.

31. Poe, quoted in Edwin Fussell, *Frontier: American Literature and the American West* (Princeton: Princeton University Press, 1965), p. 146; Emerson, "Nature," in John Elder, ed., *Nature/Walking* (Boston: Beacon Press, 1991), p. 3.

32. See Rhea; Bailey; Keith Huntress, "Another Source for Poe's *Narrative of Arthur Gordon Pym*," *American Literature* 16 (March 1944), pp. 19–25; and Randel Helms, "Another Source for Poe's *Arthur Gordon Pym*," *American Literature* 41 (January 1970), pp. 572–75.

33. See Stephen Rachman, "'Es lässt sich nicht schreiben': Plagiarism and 'The Man of the Crowd,'" in *The American Face of Edgar Allan Poe*, ed. Shawn Rosenheim and Stephen Rachman (Baltimore: Johns Hopkins University Press, 1995), pp. 51–52, 63–64, 68, 73, and 83, and Whalen, pp. 11–17, 25–27, 34–36, 45–47, and 247.

34. Emerson, "Experience," in *Selections from Ralph Waldo Emerson*, ed. Stephen E. Whicher (Boston: Houghton Mifflin, 1960; "Experience" orig. pub. 1844), p. 256.

35. Poe, *Pym*, pp. 326 and 370–71.

36. Poe, *Eureka*, in Harrison, ed., *Complete Works*, 16:185–86.

37. See Stanton, pp. 41–71; Woodbridge, "J.N. Reynolds—American," chaps. 18–22; and Julian Hawthorne, *Nathaniel Hawthorne and His Wife: A Biography* (1884: repr., Boston: Houghton Mifflin, 1893), 1:152–58 and 161–64 (the letters from Pierce to Hawthorne reveal the extent to which people acknowledged Reynolds as the key person to impress if you wanted to get a spot on the South Seas expedition). Edwin Fussell, meanwhile, argues that Hawthorne was obsessed with frontiers and was thus "at heart a Western writer"; see pp. 69–131. Certainly, Hawthorne's characters do occasionally admit to a burning wanderlust: "I felt an inexpressible longing for at least a temporary novelty. I thought of going across the Rocky Mountains, or to Europe, or up the Nile—of offering myself a volunteer on the Exploring Expedition—of taking a ramble of years, no matter in what direction, and coming back on the other side of the world." See Nathaniel Hawthorne, *The Blithedale Romance* (1852; repr., New York: Penguin, 1986), p. 140.

38. See the letter of Silliman to Dickerson, October 10, 1836, National Archives, Microfilm Publication M75, Letters Received by the Navy Department relating to the Wilkes Expedition, roll 1. Silliman quote is from the *New England Palladium and Commercial Advertiser*, June 16, 1826, p. 1; also see Mitterling, p. 86.

39. Most of the letters originally appeared in the *New-York Times* and the *New-York Courier and Enquirer* and many were collected together and published in J. N. Reynolds, *Exploring Expedition Correspondence*, undated pamphlet, copy consulted at Beinecke Library, Yale University, quote on p. 3. The letters are also reprinted in Reynolds, *Pacific and Indian Oceans; or The South Sea Surveying and Exploring Expedition: Its Inception, Progress, and Objects* (New York: Harper and Brothers, 1841), which provides an in-depth history of the political maneuverings surrounding the expedition between 1835 and 1838.

40. See, for instance, Mary P. Ryan, *Civic Wars: Democracy and Public Life in the American City during the Nineteenth Century* (Berkeley: University of California Press, 1997); Sean Wilentz, *Chants Democratic: New York City and the Rise of the American Working Class, 1788–1850* (New York: Oxford University Press, 1984); and Edward Pessen, *Jacksonian America: Society, Personality, and Politics* (Homewood, Ill.: Dorsey Press, 1969). An interesting revision of the standard picture of antebellum political culture is Glenn C. Altschuler and Stuart M. Blumin, *Rude Republic: Americans and Their Politics in the Nineteenth Century* (Princeton: Princeton University Press, 2000), pp. 3–151. Reynolds had this to say about the political climate: "Public opinion is the great arbiter in these affairs, and no department has power to reverse its final decree." Reynolds to Dickerson, letter of November 16, 1836, National Archives, Microfilm Publication M75, Letters Received by the Navy Department relating to the Wilkes Expedition, roll 1.

41. "History of Navigation in the South Seas" (review of Reynolds's *Address*), *North American Review* 45 (October 1837), pp. 361–90, quote on p. 369.

42. Quoted in Stanton, pp. 56–57.

43. Dickerson's letters are included in Reynolds, *Correspondence*, and Reynolds, *Pacific and Indian Oceans*; also see Stanton, pp. 41–47 and 57–59, and Kohlstedt, "The Nineteenth-Century Amateur Tradition."

44. Dickerson, letter 3, August 10, 1837, in Reynolds, *Correspondence*, p. 81; Reynolds, letter 11, January 1, 1838, New-York, ibid., p. 124.

45. This, in fact, was the direction in which American science would travel, for Reynolds had successfully tapped the next generation of nationally recognized scientific leaders: his choices for the expedition's geological and botanical positions, James Dwight Dana and Asa Gray, would become the most esteemed midcentury intellectuals at Yale and Harvard, respectively, with Dana inheriting Benjamin Silliman's legacy and Gray battling Louis Agassiz over the implications of Charles Darwin's researches and theories.

Quotes from Reynolds, letter 11, ibid., pp. 133 and 126.

46. Gray quoted in Stanton, p. 49; Reynolds, letter 6, July 28, 1837, New-York, *Correspondence*, pp. 39–41.

47. Reynolds, *Pacific and Indian Oceans*, p. 498.

48. This situation is documented at length in National Archives, Microfilm Publication M75, Letters Received by the Navy Department relating to the Wilkes Expedition, rolls 2–6; the letters of James Eights are particularly poignant. A group of the scientists also filed a petition with the Navy Department protesting their lack of pay: see roll 4.

49. Poinsett, in Reynolds, *Pacific and Indian Oceans*, p. 504. Also see all the documents attached by the editor of the *Southern Literary Messenger*, including newspaper editorials and a petition to the president from all the congressional members of the "Western delegation," in Reynolds, "A Leaf from an Unpublished Manuscript," *Southern Literary Messenger* 5 (June 1839), pp. 408–13. The petition can also be found in National Archives, Microfilm Publication M75, Letters Received by the Navy Department relating to the Wilkes Expedition, May 1, 1838, roll 4. Also see the letter from all the scientists insisting that Reynolds accompany the expedition, reprinted in Carroll, *Star of the West*, pp. 132–33.

50. On the legacy of the Wilkes Expedition, see especially Stanton, pp. 281–382, and William H. Goetzmann, *New Lands, New Men: America and the Second Great Age of Discovery* (1986; repr., Austin: Texas State Historical Association, 1995), pp. 273–97, quotes on pp. 284 and 278. The newest thorough treatment of the Wilkes Expedition is Philbrick, *Sea of Glory*.

51. See Goetzmann, ibid., and Stanton, pp. 247–65; on *Reynoldsia*, see Asa Gray, *United States Exploring Expedition: Botany: Phanerogamia* (New York: George P. Putnam, 1854), 1:723–26. The expedition's experience of the forest of Savaii is described in Charles Wilkes, *Narrative of the United States Exploring Expedition* (Philadelphia: Lea and Blanchard, 1845), 2:108–12.

52. One of the documents attached to Reynolds, "A Leaf from an Unpublished Manuscript," pp. 408–13.

53. First published in the *Southern Literary Messenger* 5 (April 1839), pp. 254–56; reprinted in the *New-Yorker* 7 (April 27, 1839), p. 100.

54. Poe, "Review of New Books," a review of a pamphlet reprinted from Benjamin Silliman's *American Journal of Science and Arts* called *A Brief Account of the Discoveries and Results of the United States' Exploring Expedition* (New Haven: B. L. Hamlen, 1843). The pamphlet was probably penned by Reynolds, though there is no author cited; Poe's review, in any case, is in *Graham's Magazine* 23 (though in this issue, no. 3, there is a misprint saying it's vol. 24) (September 1843), pp. 164–65.

55. Reynolds, "Mocha-Dick," p. 379.

56. Ibid., pp. 379–80.

57. See, for instance, Nathaniel Philbrick, *In the Heart of the Sea: The Tragedy of the Whaleship Essex* (New York: Viking, 2000).

58. Quote from Reynolds, "Mocha-Dick," p. 379; Emerson, journal entry of February 19, 1834,

Boston, in *The Journals and Miscellaneous Notebooks of Ralph Waldo Emerson*, ed. Alfred R. Ferguson (Cambridge, Mass.: Harvard University Press, 1964), 4:265; and see Tim Severin, *In Search of Moby Dick: Quest for the White Whale* (New York: Da Capo Press, 2001), pp. 189–207.

59. Reynolds, "Mocha-Dick," pp. 385 and 378; on authorial ambivalence in this period, see Gilmore, Whalen, Wilson, and David S. Reynolds.

60. *New-Yorker* 7 (June 22, 1839), p. 221.

61. Reynolds, "Mocha-Dick," p. 392.

62. Ibid., p. 390.

63. Poe, "The Journal of Julius Rodman," in Harrison, ed., *Complete Works*, 4:9–10 (Poe's italics), 13, and 10–11. Rodman also relished "the deep and most intense excitement with which I surveyed the wonders and majestic beauties of the wilderness. No sooner had I examined one region than I was possessed with an irresistible desire to push forward and explore another" (p. 77).

On Poe and frontiers, see Fussell, pp. 132–74, and Irwin, pp. 64–94.

Note the letter of Benjamin Rodman to Reynolds in Reynolds, *Address*, pp. 116–19.

On this archetypal American hero, who has been interpreted in many different ways, see Henry Nash Smith, *Virgin Land: The American West as Symbol and Myth* (1950; repr., Cambridge, Mass.: Harvard University Press, 1978); R. W. B. Lewis, *The American Adam: Innocence, Tragedy, and Tradition in the Nineteenth Century* (Chicago: University of Chicago Press, 1955); Richard Slotkin, *The Fatal Environment: The Myth of the Frontier in the Age of Industrialization, 1800–1890* (1985; repr., New York: HarperPerennial, 1994); and David Mogen, Mark Busby, and Paul Bryant, eds., *The Frontier Experience and the American Dream: Essays on American Literature* (College Station: Texas A&M University Press, 1989), esp. pp. 3–49, 120–31, and 205–16.

64. Reynolds, "Mocha-Dick," pp. 377, 387, and 391.

65. Robert Pogue Harrison, *Forests: The Shadow of Civilization* (Chicago: University of Chicago Press, 1992), p. 133.

66. On the link between "Mocha-Dick" and *Moby-Dick*, see, for instance, R. S. Garnett, "Moby-Dick and Mocha-Dick: A Literary Find," *Blackwood's Magazine* 226 (December 1929), pp. 841–58, and Carl Van Doren, "Mr. Melville's Moby Dick," *Bookman* 59 (April 1924), pp. 154–57.

67. Quotes are from Herman Melville, *Moby-Dick or The White Whale* (1851; repr., New York: New American Library, 1961), pp. 87, 167, 82, 21, 127–28, 310, 66, 98, 519, 398, 180, 68, and 76. There are, of course, countless interpretations of *Moby-Dick*. For a sampling of some of my favorites, see the Norton Critical Edition of *Moby-Dick*, ed. Harrison Hayford and Hershel Parker (New York: W. W. Norton, 1967); Fiedler, pp. 366–90; David S. Reynolds, pp. 288–92 and 540–51; Fussell, pp. 256–79; Gilmore, pp. 113–31; C. L. R. James, *Mariners, Renegades, and Castaways: The Story of Herman Melville and the World We Live In* (1953; repr., Detroit: Bewick/ed, 1978); Eric Wilson, *Romantic Turbulence: Chaos, Ecology, and American Space* (Houndmills, U.K.: Macmillan, 2000), pp. 71–93; Lee Clark Mitchell, *Witnesses to a Vanishing America: The Nineteenth-Century Response* (Princeton: Princeton University Press, 1981), pp. 189–212; Charles Olson, *Call Me Ishmael: A Study of Melville* (San Francisco: City Lights Books, 1947); and Harry Levin, *The Power of Blackness: Hawthorne, Poe, Melville* (New York: Knopf, 1958), pp. 196–230.

68. On the lecture series, and for quotes from Reynolds's actual lectures, see the *New York Evening Star* 6 (March 16, 1839), p. 1; *New York Daily Whig* 3 (March 18 and 20, 1839), p. 1; and Woodbridge, chap. 26, pp. 21–23. Also see the *New-Yorker* 7 (August 10, 1839), p. 330, on Reynolds's becoming a lawyer.

Reynolds was one of New York's most popular speakers between 1838 and 1844, known widely for his "eloquent and animated manner," his "irresistible language," and, more particularly, his "most forcible and beautiful similes"; "the auditory were often electrified with bursts of true genius—of original and most conclusive argument." See the *New York Evening Star* 6 (November 3, 1838), p. 2, and *New York Daily Whig* 3 (March 25, 1839), p. 2.

69. These quotes are from a stash of eight letters from Reynolds to Thomas Lyon Hamer held at the H. E. Huntington Library, Manuscripts Division; see the second letter, dated October 3, 1836, New York, and the sixth letter, December 16, 1837, Astor House [New York]. Reynolds discusses his emotional state as of 1840 in the seventh letter, dated April 22, 1840, City of New York.

70. Reynolds, *Pacific and Indian Oceans*.

71. Quoted in Woodbridge, chap. 36, pp. 28–30.

72. Reynolds and Barnard shared the podium at the Syracuse Whig State Convention of 1841: see *New York Times and Evening Star* 7 (October 9, 1841), p. 2.

"[A]ll good Whigs" is Reynolds, in a letter to William Meredith, secretary of the treasury under President Zachary Taylor, dated May 6, 1849, Brownsville, Texas, quoted in Woodbridge, chap. 37, p. 22.

73. Reynolds to Corwin, July 30, 1850, Brownsville [Texas], in Library of Congress, Manuscript Division, Thomas Corwin Correspondence, 1850–1852. Reynolds may even have been involved in some fraudulent land-claim schemes: see "Case of Judge John C. Watrous," in *Reports*, vol. 1, H. Rep., 36th Cong., 2nd sess., 1860–61, No. 2.

Excursion: Watersheds: 1859–1862

1. Louis L. Noble, *After Icebergs with a Painter: A Summer Voyage to Labrador and around New-foundland* (New York: Appleton, 1861), pp. 172–73, 165, 169, 176, and 166.

2. Church quoted in Stephen Jay Gould, "Art Meets Science in *The Heart of the Andes*: Church Paints, Humboldt Dies, Darwin Writes, and Nature Blinks in the Fateful Year of 1859," in *I Have Landed: The End of a Beginning in Natural History* (New York: Harmony Books, 2002), p. 91, and in Barbara Novak, *Nature and Culture: American Landscape and Painting, 1825–1875* (New York: Oxford University Press, 1980), p. 71; also see the *New York Times*, May 19, 1859, p. 4.

3. See Noble, p. 6; Edward Lurie, *Louis Agassiz: A Life in Science* (Chicago: University of Chicago Press, 1960), pp. 212–37; and Gould.

4. Reviewer quoted in Gerald L. Carr, *Frederic Edwin Church: The Icebergs* (Dallas: Dallas Museum of Fine Arts, 1980), p. 83. Also see Eleanor Jones Harvey, *The Voyage of the Icebergs: Frederic Church's Arctic Masterpiece* (New Haven: Dallas Museum of Art and Yale University Press, 2002).

5. Theodore Winthrop, *A Companion to the "Heart of the Andes"* (New York: Appleton, 1859), p. 9; Winthrop's review quoted in Carr, p. 73—though, as Carr notes, it's not absolutely certain that Winthrop did in fact write it. Note that *A Companion to the "Heart of the Andes"* was reprinted in Winthrop, *Life in the Open Air and Other Papers* (New York: John W. Lovell, 1862), pp. 329–74.

6. On Winthrop, see the pamphlet by Elbridge Colby, *Bibliographical Notes on Theodore Winthrop* (New York: New York Public Library, 1917); Dumas Malone, ed., *Dictionary of American Biography* (New York: Charles Scribner's Sons, 1936), 20:417; and the *New York Times*, June 14, 1861, p. 8. Colby says that Winthrop was the first officer killed in the Civil War, but the newspapers from that time indicate that several other officers were killed on the same day or even earlier.

7. Winthrop, *The Canoe and the Saddle: Adventures among the Northwestern Rivers and Forests* (1862; repr., Boston: Ticknor and Fields, 1864), p. 297.

Chapter 6: *Mountaineering in the Sierra Nevada:* The Art of Self-Exposure

1. Winthrop's birthplace and date were added to the grave later; the tomb itself is not far from those of Benjamin Silliman and James Dwight Dana. A pamphlet, *History of Grove Street Cemetery*, available at the cemetery's entrance, shows the location of the grave and counts Winthrop among the "eminent people" buried in New Haven.

On the funeral, see *The Life and Poems of Theodore Winthrop, edited by his sister* (New York: Henry Holt, 1884), pp. 299–300, and Ellsworth Eliot, *Theodore Winthrop* (New Haven: Yale University Library, 1938), p. 23. On Winthrop's life and career, also see the "Biographical Sketch" by his friend George William Curtis published in Winthrop's novel *Cecil Dreeme* (1861; repr., New York: Henry Holt, 1876), pp. 5–19, and John H. Williams and Clarence B. Bagley, *Winthrop and Curtis* (Tacoma: J. H. Williams, 1914).

2. The most thorough piece of scholarship on King is Thurman Wilkins, *Clarence King: A Biography*, rev. and enlarged ed. (Albuquerque: University of New Mexico Press, 1988); on King and Winthrop, see pp. 35, 41, and 44. Winthrop is also mentioned as a key influence on King in

Samuel Franklin Emmons, "Biographical Memoir of Clarence King," *National Academy of Sciences Biographical Memoirs* 6 (1909), p. 31.

Other perceptive studies of King include John P. O'Grady, *Pilgrims to the Wild: Everett Ruess, Henry David Thoreau, John Muir, Clarence King, Mary Austin* (Salt Lake City: University of Utah Press, 1993), pp. 87–122; and David Mazel, *American Literary Environmentalism* (Athens: University of Georgia Press, 2000), pp. 117–33.

3. And in the spirit of Winthrop's frontier novel *John Brent*, which went through sixteen editions between 1861 and 1866 (Eliot, p. 20).

4. Brewer, who came to know King quite well, explained that King later told him the story of the night when Brush read him the letter: "I wrote to a very old friend and classmate, Professor George J. Brush, an enthusiastic account of our adventure, emphasizing not only the scientific interest, but also the sublime and majestic scenery connected with it. To Clarence King, who happened to call upon him soon after the receipt, Prof. Brush read this letter; and, as King told me many times, 'that settled it.' He resolved to see California, and, in particular, Mount Shasta." Brewer's explanation, and his actual letter to Brush, are reprinted in Francis P. Farquhar, "The Whitney Survey on Mount Shasta, 1862: A Letter from William H. Brewer to Professor Brush," *California Historical Society Quarterly* 7 (June 1928), pp. 121–31, quotes on pp. 121, 124, 125, 128, and 129. Also see Rossiter W. Raymond, "Biographical Notice," in *Clarence King Memoirs*, ed. James D. Hague (New York: G. P. Putnam's Sons, 1904), pp. 310–24. Note that this chapter of *Clarence King Memoirs* is a partial reprinting of Raymond's original essay, "Biographical Notice of Clarence King," *Transactions of the American Institute of Mining Engineers* 33 (1903), pp. 619–50.

On Brewer, also see Brewer, *Up and Down California in 1860–1864: The Journal of William H. Brewer*, ed. Francis P. Farquhar (New Haven: Yale University Press, 1930).

5. King, *Mountaineering in the Sierra Nevada* (1872; repr., New York: Penguin, 1989), p. 108. All further references are to this edition unless otherwise noted.

6. Emerson, "Divinity School Address," in *Selections from Ralph Waldo Emerson*, ed. Stephen E. Whicher (Boston: Houghton Mifflin, 1960), pp. 115–16.

7. King, *Mountaineering*, p. 204.

8. An excellent history of exploration for this period, which suggests the Humboldtian potential of many early expeditions, despite their overarching colonialist goals, is William H. Goetzmann, *Army Exploration in the American West, 1803–1863* (New Haven: Yale University Press, 1959). Also see Vincent Ponko Jr., "The Military Explorers of the American West, 1838–1860," in *North American Exploration*, ed. John Logan Allen, vol. 3, *A Continent Comprehended* (Lincoln: University of Nebraska Press, 1997), pp. 332–411. This period actually saw an explosion of the use of Humboldtian surveying techniques in American exploration.

9. John D. Unruh Jr., *The Plains Across: The Overland Emigrants and the Trans-Mississippi West, 1840–60* (Urbana: University of Illinois Press, 1979), p. 120. Also see Bernard De Voto, *The Year of Decision: 1846* (Boston: Little, Brown, 1943).

10. King, "The Biographers of Lincoln," *Century Magazine* 32 (October 1886), p. 861.

11. See Henry Nash Smith, *Virgin Land: The American West as Symbol and Myth* (1950; repr., Cambridge, Mass.: Harvard University Press, 1978); John Mack Faragher, *Women and Men on the Overland Trail* (New Haven: Yale University Press, 1979), pp. 1–39; Robert M. Utley, *The Indian Frontier of the American West, 1846–1890* (Albuquerque: University of New Mexico Press, 1984), pp. 1–63; and Michael A. Morrison, *Slavery and the American West: The Eclipse of Manifest Destiny and the Coming of the Civil War* (Chapel Hill: University of North Carolina Press, 1997), esp. pp. 126–56.

12. On Frémont, see Tom Chaffin, *Pathfinder: John Charles Frémont and the Course of American Empire* (New York: Hill and Wang, 2002), pp. 433–54, and Andrew Rolle, *John Charles Frémont: Character as Destiny* (Norman: University of Oklahoma Press, 1991), pp. 162–77. The Humboldt quotes are from letters of November 21, 1856, and September 11, 1856, in Humboldt, *Letters of Alexander von Humboldt to Varnhagen von Ense, from 1827 to 1858*, translated from the second German edition by Friedrich Kapp (New York: Rudd and Carleton, 1860), pp. 339 and 324.

13. Mark Twain, *The Innocents Abroad; and Roughing It* (New York: Library of America, 1985; *Roughing It* orig. pub. 1872), pp. 629 and 635.

14. On the Homestead Act, see Robert V. Hine and John Mack Faragher, *The American West: A New Interpretive History* (New Haven: Yale University Press, 2000), pp. 33–61. On the transcontinental railroad, see Hine and Faragher, pp. 274–300, and D. W. Meinig, *The Shaping of America: A Geographical Perspective on 500 Years of History*, vol. 3, *Transcontinental America, 1850–1915* (New Haven: Yale University Press, 1998), pp. 3–28. On the military buildup and Sand Creek, see Hine and Faragher, pp. 216–33, and Utley, pp. 1–98.

15. King's reputation as an abolitionist is noted by James T. Gardiner in "Clarence King's Boyhood," a manuscript in Clarence King Papers, A-3, Huntington Library, San Marino, Calif. King refers to himself as "more than ever a Wendell Phillips man" in his letter of March 28 [1860?], New York, to Gardiner, in a folder marked "Letters from and about Clarence King," Huntington Library. The letter about King's grandmother is from the John Hay Library, Brown University, John Hay Papers, microfilm reel 8, letter of [March?] 1888.

16. "Letters from and about Clarence King," Huntington Library: letter to Gardiner of March 18, 1862.

17. King, *Mountaineering*, p. 32.

18. Henry Adams, "King," in *Clarence King Memoirs*, p. 167; also see Adams, *The Education of Henry Adams* (1907; repr., Boston: Houghton Mifflin, 1973), pp. 311–13, for Adams's character sketch of King as a Representative Man, which includes his observation that "King loved paradox; he started them like rabbits." Another friend concurred that "paradox perhaps enjoyed the hegemony of his mental states"—see William Crary Brownell, "King at the Century," in *Clarence King Memoirs*, p. 219. For a fascinating study of how paradox has in some ways defined American culture, see Michael Kammen, *People of Paradox: An Inquiry Concerning the Origins of American Civilization* (1972; repr., Ithaca: Cornell University Press, 1990).

19. Wilkins, pp. 40–44.

20. On Agassiz, see Guy Davenport, *The Intelligence of Louis Agassiz* (Boston: Beacon Press, 1963); David McCullough, "The American Adventure of Louis Agassiz," in *Brave Companions: Portraits in History* (New York: Touchstone, 1992), pp. 20–36; Louis Menand, *The Metaphysical Club: A Story of Ideas in America* (New York: Farrar, Straus, and Giroux, 2001), pp. 95–148; and Rebecca Bedell, *The Anatomy of Nature: Geology and American Landscape Painting, 1825–1875* (Princeton: Princeton University Press, 2001), pp. 114–21.

21. Thoreau and Church were quite popular among King's group of friends in 1863; see David Dickason, *The Daring Young Men: The Story of the American Pre-Raphaelites* (Bloomington: Indiana University Press, 1953), pp. 33–64 and 71–124, and Roger Stein, *John Ruskin and Aesthetic Thought in America, 1840–1900* (Cambridge, Mass.: Harvard University Press; 1967), pp. 78–185. Thoreau quotes are from "Walking," in *Nature/Walking*, ed. John Elder (Boston: Beacon Press, 1991), p. 85, and Thoreau's journal quoted in Stein, p. 92.

22. Gardiner quoted in Wilkins, p. 42; Ruskin quoted in Linda S. Ferber, " 'Determined Realists': The American Pre-Raphaelites and the Association for the Advancement of Truth in Art," in *The New Path: Ruskin and the American Pre-Raphaelites*, ed. Ferber and William H. Gerdts (New York: Schocken, 1985), p. 13.

23. Quoted in Ferber, pp. 13, 21, 28, 11, and 21; and quoted in Dickason, p. 75, and Emmons, p. 30.

24. The King-Ruskin anecdote comes from John Hay, "Clarence King," in *Clarence King Memoirs*, pp. 129–30; Ruskin quoted in Stein, p. 164; King, *Mountaineering*, pp. 252–53; Ruskin, *Modern Painters* (books 1–5 compiled in a 4-vol. set) (New York: John Wanamaker, 1905); King quoted in Rossiter Raymond, "Biographical Notice," in *Clarence King Memoirs*, p. 319.

25. Ruskin quoted in Marjorie Hope Nicolson, *Mountain Gloom and Mountain Glory: The Development of the Aesthetics of the Infinite* (1959; repr., Seattle: University of Washington Press, 1997), p. 287. Also see E. T. Cook and Alexander Wedderburn, eds., *The Complete Works of John Ruskin* (London: George Allen, 1903–12), 34:36 and 583 and 37:570.

26. King Papers, Huntington Library, A-2, #4, Notebooks, Personal: "Scientific Notes, Private"; and D-23, "Journal of a trip in the Northern Sierras and Grass Valley, September–October, 1863," entry for Sunday, September 20, p. 26.

27. King Papers, Huntington Library, "Journal of a trip in the Northern Sierras and Grass

Valley," September 7, p. 5. On the meeting with Brewer, see David H. Dickason, "Clarence King's First Westward Journey," *Huntington Library Quarterly* 7 (November 1943), pp. 82–85; Raymond, "Biographical Notice," in *Clarence King Memoirs*, pp. 310–12; and Wilkins, pp. 51–52.

28. King Papers, Huntington Library, "Journal of a trip in the Northern Sierras and Grass Valley," entries for Sunday, September 13, p. 21; Tuesday, September 22, pp. 31 and 33; Thursday, September 24, p. 35; and Saturday, September 26, p. 34 (the entry for Saturday the 26th is out of order—inserted in the journal between Wednesday the 23rd and Thursday the 24th).

29. Dickason, "Clarence King's First Westward Journey," and Wilkins, pp. 44–47.

30. J. T. Redman, "Reminiscences and experiences on my trip across the plains to California 61 years ago when I drove four mules to a covered wagon," typescript dated "June 17, 1924—Marshall, Missouri," Huntington Library.

31. Ibid. King offered another, slightly more romantic, image of a silent confrontation with Indians in an unpublished, untitled poem. Here the Indians have less agency, since they are asleep; they become almost picturesque features of the landscape. But it remains notable that King did not depict them as being inherently violent: "Too softly chant the swaying pines / To wake them. Too faintly shines / The moon's pale lamp, or yonder far / Fine jewel-dust of drifting star, / To pierce the canopy of green / And let the sleeping camp be seen. / Hid in the woodland shadows deep / Saw we the Indian hunters sleep." Bancroft Library, University of California at Berkeley, Francis Farquhar Papers, Collection of Materials Relating to Clarence King, microfilm.

32. Gardiner, "Clarence King's Boyhood," and King, "C.K.'s Notes for my [James D. Hague's] biographical sketch of him for Appleton's Encyclopedia," King Papers, Huntington Library, A-1.

King's career as an explorer is detailed not only in Wilkins but also in William H. Goetzmann, *Exploration and Empire: The Explorer and the Scientist in the Winning of the American West* (New York: W. W. Norton, 1966), pp. 355–89 and 430–66; and Richard A. Bartlett, *Great Surveys of the American West* (Norman: University of Oklahoma Press, 1962), pp. 123–215.

33. Redman, and King letters to Gardiner, "Letters from and about Clarence King," Huntington Library.

34. Adams, *Education*, p. 311.

35. "Letters from and about Clarence King," Huntington Library: see letter #12, Yale, September 19, 1861, and letter #6, New York, Saturday Evening [Spring 1860?]. Other letters are also full of this kind of language: "I am a little afraid of you. . . . I feel ever so much smaller than you and as if I was hardly a companion for you. . . . Alas! the day when we parted for . . . our individualities are so fearfully merged into one" (unnumbered letter, Yale, Sat. Nov. 23, 1861); "How are you my darling fellow. . . . Are you happy? Is your soul's sky cloudless and clear? Tell me. I am sure we will always be open. We have lived too long as soul and soul together to draw any world's, devil's, curtain of mistrust between. . . . Oh dear must we be separated?" (letter #14, Yale, Jan 30, 1862); "It all seems incomplete to me. I lack you. I find no home in any heart but yours. . . . It seems to me a sort of farce to have a friendship with anyone but yourself. My heart is taken up with you. Although I don't write much and even when I do write I don't say much but my love for you grows always and is a most absorbing passion. The deeper I feel the more it becomes an effort to express myself and at times I almost reason myself into resolving to be silent henceforth. So wait until we reach heaven then we shall see each other face to face" (letter #15, Tuesday, March 18, 1862).

36. I think exploring expeditions, in some nineteenth-century cases, may have functioned like fraternal organizations, as described, for instance, by Mark C. Carnes in *Secret Ritual and Manhood in Victorian America* (New Haven: Yale University Press, 1989) and "Middle-Class Men and the Solace of Fraternal Ritual," in *Meanings for Manhood: Constructions of Masculinity in Victorian America*, ed. Mark C. Carnes and Clyde Griffen (Chicago: University of Chicago Press, 1990), pp. 37–66. As Carnes suggests, while these ritual-based associations of men in some ways propped up the bourgeois status quo "by providing [temporary] solace from the psychic pressures of . . . social and institutional relationships," they simultaneously posed a legitimate challenge to Victorian culture. Many fraternal rituals—like Humboldt's writings—"repeatedly contravened basic tenets of capitalism," and the men who experienced them did not all go straight back to their jobs refreshed and rededicated to profitmaking (Carnes, "Middle-Class Men," 51). Moreover, the very concept of "fraternity" offered men a chance to express long-stifled emotions, and some,

like Humboldt (and Walt Whitman), decided never to accept the mantle of male stoicism, and lived out their lives in the exclusive company of affectionate men. This theme is developed in a few recent studies in the history and literature of sexuality: Jonathan Ned Katz, *Love Stories: Sex between Men before Homosexuality* (Chicago: University of Chicago Press, 2001); Caleb Crain, *American Sympathy: Men, Friendship, and Literature in the New Nation* (New Haven: Yale University Press, 2001); and Leila J. Rupp, *A Desired Past: A Short History of Same-Sex Love in America* (Chicago: University of Chicago Press, 1999). The frontier setting also clearly fostered homosocial bonding in a number of cases; see, for instance, Susan Lee Johnson, *Roaring Camp: The Social World of the Gold Rush* (New York: W. W. Norton, 2000), pp. 127–30, 159–74, and 335–37, and Adele Perry, *On the Edge of Empire: Gender, Race, and the Making of British Columbia, 1849–1871* (Toronto: University of Toronto Press, 2001), pp. 20–47 and 79–96.

37. Patricia O'Toole, in an evocative book about Henry Adams and his circle of friends, adopts the Freudian perspective that King's troubled sex life was a direct result of his relationship with his mother: see *The Five of Hearts: An Intimate Portrait of Henry Adams and His Friends, 1880–1918* (New York: Ballantine Books, 1990), esp. pp. 118–20 and 184–95. Also see Rossiter W. Raymond (who was a close associate of King's), "Biographical Notice of Clarence King," p. 620, where he notes, of King's mother, that she "devoted herself to the education of her only son, pursuing for herself many studies, that she might teach him; and becoming at the outset, as she remained always, his sympathetic and competent intellectual companion. On his part, he began as a 'mother's boy'—best of all beginnings!—and as a mother's boy, maintaining still in undiminished fervor and unstained purity the filial reverence and affection of childhood, he ended—best of all endings!" King's own view of his special relationship with his mother is revealed in a letter to John Hay: "I am the only being on earth who has never hurt her (I mean within the family) and the only one who has the power of softening the pain of wounds that never leave." Hay Papers, John Hay Library, reel 8, letter of July 30 [1892].

38. "Letters from and about Clarence King," Huntington Library: see letter #1 (Hartford, Sunday Afternoon, Oct 2, '59) and letter #11 (Aug. 19, '61, Brattleboro, Vermont).

39. As I argued in chapter 2, with regard to Humboldt; see E. Anthony Rotundo, *American Manhood: Transformations in Masculinity from the Revolution to the Modern Era* (New York: Basic Books, 1993), and Rupp.

40. Hay Papers, John Hay Library, reel 7, letter of King to Hay, March 8, 1883. King often wrote "Burn this!" on his letters to Hay. Obviously, Hay didn't.

King also sometimes equated his relationship to Hay with his relationship to his mother: "Letter writing has been near to impossible to me during all my illness, and strange to say most impossible to those for whom I have the most feeling—my mother and you. It seemed as if the stirring of a sentiment overcame me hopelessly." Ibid., reel 8, letter of Feb. 3, [1894].

41. "Letters from and about Clarence King," Huntington Library: letter #13, Yale, October 10, 1861.

42. Hay Papers, John Hay Library, reel 8, letter of May 30, 1885. The "seal" of the Five of Hearts was a simple, plain playing card in the upper left corner of the stationery, with no numeral, just five hearts arranged in the traditional pattern: two pips on top, one in the center, and two on the bottom. On the cover page of the quoted letter, King circled the upper right heart and wrote in a question mark next to it.

43. Ibid., letters of July 28, 1887; May 30, 1885; Sept. 8, 1891; and Aug. 4, 1887.

44. King, *Mountaineering*, pp. 228 and 53.

45. Ibid., p. 66.

46. "Masculine primitive ethos": Gail Bederman quoted in Mazel, p. 118—see Bederman, *Manliness and Civilization: A Cultural History of Gender and Race in the United States, 1880–1917* (Chicago: University of Chicago Press, 1995), p. 22; "blustering": Mazel, p. 118; "romantic ideal of self-reliant heroism": William Howarth, introduction to *Mountaineering*, p. xii; "risk-taking": Kevin Starr, *Americans and the California Dream, 1850–1915* (New York: Oxford University Press, 1973), p. 187; "adversarial": Michael Smith, *Pacific Visions: California Scientists and the Environment, 1850–1915* (New Haven: Yale University Press, 1987), p. 82.

47. See, for instance, Bederman; Lisa Bloom, *Gender on Ice: American Ideologies of Polar*

Expeditions (Minneapolis: University of Minnesota Press, 1993); and Dana D. Nelson, *National Manhood: Capitalist Citizenship and the Imagined Fraternity of White Men* (Durham, N.C.: Duke University Press, 1998).

48. King, *Mountaineering*, pp. 42, 47, and 51.

49. Henry Adams, "Review of *Mountaineering in the Sierra Nevada*," *North American Review* 114 (April 1872), p. 446.

50. Quoted in Wilkins, p. 150.

51. Brewer quoted in Wilkins, p. 71. Brewer, for one, believed that King "deserves a gold medal for his pluck," and noted the extent to which King embraced both the mental and physical challenges of mountaineering: "King is enthusiastic, is wonderfully *tough*, has the greatest endurance I have ever seen, and is withal very muscular. He is a most perfect specimen of health." Brewer, *Such a Landscape! A Narrative of the 1864 California Geological Survey Exploration of Yosemite, Sequoia and Kings Canyon from the Diary, Fieldnotes, Letters and Reports of William Henry Brewer*, ed. William Alsup (Yosemite National Park: Yosemite Associates and Sequoia Natural History Association, 1987), pp. 56 and 50.

52. King, *Mountaineering*, pp. 54 and 251, and King Papers, Huntington Library, "Journal of a trip in the Northern Sierras and Grass Valley," Saturday, Sept. 19, p. 26.

53. Thoreau, *The Maine Woods* (1864; repr., New York: W. W. Norton, 1950; version arranged with notes by Dudley C. Lunt), p. 278.

54. Brewer, *Such a Landscape!*, p. 49; King, *Mountaineering*, pp. 51 and 54; and Richard Wilson, "American Vision and Landscape: The Western Images of Clarence King and Timothy H. O'Sullivan" (Ph.D. diss., University of New Mexico, 1979), p. 54. Also see Perry, pp. 20–47 and 79–96, and Mazel, pp. 130–32.

55. King, *Mountaineering*, p. 58; Herman Melville, *Moby Dick or The White Whale* (1851; repr., New York: New American Library, 1961), p. 310; "Letters from and about Clarence King,"Huntington Library, letter #13, Yale, October 10, 1861.

56. "Letters from and about Clarence King," Huntington Library: letter #6, NY Sat Eve [spring 1860?]; letter #13, Yale, October 10, 1861; and letter #1, Hartford, Sunday Afternoon, Oct 2, '59.

57. King Papers, Huntington Library, "Journal of a trip in the Northern Sierras and Grass Valley," Tuesday, Sept. 29, p. 60.

58. Brewer in *Clarence King Memoirs*, p. 320; Brewer in *Such a Landscape!*, p. 18. The full citation has Brewer commenting that King "is a good fellow, tough, but doesn't like to do unpleasant work—yet I like him."

59. James Gardiner quoted in Wilkins, p. 53.

60. See Smith, p. 60.

61. King, *Mountaineering*, p. 153.

62. Ibid.

63. "Letters from and about Clarence King," Huntington Library: letter #13, Yale, October 10, 1861; King Papers, D-17, Scientific Notebooks, "Miscellaneous Notes, 1869," journal entry for 5/19/69.

64. King, *Mountaineering*, pp. 153–55.

65. See Wilkins, pp. 57–59; Mazel, pp. 104–6 and 110–17; and Witold Rybczynski, *A Clearing in the Distance: Frederick Law Olmsted and America in the 19th Century* (New York: Touchstone, 2000), pp. 225–43. Whitney himself was somewhat conflicted about these kinds of applications of his survey work; see Gerald D. Nash, "The Conflict between Pure and Applied Science in 19th-Century Public Policy: The California Geological Survey, 1860–1874," *Isis* 54 (June 1963), pp. 217–28.

66. Mazel, p. 132. Bartlett, Goetzmann, Smith, and Starr draw similar conclusions.

67. King, "Artium Magister," *North American Review* 147 (October 1888), p. 382; and King Papers, Huntington Library, "Journal of a trip in the Northern Sierras and Grass Valley," Monday, Oct. 5, pp. 72–74.

68. National Archives, Record Group 77, "King Survey," King to Humphreys, August 22, 1875; King, *Mountaineering*, p. 160.

69. Friedrich Nietzsche, *On the Genealogy of Morals*, in *On the Genealogy of Morals and Ecce*

Homo, ed. Walter Kaufmann (New York: Vintage, 1967; *Genealogy* orig. pub. 1887), pp. 155–56 and 147.

70. King [unsigned], "Style and the Monument," *North American Review* 141 (November 1885), p. 449; "Artium Magister," p. 371; *Mountaineering*, p. 242.

71. King quoted in Edgar Beecher Bronson, *Reminiscences of a Ranchman* (New York: McClure, 1908), p. 3.

72. Walt Whitman, *Democratic Vistas*, in *The Portable Walt Whitman*, ed. Mark Van Doren (New York: Penguin, 1973), pp. 349, 351, and 363; Howells, "Review of *Mountaineering*," *Atlantic Monthly* 29 (April 1872), pp. 500–501; and King, *Mountaineering*, p. 175. On Howells and literary realism in this period, see, for instance, Alan Trachtenberg, *The Incorporation of America: Culture and Society in the Gilded Age* (New York: Hill and Wang, 1982), pp. 182–207.

73. King, *Mountaineering*, pp. 127 and 69.

74. Ibid., pp. 242, 245, and 24; "Style and the Monument," p. 449; and "Artium Magister," p. 377.

75. See Robert Cahn, *American Photographers and the National Parks* (New York: Viking, 1981), pp. 124–27; Rybczinski, pp. 236–39 and 256–61; Wilkins, pp. 59–90; Roderick Nash, *Wilderness and the American Mind*, 4th ed. (1967; repr., New Haven: Yale University Press, 2001), pp. 96–121; and Weston J. Naef and James N. Wood, *Era of Exploration: The Rise of Landscape Photography in the American West, 1860–1885* (Buffalo: Albright-Knox Art Gallery and the Metropolitan Museum of Art, 1975), pp. 34–41 and 79–124.

The maps King produced for the 40th Parallel Survey were particularly stunning, as noted by Wallace Stegner in his novel *Angle of Repose*, whose main female character, Susan Ward, decorates her cabin with them. Stegner also includes a long section on King himself: see *Angle of Repose* (1971; repr., New York: Fawcett Crest, 1989), pp. 219–35.

76. See Alfred Runte, *National Parks: The American Experience*, 3rd ed. (1979; repr., Lincoln: University of Nebraska Press, 1997), as well as Runte, *Yosemite: The Embattled Wilderness* (Lincoln: University of Nebraska Press, 1990); Roderick Nash, pp. 106–7; Hans Huth, *Nature and the American: Three Centuries of Changing Attitudes* (Berkeley: University of California Press, 1957); and Max Oelschlaeger, *The Idea of Wilderness: From Prehistory to the Age of Ecology* (New Haven: Yale University Press, 1991).

77. See Solnit's magnificent book *Savage Dreams: A Journey into the Landscape Wars of the American West* (1994; repr., Berkeley: University of California Press, 1999), pp. 215–364, quotation on p. 222; William Cronon, "The Trouble with Wilderness; or, Getting Back to the Wrong Nature," in *Uncommon Ground: Rethinking the Human Place in Nature* (New York: W. W. Norton, 1996), pp. 69–90; Robert H. Keller and Michael F. Turek, *American Indians and National Parks* (Tucson: University of Arizona Press, 1998), pp. 19–22; and Mark Spence, *Dispossessing the Wilderness: Indian Removal and the Making of the National Parks* (New York: Oxford University Press, 1999), pp. 101–39.

78. King, *Mountaineering*, p. 116.

79. Ibid., pp. 114–15, 117, 121, 123–24, and 130–31.

80. See the plates in Geological Survey of California (J. D. Whitney, State Geologist), *The Yosemite Book; A Description of the Yosemite Valley and the Adjacent Region of the Sierra Nevada, and of the Big Trees of California* (New York: Julius Bien, 1868). Not all of Watkins's 1866 photos made it into the book, though; his prints are scattered in various archival collections. I looked at some at the Beinecke Library, Yale University, "Mammoth-plate Photographs of California." Also see Naef and Wood, pp. 34–41, 79–86, and 91–112.

81. Quoted in Robert Taft, *Photography and the American Scene: A Social History, 1839–1889* (1938; repr., New York: Dover, 1964), p. 278. For other accounts of Durant's excursion, see Barry B. Combs, *Westward to Promontory* (New York: Garland Books, 1969), pp. 14–16, and Naef and Wood, p. 43.

82. King quoted in Wilkins, p. 101; National Archives, Record Group 77, Office of the Chief of Engineers, Records of the U.S. Geological Exploration of the 40th Parallel ("King Survey"), 1867–81, Humphreys to King, March 21, 1867; Stanton quoted in James D. Hague, "Memorabilia," in *Clarence King Memoirs*, p. 385.

83. *New York Times*, May 2, 1867, p. 8; Henry Adams, *Education*, p. 312. George P. Merrill, in his book *The First One-Hundred Years of American Geology*, has asserted that the crew of the 40th Parallel Survey "was beyond question the best equipped by training of any that had thus far entered the field of American geology" (1924; repr., New York: Hafner, 1964), p. 531. For histories of the 40th Parallel Survey, see Goetzmann, *Exploration and Empire*, pp. 430–66, and Bartlett, pp. 123–215.

84. King Papers, Huntington Library, D-7, Scientific Notebooks, 1866; D-17, Miscellaneous Notes, 4/20/69; and A-2, #5, Personal Notebooks.

Another notable reference to Humboldt in King's field notebooks: in 1864, King recorded his strategy of obtaining a specimen of the springtime bloom of the sequoia tree by shooting down a branch with his gun, at which point he remembered reading that Humboldt had "complained of being years in the tropics without being able to examine the flowers of the palm, which hung 60 feet overhead, because he could hire no native to climb for him." King's conclusion was "that the great Humboldt would have been better off with a six-shooter." See King Papers, Huntington Library, B-1, "Field Notes of Campaign in High Sierra Nevada, summer of 1864," p. 5.

85. King, *Mountaineering*, p. 78. An illuminating exploration of Dante's dark view of wilderness is in Robert Pogue Harrison, *Forests: The Shadow of Civilization* (Chicago: University of Chicago Press, 1992), pp. 81–87.

86. The Poe quote is from "The Fall of the House of Usher" (1839) in *"The Fall of the House of Usher" and Other Tales* (New York: New American Library, 1980), p. 113; the King quotes are from *Mountaineering*, pp. 168 and 165.

On O'Sullivan, see Naef and Wood, pp. 125–66; Joel Snyder, *American Frontiers: The Photographs of Timothy H. O'Sullivan, 1867–74* (Millerton, N.Y.: Aperture, 1981); James D. Horan, *Timothy O'Sullivan: America's Forgotten Photographer* (Garden City, N.Y.: Doubleday, 1966); and Rick Dingus, *The Photographic Artifacts of Timothy H. O'Sullivan* (Albuquerque: University of New Mexico Press, 1982). Dingus, as a member of the Rephotographic Survey Project, found several of the locations of O'Sullivan's frontier photographs and includes in his book some of his own photographs of the same scenes, in their modern context. Also see Alan Trachtenberg, *Reading American Photographs: Images as History, Mathew Brady to Walker Evans* (New York: Hill and Wang, 1989), pp. 119–63.

87. John Samson, "Photographs from the High Rockies," *Harper's Monthly Magazine* 39 (September 1869), pp. 470–71. Samson is probably O'Sullivan's pseudonym. In any case, O'Sullivan is quoted directly in the article, and the narrative follows the photographer's adventures all the way through. If Samson did exist, he either interviewed O'Sullivan or received letters from him.

88. See, for instance, Trachtenberg, *Reading American Photographs*, pp. 99, 102, and 129.

89. Naef and Wood, p. 51.

90. King, *Mountaineering*, pp. 179–80.

91. See Snyder, p. 37; "The Cross of Snow" quoted in Taft, p. 305. On Jackson, also see Naef and Wood, pp. 219–50.

92. See, for instance, Rebecca Solnit's essay "Uplift and Separate: The Aesthetics of Nature Calendars," in her book *As Eve Said to the Serpent: On Landscape, Gender, and Art* (Athens: University of Georgia Press, 2001), pp. 200–205.

93. See Rebecca Solnit, *River of Shadows: Eadweard Muybridge and the Technological Wild West* (New York: Viking, 2003), pp. 75–124.

94. King Papers, Huntington Library, "Journal of a trip in the Northern Sierras and Grass Valley," Tuesday, Sept. 29, pp. 67–68; and see Brewer, *Up and Down California*, p. 465.

95. King, *Mountaineering*, pp. 194, 209, 224, 208, 220, and 199–200.

96. King, "On the Discovery of Actual Glaciers in the Mountains of the Pacific Slope," *American Journal of Science and Arts*, 3rd ser., 1 (March 1871), pp. 157–67. Letter to Humphreys is in National Archives, Record Group 57.2.1, King Survey, Microfilm Publication M622, roll 3, letter of 10/10/70.

97. King, *Mountaineering*, pp. 198, 216–18, and 221; and note King, "Active Glaciers within the United States," *Atlantic Monthly* 27 (March 1871), pp. 371–77.

98. King, *Mountaineering*, p. 202.

Chapter 7: "Catastrophism and the Evolution of Environment":
A Science of Humility

1. Clarence King, "On the Discovery of Actual Glaciers in the Mountains of the Pacific Slope," *American Journal of Science and Arts*, 3rd ser., 1 (March 1871), p. 158; National Archives, Record Group 57.2.1, King Survey, Microfilm Publication M622, roll 3, Copies of Letters and Reports sent to the Chief of Engineers by Clarence King, March 1867–January 1879, letter of Clarence King to General A. A. Humphreys, September 2, 1870, Camp at Mount Shasta, p. 195; and John Samson, "Photographs from the High Rockies," *Harper's Monthly Magazine* 39 (September 1869), p. 470.
2. National Archives, RG 57.2.1, King Survey Microfilm, roll 3, letter of King to Humphreys, May 11, 1870, p. 178.
3. Clarence King Papers, H. E. Huntington Library, B-1, "Field Notes of Campaign in High Sierra Nevada": "Attempt to Climb Mount Whitney," pp. 9–10.
4. King, *Mountaineering in the Sierra Nevada* (1872; repr., New York: Penguin, 1989), p. 240. Unless otherwise noted, further references will be to this edition.
5. Ibid., pp. 226, 238, 232, 239, 237, 240, and 239. On the question of being first, see Thomas P. Slaughter, *Exploring Lewis and Clark: Reflections on Men and Wilderness* (New York: Knopf, 2003), pp. 27–64.
6. Henry Adams, *The Education of Henry Adams* (1907; repr., Boston: Houghton Mifflin, 1973), pp. 311–13 and 328. Also see p. 346, where Adams describes the devastating impact of the Depression of 1893 and notes that "among the earliest wreckage had been the fortunes of Clarence King. The lesson taught whatever the bystander chose to read in it; but to Adams it seemed singularly full of moral, if he could but understand it. In 1871 he had thought King's education ideal, and his personal fitness unrivalled. No other young American approached him for the combination of chances—physical energy, social standing, mental scope and training, wit, geniality, and science, that seemed superlatively American and irresistibly strong."
7. See William A. Goodyear, "On the Situation and Altitude of Mount Whitney," *Proceedings of the California Academy of Sciences* 5 (August 4, 1873), pp. 139–44; Thurman Wilkins, *Clarence King: A Biography*, rev. and enlarged ed. (Albuquerque: University of New Mexico Press, 1988), pp. 154–55; and King, *Mountaineering*, pp. 240–41.
8. Quoted in Wilkins, p. 155.
9. King, *Mountaineering*, p. 242.
10. This preface is not included in the Penguin edition of *Mountaineering*, so for this quote I've used an edition by Charles Scribner's Sons (1874; repr., New York, 1902), p. x.
11. The official certificate sent to King from the Boston Society of Natural History can be found in the King Papers, Huntington Library, box 2; it is dated March 1881, and signed by the society's honorary secretary, a Mr. Abbot.
12. King, *Mountaineering*, pp. 252–53. Many literary critics conclude their analysis of this passage at this point, asserting that King was simply flexing his scientific muscles: see, for instance, John P. O'Grady, *Pilgrims to the Wild: Everett Ruess, Henry David Thoreau, John Muir, Clarence King, Mary Austin* (Salt Lake City: University of Utah Press, 1993), pp. 120–22, and David Mazel, *American Literary Environmentalism* (Athens: University of Georgia Press, 2000), pp. 128–30.
13. King, *Mountaineering*, pp. 244, 249, and 246. The historian Michael Smith views King's mistake on Sheep Rock as pure irresponsibility, arguing that "careful attention to the terrain or a verification of his barometric readings could have corrected his error"; see Smith, *Pacific Visions: California Scientists and the Environment, 1850–1915* (New Haven: Yale University Press, 1987), p. 90.
14. King, *Mountaineering*, pp. 242, 252–53, and 250–51.
15. See George H. Daniels, *Science in American Society: A Social History* (New York: Knopf, 1971), esp. pp. 174–287; William H. Goetzmann, *New Lands, New Men: America and the Second Great Age of Discovery* (1986; repr., Austin: Texas State Historical Association, 1995); Burton J. Bledstein, *The Culture of Professionalism: The Middle Class and the Development of Higher Education in America* (New York: W. W. Norton, 1976), esp. pp. 318–31; and Dorothy Ross, *The Origins of American Social Science* (New York: Cambridge University Press, 1991), p. 37. Theodore Roosevelt

is quoted in Keir B. Sterling, *Last of the Naturalists: The Career of C. Hart Merriam*, rev. ed. (New York: Arno Press, 1977), p. 392.

16. See Alan Trachtenberg, *The Incorporation of America: Culture and Society in the Gilded Age* (New York: Hill and Wang, 1982), pp. 38–100; Daniels, pp. 174–287; Rodman W. Paul, *Mining Frontiers of the Far West, 1848–1880* (1963; repr., Albuquerque: University of New Mexico Press, 2001); and David F. Noble, *America by Design: Science, Technology, and the Rise of Corporate Capitalism* (New York: Knopf, 1977).

17. See Daniels, esp. pp. 265–87, and Christopher Lasch, *The True and Only Heaven: Progress and Its Critics* (New York: W. W. Norton, 1991), pp. 51 and 195.

18. Merriam quoted in Daniels, p. 279. On Haeckel, see Donald Worster, *Nature's Economy: A History of Ecological Ideas* (New York: Cambridge University Press, 1977), p. 192, and Anna Bramwell, *Ecology in the Twentieth Century: A History* (New Haven: Yale University Press, 1989), pp. 39–63. And see Rowland, "A Plea for Pure Science," *Proceedings of the American Association for the Advancement of Science* 32 (August 1883), pp. 105–26.

19. King Papers, Huntington Library, D-16, "Memoranda, Virginia City, Nevada, 1868."

20. William Dean Howells, "Meetings with King," in *Clarence King Memoirs* (New York: G. P. Putnam's Sons, 1904), p. 140.

21. Adams, "Review of *Systematic Geology*," *The Nation* 28 (January 23, 1879), p. 74.

22. Quoted in Wilkins, p. 187.

23. King [unsigned], "Current Literature" (Book Reviews), *Overland Monthly* 5 (December 1870), p. 580.

24. King, *Mountaineering*, p. 206.

25. National Archives, Record Group 77, "King Survey," King to Humphreys, February 17, 1873.

26. Quoted in Edgar Beecher Bronson, "A Man of East and West: Clarence King, Geologist, Savant, and Wit," *Century Magazine* 80 (July 1910), p. 382.

27. See Daniels, pp. 223–64, esp. 244–45; Louis Menand, *The Metaphysical Club: A Story of Ideas in America* (New York: Farrar, Straus, and Giroux, 2001), pp. 120–28; Cynthia Eagle Russett, *Darwin in America: The Intellectual Response, 1865–1912* (San Francisco: W. H. Freeman, 1976), pp. 1–45; and Edward J. Larson, *Summer for the Gods: The Scopes Trial and America's Continuing Debate over Science and Religion* (New York: Basic Books, 1997).

28. "The struggle for existence" is Darwin's phrase; on this element of his thought, see Worster, *Nature's Economy*, pp. 114–87, esp. pp. 147–69. Darwin also wrote that "[a]ll organic beings are striving to seize on each place in the economy of nature." See Worster, *Nature's Economy*, p. 158.

29. Darwin quoted in Menand, p. 210; Darwin quoted in Worster, *Nature's Economy*, pp. 164–65. Also see Menand, pp. 120–27. When Darwin discussed this process in terms of his famous tree-of-life metaphor, he explained that new branches had to destroy "the less vigorous ones," with "the dead and lost branches rudely representing extinct genera and families." See Worster, *Nature's Economy*, p. 164.

30. Darwin quoted in Worster, *Nature's Economy*, p. 180. Note that Darwin referred to nature as "a web of complex relations": see Worster, p. 156. Haeckel quoted in Worster, p. 192. Also see Robert M. Young, *Darwin's Metaphor: Nature's Place in Victorian Culture* (New York: Cambridge University Press, 1985).

31. Hudson quoted in Richard Haymaker, *From Pampas to Hedgerows and Downs: A Study of W.H. Hudson* (New York: Bookman Associates, 1954), p. 74.

32. See especially Menand, pp. 140–46, and Daniels, pp. 262–64.

33. On Spencer and his influence, see Daniels, pp. 248–64; Menand, pp. 121 and 140–44; Russett, pp. 14–19; and Richard Hofstadter, *Social Darwinism in American Thought*, rev. ed. (1944; repr., New York: George Braziller, 1959), esp. pp. 13–50. Carnegie quoted in Trachtenberg, *Incorporation of America*, p. 81.

34. See especially William Stanton, *The Leopard's Spots: Scientific Attitudes toward Race in America, 1815–1859* (Chicago: University of Chicago Press, 1960), and Stephen Jay Gould, *The Mismeasure of Man* (New York: W. W. Norton, 1981), pp. 30–74, as well as Menand, pp. 97–148.

Note also George W. Stocking Jr., *Race, Culture, and Evolution: Essays on the History of Evolution* (New York: Free Press, 1968), especially the essays "The Persistence of Polygenist Thought in Post-Darwinian Anthropology" (pp. 42–68) and "The Dark-Skinned Savage: The Image of Primitive Man in Evolutionary Anthropology" (pp. 110–32).

35. See Menand, pp. 12–16 and 144–45.

36. See Reginald Horsman, "Scientific Racism and the American Indian in the Mid-Nineteenth Century," *American Quarterly* 27 (May 1975), pp. 152–68; Frederick C. Hoxie, *A Final Promise: The Campaign to Assimilate the Indians, 1880–1920* (Lincoln: University of Nebraska Press, 1984); Robert F. Berkhofer Jr., *The White Man's Indian: Images of the American Indian from Columbus to the Present* (New York: Vintage Books, 1979), esp. pp. 44–69 and 153–75; and Roy Harvey Pearce, *Savagism and Civilization: A Study of the Indian and the American Mind* (Baltimore: Johns Hopkins University Press, 1965).

37. Wilkins, pp. 81–83, and King, *Mountaineering*, pp. 33–34. Wilkins, then, was clearly overstating the case when he called King "a staunch defender of the native American" (p. 123). One comment he might have had in mind, though, was King's assertion in a letter to John Hay that "enemies are impossible to me among archaic peoples": John Hay Library, Brown University, John Hay Papers, microfilm reel 8, letter of Aug. 12, 1888.

38. Powell quoted in Donald Worster, *A River Running West: The Life of John Wesley Powell* (New York: Oxford University Press, 2001), p. 284. Worster's book can't be topped for thoroughness, but the earlier biography of Powell by Wallace Stegner is also a classic work of western history: *Beyond the Hundredth Meridian: John Wesley Powell and the Second Opening of the West* (1953; repr., Lincoln: University of Nebraska Press, 1982). Note also that Powell essentially founded the Bureau of Ethnology.

39. Bronson, "A Man of East and West," p. 376; Hay Papers, John Hay Library, reel 8, letter of King to Hay, May 30, 1885. On the subject of Helen Hunt Jackson, King wrote to Hay that "I have just read Ramona. . . . The story of the wrongs of the poor Mission Indians is gospel truth and the character of [the novel's Indian hero] Alessandro not much overdone. . . . I am inclined to like it very much."

40. National Archives, Record Group 77, "King Survey," King to Humphreys, February 11, 1873; King, *Mountaineering*, p. 168; and Wilkins, p. 117. The agent was named H. G. Parker and was friendly with the survey crew; he did the taxidermy on some of the specimens collected by King's ornithologist, Robert Ridgway.

On naming, also see Stephen Greenblatt, *Marvelous Possessions: The Wonder of the New World* (Chicago: University of Chicago Press, 1991), pp. 80–85, and Alan Trachtenberg, "Naming the View," in *Reading American Photographs: Images as History, Mathew Brady to Walker Evans* (New York: Hill and Wang, 1989), pp. 119–63.

41. Trachtenberg, *Reading American Photographs*, p. 127.

42. In *Mountaineering*, King referred to the mountain as "that firm peak with titan strength and brow so square and solid, it seems altogether natural we should have named it for California's statesman, John Conness" (p. 228). Also see Wilkins, p. 131.

43. Adams, *Education*, p. 321.

44. Hague, "Memorabilia," in *Clarence King Memoirs*, pp. 406–7; Wilkins, p. 356; and Hay Papers, John Hay Library, reel 8, letter of King to Hay, November 12 [1888?]. King referred to Lancaster, in a letter to Henry Adams, as "a trained nurse and a monument of medical wisdom." Massachusetts Historical Society (MHS), Clarence King Papers, letter of December 31, 1893.

45. Hay Papers, John Hay Library, reel 8, letter of Sept. 8, 1891; letter marked "Saturday" [1890?]; undated letter on Union League Club stationery [1891?]; and Adams, "King," in *Clarence King Memoirs*, p. 172.

46. Quoted in Wilkins, p. 362.

47. When King published his speech, he simplified its title to "Catastrophism and Evolution." I continue to use the original title in the text, but I will refer to the published title in the notes: *American Naturalist* 11 (August 1877), p. 464. Also see Wilkins, p. 220.

48. Quoted in Worster, *Nature's Economy*, p. 133.

49. King, "Catastrophism and Evolution," pp. 467 and 452.

50. Leopold, "Wilderness" (unpublished manuscript, c. 1935), in *The Essential Aldo Leopold: Quotations and Commentaries*, ed. Curt Meine and Richard L. Knight (Madison: University of Wisconsin Press, 1999), p. 245.

51. See Smith, p. 2, and Stephen Jay Gould, *Time's Arrow, Time's Cycle: Myth and Metaphor in the Discovery of Geological Time* (Cambridge, Mass.: Harvard University Press, 1987), pp. 1–3. Freud, patting himself on the back, initially made this observation about science in 1917. On the "deep time" of geology and the Darwinian time frame, also see John McPhee, *Basin and Range* (New York: Noonday Press, 1981), pp. 20 and 120–29, and Stephen Kern, *The Culture of Time and Space, 1880–1918* (Cambridge, Mass.: Harvard University Press, 1983), pp. 37–38.

52. King, "Catastrophism and Evolution," pp. 449, 455, and 463.

53. Ibid., pp. 461, 451, and 462. King not only criticized uniformitarians for their lack of imagination but also questioned whether they even employed the scientific method: "they start with a gratuitous assumption (vast time), fortify it by an analogy of unknown relevancy (the present rate), and serenely appeal to the absence of evidence against them as proof in their favor" (p. 462).

Humboldt quotation is from his *Researches, Concerning the Institutions and Monuments of the Ancient Inhabitants of America, with Descriptions and Views of Some of the Most Striking Scenes in the Cordilleras* (London: Longman, Hurst, 1814), 1:12.

On the debate between uniformitarians and catastrophists, see Stephen Jay Gould, *Ever Since Darwin: Reflections in Natural History* (New York: W. W. Norton, 1977), pp. 147–52, and Gould, *Time's Arrow, Time's Cycle*, pp. 112–42.

54. King, "Catastrophism and Evolution," pp. 463, 464, 461, 457, 464, and 453.

55. Ibid., pp. 450, 469, and 451; King, *Systematic Geology*, vol. 1, *Report of the U.S. Geological Exploration of the Fortieth Parallel* (Washington, D.C.: Government Printing Office, 1878), p. 727; King, *Mountaineering*, p. 126; and King, "The Biographers of Lincoln," *Century Magazine* 32 (October 1886), p. 868.

56. King, "Catastrophism and Evolution," pp. 461, 463, 451–52, and 454, and *Mountaineering*, pp. 262 and 251.

57. King, "Artium Magister," *North American Review* 147 (October 1888), pp. 371 and 383–84.

58. See Adams, *Education*; Trachtenberg, *Incorporation of America*; and Ray Ginger, *Age of Excess: The U.S. from 1877 to 1914* (New York: Macmillan, 1965).

59. King, "The Age of the Earth," *American Journal of Science* 45 (January 1893), pp. 1–20.

60. Gardiner quoted in Wilkins, p. 237; on King's involvement in the cattle industry in general, see Wilkins, pp. 230–42.

61. Adams, "Review of *Systematic Geology*," p. 74.

62. King, *Systematic Geology*, p. 5.

63. William H. Goetzmann, *Exploration and Empire: The Explorer and the Scientist in the Winning of the American West* (1966; repr., New York: W. W. Norton, 1978), p. 464.

64. Gould, *Ever Since Darwin*, p. 152; Gould, quoted in Richard Huggett, *Catastrophism: Systems of Earth History* (London: Edward Arnold, 1990), p. 8; the British geologist Derek Ager, quoted in John D. Cooper, Richard H. Miller, and Jacqueline Patterson, *A Trip Through Time: Principles of Historical Geology* (Columbus: Merrill, 1986), p. 75; and King, "Catastrophism and Evolution," p. 457. Also see Brian J. Skinner and Stephen C. Porter, *Physical Geology* (New York: John Wiley and Sons, 1987), pp. 35–36.

Of course, King's theory of "modified catastrophism," which was parallel to what King referred to as his colleague Thomas Henry Huxley's "proposed environmental geology," did contain errors: King suspected, for instance, that catastrophes could actually cause "morphological change on the part of plastic species" ("Catastrophism and Evolution," p. 469). And his rhetoric generally exaggerated the speed with which catastrophes occurred: the "sudden physical change" noted above may indeed have been sudden in geological terms, but that could mean years or even millennia.

It is also worth noting that geological science is continually evolving. The theory of continental drift, for instance, now taken for granted, did not erupt onto the scene until the 1960s.

65. See Goetzmann, *Exploration and Empire;* Weston J. Naef and James N. Wood, *Era of Exploration: The Rise of Landscape Photography in the American West, 1860–1885* (Buffalo: Albright-Knox Art Gallery and the Metropolitan Museum of Art, 1975); Richard A. Bartlett, *Great Surveys of the American West* (Norman: University of Oklahoma Press, 1962); Bartlett, "Scientific Exploration of the American West, 1865–1900," in *North American Exploration,* vol. 3, *A Continent Comprehended,* ed. John Logan Allen (Lincoln: University of Nebraska Press, 1997), pp. 461–520; and Howard D. Kramer, "The Scientist in the West, 1870–1880," *Pacific Historical Review* 12 (September 1943), pp. 239–51.

66. See Henry Nash Smith, "Clarence King, John Wesley Powell, and the Establishment of the United States Geological Survey," *Mississippi Valley Historical Review* 34 (June 1947), pp. 37–58.

67. Ibid., and also see Samuel Franklin Emmons, "Clarence King—Geologist," in *Clarence King Memoirs,* pp. 272–73; and D. H. Dickason, "Henry Adams and Clarence King, The Record of a Friendship," *New England Quarterly* 17 (June 1944), p. 236.

Hewitt's quote is from 45th Cong., 3rd sess., *Congressional Record,* 8, pt. 2 (Feb. 11, 1879), col. 1203. King is quoted in Wilkins, pp. 262 and 277. The "landed interests" spoke especially vehemently through Representative Thomas Patterson of Colorado (nicknamed "Old Perplexity"): see 45th Cong., 3rd sess., *Congressional Record* 8, pt. 3, appendix, col. 217.

68. George Perkins Marsh, *Man and Nature; or, Physical Geography as Modified by Human Action* (New York: Charles Scribner, 1864), p. 18. Marsh, now considered a founding father of conservation, was in fact a devoted Humboldtian; he wrote an essay in Humboldt's honor the year after the Prussian died: "The Study of Nature," *Christian Examiner* 68 (January 1860), pp. 33–62.

69. Henry Nash Smith's article on the founding of the USGS (see note 66) emphasizes Powell's key role; also see Henry Nash Smith, *Virgin Land: The American West as Symbol and Myth* (1950; repr., Cambridge, Mass.: Harvard University Press, 1978), esp. pp. 179–83 and 287, on the myth of rain following the plow, as well as Stegner, pp. 202–51, and Worster, *A River Running West,* pp. 337–96.

John Wesley Powell, *Report on the Lands of the Arid Region of the United States* (1878; repr., Cambridge, Mass.: Harvard University Press, 1962), p. 33; Powell's first quote is from Mary C. Rabbitt, "John Wesley Powell: Pioneer Statesman of Federal Science," in *The Colorado River Region and John Wesley Powell,* Geological Survey Professional Paper 669 (Washington, D.C.: Government Printing Office, 1969), p. 17.

On the Cosmos Club, see Wilcomb E. Washburn, *The Cosmos Club of Washington: A Centennial History, 1878–1978* (Washington, D.C.: Cosmos Club, 1978), and the Cosmos Club, *The Twenty-fifth Anniversary of the Founding of the Cosmos Club of Washington* (Washington, D.C.: Cosmos Club, 1904). Early members included King and his USGS colleague Grove Karl Gilbert, who commented that one of the key purposes of the club was "to bind the scientific men of Washington by a social tie and thus promote the solidarity which is important to their proper work and influence"—a goal J. N. Reynolds surely would have applauded (Cosmos Club, p. 40).

70. King, "Catastrophism and Evolution," p. 469; King, "The Education of the Future," *Forum* 13 (March 1892), p. 29. On Powell and redemption and reclamation, see Stegner, pp. 294–350, and Worster, *A River Running West,* pp. 467–532.

71. See the *New-York Daily Tribune,* October 31, 1893, p. 4, and November 4, 1893, p. 7; the *New York World,* October 31, 1893, p. 3; and the *New York Sun,* November 3, 1893, p. 1. King's friend John La Farge quoted in Wilkins, p. 388. On this episode, also see Patricia O'Toole, *The Five of Hearts: An Intimate Portrait of Henry Adams and His Friends, 1880–1918* (New York: Ballantine Books, 1990), pp. 267–68.

72. "Scientific imagination" is from a series of resolutions adopted by a group of King's friends and colleagues at a memorial meeting shortly after his death. The resolutions were reprinted in a special memorial pamphlet: S. F. Emmons et al., *Clarence King: A Memorial* (New York: The Engineering and Mining Journal, 1902), p. 4.

"Scientific education" is from the *San Francisco Morning Bulletin,* November 27, 1872, quoted in Wilkins, p. 184.

73. Note the comments of Samuel Franklin Emmons relating to King's Humboldtian

approach to fieldwork, in Emmons, "Biographical Memoir of Clarence King," *National Academy of Sciences Biographical Memoirs* 6 (1909), p. 38: "As far as it was possible to human foresight, King had provided means to overcome the difficulties liable to be encountered. Guided by his previous experience in such work, he had personally supervised the preparation of every article of the party's equipment, from the scientific instruments, many of which were specially constructed for the purpose of his own special designs, down to the minor details of construction of wagons and pack saddles."

74. Janin paraphrased by King in his testimony for a lawsuit stemming from the diamond hoax, reprinted in San Francisco's *Daily Morning Call*, January 21, 1875, p. 1. According to Peter Farquhar, if you go looking today for the site of the fake diamond deposit, you'll actually find it in Colorado, just a few miles south of the Wyoming border; see "Site of the Diamond Swindle of 1872," *California Historical Society Quarterly* 42 (March 1963), pp. 49–53. The story originally broke in San Francisco's *Daily Alta California*, November 26, 1872, p. 1: "A Swindle: Exposure of the Diamond Fraud." For in-depth narratives of the whole scandal, see Wilkins, 170–85; Harry H. Crosby, "The Great Diamond Fraud," *American Heritage* 7 (February 1956), pp. 58–63 and 100; King, "Report of Clarence King, United States Geologist, To the Board of Directors of the San Francisco and New York Mining and Commercial Company," reprinted in *Engineering and Mining Journal* 14 (December 10, 1872), pp. 377 and 379–80; Allen D. Wilson, "The Great California Diamond Mines: A True Story," *Overland Monthly*, 2nd ser., 43 (April 1904), pp. 291–96; "The Diamond Bubble and Its Bursting," *Nation* 15 (December 12, 1872), pp. 379–80; Bartlett, *Great Surveys*, pp. 187–205; S. F. Emmons, "The Diamond Discovery of 1872," typescript in box 32 of the S. F. Emmons Papers, Manuscript Division, Library of Congress, Washington, D.C.; and National Archives, Record Group 77, "King Survey," King to Humphreys, Nov. 27, 1872, roll 3, pp. 378–82.

Also see Mazel's epistemological interpretation of King's role in the Great Diamond Hoax, pp. 123–27.

75. National Archives, Record Group 77, "King Survey," King to Humphreys, Nov. 27, 1872, roll 3, pp. 379–80; King, testimony reprinted in the *Daily Morning Call*, January 21, 1875, p. 1; King, "Report of Clarence King," reprinted in the *Daily Alta California*, November 26, 1872, p. 1 (also in *Engineering and Mining Journal* 14 [December 10, 1872], pp. 379–80).

For King obituaries, see, for example, James D. Hague, "Clarence King: An Appreciation," *New York Evening Post*, December 28, 1901, p. 5; Emmons et al., *Clarence King: A Memorial*, p. 9.

76. On this period in King's life, a thorough source is Wilkins, pp. 288–389; also see O'Toole.

77. Adams, *Education*, p. 328.

78. MHS, King Papers, letter of September 25, 1889.

79. King, "Artium Magister," pp. 380 and 382, and "Education of the Future," p. 22.

80. MHS, King Papers, letter of September 25, 1889 (to Henry Adams). The letter continues, referring to his family members: "Not merely their maintenance but their whole affairs have rested on my shoulders ever since. As a consequence, the quarter of a million I have earned professionally has gone to them." The financial burden King bore in caring for his Rhode Island family is one of the most prevalent themes in his correspondence with Hay and Adams in the 1880s and '90s.

81. Emmons, "Biographical Memoir," p. 30; King's friends' official resolutions, in Emmons et al., *Clarence King: A Memorial*, p. 4; Emmons, in *Clarence King: A Memorial*, p. 5.

82. Stegner, p. 21. Also see Michael Smith, and Bartlett, *Great Surveys*.

83. Hay and James quoted in Wilkins, p. 325.

84. Emmons, "Biographical Memoir," pp. 38–39.

James D. Hague agreed: "King, always a delightful companion, was especially so in camp. Everybody missed him when he went away and was glad when he came back. If any discontenting grievances, dissensions or difficulties had arisen during King's absence, they all vanished before his genial presence and cheerful spirit as soon as he returned." See Hague, "Memorabilia," in *Clarence King Memoirs*, pp. 401–2.

King himself, looking back at his life, highlighted his "qualities as a field leader. If I succeeded in anything, it was in personally impressing the whole corps and making it uniformly harmonious and patient; and I think I did that as much as anything else by a sort of natural spirit of command and personal sympathy with all hands and conditions, from geologists to mules. 'Tis but a step from the sublime to the ridiculous." See his notes to James D. Hague for Hague's biographical sketch of King for Appleton's Encyclopedia, King Papers, Huntington Library, A-1, p. 3.

85. John Hay, "Clarence King," and William Dean Howells, "Meetings with King," in *Clarence King Memoirs*, pp. 120, 152, and 155.

86. Hay Papers, John Hay Library, reel 7, letter of Sunday [September 1884?].

87. Ibid., reel 8, letter of King to Hay, May 16 [1894?].

88. Quoted in Wilkins, p. 294.

89. Quoted in Trachtenberg, *Incorporation of America*, p. 47.

90. This letter was printed in the *New York Daily Mirror*, November 22, 1933, p. 9 (in the context of a trial in which Ada tried to sue James Gardiner's estate for money that she claimed King had promised to her).

91. MHS, King Papers, letter of King to Adams, Sept. [1887] and letter of Sept. 25 [1889]. On attitudes toward miscegenation in the United States, see Werner Sollors, ed., *Interracialism: Black-White Intermarriage in American History, Literature, and Law* (New York: Oxford University Press, 2000).

King's relationship with Ada becomes even more interesting in the context of King's generally positive feelings about miscegenation and his fondness for the rhetoric of abolitionist Wendell Phillips. In 1862, King had declared himself a firm Phillips supporter (see chapter 6), and in 1863 Phillips spoke of "the melting of the Negro into the various races that congregate on the continent" in "gradual and harmonizing union, in honorable marriage. In my nationality, there is but one idea—the harmonious and equal mingling of all races." Quoted in Gary B. Nash, "The Hidden History of Mestizo America," in *Sex, Love, Race: Crossing Boundaries in North American History*, ed. Martha Hodes (New York: New York University Press, 1999), p. 22.

92. Emmons, "Biographical Memoir," p. 30.

93. See *New York Daily Mirror*, November 22, 1933, pp. 3, 8, and 9, and Wilkins, p. 380.

King's willingness to go into debt to support Ada is notable, since his debt to Hay, especially, caused him a great deal of suffering. "Dear Hay," he wrote, in 1894, "A few dozen times this summer I have begun to write you and every time, after a few lines, I broke down. The one great sin of my life is to have blindly got into such a miserable position with you who are the dearest friend I ever had or shall ever have on earth. I never should have recovered reason and dared to begin life again had there not been a germ of faith that refused to perish. That little mustard seed grows within me and its still small voice keeps me up with a promise of remaking my life and of righting myself so far as mere debit and credit goes with you." Hay Papers, John Hay Library, reel 8, letter of Aug. 30, 1894.

94. Quoted in Wilkins, p. 397.

95. The economic historian David Landes quoted in Trachtenberg, *Incorporation of America*, p. 39; Adams, *Education*, p. 338; and King quoted in Wilkins, p. 387.

96. Hay Papers, John Hay Library, reel 8, letter of December 1897. On King's relationship to specialization, his friend Rossiter Raymond commented: "Perhaps we might say that, in this age, scientific distinction must be won, as a rule, in some specialty, and at the cost of an exclusive devotion to that one department; so that the great specialist, however versatile he might have become, if all his original endowments had been utilized, is at last, to the eyes of men, simply the impersonal representative of one idea or sphere. . . . Clarence King did not thus sacrifice himself to his work." See Raymond, "Biographical Notice," in *Clarence King Memoirs*, pp. 370–71.

97. See, for instance, John Mack Faragher, ed., *Rereading Frederick Jackson Turner: "The Significance of the Frontier in American History" and Other Essays* (New York: Henry Holt, 1994).

98. Roosevelt explicitly upheld "the stern manly qualities that are invaluable to a nation"—see *Ranch Life and the Hunting-Trail* (Ann Arbor: University of Michigan Press, 1966; facsimile of the 1888 ed.), p. ii. And he lamented the passing of the frontier: "We who have felt the charm of the life, and have exulted in its abounding vigor and its bold, restless freedom, will not only regret its passing for our own sakes, but must also feel real sorrow that those who come after us are

not to see, as we have seen, what is perhaps the pleasantest, the healthiest, and most exciting phase of American existence" (p. 24). On Roosevelt's Rough-Riding antimodernism, also see Gail Bederman, *Manliness and Civilization: A Cultural History of Gender and Race in the United States, 1880–1917* (Chicago: University of Chicago Press, 1995).

On U.S. expansionism and imperialism, see Thomas Schoonover, *Uncle Sam's War of 1898 and the Origins of Globalization* (Lexington: University Press of Kentucky, 2003); Walter LaFeber, *The New Empire: An Interpretation of American Expansion, 1860–1898* (1963; repr., Ithaca: Cornell University Press, 1967); and Amy Kaplan and Donald E. Pease, eds., *Cultures of United States Imperialism* (Durham, N.C.: Duke University Press, 1993).

99. King, "Shall Cuba Be Free?," *Forum* 20 (September 1895), p. 51.

100. Ibid., pp. 51, 50, and 57.

101. Ibid., p. 50; King, "Fire and Sword in Cuba," *Forum* 22 (September 1896), pp. 36 and 52. For context on U.S. involvement in Cuba, see Schoonover; Philip S. Foner, *The Spanish-Cuban-American War and the Birth of American Imperialism, 1895–1902*, 2 vols. (New York: Monthly Review Press, 1972), esp. 1:71 and 90 on King; and Louis A Pérez, *The War of 1898: The United States and Cuba in History and Historiography* (Chapel Hill: University of North Carolina Press, 1998).

King also admitted in a letter that he considered himself "diametrically opposed to the United States" (quoted in Wilkins, p. 351). Back in 1884, immediately after the election of Grover Cleveland, King had written to John Hay: "It is a serious result of our American system, that a grovelling ignoramus should be forced on us. . . . A political system, my dear Hay, which gave us Andy Johnson, Grant, Hayes, and Arthur is a failure. Cleveland seals it. I have always maintained that America was an utter, absolute failure socially, but a considerable success politically. It looks as if that last saving feature might break down too and then there would be left only cooking and perennial friendship to cheer us up." Hay Papers, John Hay Library, reel 8, letter of Nov. 5, 1884.

102. King, *Mountaineering*, p. 202.

103. Quoted in O'Toole, p. 356.

104. Hay Papers, John Hay Library, reel 7, letter of Aug. 24 [1882]; reel 7, letter of Jan. 12 [1876]; and reel 8, letter of July 28, 1887. The "lonely moor" letter is perhaps the best evidence I've seen of King's homosexual tendencies. It should be noted that Gore Vidal and Lionel Trilling have gone on record as asserting that King, Adams, and Hay may well have been lovers; see the summary of their position in Wilkins, p. 23.

105. See O'Toole, pp. 401–3, and Wilkins, pp. 412–13, as well as the *New York Daily Mirror*, November 22, 1933, pp. 3, 8, and 9.

106. Quoted in O'Toole, p. 358.

107. Edward Cary, "Century Necrological Notice," in *Clarence King Memoirs*, p. 236. Also see Brownell, "King at the Century," ibid., p. 223: "At moments assuredly it [his imagination] held him quite enthralled within an almost hypnotic control, and he followed its beckoning with the confident eagerness of ecstasy."

108. King, "Shall Cuba Be Free?," p. 57.

Excursion: Yreka: Just North of Mount Shasta

1. Clarence King, *Mountaineering in the Sierra Nevada* (1872; repr., New York: Penguin, 1989), p. 200.

2. Quoted in Thurman Wilkins, *Clarence King: A Biography*, rev. and enlarged ed. (Albuquerque: University of New Mexico Press, 1988), p. 407.

3. King actually made it just into Oregon on this 1863 trip before turning back southward. The quote is from the Clarence King Papers, Huntington Library, D-21, Notebooks, Scientific: Northern Sierras and Coast Ranges, p. 1.

4. Muir, "Snow-Storm on Mount Shasta," *Harper's New Monthly Magazine* 55 (September 1877), p. 521.

5. This citation of Ogden derives from Francis P. Farquhar's extensive research on Mount Shasta; see the Farquhar Papers at the Bancroft Library, Berkeley, California, carton 4, folder 1.

6. Horatio Hale, *United States Exploring Expedition: Ethnography and Philology* (Philadelphia: Lea and Blanchard, 1846), 6:218; also see 6:199. Geologist James Dwight Dana called the peak "Mount Shasty": "A heavy mist covered the region as we approached it. Gazing up intently for the peak, visible in the earlier part of the day, we barely discovered some lights and shades far above us, which produced, through the indefiniteness of the view, a vision of immensity such as pertains to the vast universe rather than to our own planet." Dana, *United States Exploring Expedition: Geology* (Philadelphia: C. Sherman, 1849), 10:615; also see 10:641–43. Charles Wilkes, though he did not actually see Shasta himself, made use of the notes compiled by the men who did explore the area, and in his narrative called the mountain Shaste Peak: see Wilkes, *Narrative of the United States Exploring Expedition* (Philadelphia: Lea and Blanchard, 1845), 5:240.

7. John Charles Frémont, *Geographical Memoir upon Upper California* (1848; repr., San Francisco: Book Club of California, 1964), p. 22; also see the map enclosed with this edition.

8. Hubert Howe Bancroft, *The Works of Hubert Howe Bancroft*, vol. 23, *History of California, Vol. 6, 1848–1859* (San Francisco: History Co., 1888), p. 494.

Chapter 8: *In the Lena Delta:* Arctic Tragedy and American Imperialism

1. Alexander von Humboldt, *Cosmos: A Sketch of a Physical Description of the Universe*, trans. E. C. Otté (New York: Harper and Brothers, 1859), 5:389–90. Note that Humboldt cites James Dwight Dana, *United States Exploring Expedition: Geology* (Philadelphia: C. Sherman, 1849), 10:615–20, 640, and 643–45.

2. Humboldt, *Cosmos*, 5:403.

3. Humboldt, *Cosmos* (Baltimore: Johns Hopkins University Press, 1997), 2:97. Also note Humboldt's comment that "[e]ven the icy north is cheered for months together by the presence of herbs and large Alpine blossoms covering the earth, and by the aspect of a mild azure sky" (p. 95).

4. Humboldt, *Cosmos*, 5:349.

5. George W. Melville, *In the Lena Delta* (London: Longmans, Green, 1885), pp. 1 and 6.

6. The most thorough treatment of the *Jeannette* expedition is Leonard F. Guttridge, *Icebound: The* Jeannette *Expedition's Quest for the North Pole* (Annapolis: Naval Institute Press, 1986). My references are to this edition, but note that there is a newer, more accessible edition published as a trade paperback by Berkley Books (New York) in December 2001. An earlier history is A. A. Hoehling, *The* Jeannette *Expedition: An Ill-Fated Journey to the Arctic* (New York: Abelard-Schuman, 1967). There is also a fictionalized account of the expedition worth noting, by Commander Edward Ellsberg, *Hell on Ice: The Saga of the* Jeannette (New York: Dodd, Mead, 1938). Ellsberg, a naval engineer, told the story in the first person from Melville's point of view.

7. Letter from J. L. O'Sullivan to Bennett, March 3, 1878, National Archives, Record Group 45, Naval Records Collection, Subject File, U.S. Navy 1775–1910, box no. 373, 1877–1879, OC—Cruises and Voyages (Special), *Jeannette* Expedition.

8. The scene of the *Jeannette's* departure is described in various newspapers; see especially Bennett's paper, the *New York Herald*, July 9, 1879, p. 1 (the blithe headline reads "OFF TO THE POLE"), and the *San Francisco Chronicle* of the same date. The headline in the *Herald* on the following day reads: "THE JEANNETTE'S MISSION: Scientists Deeply Interested in the Hazardous Mission"; in other words, even the purely publicity-minded Bennett depended on the scientific angle as well as the danger angle to draw attention to the expedition.

"Geographical adventure" is from the *New York Commercial Advertiser*, quoted in Guttridge, p. 3; the second quote is from a resolution of the San Francisco Chamber of Commerce, adopted on June 30, 1879, which can be read in full in the National Archives, Record Group 45, Naval Records Collection, Subject File, U.S. Navy 1775–1910, box no. 375, 1879–1884, OC—Cruises and Voyages (Special), *Jeannette* Expedition.

9. A list of the relics can be found in Geographical Society of the Pacific, *An Examination into the Genuineness of the "Jeannette" Relics* (San Francisco: John Partridge, 1896), pp. 10–11. The discoveries made by the native seal hunters and Lytzen's subsequent article are most commonly discussed in histories of the 1893 Arctic expedition led by the Norwegian explorer Fridtjof Nansen.

Alerted to the drift of the *Jeannette*'s relics by a newspaper article—which summarized a Norwegian Academy of Sciences lecture by Professor Henrik Mohn, who had read Lytzen's article— Nansen hit upon the idea of building a ship designed to withstand the crush of the ice pack and then simply allowing it to drift with the ice toward the North Pole. See, for example, Roland Huntford, *Nansen: The Explorer as Hero* (London: Duckworth, 1997), p. 148, and L. H. Neatby, *Discovery in Russian and Siberian Waters* (Athens: Ohio University Press, 1973), p. 147. An interesting nineteenth-century account of Arctic drift is Henry Mellen Prentiss, *The Great Polar Current* (New York: F. A. Stokes, 1897).

10. Even recent histories of polar exploration seem to embrace this interpretation. See Walter A. Wood, "United States Arctic Exploration Through 1939," in *United States Polar Exploration*, ed. Herman R. Friis and Shelby G. Bale Jr. (Athens: Ohio University Press, 1970), p. 12, and John Maxtone-Graham, *Safe Return Doubtful: The Heroic Age of Polar Exploration* (1988; repr., London: Constable, 2000), pp. 77–89 and 107–64. It's also significant that the *Jeannette* expedition definitively disproved the theory of the open polar sea.

11. There are four fascinating firsthand accounts of the *Jeannette* expedition: George Washington De Long, *The Voyage of the* Jeannette: *The Ship and Ice Journals of George W. De Long, Lieutenant-Commander U.S.N., and Commander of the Polar Expedition of 1879–1881,* 2 vols. (Boston: Houghton Mifflin, 1883; heavily edited by De Long's wife, Emma De Long); Melville, *In the Lena Delta*; John W. Danenhower, *Narrative of the "Jeannette"* (Boston: J. R. Osgood, 1882); and Raymond L. Newcomb and Richard W. Bliss, *Our Lost Explorers: The Narrative of the* Jeannette *Arctic Expedition* (Hartford, Conn.: American Publishing, 1882). See also Guttridge, p. 303 (on the De Long Islands). All of these serve as correctives to the inaccurate secondhand account written by H. L. Williams, *History of the Adventurous Voyage and Terrible Shipwreck of the U.S. Steamer "Jeannette"* (New York: A. T. B. De Witt, 1882).

12. The most detailed account of the two inquiries is in Guttridge, pp. 298–326.

13. See Guttridge, pp. 24–41; De Long quoted on p. 147.

14. There were many reasons why both investigative committees decided to mount a cover-up. Primary among them, though, was probably the simple desire to maintain the navy's reputation and support among the American public. See Guttridge, pp. 291–94 and 313–14.

15. De Long receives no mention in even so astute a history of nineteenth-century exploration as William H. Goetzmann, *New Lands, New Men: America and the Second Great Age of Discovery* (1986; repr., Austin: Texas State Historical Association, 1995). Over the years, histories focusing on polar exploration have occasionally devoted a chapter or half-chapter to the *Jeannette* expedition. Aside from the texts already mentioned, see, for example, Jeannette Mirsky, *Northern Conquest: The Story of Arctic Exploration from Earliest Times to the Present* (London: H. Hamilton, 1934), pp. 218–22; Clive Holland, ed., *Farthest North: The Quest for the Pole* (London: Robinson, 1994), pp. 87–106; Clive Johnson, *The Devil's Labyrinth: Encounters with the Arctic* (Shrewsbury, U.K.: Swan Hill Press, 1995), pp. 28–39; and Beau Riffenburgh, *The Myth of the Explorer* (New York: Oxford University Press, 1994), pp. 76–81 and 84–85.

16. See Guttridge, pp. 271–97. Press coverage was particularly strong because of Bennett's involvement. Bennett also dispatched two of his journalists to Siberia once it was reported that the *Jeannette* had gone down and the survivors had landed on the Russian coast. John P. Jackson sent back the first scandalous articles about the disaster, and William Gilder published a popular book, *Ice-Pack and Tundra: An Account of the Search for the* Jeannette *and a Sledge Journey Through Siberia* (New York: Scribner's, 1883).

17. See John W. Danenhower, *The Polar Question: Paper to be Read and Discussed at the Naval Institute, Annapolis, Md., October 9, 1885,* pamphlet published by the United States Naval Institute (August 15, 1885), p. 53; and Guttridge, p. 328.

18. See Hoehling, pp. 209–12.

19. Good summaries of Melville's career can be found in William Ledyard Cathcart, "George Wallace Melville, Engineer-in-Chief of the United States Navy," *Cassier's Magazine* 11 (April 1897), pp. 459–80, and Dumas Malone, ed., *Dictionary of American Biography* (New York: Charles Scribner's Sons, 1933), 12:521–22.

20. Elizabeth Johns, *Thomas Eakins: The Heroism of Modern Life* (Princeton: Princeton University Press, 1983), pp. 3 and 150; see also John Wilmerding, ed., *Thomas Eakins (1844–1916) and the Heart of American Life* (London: National Portrait Gallery, 1993), p. 18, and Martin A. Berger, *Man Made: Thomas Eakins and the Construction of Gilded Age Manhood* (Berkeley: University of California Press, 2000).

21. See Lloyd Goodrich, *Thomas Eakins* (Cambridge, Mass.: Harvard University Press, 1982), 2:137–38.

22. On the general tone of Eakins's portraits of professionals, see David E. Shi, *Facing Facts: Realism in American Thought and Culture, 1850–1920* (New York: Oxford University Press, 1995), p. 146.

23. Goodrich, pp. 208–11.

24. As Nicolai Cikovsky Jr. points out, in his brief essay about the 1905 portrait in Wilmerding, p. 167.

25. "Literature of the Day," *Lippincott's Magazine* 35 (April 1885), p. 421.

26. Quoted in Guttridge, p. 82.

27. Dartmouth College Special Collections, Rauner Library, "Ice Journals" of George Wallace Melville.

28. The original source suggesting that this medal was the Order of St. Stanislaus appears to be Margaret McHenry, *Thomas Eakins Who Painted* (Oreland, Penn.: privately printed, 1945), pp. 117–19. (Art historians who comment on this painting generally echo McHenry: see Cikovsky in Wilmerding, and Goodrich.) I've been able to confirm McHenry's claim by means of letters I found in the Hoover Institution Archives at Stanford University. Melville was received by Czar Alexander III when he passed through St. Petersburg in 1882, but he did not actually get the medal until twenty-two years later, as explained in letters exchanged by Melville and Captain A. G. Boutakoff of the Imperial Russian Navy in the summer of 1904. These letters identify Melville's medal as being the "Imperial Order of St. Stanislas, first class, with Grand Cordon," and Melville thanked the captain for the honor profusely, attesting to his "appreciation of the kindness and hospitality which I received from your officials more than twenty years ago, when I landed in the Lena Delta, after the loss of the Jeannette, and when my men and myself were practically helpless and starving." The Russians' "sincerity and brotherly kindness . . . filled my heart with gratitude and left an indelible impression." See the letters of August 31, 1904, and September 15, 1904, in the Hoover Institution, Stanford University, Russia Posol'stvo (U.S.) Records, 1897–1947, Naval Agent Office File, box 285.

29. All the descriptions I've seen of the 1905 portrait put exclusive emphasis on its romantic heroism. See Cikovsky in Wilmerding, pp. 166–67; Goodrich, pp. 208–11; and McHenry, pp. 117–19.

30. See, for instance, Cikovsky in Wilmerding, p. 167; the real Melville inspired a similar description in Malone, p. 522.

31. Elisha Kent Kane, *Arctic Explorations: The Second Grinnell Expedition in Search of Sir John Franklin, 1853, 1854, 1855* (Philadelphia: Childs and Peterson, 1856), p. 228. Also note that Kane served as the senior medical officer aboard the First Grinnell Expedition in search of Franklin (1850–51), commanded by Edwin J. De Haven, who filed the expedition's official report but asked Kane to write up a history of their efforts for popular consumption: Kane, *The United States Grinnell Expedition in Search of Sir John Franklin: A Personal Narrative* (New York: Harper and Brothers, 1854). On Kane, also see Larzer Ziff, "Arctic Exploration and the Romance of Failure," *Raritan* 23 (Fall 2003), pp. 58–79; William Elder, *Biography of Elisha Kent Kane* (Philadelphia: Childs and Peterson, 1858); Jeannette Mirsky, *Elisha Kent Kane and the Seafaring Frontier* (Boston: Little, Brown, 1954); and David Alexander Chapin, "Exploring Other Worlds: Margaret Fox, Elisha Kane, and the Antebellum Culture of Curiosity" (Ph.D. diss., University of New Hampshire, 2000). Sales figures from Richard Vaughan, *The Arctic: A History* (Phoenix Mill: Alan Sutton Publishing, 1994), p. 174.

32. Basic overviews of American Arctic exploration for this period can be found in Goetzmann, pp. 359–62 and 423–44 (despite the fact that he doesn't cover the *Jeannette*); Vaughan,

pp. 168–92; Charles Officer and Jake Page, *A Fabulous Kingdom: The Exploration of the Arctic* (New York: Oxford University Press, 2001), pp. 92–130; Jean Malaurie, *Ultima Thule: Explorers and Natives in the Polar North* (New York: W. W. Norton, 2003), pp. 74–127; and Michael Frederick Robinson, "The Coldest Crucible: Arctic Exploration and American Culture, 1850–1910" (Ph.D. diss., University of Wisconsin—Madison, 2002). Also see W. Gillies Ross, "Nineteenth-Century Exploration of the Arctic," in *North American Exploration*, vol. 3, *A Continent Comprehended*, ed. John Logan Allen (Lincoln: University of Nebraska Press, 1997), pp. 244–331, which covers both American and European Arctic expeditions. Robinson's excellent dissertation is scheduled to be published (with the same title) in 2006 by the University of Chicago Press.

33. William Healey Dall, *Alaska and Its Resources* (Boston: Lee and Shepard, 1870). Also see Goetzmann, pp. 423–33, and Vaughan, pp. 178–84. On Dall's work for the Coast and Geodetic Survey, see J. E. Nourse, *American Explorations in the Ice Zones* (Boston: D. Lothrop, 1884), pp. 367–69.

34. The purchase of Alaska was generally known as "Seward's Folly" and the territory itself as "Seward's Icebox."

35. Quoted in Riffenburgh, p. 131.

36. On Maury, see Goetzmann, pp. 315–30 and 359–62, and Chester G. Hearn, *Tracks in the Sea: Matthew Fontaine Maury and the Mapping of the Oceans* (Camden, Maine: International Marine/McGraw Hill, 2002).

37. On the search for Franklin, see Pierre Berton, *The Arctic Grail: The Quest for the North West Passage and the North Pole, 1818–1909* (Toronto: McClelland and Stewart, 1988), pp. 150–269; Vaughan, pp. 154–60 and 168–78; Officer and Page, pp. 83–91; and Mirsky, *Northern Conquest*, pp. 153–79. The final American search for Franklin relics, a highly successful expedition using Inuit travel techniques, was conducted by Lieutenant Frederick Schwatka between 1878 and 1880.

38. See, for instance, Edmund Blair Bolles, *Ice Finders: How a Poet, a Professor, and a Politician Discovered the Ice Age* (Washington, D.C.: Counterpoint, 1999).

39. On Franklin and his baggage, see, Richard King, *The Franklin Expedition from First to Last* (London: J. Churchill, 1855); Mirsky, *Northern Conquest*, pp. 145–52; and Annie Dillard, "An Expedition to the Pole," in *Teaching a Stone to Talk: Expeditions and Encounters* (New York: Harper and Row, 1982), pp. 17–52.

40. On Hayes, see Malaurie, pp. 94–105; Berton, pp. 256–57, 274–75, 280–86, and 353–64; and Wendell H. Oswalt, *Eskimos and Explorers* (1979; repr., Lincoln: University of Nebraska Press, 1999), pp. 112–20. Hayes wrote a book based on experiences he had on Kane's expedition: *An Arctic Boat Journey in the Autumn of 1854* (Boston: Brown, Taggard, and Chase, 1860); and then, after the Civil War, he penned *The Open Polar Sea: A Narrative of a Voyage of Discovery Towards the North Pole in the Schooner* United States (New York: Hurd and Houghton, 1867). A recent novel, *The Voyage of the* Narwhal, by Andrea Barrett (New York: W. W. Norton, 1998), is based largely on the expeditions of Kane and Hayes.

41. Thomas Hickey quoted in Charles Francis Hall, *Arctic Researches and Life among the Esquimaux: Being the Narrative of an Expedition in Search of Sir John Franklin in 1860, 1861, and 1862* (New York: Harper and Brothers, 1865), pp. xxi–xxii.

42. See Shari M. Huhndorf, *Going Native: Indians in the American Cultural Imagination* (Ithaca, N.Y.: Cornell University Press, 2001), esp. pp. 79–128.

43. On Hall's relationship to native cultures, see Malaurie, pp. 111–27; Vaughan, pp. 174–77; and Goetzmann, pp. 433–41.

44. Hall, pp. 131–32. Hall also frequently makes such comments as "The honesty of this people is remarkable" (p. 104) and has section titles like "Native Sagacity in studying Natural History" and "Inuit Ingenuity" (see p. 566).

45. Quoted in Riffenburgh, p. 40.

46. The classic biography of Hall, whose author received permission to exhume Hall's remains and test them for arsenic, is Chauncey C. Loomis, *Weird and Tragic Shores: The Story of Charles Francis Hall, Explorer* (New York: Knopf, 1971). Two recent studies of the *Polaris* expedition are Richard Parry, *Trial by Ice: The True Story of Murder and Survival on the 1871 Polaris*

Expedition (New York: Ballantine Books, 2001), and Bruce Henderson, *Fatal North: Adventure and Survival Aboard USS* Polaris, *the First U.S. Expedition to the North Pole* (New York: New American Library, 2001).

47. On the evolution of Arctic exploration, see especially Riffenburgh and Robinson, passim.

48. See Danenhower, pp. 84, 88–91, and 93; on Jackson, see Guttridge, pp. 271, 273, and 345.

49. Quoted in Guttridge, p. 326.

50. Melville, *In the Lena Delta*, pp. 7 and 14.

51. Quoted in Guttridge, pp. 149 and 114. Note that the administering of lime juice as an anti-scorbutic was a brand-new practice.

52. Quoted in Holland, p. 91.

53. Homosocial, and possibly homosexual, bonding was quite common on long voyages of exploration, though when such expeditions got caught in the ice tempers tended to flare and relationships became strained.

54. Melville, *In the Lena Delta*, p. 13.

55. All these quotes are from the ship's official log, in the National Archives, Record Group 45, Naval Records Collection, Microfilm Publication T297 ("Log of the Jeannette"), entries for Feb. 16/17, 1881; May 17/18, 1881; and June 7, 1881. Note that Melville most likely kept a log as well, but only his Lena Delta journals have come to light.

The most detailed rendering of the expedition's day-to-day travails is Guttridge, drawing on all the different firsthand accounts. I've generally followed his narrative, which squares with De Long's log, for the basic facts.

56. Melville, *In the Lena Delta*, p. 18.

57. The findings of the committee were published as the *Proceedings of an Investigation into the* Jeannette *by the Naval Affairs Subcommittee, 48th Congress, 1st Session, House Miscellaneous Document 66* (Washington, D.C.: Government Printing Office, 1885). The navy was clearer about exonerating all the survivors: See U.S. Navy, Courts of Inquiry, Jeannette (ship) 1882, *Proceedings of a Court of Inquiry Convened at the Navy Department . . . to Investigate the Circumstances of the Loss . . . of the Exploring Steamer "Jeannette"* (Washington, D.C.: Government Printing Office, 1883), pp. 260–66.

58. In many parts of the book, Melville actually alerts the reader to the fact that he is simply copying from his journals or letters, but even in other places the texts are markedly similar. Compare, for instance, his letter of January 6, 1882, Yakutz, Siberia, to the secretary of the navy, in the Letter Book of Dartmouth's "Ice Journals" collection, with *In the Lena Delta*, pp. 77–125, on the subject of Melville's arrival in the delta and first attempt to search for De Long. Direct journal transcriptions can be found on pp. 297–366 (Feb. 26–April 24, 1882). Also note that Melville did file an official report: "Report of Chief Engineer George Wallace Melville in Connection with the Jeannette Expedition (Oct. 17, 1882)," included in the *Report of the Secretary of the Navy*, 47th Cong., 2nd sess., 1882–3, H. Ex. Doc. 1, pt. 3, pp. 57–70; also see pp. 15–19 for the secretary's overview, and pp. 71–89 for additional related documents, including an official report filed by Danenhower.

59. Quoted in Guttridge, p. 330.

60. Melville, "Ice Journals," entry for November 23, 1881.

61. Ibid., entries for November 26, 23, and 18, 1881. Also see *In the Lena Delta*, p. 98: "My own, and the limbs of Leach, Manson, and Lauterbach, were so badly frozen that we were forced to crawl on our hands and knees."

62. Melville, *In the Lena Delta*, p. 147.

63. Melville, "Ice Journals," entry for February 27, 1882.

64. National Archives, Record Group 45, Naval Records Collection, "Letters from Officers Commanding Expeditions, Jan. 1881–Dec. 1885," vol. 24 of 27, Letters and Telegrams Concerning the Loss of the U.S.S. Jeannette and Rodgers, 1881–82–83, Document 61, letter of March 24, 1882.

65. Melville, *In the Lena Delta*, pp. 179 and 159.

66. Melville, "Ice Journals," Letter Book, letter of Jan. 5, 1882 (also see letter of Jan. 2).

67. Melville, *In the Lena Delta*, pp. 128, 267, 406, 185, 318, 308–9, and 327.

68. Melville, "Ice Journals," entries for July 14–25, 1882, and December 8, 1881, and *In the Lena Delta*, p. 250.

69. Melville, *In the Lena Delta*, pp. 250, 19, 262, 265, and 344.

70. For instance, Melville makes direct reference to Ruskin in Melville, "The Military Duty of the Engineering Institutions," *Forum* 32 (January 1902), p. 521.

71. Melville, *In the Lena Delta*, p. 147.

72. Ibid., pp. 471–83 and v; see also Melville, "Remarks on Polar Expedition," *Proceedings of the American Philosophical Society* 36 (October 29, 1897), pp. 454–61.

73. McHenry, p. 117.

74. Melville, "The Engineer's Duty as a Citizen," *Journal of the American Society of Mechanical Engineers* 32 (July–December 1910), pp. 1128 and 1132.

75. See Burton J. Bledstein, *The Culture of Professionalism: The Middle Class and the Development of Higher Education in America* (New York: W. W. Norton, 1976); Edwin T. Layton, *The Revolt of the Engineers: Social Responsibility and the American Engineering Profession* (Baltimore: Johns Hopkins University Press, 1986), pp. 1–133; and Terry S. Reynolds, ed., *The Engineer in America* (Chicago: University of Chicago Press, 1991), pp. 1–41.

76. Melville, *In the Lena Delta*, pp. 385 and 403.

77. Melville, "The Engineer's Duty as a Citizen," p. 1131.

78. Stegner's quote is from "Crossing into Eden," in *Where the Bluebird Sings to the Lemonade Springs: Living and Writing in the West* (New York: Penguin, 1992), p. 41.

79. As T. J. Jackson Lears has argued, men of their ilk "longed to rekindle possibilities for authentic experience, physical or spiritual—possibilities they felt had existed once before, long ago." See Lears, *No Place of Grace: Antimodernism and the Transformation of American Culture, 1880–1920* (New York: Pantheon, 1981), p. 57.

80. And with Washington Irving, Daniel Webster, Albert Gallatin, the historian William Prescott, and the Humboldtian Whig (and opponent of the Mexican War) D. D. Barnard. See the *Columbia University Alumni Register, 1754–1931* (New York: Columbia University Press, 1932), pp. 594, 723, and 1182–83.

81. Cathcart, p. 478.

82. Michael Fried, *Reading, Writing, Disfiguration: On Thomas Eakins and Stephen Crane* (Chicago: University of Chicago Press, 1987), pp. 46–52.

83. See *The Champion Single Sculls (Max Schmitt in a Single Scull)* (1871), in Gordon Hendricks, *The Life and Work of Thomas Eakins* (New York: Grossman, 1974), p. 172, plate 10.

84. Melville, *In the Lena Delta*, p. 471.

85. See Guttridge, p. 328.

86. Melville, preface to *A Tenderfoot with Peary*, by George Borup (New York: Frederick A. Stokes, 1911), pp. viii–ix.

87. Cathcart, p. 480.

88. Quoted ibid., p. 478.

89. Ibid., p. 465.

90. Ibid. "We have laid others of our shipmates away in many a land," Melville said in 1896. "Aye, and buried them, too, beneath the lap and roll of every crested sea, from the equator, with its festering fever, to the poles, where the snow-god and the ice king hold everlasting sway."

91. Melville, "The Engineer's Duty as a Citizen," p. 1129.

92. On this whole episode, see Guttridge, pp. 274, 281–82, and 295–97.

93. Cathcart, p. 475.

94. Malone, pp. 521–22.

95. Guttridge, p. 325.

96. Quoted in Cathcart, p. 467. Another of Melville's commanders had confirmed "the high reputation this gentleman has throughout the service for professional skill, executive ability, energy, and zeal. In all these qualities, as well as in those other essential ones that go to make up true manhood, it is no disparagement to his fellows to say that I believe he has not his superior in his corps" (ibid., p. 466).

97. Melville, "The Engineer's Duty as a Citizen," p. 1129.

98. On Greely, see Leonard F. Guttridge, *Ghosts of Cape Sabine: The Harrowing True Story of the Greely Expedition* (New York: G. P. Putnam's Sons, 2000); Alden Todd, *Abandoned: The Story of the Greely Arctic Expedition, 1881–1884* (1961; repr., Fairbanks: University of Alaska Press, 2001); Maxtone-Graham, pp. 89–106; and Mirsky, *Northern Conquest*, pp. 206–17. Greely's own account of his adventures is *Three Years of Arctic Service*, 2 vols. (New York: Charles Scribner's Sons, 1886).

99. Melville, *In the Lena Delta*, p. 415.

100. See Robert L. Beisner, *Twelve against Empire: The Anti-Imperialists, 1898–1900* (1968; repr., Chicago: Imprint Publications, 1992), and Thomas Schoonover, *Uncle Sam's War of 1898 and the Origins of Globalization* (Lexington: University Press of Kentucky, 2003).

101. Melville, *Views of Commodore George W. Melville, Chief Engineer of the Navy, as to the Strategic and Commercial Value of the Nicaraguan Canal, the Future Control of the Pacific Ocean, the Strategic Value of Hawaii, and its Annexation to the United States* (Washington, D.C.: Government Printing Office, 1898), passim.

102. Ibid., pp. 7, 30, 28, and 4.

103. Melville, *Abandon the Philippines* (Boston, 1905). Unfortunately, I have not been able to locate this pamphlet. I know it exists because it is listed in the card catalog of Sterling Memorial Library, Yale University, but it is missing from the stacks. I've done an exhaustive electronic search and physical searches in several East Coast libraries but still haven't found a copy. Fortunately, Melville referred to his stance against U.S. control of the Philippines in another 1905 publication, "The Important Elements in Naval Conflicts," in "The United States as a World Power," special annual meeting number, *Annals of the American Academy of Political and Social Science* 26 (July 1905), pp. 123–36, esp. 130–32. Here, again, Melville's position comes across as strategic and defensive rather than radical and idealistic, but it is still clearly anti-imperialist.

104. The story of St. Stanislaus (also called St. Stanislaw) is from Saint Stanislaw Publications Committee, *Saint Stanislaw, Bishop of Krakow: In Commemoration of the 900th Anniversary of His Martyrdom in 1079* (Santa Barbara, Calif.: Saint Stanislaw Publications Committee, 1979), pp. 15–36, and from *The New Encyclopaedia Britannica, Volume 11, Micropaedia*, 15th ed. (Chicago: Encyclopaedia Britannica, 1998), p. 211.

Chapter 9: *The Cruise of the* Corwin: Nature, Natives, Nation

1. John Muir, "Living Glaciers of California," *Harper's New Monthly Magazine* 51 (November 1875), p. 770.

2. The only bits of evidence I've seen that suggest they even met are in two letters written by John Muir to Jeanne C. Carr, both sent from Yosemite, on August 13 and September 8, 1871, and reprinted in Bonnie Johanna Gisel, ed., *Kindred and Related Spirits: The Letters of John Muir and Jeanne C. Carr* (Salt Lake City: University of Utah Press, 2001), pp. 146–47.

3. Clarence King, *Systematic Geology*, vol. 1, *Report of the U.S. Geological Exploration of the Fortieth Parallel* (Washington, D.C.: Government Printing Office, 1878), pp. 477–48. An excellent explanation of the disagreement between Muir and King over the geological history of Yosemite can be found in Keith Burich, "Josiah Dwight Whitney, John Muir and Clarence King, and the 'Chasm of the Yosemite,'" in *John Muir in Historical Perspective*, ed. Sally M. Miller (New York: Peter Lang, 1999), pp. 165–84. Burich finds that, on balance, King's perspective is closest to the one accepted by geologists today. Also see Dennis R. Dean, "Muir and Geology," in *John Muir: Life and Work*, ed. Sally M. Miller (Albuquerque: University of New Mexico Press, 1993), pp. 168–93.

4. Muir, "Exploring in the Great Tuolumne Canyon," *Overland Monthly* 11 (August 1871), pp. 140–41.

5. Muir, "Living Glaciers of California," pp. 772 and 775–76; also see Muir, "Snow-Storm on Mount Shasta," *Harper's New Monthly Magazine* 55 (September 1877), p. 522.

6. Muir, "Living Glaciers of California," pp. 770–72.

7. Muir, "Geological and Botanical Characteristics of Shasta," paper written May 6, 1875, San Francisco, California, and sent to A. F. Rodgers, Assistant, U.S. Coast Survey; copy consulted on microfilm in the Bancroft Library, Berkeley, Calif.

8. All of the quotes in this section are from Muir, "Snow-Storm on Mount Shasta," pp. 521–30.

9. See Frederick Turner, *Rediscovering America: John Muir in His Time and Ours* (New York: Viking, 1985), p. 229. This is one of the more important of the many studies of Muir, along with Michael P. Cohen, *The Pathless Way: John Muir and American Wilderness* (Madison: University of Wisconsin Press, 1984). Also note that Donald Worster is currently working on a new biography of Muir; see his article, "John Muir and the Modern Passion for Nature," *Environmental History* 10 (January 2005), pp. 8–19. This article promises a trenchant reevaluation of Muir; I completely agree with Worster when he says that Muir's environmental radicalism became "more conservative and more compromised" in his later years (p. 13).

10. Muir, *Our National Parks* (1901; repr., Boston: Houghton Mifflin, 1916), p. 63.

11. As much as I respect the terms of Michael Smith's argument on this subject, my conclusions are almost exactly opposite to his. See Michael Smith, *Pacific Visions: California Scientists and the Environment, 1850–1915* (New Haven: Yale University Press, 1987), pp. 71–103.

12. Muir, *Letters from Alaska*, ed. Robert Engberg and Bruce Merrell (Madison: University of Wisconsin Press, 1993), pp. 5, 14–15, 23, and 27. On Indian summer, see Harriette Simpson Arnow, *Seedtime on the Cumberland* (New York: Macmillan, 1960), pp. 138–39.

13. The difference between Muir's final version of *Travels in Alaska* (Boston: Houghton Mifflin, 1915), probably his most thoroughly revised text, and his journals and letters from 1879 and 1880, is striking. A simple comparison can be made by consulting *Travels in Alaska; Letters from Alaska*; Muir, *John of the Mountains: The Unpublished Journals of John Muir*, ed. Linnie Marsh Wolfe (Boston: Houghton Mifflin, 1938), pp. 245–80; and William Frederic Badè, *The Life and Letters of John Muir* (Boston: Houghton Mifflin, 1924), 2:123–61, which reprints several of the personal letters he wrote during the two trips.

14. On this point, see especially Ronald H. Limbaugh, *John Muir's "Stickeen" and the Lessons of Nature* (Fairbanks: University of Alaska Press, 1996), p. xii.

15. Muir, *Travels in Alaska*, p. 96.

16. Robert Bruce Campbell, in his Ph.D. dissertation, "Inside Passage: Alaskan Travel, American Culture, and the Nature of Empire, 1867–1898" (Yale University, 2003), argues that Muir's writings about Alaska, especially those connected to his touristic trips of 1879 and 1880, were consistent with the general Alaska travel literature of the day, which portrayed this relatively new territory as a beautiful, pristine landscape to be enjoyed and exploited by worthy pioneers (of both travel and development) from the States. See esp. pp. 103–27 and 324–80.

17. Strangely, it is even omitted from the list of "Books by John Muir" compiled by Stephen Fox, in *John Muir and His Legacy: The American Conservation Movement* (Boston: Little, Brown, 1981), p. 379. Fox includes all the other posthumous publications, so he seems simply to have overlooked this one. Almost none of Muir's biographers or critics, even if they mention this text, make any attempt to analyze it or suggest its significance to Muir's career.

18. It is unclear why Muir ignored the 1881 trip when he was working on the final manuscript for *Travels in Alaska* during the last two years of his life. It seems possible that he had already planned to do a separate volume devoted solely to the 1881 voyage, so Badè's work on *The Cruise of the Corwin* may well have fulfilled Muir's own desires. Thanks to Badè's introduction and editorial notes, it is possible in many places, though not all, to tell whether the text of *The Cruise of the Corwin* has been drawn from Muir's journal or from a letter to the *Evening Bulletin*. Where significant, I've cross-checked passages and specified the source (journal or newspaper) in the notes, but the vast majority of the text is drawn from Muir's letters to the newspaper. The twenty-one letters appeared as articles on the following dates (all in 1881): June 20 (p. 1); July 13 (five letters, all on p. 1); July 25 (p. 1); August 15 (three letters, all on p. 3); August 16 (three letters, all on p. 1); September 28 (p. 4); September 29 (p. 1); October 22 (p. 4); October 24 (p. 3); October 25 (p. 1); October 26 (p. 1); October 27 (p. 3); and October 31 (p. 1). The complete journals can be read in the John Muir Papers, microfilm reel 26 ("Journals and Sketchbooks, 1879–1881") and reel 27 ("Journals and Sketchbooks, 1881–1890"); the best sketches are on reel 26, but there are very useful transcriptions of all the journals on reel 27.

19. Muir, *The Cruise of the Corwin* (1917; repr., San Francisco: Sierra Club Books, 1993), p. 62.

I'll be using this edition, which comes with a foreword by Roderick Nash as well as the original introduction by Badè. This letter, of June 20, was published in the *Evening Bulletin* on August 16, p. 1.

20. Yosemite was still just a state park, of course, the national park system not yet having been inaugurated. For an account of wilderness preservation in this era, both in California and on a national scale, see Roderick Nash, *Wilderness and the American Mind*, 4th ed. (1967; repr., New Haven: Yale University Press, 2001), pp. 96–121.

21. The quote is from a letter Muir wrote to his wife from Siberia in 1881, noted in Badè's introduction to *Cruise of the* Corwin, p. xi (letter to Louie Strentzel Muir, June 14, 1881, Plover Bay, Siberia, Muir Papers, microfilm reel 4, Correspondence and Related Documents, 1880–1883). Note also that some of Muir's letters to his wife from aboard the *Corwin* are reprinted in Badè, *Life and Letters of John Muir*, 2:163–91.

22. Muir, *Cruise of the* Corwin, p. 156.

23. Ibid., p. 86. (*Evening Bulletin*, August 15, p. 3.)

24. Useful biographical accounts of this portion of Muir's life can be found in Thurman Wilkins, *John Muir: Apostle of Nature* (Norman: University of Oklahoma Press, 1995), pp. 134–53; Fox, pp. 64–71; Turner, pp. 252–63; Linnie Marsh Wolfe, *Son of the Wilderness: The Life of John Muir* (New York: Knopf, 1945), pp. 188–227; Badè, *Life and Letters of John Muir*, 2:100–192; and the foreword (by Roderick Nash) and introduction (by Badè) to *Cruise of the* Corwin, pp. vii–xxix.

25. Letter to Louie Strentzel Muir, May 4, 1881, San Francisco, Muir Papers, microfilm reel 4, Correspondence and Related Documents, 1880–1883. The letter goes on at length, in a rationalizing sort of way, about Muir's health issues: "I have been losing all appetite during the last few months and am now as fleshless as I ever was. All my friends look in my face as if wondering at its whiteness and ask if I am well. So this trip is a necessity for health—if no more."

26. It is generally not recognized that Muir contributed substantially to the nascent tradition of American ethnographic writing. The only scholar who has made any attempt to examine this aspect of Muir's career, to my knowledge, is Richard F. Fleck, in *Henry Thoreau and John Muir among the Indians* (Hamden, Conn.: Archon Books, 1985).

27. Quoted in Cohen, p. 189.

28. Muir, *Our National Parks*, pp. 3–4.

29. Ibid., p. 32.

30. Quoted in Robert L. Dorman, *A Word for Nature: Four Pioneering Environmental Advocates, 1845–1913* (Chapel Hill: University of North Carolina Press, 1998), p. 147. Dorman's chapter on Muir (pp. 103–71) does a good job of both placing Muir in his historical context and explaining the consensus scholarly opinion of Muir's relevance to modern environmentalism.

31. Muir, *The Mountains of California* (1894; repr., New York: Penguin, 1985), p. 42.

32. Quoted in Dorman, p. 155.

33. Roderick Nash titled his chapter on Muir (pp. 122–40), "John Muir: Publicizer."

34. Muir, *Mountains*, pp. 45–48 and 40. In 1888–89, Muir edited and coauthored part of a work (published in stages) called *Picturesque California* (San Francisco: J. Dewing, 1888–91), which was certainly a piece of propaganda, though not for wilderness per se.

This passage from *Mountains* much later became the inspiration behind the title and a key scene in James Dickey's novel *Deliverance* (Boston: Houghton Mifflin, 1970), as well as the bedrock for a poem by Gary Snyder (section 8 of "Burning," in *Myths and Texts* [Totem Press, 1960], p. 39).

35. Muir, *My First Summer in the Sierra* (Boston: Houghton Mifflin, 1911), p. 16.

36. See William James, *The Varieties of Religious Experience* (1902; repr., New York: Penguin, 1982). On the link between Muir and James, see Don Weiss, "John Muir and the Wilderness Ideal," in Miller, ed., *John Muir: Life and Work*, pp. 118–34, esp. 121–22.

37. Muir quoted in Turner, p. 222. For a deeper analysis of Muir's relationship to Ruskin, see Terry Gifford, "Muir's Ruskin: John Muir's Reservations about Ruskin Reviewed," in Miller, ed., *John Muir in Historical Perspective*, pp. 137–50.

38. See Dorman, pp. 119–22, and James D. Heffernan, "Why Wilderness? John Muir's 'Deep Ecology,'" in Miller, ed., *John Muir: Life and Work*, pp. 102–16. Also see Arne Naess, "The Shallow

and the Deep, Long-Range Ecology Movement," *Inquiry* 16 (Spring 1973), pp. 95–100, and Bill Devall and George Sessions, *Deep Ecology: Living as if Nature Mattered* (Layton, Utah: Peregrine Smith Books, 1985). The Deep Ecologists argue for what they call "biospherical egalitarianism" (Naess quoted in Heffernan, p. 104), which is in some ways a Humboldtian concept but which also denies one of Humboldt's key ideas about the necessity of utilizing certain resources and, in turn, ignores Humboldt's emphasis on issues of social justice.

39. Muir, *A Thousand-Mile Walk to the Gulf* (1916; repr., San Francisco: Sierra Club Books, 1991), p. 56. This text, like *The Cruise of the* Corwin, was published posthumously but was actually written fairly early in Muir's life, long before he embraced his role in politics.

40. Muir, *Our National Parks*, pp. 57–58 and 206–8; Muir, *My First Summer in the Sierra*, pp. 34–35; Muir, *Thousand-Mile Walk*, pp. 78–79.

41. Muir, *Thousand-Mile Walk*, pp. 78–79; and Muir quoted in Dorman, p. 116. Muir also wrote: "I have never yet happened upon a trace of evidence that seemed to show that any one animal was ever made for another as much as it was made for itself" (*Steep Trails* [Boston: Houghton Mifflin, 1918], p. 12).

42. Muir, *Steep Trails*, pp. 11–12.

43. Muir, *Cruise of the* Corwin, pp. 204 and 206.

44. See Wilkins, pp. 149 and 159, and Badè, *Life and Letters of John Muir*, 2:191. Muir met Gray when the botanist visited Yosemite in 1872; the curmudgeonly mountain man did not always take kindly to academics, but in this case the meeting went delightfully, and Muir found Gray to be "a great, progressive, unlimited man like Darwin." See ibid., 1:335 and 337.

45. Muir, *Cruise of the* Corwin, pp. 207 and 213. Muir describes his botanizing throughout the text; for some of the lists of plants, see his "Botanical Notes," appended to *Cruise of the* Corwin, pp. 205–19. This document was originally published as part of the two governmental reports on the *Corwin*'s expedition, as was Muir's article on "The Glaciation of the Arctic and Sub-Arctic Regions Visited During the Cruise" (also appended to *Cruise of the* Corwin, pp. 185–205). See *Treasury Department Document No. 429* (Washington, D.C.: Government Printing Office, 1883) for the botany and *Senate Executive Document No. 204* (Washington, D.C.: Government Printing Office, 1884) for the glaciology. Note that the 1883 document is also catalogued as Irving C. Rosse, John Muir, and E. W. Nelson, *The Cruise of the Revenue-Steamer Corwin in Alaska and the N.W. Arctic Ocean in 1881; Notes and Memoranda—Medical and Anthropological, Botanical, Ornithological* (Washington, D.C.: Government Printing Office, 1883).

46. About half of the books John Muir owned are housed in the John Muir Papers and Related Collections, Holt-Atherton Special Collections, University of the Pacific, Stockton, Calif. I looked through Muir's Humboldt stash on October 16, 2001. There is a complete five-volume set of *Cosmos*, published in London in the 1870s (the Otté translation); a London edition of *Views of Nature* translated by Otté and H. G. Bohn (1896); and the three-volume Thomasina Ross translation of the *Personal Narrative*, published in London in 1907–8. All were published by George Bell and Sons except for volumes 1 and 3 of *Cosmos*, which were published by Bell and Daldy. The emphasis on Indians is especially noticeable in the *Personal Narrative*.

47. Muir, in Wolfe, *John of the Mountains*, p. 261.

48. Muir, *Cruise of the* Corwin, pp. 134–35 and 51.

49. Ibid., pp. 51, 155–56, and 103.

50. Muir, in Wolfe, *John of the Mountains*, pp. 273–74.

51. Muir, *Cruise of the* Corwin, pp. 50, 14, and 16. These kinds of observations appear in both the journals and the newspaper articles; see, for instance, the letters published in the *Evening Bulletin* on July 13, p. 1, especially "At St. Paul."

On the seal and sea otter trade, see Donald Craig Mitchell, *Sold American: The Story of Alaska Natives and Their Land, 1867–1959* (Hanover, N.H.: Dartmouth College/University Press of New England, 1997), pp. 30–34, and Ted C. Hinckley, *The Canoe Rocks: Alaska's Tlingit and the Euramerican Frontier, 1800–1912* (Lanham, Md.: University Press of America, 1996), pp. 28–38.

52. Muir, *Cruise of the* Corwin, pp. 96, 101, 53, and 88. The last quotes come from the *Evening*

Bulletin of August 15, p. 3. Muir's original journal entry said of the Eskimos that they were *"in every way infinitely* better behaved than white men." Muir Papers, microfilm reel 27, "Journals and Sketchbooks, 1881–1890," entry for June 7, emphasis in original.

53. See Brian W. Dippie, *The Vanishing American: White Attitudes and U.S. Indian Policy* (Middletown, Conn.: Wesleyan University Press, 1982).

54. Muir, *Cruise of the* Corwin, pp. 112–13, 127–28, 123, 162, and 20.

55. On Rousseau and the cult of the noble savage, see Robert F. Berkhofer Jr., *The White Man's Indian: Images of the American Indian from Columbus to the Present* (New York: Vintage Books, 1979), pp. 72–80.

56. See Campbell, and compare, for instance, the perspective expressed in George Wardman, *A Trip to Alaska: A Narrative of What Was Seen and Heard during a Summer Cruise in Alaskan Waters* (San Francisco: Samuel Carson, 1884).

57. Muir, *Cruise of the* Corwin, pp. 171–72. This passage comes from the *Evening Bulletin* of October 27, p. 3.

58. Ibid., pp. 35, 126, 73, and 56.

59. Ibid., p. 105. This passage comes from the *Evening Bulletin* of October 26, p. 1.

60. Muir, *Cruise of the* Corwin, pp. 54, 37, and 19. These passages all come from the *Evening Bulletin* of July 13, p. 1.

61. A copy of the message under the cairn can be examined in the National Archives, Record Group 45, Naval Records Collection, "Letters from Officers Commanding Expeditions, Jan. 1881–Dec. 1885," vol. 24 of 27, Letters and Telegrams Concerning the Loss of the U.S.S. Jeannette and Rodgers, 1881–82–83, Document 8. It reads as follows: "U.S. Str. Corwin. Wrangel's Land, Aug. 12, 1881. The U.S. Str. Corwin, Captain C.L. Hooper commanding, visited this Land in search of tidings from the U.S. Exploring Steamer Jeannette. A cask of provisions will be found on the second cliff to the northward. All well on board." For Muir's comments on the landing and some extended musings on De Long and the *Jeannette* expedition, see *Cruise of the* Corwin, pp. 133–43.

62. Ibid., pp. 93 and 62.

63. Letter to Louie Strentzel Muir, Monday, May 16, 1881, 10A.M., aboard the *Corwin*, two miles from shore, Muir Papers, microfilm reel 4, Correspondence and Related Documents, 1880–1883.

64. Jonathan Auerbach, *Male Call: Becoming Jack London* (Durham, N.C.: Duke University Press, 1996), p. 48.

65. Muir, *Cruise of the* Corwin, p. 92.

66. See Shepard Krech III, *The Ecological Indian: Myth and History* (New York: W. W. Norton, 1999).

67. William Cronon, in *Nature's Metropolis: Chicago and the Great West* (New York: W. W. Norton, 1991), has argued that in the mid-to-late-nineteenth century, market forces both forged intimate connections between city and country and simultaneously concealed these connections, so that most Americans no longer knew where their food—or their clothing, or their furniture—came from.

68. Muir, *Cruise of the* Corwin, pp. 168–69, 102–3, 32, 25, and 20.

69. Ibid., pp. 51 and 30.

70. See Mitchell, pp. 132–36, and John R. Bockstoce, *Whales, Ice, and Men: The History of Whaling in the Western Arctic* (Seattle: University of Washington Press, 1986), esp. pp. 143–230. For a cultural analysis of portrayals of the Alaskan environment, see Susan Kollin, *Nature's State: Imagining Alaska as the Last Frontier* (Chapel Hill: University of North Carolina Press, 2001).

71. Muir, *Cruise of the* Corwin, pp. 150, 146, 35, and 46–47. Muir's comments about whaling appeared in the *Evening Bulletin* on October 24: "Perils of Whaling," p. 3.

72. Muir, *Cruise of the* Corwin, pp. 18, 41, 43–44, 63, 95, and 80.

73. Ibid., pp. 55 and 26. These actual phrases come from the journals, but they refer to scenes described in the letters; see the *Evening Bulletin* of July 13, p. 1.

74. Boas's book, *The Central Eskimo* (1888; repr., Lincoln: University of Nebraska Press,

1964), is generally considered to be the first major scientific monograph on the Eskimos. See also Ludger Müller-Wille, ed., *Franz Boas among the Inuit of Baffin Island, 1883–1884: Journals and Letters*, trans. William Barr (Toronto: University of Toronto Press, 1998).

75. Boas, "A Year among the Eskimo," in *The Shaping of American Anthropology, 1883–1911: A Franz Boas Reader*, ed. George W. Stocking Jr. (New York: Basic Books, 1974), p. 55; and Boas quoted by Stocking, p. 23. On Boas's career in general and his debt to Humboldt, see Claudia Roth Pierpont, "The Measure of America: The Anthropologist Who Fought Racism," *New Yorker* 80 (March 8, 2004), pp. 48–63, esp. pp. 50–51 on Humboldt. Also see Boas, "The Study of Geography," *Science* 9 (February 11, 1887), pp. 137–41.

76. See, for instance, Muir, *Cruise of the Corwin*, p. 88.

77. Quoted in Turner, p. 232.

78. See Cohen, p. 189.

79. John Muir Papers and Related Collections, University of the Pacific. Muir noted passages about Indians in this volume of *Personal Narrative* (London: George Bell and Sons, 1907) on the following pages: 27, 37, 56, 152, 161, 202–6, 213, 219, 228, 232, 241–48, 266, 296, 306–8, 334–37, 349, 354, 360–62, 372, 455, 461–66, 473, 481, and 495—for a total of forty-seven pages. He noted passages about trees on twenty-six pages.

80. Cohen, p. 188.

81. See, for instance, Wilkins, p. 159.

82. On this period of his life, see Wilkins, pp. 159–68, and Badè, *Life and Letters of John Muir*, 2:193–251.

83. Muir, letter of Sept. 13, 1889 (transcript), in Francis P. Farquhar Papers, Bancroft Library, Berkeley, California, carton 4, John Muir folder 1.

84. John Muir Collection, Huntington Library, letter of Aug. 14, 1912, Martinez, California, Muir to Enos Abijah Mills.

85. C. Hart Merriam, "To the Memory of John Muir," *Sierra Club Bulletin* 10 (January 1917); I read this article on the Web at www.sierraclub.org/john_muir_exhibit/life/memory_jm_c_hart_merriam.html.

86. Ibid.

Excursion: Home: The Harriman Expedition

1. An excellent history of this expedition is William H. Goetzmann and Kay Sloan, *Looking Far North: The Harriman Expedition to Alaska, 1899* (New York: Viking Press, 1982). For a more personal history, see Nancy Lord, *Green Alaska: Dreams from the Far Coast* (Washington, D.C.: Counterpoint, 1999).

2. See C. Hart Merriam, *Results of a Biological Survey of Mount Shasta, California* (Washington, D.C.: U.S. Department of Agriculture, Division of Biological Survey, 1899).

3. See, for instance, Donald Worster, *Nature's Economy: A History of Ecological Ideas* (New York: Cambridge University Press, 1977), pp. 194–204.

4. Merriam quoted in Keir B. Sterling, *Last of the Naturalists: The Career of C. Hart Merriam*, rev. ed. (New York: Arno Press, 1977), p. 213.

5. See, for instance, the posthumous volume *Studies of California Indians* (Berkeley: University of California Press, 1955), which is based on the collection of Merriam's papers at the Bancroft Library, Berkeley, California (there is also an important collection of Merriam's Indian materials at the Library of Congress). His article on Shasta is "ETHNOGRAPHY—Source of the name Shasta," *Journal of the Washington Academy of Sciences* 16 (November 18, 1926), transcription consulted in the Francis P. Farquhar Papers, carton 4, Shasta folder 1, Bancroft Library, Berkeley, California.

6. Quoted in Goetzmann and Sloan, p. ix.

7. Quoted ibid., p. 152.

8. Quoted ibid., p. 124.

9. Quoted in S. Hall Young, *Alaska Days with John Muir* (New York: Fleming H. Revell, 1915), p. 220.

10. George Bird Grinnell, "Natives of the Alaska Coast Region," in *Narrative, Glaciers, Natives,* vol. 1, Harriman Alaska Series, by John Burroughs, John Muir, and George Bird Grinnell (1902; repr., Washington, D.C.: Smithsonian Institution, 1910), p. 183.

11. C. Hart Merriam, "Bogoslof, Our Newest Volcano," in *History, Geography, Resources,* vol. 2, Harriman Alaska Series, by William H. Dall, Charles Keeler, et al. (1902; repr., Washington, D.C.: Smithsonian Institution, 1910), pp. 291–336.

12. See the rare Harriman Alaska Expedition Souvenir Book (copy consulted at Beinecke Library, Yale University).

On Curtis, see Edward S. Curtis, *The North American Indian: The Complete Portfolios* (Köln: Taschen, 1997), and Anne Makepeace, *Edward S. Curtis: Coming to Light* (Washington, D.C.: National Geographic, 2001).

13. See Lisa Bloom, *Gender on Ice: American Ideologies of Polar Expeditions* (Minneapolis: University of Minnesota Press, 1993); this book overstates its case by trying to fit all polar exploration into the category of colonialist fantasy, but I generally agree with Bloom's analysis of Peary and Cook (it would be a stretch to consider either one of them a Humboldtian explorer).

Chapter 10: The Grounding of American Environmentalism

1. Thoreau, *Walden* (1854; repr., Boston: Beacon Press, 2004), p. 195; Stephen Greenblatt, *Marvelous Possessions: The Wonder of the New World* (Chicago: University of Chicago Press, 1991), p. 127. Also see John Aldrich Christie, *Thoreau as World Traveler* (New York: Columbia University Press, 1965).

2. Alain de Botton is particularly eloquent on the topic of receptivity and metaphorical field-work in his lovely book *The Art of Travel* (New York: Pantheon, 2002), esp. pp. 239–49. Another wonderfully meandering meditation is Rebecca Solnit, *A Field Guide to Getting Lost* (New York: Viking, 2005), and if you're still in the mood to explore, open up to any page of John R. Stilgoe, *Landscape and Images* (Charlottesville: University of Virginia Press, 2005).

3. Carey McWilliams, *Ill Fares the Land* (Boston: Little, Brown, 1942), p. 386, and see *Factories in the Field* (Boston: Little, Brown, 1939). McWilliams especially celebrated the multifarious landscapes he experienced in California: the "coast line, mountain ranges, desert areas and lush valleys": McWilliams, "Personal Note," *Nation* 201 (September 20, 1965), p. 25. Also see Aaron Sachs, "Civil Rights in the Field: Carey McWilliams as a Public-Interest Historian and Social Ecologist," *Pacific Historical Review* 73 (May 2004), pp. 215–48.

4. See Mary Austin, *The Land of Little Rain,* in *Stories from the Country of Lost Borders,* ed. Marjorie Pryse (New Brunswick, N.J.: Rutgers University Press, 1987), pp. 13 and 11; and Austin, *Earth Horizon* (Boston: Houghton Mifflin, 1932), pp. 198 and 266.

Another nature writer with Humboldtian tendencies is Joseph Wood Crutch, who seized on the quickly developing science of ecology to urge people to work for the unity embodied by nature's interdependent web of life, to reach across the barriers of race, class, and country while at the same time transcending their seeming separation from their environment. "We must be a part not only of the human community," he wrote in 1955, "but of the whole community; we must acknowledge not only some sort of oneness with our neighbors, our countrymen, and our civilization but also some respect for the natural." Quoted in Donald Worster, *Nature's Economy: A History of Ecological Ideas* (New York: Cambridge University Press, 1977), p. 334.

5. See David N. Livingstone, *The Geographical Tradition: Episodes in the History of a Contested Enterprise* (Oxford: Blackwell, 1992), pp. 294–303. I am grateful to Kent Mathewson for alerting me to the connection between Humboldt and Sauer.

6. See Henry J. Bruman, "Carl Sauer in Midcareer: A Personal View by One of his Students," in *Carl O. Sauer: A Tribute,* ed. Martin S. Kenzer (Corvallis: Oregon State University Press, 1987), p. 128.

7. See Carl O. Sauer et al., "The Relation of Man to Nature in the Southwest: A Conference," *Huntington Library Quarterly* 8 (February 1945), pp. 115–51, and Kent Mathewson, "Sauer South by Southwest: Antimodernism and the Austral Impulse," in Kenzer, pp. 90–111.

8. Cited in Livingstone, p. 301.

9. George Perkins Marsh, *Man and Nature; or, Physical Geography as Modified by Human Action* (New York: Charles Scribner, 1864), p. 36. On the links between Sauer, Mumford, Marsh, and Humboldt, see David Lowenthal, *George Perkins Marsh: Prophet of Conservation* (1958; repr., Seattle: University of Washington Press, 2000), esp. pp. xv–xxi and 400–431, and William L. Thomas, ed., *Man's Role in Changing the Face of the Earth* (Chicago: University of Chicago Press, 1955), based on an international symposium dedicated to George P. Marsh and organized by Carl Sauer, Marston Bates, and Lewis Mumford.

10. Mumford quoted in John P. Clark, "A Social Ecology," *Capitalism, Nature, Socialism* 8 (September 1997), p. 7; also see Mark Luccarelli, *Lewis Mumford and the Ecological Region: The Politics of Planning* (New York: Guilford Press, 1995), and Ramachandra Guha, "Lewis Mumford, the Forgotten American Environmentalist: An Essay in Rehabilitation," in *Minding Nature: The Philosophers of Ecology*, ed. David Macauley (New York: Guilford Press, 1996), pp. 209–28. Note that Clark's excellent essay is one of the very few published attempts at a historical reconstruction of social ecologist ideas. Clark is especially interested in Reclus, a truly inspiring Humboldtian of French origin: see John P. Clark and Camille Martin, eds., *Anarchy, Geography, Modernity: The Radical Social Thought of Elisée Reclus* (Lanham, Md.: Lexington Books, 2004).

11. Lewis Mumford, *The Culture of Cities* (1938; repr., San Diego: Harcourt Brace Jovanovich, 1970), p. 392.

12. Ibid., pp. 303 and 382; Thoreau, p. 195; Kropotkin quoted in David Macauley, "Evolution and Revolution: The Ecological Anarchism of Kropotkin and Bookchin," in *Social Ecology After Bookchin*, ed. Andrew Light (New York: Guilford Press, 1998), p. 302.

13. Murray Bookchin, *The Modern Crisis* (Philadelphia: New Society Publishers, 1986), p. 59.

14. Bookchin, *Post-Scarcity Anarchism* (Berkeley: Ramparts Press, 1971), p. 70.

15. Ibid., p. 63.

16. See, for instance, Andrew Light, "Introduction: Bookchin as/and Social Ecology," in Light, p. 9; Mark Dowie, *Losing Ground: American Environmentalism at the Close of the Twentieth Century* (Cambridge, Mass.: MIT Press, 1995), pp. 228–29; and Werner Hülsberg, *The German Greens: A Social and Political Profile* (London: Verso, 1988).

17. See, for instance, J. Baird Callicott and Michael P. Nelson, eds., *The Great New Wilderness Debate* (Athens: University of Georgia Press, 1998), esp. Carl Talbot, "The Wilderness Narrative and the Cultural Logic of Capitalism," pp. 325–33.

18. Robert Merideth, *The Environmentalist's Bookshelf: A Guide to the Best Books* (New York: G. K. Hall, 1993).

19. Dowie, p. 18. Leopold's classic work is *A Sand County Almanac* (1949; repr., New York: Ballantine, 1988).

20. Rachel Carson, *Silent Spring* (1962; repr., Boston: Houghton Mifflin, 1987), p. 32; also see pp. 1–3 and 103–28.

21. Mark Dowie has suggested that environmental justice may well wind up being at the forefront of the so-called fourth wave of American environmentalism (see Dowie, pp. 125–74 and 205–63), and I agree. For even deeper analyses of the environmental justice movement, see Stephen Sandweiss, "The Social Construction of Environmental Justice," in *Environmental Injustices, Political Struggles: Race, Class, and the Environment*, ed. David E. Camacho (Durham, N.C.: Duke University Press, 1998), pp. 31–57; Rodger C. Field, "Risk and Justice: Capitalist Production and the Environment," in *The Struggle for Ecological Democracy: Environmental Justice Movements in the United States*, ed. Daniel Faber (New York: Guilford Press, 1998), pp. 81–103; Giovanna Di Chiro, "Nature as Community: The Convergence of Environment and Social Justice," in *Uncommon Ground: Rethinking the Human Place in Nature*, ed. William Cronon (New York: W. W. Norton, 1996), pp. 298–320; and Luke W. Cole and Sheila R. Foster, *From the Ground Up: Environmental Racism and the Rise of the Environmental Justice Movement* (New York: New York University Press, 2001).

22. See, for instance, Samuel P. Hays, *Beauty, Health, and Permanence: Environmental Politics in the United States, 1955–1985* (New York: Cambridge University Press, 1987), and Hal K. Rothman, *Saving the Planet: The American Response to the Environment in the Twentieth Century* (Chicago: Ivan R. Dee, 2000).

23. See Mark W. T. Harvey, *A Symbol of Wilderness: Echo Park and the American Conservation Movement* (Albuquerque: University of New Mexico Press, 1994), and Eliot Porter, *The Place No One Knew: Glen Canyon on the Colorado,* ed. David R. Brower (San Francisco: Sierra Club, 1963).

24. See Dowie, and Anna Bramwell, *The Fading of the Greens: The Decline of Environmental Politics in the West* (New Haven: Yale University Press, 1994). And note Michael Shellenberger and Ted Nordhaus, *The Death of Environmentalism: Global Warming Politics in a Post-Environmental World* (pamphlet published by the Nathan Cummings Foundation, 2004).

25. A typical viewpoint (this attitude pervades the literature) is that expressed by Rothman (p. 5): "Until the 1890s only an occasional voice—Henry David Thoreau, George Perkins Marsh, or Ralph Waldo Emerson—had seen the American land as anything other than a source of unlimited wealth. For most Americans during the first three hundred years of the European experience in the New World, the response to the sight of a tree had been to cut it down and make it useful: into shelter, into transportation, into fuel, into an article for storing food."

26. For an illuminating exploration of this problem, see Jennifer Price, *Flight Maps: Adventures with Nature in Modern America* (New York: Basic Books, 1999).

27. Reynolds to Corwin, Sept. 30, 1850, Brownsville [Texas], in Library of Congress, Manuscript Division, Thomas Corwin Correspondence, 1850–1852.

28. Letter of Clarence King to Henry Adams, Sept. 1887, New York, Massachusetts Historical Society, Clarence King Papers; also see King's letter to John Hay of Nov. 14, 1895, The Century Club (John Hay Library, Brown University, John Hay Papers, microfilm reel 8), where he writes about "Cervantes, whose temperament mine closely resembles."

29. Edmund Clarence Stedman, "King—'The Frolic and the Gentle,'" and William Crary Brownell, "King at the Century," in *Clarence King Memoirs,* ed. James D. Hague (New York: G. P. Putnam's Sons, 1904), pp. 203 and 222.

30. Letter of King to Hay, Sept. 29 [1894?], The Century Club, Hay Papers, John Hay Library, reel 8.

31. Clarence King, "The Education of the Future," *Forum* 13 (March 1892), p. 26.

32. William Ledyard Cathcart, "George Wallace Melville, Engineer-in-Chief of the United States Navy," *Cassier's Magazine* 11 (April 1897), p. 480.

33. See John Muir, *John Muir's Last Journey: South to the Amazon and East to Africa, Unpublished Journals and Selected Correspondence,* ed. Michael P. Branch (Washington, D.C.: Island Press, 2001), and C. Michael Hall and Stephen R. Mark, "The Botanist's Last Journey: John Muir in South America and Southern Africa, 1911–12," in *John Muir in Historical Perspective,* ed. Sally M. Miller (New York: Peter Lang, 1999), pp. 217–32.

34. Letter of Jan. 7, 1868, Cedar Keys, Muir Family Papers, H. E. Huntington Library, San Marino, Calif.

35. John Muir Papers and Related Collections, Holt-Atherton Special Collections, University of the Pacific, Stockton, Calif.: Humboldt, *Personal Narrative of Travels to the Equinoctial Regions of America, 1799–1804* (London: George Bell and Sons, 1907), p. ix.

36. See, for instance, Rebecca Solnit, *Wanderlust: A History of Walking* (New York: Viking, 2000), pp. 148–68.

37. Bruce Chatwin, *The Songlines* (New York: Viking, 1987), pp. 2 and 70.

38. Ibid., p. 272.

39. See, for instance, Daniel Boyarin and Jonathan Boyarin, "Diaspora: Generation and the Ground of Jewish Identity," in *Identities,* ed. Kwame Anthony Appiah and Henry Louis Gates Jr. (Chicago: University of Chicago Press, 1995), pp. 305–37; Paul Gilroy, *Against Race: Imagining Political Culture beyond the Color Line* (Cambridge, Mass.: Harvard University Press, 2000), esp. pp. 97–133 and 279–356; James Clifford, *Routes: Travel and Translation in the Late Twentieth Century* (Cambridge, Mass.: Harvard University Press, 1997), pp. 244–77; and Arif Dirlik, "The Global in the Local," in *Global/Local: Cultural Production and the Transnational Imaginary,* ed. Rob Wilson and Wimal Dissayanke (Durham, N.C.: Duke University Press, 1996), pp. 21–45.

40. Stein quoted in Wallace Stegner, "Living Dry," in *Where the Bluebird Sings to the Lemonade Springs: Living and Writing in the West* (New York: Penguin, 1992), p. 72. Stegner argued quite

eloquently that what America needs most is to settle down and develop a sense of place: see pp. xv–xxiii, 3–21, 57–75, 99–116, and 199–206.

41. See Stegner, ibid., esp. p. 71; Richard Slotkin, *The Fatal Environment: The Myth of the Frontier in the Age of Industrialization, 1800–1890* (1985; repr., New York: HarperPerennial, 1994), and *Gunfighter Nation: The Myth of the Frontier in Twentieth-Century America* (New York: HarperPerennial, 1993); Justin D. Edwards, *Exotic Journeys: Exploring the Erotics of U.S. Travel Literature, 1840–1930* (Hanover, N.H.: University Press of New England, 2001); and Terry Caesar, *Forgiving the Boundaries: Home as Abroad in American Travel Writing* (Athens: University of Georgia Press, 1995).

42. On the perennial idea of gardening as a panacea, most famously proposed by Voltaire, see Verlyn Klinkenborg, "Candide's Advice," editorial, June 18, 2004, *New York Times*, p. A30.

43. Yi-Fu Tuan, *Cosmos and Hearth: A Cosmopolite's Viewpoint* (Minneapolis: University of Minnesota Press, 1996), pp. 1–2 and 14. Also see De Botton. A fairly typical plea for the cultivation of placedness in America is Harold P. Simonson, *Beyond the Frontier: Writers, Regionalism, and a Sense of Place* (Fort Worth: Texas Christian University Press, 1989), esp. pp. 170–75. A helpful exploration of the pros and cons of placedness, using Thoreau as a kind of touchstone, is Richard J. Schneider, ed., *Thoreau's Sense of Place: Essays in American Environmental Writing* (Iowa City: University of Iowa Press, 2000).

44. Quoted in Robert Pogue Harrison, *Forests: The Shadow of Civilization* (Chicago: University of Chicago Press, 1992), p. 265. Harrison himself elaborates: "We call it the loss of nature, or the loss of wildlife habitat, or the loss of biodiversity, but underlying the ecological concern is perhaps a much deeper apprehension about the disappearance of boundaries, without which the human abode loses its grounding. Somewhere we still sense—who knows for how much longer?—that we make ourselves at home only in our estrangement" (p. 247). Also see Yi-Fu Tuan's reinterpretation of escapism: *Escapism* (Baltimore: Johns Hopkins University Press, 1998).

45. Thoreau, p. 297.

46. Ibid., pp. 6 and 305.

47. My thinking about today's evolving environmental situation and the United States' ongoing response to it has been influenced by a number of works, in addition to those already cited; for instance: Michael E. Zimmerman, *Contesting Earth's Future: Radical Ecology and Postmodernity* (Berkeley: University of California Press, 1994); Arran E. Gare, *Postmodernism and the Environmental Crisis* (London: Routledge, 1995); Allan Schnaiberg and Kenneth Alan Gould, *Environment and Society: The Enduring Conflict* (New York: St. Martin's, 1994); Robert J. Brulle, *Agency, Democracy, and Nature: The U.S. Environmental Movement from a Critical Theory Perspective* (Cambridge, Mass.: MIT Press, 2000); Dana Phillips, *The Truth of Ecology: Nature, Culture, and Literature in America* (New York: Oxford University Press, 2003); and Alexander Wilson, *The Culture of Nature: North American Landscape from Disney to the Exxon Valdez* (Toronto: Between the Lines, 1991).

48. As Elisée Reclus put it: "Certainly, man must take possession of the earth's surface and know how to utilize its forces. However, one cannot help lamenting the brutality with which this process is carried out"—and he meant social brutality as well as environmental. See Clark and Martin, p. 125.

For powerful histories of the potentially beautiful human attachment to the world, of our shaping of nature in ways that allow nature also to shape us, see Anne Whitson Spirn, *The Language of Landscape* (New Haven: Yale University Press, 1998); Simon Schama, *Landscape and Memory* (New York: Knopf, 1995); and Yi-Fu Tuan, *Passing Strange and Wonderful: Aesthetics, Nature, and Culture* (Washington, D.C.: Island Press, 1993). Also useful, as an overview, is Keith Thomas, *Man and the Natural World: A History of the Modern Sensibility* (New York: Pantheon, 1983).

49. Thoreau, p. 266.

Epilogue: Humboldt on Chimborazo

1. Humboldt wrote these journal entries in French, with a few Spanish words thrown in; the text above is my translation. For the original text, see Margot Faak, ed., *Reise auf dem Río Magdalena, durch die Anden und Mexico* (Berlin: Akademie Verlag, 1986), 1:219–20.

2. Ibid., p. 219, my translation. Montúfar was of course a creole; Chagra de San Juan seems to have been half-Indian. Interestingly, Humboldt contextualized his climb of Chimborazo with some social commentary about the surrounding settlements. Walking along the main road, he noted that "one never meets an Indian here . . . who is not engaged in cleaning or spinning cotton. How much would this industry grow if only those who worked at it (the Indians) were stimulated by the opportunity to enjoy the fruits of their own labors! But alas—they are slaves, without liberty, without property, and without equipment!" My translation, p. 217.

3. Albert Camus, *The Myth of Sisyphus and Other Essays* (New York: Vintage Books, 1955; orig. pub. in French in 1942), pp. 88–91.

4. Humboldt, *Cosmos: A Sketch of a Physical Description of the Universe* (Baltimore: Johns Hopkins University Press, 1997), 1:73.

5. Humboldt, *Personal Narrative of Travels to the Equinoctial Regions of America, 1799–1804* (London: Henry G. Bohn, 1852–53), 1:90–91.

6. Humboldt, *Cosmos*, 1:40–41.

7. Ibid., 2:294—though this was in fact a direct echo of something he had written much earlier: in his *Critical Examination of the History of the Geography of the New Continent* (5 vols., 1836–39), he wrote, about "the extension of the sphere of knowledge": "Feeble spirits at each epoch believe that humanity has arrived at its culminating point of its progressive march; they forget that, by the intimate connection of all truths, with each step that we advance, the field to traverse reveals itself to be that much vaster, bordered by a horizon that endlessly retreats." See Humboldt, *Examen Critique de l'Histoire de la Géographie du Nouveau Continent* (Paris: Gide, 1837), 3:154, my translation.

For a slightly more extensive bibliography, see Aaron Jacob Sachs, "The Humboldt Current: Avant-Garde Exploration and Environmental Thought in 19th-Century America" (Ph.D. diss., Yale University, 2004).

PRIMARY SOURCES

Humboldt

Manuscripts and Archives

American Philosophical Society, Philadelphia: Humboldt Collection.
Archivo Histórico Nacional, Madrid, Sección Estado, legajo 4709 ("Memoria" from Humboldt to the king of Spain, 1799).
Boston Public Library, Rare Books Department: "Alexander von Humboldt on the Fugitive Slave Law."
H. E. Huntington Library, San Marino, Calif.: Manuscripts Division: Rhees Collection.

Works by Humboldt

Asie Centrale: Recherches sur les Chaînes de Montagnes et la Climatologie Comparée. 3 vols. Paris: Gide, 1843.
Aspects of Nature. Philadelphia: Lea and Blanchard, 1849.
Cosmos: A Sketch of a Physical Description of the Universe. Translated by E. C. Otté. 5 vols. London: Henry G. Bohn, 1849, 1851, 1852, and 1870; orig. pub. in German between 1845 and 1861.
Cosmos. Translated by E. C. Otté. 2 vols. Baltimore: Johns Hopkins University Press, 1997. Introductions by Nicolaas A. Rupke (vol. 1) and Michael Dettelbach (vol. 2).
Essai Politique sur le Royaume de la Nouvelle Espagne. 2 vols. and atlas. Paris: F. Schoell, 1811.
Essai sur la Géographie des Plantes. Paris: F. Schoell, 1807.
Examen Critique de l'Histoire de la Géographie du Nouveau Continent. 5 vols. Paris: Gide, 1836–39.
Faak, Margot, ed. *Lateinamerike am Vorabend der Unabhärigigkeitsrevolution: eine Anthologie von Impressionen und Urteilen, aus seinen Reisetagesbüchern.* Berlin: Akademie Verlag, 1982.
———. *Reise auf dem Río Magdalena, durch die Anden und Mexico.* Vol. 1. Berlin: Akademie Verlag, 1986.
Fragmens de Géologie et de Climatologie Asiatiques. 2 vols. Paris: A. Pihan Delaforest, 1831.
Hamy, E. T., ed. *Lettres Américaines d'Alexandre de Humboldt (1798–1807).* Paris: Librairie Orientale et Américaine, 1904.
Jahn, Ilse, and Fritz G. Lang, eds. *Die Jugendbriefe Alexander von Humboldts: 1787–1799.* Berlin: Akademie Verlag, 1973.
Letters of Alexander von Humboldt to Varnhagen von Ense, from 1827 to 1858. Translated from the second German edition by Friedrich Kapp. New York: Rudd and Carleton, 1860.
"Des Lignes Isothermes et de la Distribution de la Chaleur sur le Globe." *Mémoires de Physique et de Chimie de la Société d'Arcueil* 3 (1817), pp. 462–602.
"Mes Confessions, à lire et à me renvoyer un jour." In *Lettres Américaines d'Alexandre de Humboldt (1798–1807),* ed. E. T. Hamy (see above), pp. 236–44.

Moheit, Ulrike, ed. *Alexander von Humboldt: Briefe aus Amerika, 1799–1804*. Berlin: Akademie Verlag, 1993.

Personal Narrative of a Journey to the Equinoctial Regions of the New Continent. Abridged ed. London: Penguin, 1995. Introduction by Jason Wilson, Historical Introduction by Malcolm Nicolson.

Personal Narrative of Travels to the Equinoctial Regions of America, 1799–1804. Translated by Thomasina Ross. 3 vols. London: Henry G. Bohn, 1852–53; orig. pub. in French between 1814 and 1825.

Political Essay on the Kingdom of New Spain. Translated by John Black, edited by Mary Maples Dunn. Abridged ed. New York: Knopf, 1972; orig. pub. in French in 1811.

Researches, Concerning the Institutions and Monuments of the Ancient Inhabitants of America, with Descriptions and Views of Some of the Most Striking Scenes in the Cordilleras. Translated by Helen Maria Williams. 2 vols. London: Longman, Hurst, 1814.

Schwarz, Ingo, ed. *Alexander von Humboldt und die Vereinigten Staaten von America: Briefwechsel*. Berlin: Akademie Verlag, 2004.

Umrisse von Vulkanen aus den Cordilleran von Quito und Mexico. Stuttgart und Tübingen: J. G. Cotta, 1853.

Views of Nature: Or Contemplations on the Sublime Phenomena of Creation; with Scientific Illustrations. Translated by E. C. Otté and H. G. Bohn. London: Henry G. Bohn, 1850; orig. pub. in German in 1808.

Vues des Cordillères et Monumens des Peuples Indigènes de l'Amérique, 2 vols. Paris: Chez L. Bourgeois-Maze, 1840; orig. pub. 1810.

Reynolds

Manuscripts and Archives

Boston Museum of Science: Archives of the Boston Society of Natural History.

Columbia University, New York: Columbiana Archives, *Columbia University Annual Commencement Scrapbook, 1830–1849*, and *Minutes of the Trustees of Columbia College*, vol. 3, pt. 2 (typescript); and College Papers, Special Manuscript Division.

H. E. Huntington Library, San Marino, Calif., Manuscripts Division: Thomas Lyon Hamer Papers.

Library of Congress, Washington, D.C., Manuscript Division: Thomas Corwin Correspondence, 1850–1852.

National Archives, Washington, D.C.: Microfilm Publication M75, Letters Received by the Navy Department relating to the Wilkes Expedition, Letters Relating to Preparations for the Exploring Expedition; and Microfilm Publication M124, Miscellaneous Letters Received by the Secretary of the Navy, 1801–1884.

Princeton University Library, Princeton, N.J., Special Collections: Samuel L. Southard Papers; Richard G. Woodbridge III, "J.N. Reynolds—American," typescript.

Works by Reynolds

Address on the Subject of a Surveying and Exploring Expedition to the Pacific Ocean and South Seas. Delivered in the Hall of Representatives on the Evening of April 3, 1836. New York: Harper and Brothers, 1836.

"Bearding a Sea-Lion in His Den." *Knickerbocker* 13 (June 1839), pp. 524–26.

Exploring Expedition Correspondence. New York, [1838?] (undated pamphlet).

"A Leaf from an Unpublished Journal." *New-Yorker* 7 (June 29, 1839), pp. 228–29.

"A Leaf from an Unpublished Manuscript." *Southern Literary Messenger* 5 (June 1839), pp. 408–13.

"Leaves from an Unpublished Journal." *New-York Mirror* 15 (April 21, 1838), pp. 340–41.

A Life of George Washington, in Latin Prose: by Francis Glass, A.M., of Ohio. Edited by J. N. Reynolds. New York: Harper and Brothers, 1835.

"Mocha-Dick; or, The White Whale of the Pacific: A Leaf from a Manuscript Journal." *Knicker-bocker* 13 (May 1839), pp. 377–92.

Pacific and Indian Oceans; or The South Sea Surveying and Exploring Expedition: Its Inception, Progress, and Objects. New York: Harper and Brothers, 1841.

Remarks on a Review of Symmes' Theory, which appeared in the American Quarterly Review. Washington, D.C.: Gales and Seaton, 1827.

"A Report in relation to Islands, Reefs, and Shoals in the Pacific Ocean &c.," referred to Committee on Commerce January 27, 1835, H. Doc. 105, 23rd Cong., 2nd sess., 28 pp., serial 273. Also see the revised version, March 21, 1836, S. Doc. 262, 24th Cong., 1st sess., 87 pp., serial 281 (which includes memorials favoring an exploring expedition). Originally printed in 1828 and titled *Pacific Ocean and South Seas.*

"Rough Notes of Rough Adventures." *Southern Literary Messenger* 9 (December 1843), pp. 705–15.

Untitled article. *New-York Mirror* 3 (May 27, 1826), p. 351.

Voyage of the United States Frigate Potomac, *under the Command of Commodore John Downes, during the Circumnavigation of the Globe, in the Years 1831, 1832, 1833, and 1834.* New York: Harper and Brothers, 1835.

King

Manuscripts and Archives

American Philosophical Society, Philadelphia: Clarence King Papers.

Bancroft Library, University of California at Berkeley: Francis P. Farquhar Collection of Materials Relating to Clarence King.

Beinecke Library, Yale University, New Haven, Conn.: Carleton Watkins Collection ("Mammothplate Photographs of California" and "Photographs of California from the U.S. Geological Exploration of the 40th Parallel [1870]").

H. E. Huntington Library, San Marino, Calif., Manuscripts Division: W. W. Bailey Papers; James D. Hague Papers; Clarence King Papers; J. T. Redman typescript, "Reminiscences and experiences on my trip across the plains to California 61 years ago when I drove four mules to a covered wagon."

John Hay Library, Brown University, Providence, R.I.: John Hay Papers, microfilm reels 7–8.

Library of Congress, Washington, D.C.: Prints and Photographs Collection; Manuscript Division, S. F. Emmons Papers.

Massachusetts Historical Society, Boston, Mass.: Clarence King Papers.

National Archives, Washington, D.C.: Record Group 57, Geological Survey Records, Copies of Letters and Reports sent to the Chief of Engineers; Record Group 77, Records of the U.S. Geological Exploration of the 40th Parallel ("King Survey"), 1867–81, Office of the Chief of Engineers; The Still Pictures Section, collection of the U.S. Geological Exploration of the 40th Parallel.

Works by King

"Active Glaciers within the United States." *Atlantic Monthly* 27 (March 1871), pp. 371–77.

"The Age of the Earth." *American Journal of Science* 45 (January 1893), pp. 1–20.

"Artium Magister." *North American Review* 147 (October 1888), pp. 369–84.

"Bancroft's *Native Races of the Pacific States.*" *Atlantic Monthly* 35 (February 1875), pp. 163–73.

"The Biographers of Lincoln." *Century Magazine* 32 (October 1886), pp. 861–69.

"Catastrophism and Evolution." *American Naturalist* 11 (August 1877), pp. 449–70.

[Unsigned]. "Current Literature" (Book Reviews). *Overland Monthly* 5 (December 1870), pp. 578–83.

"The Education of the Future." *Forum* 13 (March 1892), pp. 20–33.

"The Falls of the Shoshone." *Overland Monthly* 5 (October 1870), pp. 379–85.

"Fire and Sword in Cuba." *Forum* 22 (September 1896), pp. 31–52.

"The Helmet of Mambrino." *Century Magazine* 32 (May 1886), pp. 154–59.
[Unsigned]. "John Hay." *Scribner's Monthly* 7 (April 1874), pp. 736–39.
Mountaineering in the Sierra Nevada. 1872; repr., New York: Penguin, 1989. The new preface appears in the reprint of the 1874 ed. (New York: Scribner's, 1902).
"Note on the Uinta and Wasatch Ranges." *American Journal of Science*, 3rd ser., 11 (1876), p. 494.
"On the Discovery of Actual Glaciers in the Mountains of the Pacific Slope." *American Journal of Science and Arts*, 3rd ser., 1 (March 1871), pp. 157–67.
"Paleozoic Subdivisions of the Fortieth Parallel." *American Journal of Science*, 3rd ser., 11 (1876), pp. 475–82.
"Report of Clarence King, United States Geologist, To the Board of Directors of the San Francisco and New York Mining and Commercial Company." Reprinted in the *Daily Alta California*, November 26, 1872, p. 1.
"Shall Cuba Be Free?" *Forum* 20 (September 1895), pp. 50–65.
[Unsigned]. "Style and the Monument." *North American Review* 141 (November 1885), pp. 443–53.
Systematic Geology. vol. 1, *Report of the U.S. Geological Exploration of the Fortieth Parallel.* Washington, D.C.: Government Printing Office, 1878.
The Three Lakes: Marian, Lall, and Jan, and How They Were Named. San Francisco: privately printed, 1870. Copy available at the Boston Public Library.

Melville

Manuscripts and Archives

Dartmouth College Library, Hanover, N.H.: George Wallace Melville Papers, 1881–1882, including his Arctic journal.
Hoover Institution, Stanford University, Stanford, Calif.: Russia Posol'stvo (U.S.) Records, 1897–1947, Naval Agent Office File.
National Archives, Washington, D.C.: Record Group 45, Naval Records Collection (Letters from Officers Commanding Expeditions; Microfilm Publication T297, "Log of the Jeannette"; Subject File on the *Jeannette* Expedition).

Works by Melville

Abandon the Philippines. Boston, 1905.
"The Engineer's Duty as a Citizen." *Journal of the American Society of Mechanical Engineers* 32 (July–December 1910), pp. 1127–32.
Gems of Wonderful Chicago and the World's Fair. Chicago: Geo. W. Melville, 1893.
"The Important Elements in Naval Conflicts." In "The United States as a World Power." Special annual meeting number, *Annals of the American Academy of Political and Social Science* 26 (July 1905), pp. 123–36.
In the Lena Delta. London: Longmans, Green, 1885.
"The Military Duty of the Engineering Institutions." *Forum* 32 (January 1902), pp. 515–27.
"Preface" to *A Tenderfoot with Peary*, by George Borup. New York: Frederick A. Stokes, 1911, pp. viii–ix.
"Remarks on Polar Expedition." *Proceedings of the American Philosophical Society* 36 (October 29, 1897), pp. 454–61.
"Report of Chief Engineer George Wallace Melville in Connection with the Jeannette Expedition (Oct. 17, 1882)." *Report of the Secretary of the Navy.* 47th Cong., 2nd sess., 1882–3. H. Ex. Doc. 1, pt. 3, pp. 57–70.
Views of Commodore George W. Melville, Chief Engineer of the Navy, as to the Strategic and Commercial Value of the Nicaraguan Canal, the Future Control of the Pacific Ocean, the Strategic Value of Hawaii, and its Annexation to the United States. Washington, D.C.: Government Printing Office, 1898.

Muir

Manuscripts and Archives

Bancroft Library, University of California at Berkeley: Francis P. Farquhar Papers; John Muir Papers, 1860–1914, correspondence on Yosemite and conservation in California, plus a report on the geology and botany of Mount Shasta.

H. E. Huntington Library, San Marino, Calif., Manuscripts Division: John Muir Collection, Muir Family Papers.

University of the Pacific, Holt-Atherton Special Collections, John Muir Center, Stockton, Calif.: John Muir Papers and Related Collections.

Yale University, Microfilm Collections: John Muir Papers.

Works by Muir

"The Ancient Glaciers of the Sierra." *The Californian* 2 (December 1880), pp. 550–57.

Badè, William Frederic, ed. *The Life and Letters of John Muir*. 2 vols. Boston: Houghton Mifflin, 1923–24.

Branch, Michael P., ed. *John Muir's Last Journey: South to the Amazon and East to Africa, Unpublished Journals and Selected Correspondence*. Washington, D.C.: Island Press, 2001.

The Cruise of the Corwin. 1917. Reprint, San Francisco: Sierra Club Books, 1993.

Engberg, Robert, and Bruce Merrell, eds. *Letters from Alaska*. Madison: University of Wisconsin Press, 1993.

"Exploring in the Great Tuolumne Canyon." *Overland Monthly* 11 (August 1871), pp. 140–41.

Gisel, Bonnie Johanna, ed. *Kindred and Related Spirits: The Letters of John Muir and Jeanne C. Carr*. Salt Lake City: University of Utah Press, 2001.

Letters to a Friend. Boston: Houghton Mifflin, 1915.

"Living Glaciers of California." *Harper's New Monthly Magazine* 51 (November 1875), pp. 769–76.

The Mountains of California. 1894. Reprint, New York: Penguin, 1985.

My First Summer in the Sierra. Boston: Houghton Mifflin, 1911.

Our National Parks. 1901. Reprint, Boston: Houghton Mifflin, 1916.

Picturesque California. San Francisco: J. Dewing, 1888–91.

Rosse, Irving C., John Muir, and E. W. Nelson. *The Cruise of the Revenue-Steamer Corwin in Alaska and the N.W. Arctic Ocean in 1881; Notes and Memoranda—Medical and Anthropological, Botanical, Ornithological*. Washington, D.C.: Government Printing Office, 1883.

"Snow-Storm on Mount Shasta." *Harper's New Monthly Magazine* 55 (September 1877), pp. 521–30.

Steep Trails. Boston: Houghton Mifflin, 1918.

A Thousand-Mile Walk to the Gulf. 1916. Reprint, San Francisco: Sierra Club Books, 1991.

Travels in Alaska. 1915. Reprint, New York: AMS Press, 1978.

Wolfe, Linnie Marsh, ed. *John of the Mountains: The Unpublished Journals of John Muir*. Boston: Houghton Mifflin, 1938.

Other

Manuscripts and Archives

American Philosophical Society, Philadelphia: Elisha Kent Kane Papers.

Beinecke Rare Book and Manuscripts Library, New Haven, Conn.: Western Americana Collection.

Published Works

Abert, J. W. *Western America in 1846–1847: The Original Travel Diary of Lieutenant J.W. Abert, who mapped New Mexico for the United States Army*. Edited by John Galvin. San Francisco: John Howell Books, 1966.

Adams, Henry. *The Education of Henry Adams*. 1907. Reprint, Boston: Houghton Mifflin, 1973.

———. "Review of *Mountaineering in the Sierra Nevada*." *North American Review* 114 (April 1872), p. 446.

———. "Review of *Systematic Geology*." *The Nation* 28 (January 23, 1879), p. 74.

Agassiz, Louis. *Address Delivered on the Centennial Anniversary of the Birth of Alexander von Humboldt; Under the Auspices of the Boston Society of Natural History; with an account of the evening reception.* Boston: Boston Society of Natural History, 1869.

"Alexander von Humboldt and his Cosmos." *Methodist Quarterly Review*, 4th ser., 12 (July 1860), pp. 414–32.

American Geographical and Statistical Society. *Tribute to the Memory of Humboldt.* New York: H. H. Lloyd, June 15, 1859.

Bancroft, Hubert Howe. *The Works of Hubert Howe Bancroft.* Vol. 23, *History of California, Vol. 6, 1848–1859.* San Francisco: History Co., 1888.

Barnard, D. D. "The War with Mexico." *American Review: A Whig Journal of Politics, Literature, Art, and Science* 3 (June 1846), pp. 571–80.

"Baron Humboldt's Cosmos." *Christian Examiner and Religious Miscellany*, 48, 4th series Vol. 13 (January 1850), pp. 53–88.

"Baron Humboldt's *Kosmos: A General Survey of the Physical Phenomena of the Universe.*" *North British Review* 4 (November 1845–February 1846), pp. 202–54.

Blodget, Lorin. *Climatology of the United States.* Philadelphia: J. B. Lippincott, 1857.

Boas, Franz. *The Central Eskimo.* 1888. Reprint, Lincoln: University of Nebraska Press, 1964.

———. "The Study of Geography." *Science* 9 (February 11, 1887), pp. 137–41.

Bouvé, T. T. "Address." *Proceedings of the Boston Society of Natural History* 18 (Boston: Boston Society of Natural History, 1877), pp. 242–51.

Brewer, William H. *Such a Landscape! A Narrative of the 1864 California Geological Survey Exploration of Yosemite, Sequoia and Kings Canyon from the Diary, Fieldnotes, Letters and Reports of William Henry Brewer.* Edited by William Alsup. Yosemite National Park: Yosemite Associates and Sequoia Natural History Association, 1987.

———. *Up and Down California in 1860–1864: The Journal of William H. Brewer.* Edited by Francis P. Farquhar. New Haven: Yale University Press, 1930.

Bronson, Edgar Beecher. "A Man of East and West: Clarence King, Geologist, Savant, and Wit." *Century Magazine* 80 (July 1910), pp. 376–82.

———. *Reminiscences of a Ranchman.* New York: McClure, 1908.

Burroughs, John, John Muir, and George Bird Grinnell. *Narrative, Glaciers, Natives.* Vol. 1, Harriman Alaska Series. 1902. Reprint, Washington, D.C.: Smithsonian Institution, 1910.

Carrell [*sic;* proper spelling is Carroll], A. E. "The First American Exploring Expedition." *Harper's New Monthly Magazine* 44 (December 1871), pp. 60–64.

Carroll, Anna Ella. *The Star of the West; or, National Men and National Measures.* 3rd ed. 1856. Reprint, New York: Miller, Orton, 1857.

"Church's Heart of the Andes." *Harper's Weekly* 3 (May 7, 1859), p. 291.

Clark, P. "The Symmes Theory of the Earth." *Atlantic Monthly* 31 (April 1873), pp. 471–80.

Dall, William Healey. *Alaska and Its Resources.* Boston: Lee and Shepard, 1870.

Dall, William H., Charles Keeler, et al. *History, Geography, Resources.* Vol. 2, Harriman Alaska Series. 1902. Reprint, Washington, D.C.: Smithsonian Institution, 1910.

Dana, James Dwight. *United States Exploring Expedition: Geology.* Vol. 10. Philadelphia: C. Sherman, 1849.

Dana, Richard Henry. *Two Years before the Mast.* 1840. Reprint, New York: New American Library, 1964.

Danenhower, John W. *Narrative of the "Jeannette."* Boston: J. R. Osgood, 1882.

———. *The Polar Question: Paper to be Read and Discussed at the Naval Institute, Annapolis, Md., October 9, 1885.* Annapolis: United States Naval Institute, August 15, 1885.

De Long, George Washington. *The Voyage of the* Jeannette: *The Ship and Ice Journals of George W. De Long, Lieutenant-Commander U.S.N., and Commander of the Polar Expedition of 1879–1881.* 2 vols. Boston: Houghton Mifflin, 1883.

Diderot, Denis. "Supplement to Bougainville's *Voyage.*" In *Rameau's Nephew and Other Works.* Translated by Jacques Barzun and Ralph H. Bowen. Indianapolis: Bobbs-Merrill, 1964; orig. pub. 1796.

Donaldson, Thomas. *The George Catlin Indian Gallery in the U.S. National Museum.* Washington, D.C.: United States National Museum, 1886.

Eights, James. "Description of a New Animal Belonging to the Arachnides of Latreille; Discovered in the Sea Along the Shores of the New South Shetland Islands." *Boston Journal of Natural History* 1 (Boston: Hilliard, Gray, 1837), pp. 203–6.

———. "On the Icebergs of the Antarctic Sea." *American Journal of Agriculture and Science* 4 (July 1846), pp. 20–24.

Emerson, Ralph Waldo. "Experience" and "Divinity School Address." In *Selections from Ralph Waldo Emerson*, edited by Stephen E. Whicher. Boston: Houghton Mifflin, 1960.

———. *The Journals and Miscellaneous Notebooks of Ralph Waldo Emerson*. Edited by Alfred P. Ferguson. Vol. 4. Cambridge, Mass.: Harvard University Press, 1964.

———. "Nature." In *Nature/Walking*, edited by John Elder. Boston: Beacon Press, 1991.

Evans, Estwick. *A Pedestrious Tour, of Four Thousand Miles, through the Western States and Territories, during the Winter and Spring of 1818*. Concord, N.H.: Joseph C. Spear, 1819.

Fanning, Edmund. *Voyages and Discoveries in the South Seas, 1792–1832*. Compilation ed. Salem, Mass.: Marine Research Society, 1924.

———. *Voyages Round the World*. New York: Collins and Hannay, 1833.

———. *Voyages to the South Seas*. New York: William H. Vermilye, 1838.

Forbes, James David. "Humboldt's *Cosmos*." *Eclectic Magazine* 7 (March 1846), pp. 353–75.

Forster, Georg. *Ansichten vom Niederrhein, von Brabant, Flandern, Holland, England, und Frankreich*. 1792. Reprint, Leipzig: F. A. Brodhaus, 1868.

———. *A Voyage Round the World*. 1777. Reprint, Honolulu: University of Hawaii Press, 2000.

Frémont, John Charles. *Geographical Memoir upon Upper California*. 1848. Reprint, San Francisco: Book Club of California, 1964.

———. *Report of the Exploring Expedition to the Rocky Mountains in the year 1842, and to Oregon and north California in the years 1843–'44*. Washington, D.C.: Blair and Rives, 1845.

Gallatin, Albert. *A Synopsis of the Indian Tribes within the United States*. Cambridge, Mass.: American Antiquarian Society, 1836.

Geographical Society of the Pacific. *An Examination into the Genuineness of the "Jeannette" Relics*. San Francisco: John Partridge, 1896.

Geological Survey of California (J. D. Whitney, State Geologist). *The Yosemite Book; A Description of the Yosemite Valley and the Adjacent Region of the Sierra Nevada, and of the Big Trees of California*. New York: Julius Bien, 1868.

Gilder, William. *Ice-Pack and Tundra: An Account of the Search for the* Jeannette *and a Sledge Journey Through Siberia*. New York: Scribner's, 1883.

Goodyear, William A. "On the Situation and Altitude of Mount Whitney." *Proceedings of the California Academy of Sciences* 5 (August 4, 1873), pp. 139–44.

Gould, Augustus A. "Notice on the Origin, Progress and Present Condition of the Boston Society of Natural History." *American Quarterly Register* 14 (February 1842), pp. 236–41.

Gray, Asa. *United States Exploring Expedition: Botany: Phanerogamia*. Vol. 1. New York: George P. Putnam, 1854.

Greely, Adolphus. *Three Years of Arctic Service*. 2 vols. New York: Charles Scribner's Sons, 1886.

Greenwood, Rev. F. W. P. "An Address Delivered before the Boston Society of Natural History, at the opening of their New Hall in Tremont Street." *Boston Journal of Natural History* (Boston: Hilliard, Gray, 1837), pp. 5–14.

Guyot, Arnold. *Earth and Man: Lectures on Comparative Physical Geography, in its Relation to the History of Mankind*. Boston: Gould and Lincoln, 1854.

Hale, Horatio. *United States Exploring Expedition: Ethnography and Philology*. Vol. 6. Philadelphia: Lea and Blanchard, 1846.

Hall, Charles Francis. *Arctic Researches and Life among the Esquimaux: Being the Narrative of an Expedition in Search of Sir John Franklin in 1860, 1861, and 1862*. New York: Harper and Brothers, 1865.

Hayes, Isaac Israel. *An Arctic Boat Journey in the Autumn of 1854*. Boston: Brown, Taggard, and Chase, 1860.

————. *The Open Polar Sea: A Narrative of a Voyage of Discovery Towards the North Pole in the Schooner* United States. New York: Hurd and Houghton, 1867.

Heinzen, Karl. *The True Character of Humboldt; Oration Delivered at the German Humboldt Festival, in Boston*. Indianapolis: Association for the Propagation of Radical Principles, 1869.

"History of Navigation in the South Seas" (review of Reynolds's *Address*). *North American Review* 45 (October 1837), pp. 361–90.

Howe, Henry. "The Romantic History of Jeremiah N. Reynolds." In *Historical Collections of Ohio in Two Volumes: An Encyclopedia of the State*. Vol. 1. Columbus: Henry Howe and Son, 1889, pp. 430–32.

Howells, William Dean. *"Criticism and Fiction" and Other Essays*. Edited by Clara Marburg Kirk and Rudolf Kirk. New York: New York University Press, 1959.

————. "Review of *Mountaineering in the Sierra Nevada*," *Atlantic Monthly* 29 (April 1872), pp. 500–501.

"Humboldt, Ritter, and the New Geography." *New Englander* 18 (May 1860), pp. 277–306.

Ingersoll, Robert. *The Gods and Other Lectures*. Peoria, Ill.: Religio Philosophical Publishing House, 1874.

Irving, Washington. *Astoria*. 1836. Reprint, Portland, Ore.: Binfords and Mort, 1967.

————. "Rip Van Winkle." In *The Sketch-Book of Geoffrey Crayon, Gent*. 1819. Reprint, New York: G. P. Putnam's Sons, 1888.

Jackson, Helen Hunt. *A Century of Dishonor: A Sketch of the United States Government's Dealings with Some of the Indian Tribes*. New York: Harper and Brothers, 1881.

————. *Ramona: A Story*. Boston: Roberts Brothers, 1884.

James, William. *The Varieties of Religious Experience*. 1902. Reprint, New York: Penguin, 1982.

Kane, Elisha Kent. *Arctic Explorations: The Second Grinnell Expedition in Search of Sir John Franklin, 1853, 1854, 1855*. Philadelphia: Childs and Peterson, 1856.

————. *The United States Grinnell Expedition in Search of Sir John Franklin: A Personal Narrative*. New York: Harper and Brothers, 1854.

Keating, William. *Narrative of an Expedition to the Source of the St. Peter's River*. Vol. 1. London: Geo. B. Whittaker, 1825.

Klencke and Schleiser. *Lives of the Brothers Humboldt, Alexander and William*. Translated by Juliette Bauer. London: Ingram, Cooke, 1852.

Madden, E. F. "Symmes and His Theory." *Harper's New Monthly Magazine* 65 (October 1882), pp. 740–44.

Malthus, Thomas Robert. *An Essay on Population*. London: J. Johnson, 1798.

Marsh, George Perkins. *Man and Nature; or, Physical Geography as Modified by Human Action*. New York: Charles Scribner, 1864.

————. "The Study of Nature." *Christian Examiner* 68 (January 1860), pp. 33–62.

Matthews, Thomas J. *Lecture on Symmes' Theory of Concentric Spheres, Read at the Western Museum*. Cincinnati: A. N. Deming, 1824.

McBride, James. *Symmes's Theory of Concentric Spheres*. Cincinnati: Morgan, Lodge, and Fisher, 1826.

Melville, Herman. *Moby Dick or The White Whale*. 1851. Reprint, New York: New American Library, 1961.

————. *Typee: A Peep at Polynesian Life during a Four Months' Residence in a Valley of the Marquesas*. 1846. Reprint, New York: New American Library, 1964.

Merriam, C. Hart. *Results of a Biological Survey of Mount Shasta, California*. Washington, D.C.: U.S. Department of Agriculture, Division of Biological Survey, 1899.

————. *Studies of California Indians*. Berkeley: University of California Press, 1955.

Möllhausen, Baldwin. *Diary of a Journey from the Mississippi to the Coasts of the Pacific*. 2 vols. London: Longman, Brown, Green, Longmans, and Roberts, 1858.

Newcomb, Raymond L., and Richard W. Bliss. *Our Lost Explorers: The Narrative of the* Jeannette *Arctic Expedition*. Hartford, Conn.: American Publishing, 1882.

Nicollet, J. N. *Report intended to illustrate a map of the hydrographical basin of the Upper Mississippi River*. Washington, D.C.: Blair and Rives, 1843.

Noble, Louis L. *After Icebergs with a Painter: A Summer Voyage to Labrador and around Newfound-land*. New York: D. Appleton, 1861.

Nourse, J. E. *American Explorations in the Ice Zones*. Boston: D. Lothrop, 1884.

Parkman, Francis, Jr. *The Oregon Trail*. 1849. Reprint, New York: Penguin, 1985.

Poe, Edgar Allan. *The Complete Works of Edgar Allan Poe*. Edited by James A. Harrison. Vols. 4, 9, 15, 16, and 17. New York: AMS Press, 1965.

Powell, John Wesley. *The Exploration of the Colorado River and Its Canyons*. 1875. Reprint, New York: Penguin, 1987.

———. *Report on the Lands of the Arid Region of the United States*. 1878. Reprint, Cambridge, Mass.: Harvard University Press, 1962.

———. *Selected Prose of John Wesley Powell*. Edited by George Crossette. Boston: David R. Godine, 1970.

Prentiss, Henry Mellen. *The Great Polar Current*. New York: F. A. Stokes, 1897.

Proceedings of a Court of Inquiry Convened at the Navy Department . . . to Investigate the Circumstances of the Loss . . . of the Exploring Steamer "Jeannette." Washington, D.C.: Government Printing Office, 1883.

Proceedings of an Investigation into the Jeannette *by the Naval Affairs Subcommittee, 48th Congress, 1st Session, House Miscellaneous Document 66*. Washington, D.C.: Government Printing Office, 1885.

Raymond, Rossiter. "Biographical Notice of Clarence King." *Transactions of the American Institute of Mining Engineers* 33 (1903), pp. 619–50.

"Reynolds' Lecture on Maritime Discovery." *New-York Mirror* 13 (April 30, 1836), p. 350.

Roosevelt, Theodore. *Ranch Life and the Hunting-Trail*. Ann Arbor: University of Michigan Press, 1966; facsimile of the 1888 ed.

Rose, Gustav. *Mineralogisch-geognostische Reise nach dem Ural, dem Altai, und dem Kapischen Meere*. 2 vols. Berlin: Verlag der Sanderschen Buchhandlung, 1837.

Rousseau, Jean-Jacques. *A Discourse on Inequality*. Translated by Maurice Cranston. London: Penguin, 1984; orig. pub. in French in 1755.

Rowland, Henry. "A Plea for Pure Science." *Proceedings of the American Association for the Advancement of Science* 32 (August 1883), pp. 105–26.

Ruskin, John. *The Complete Works of John Ruskin*. Edited by E. T. Cook and Alexander Wedderburn. Vols. 34–37 (Journals). London: George Allen, 1903–12.

———. *Modern Painters* (books 1–5, compiled in a 4-vol. set). New York: John Wanamaker, 1905; orig. pub. 1843, 1846, 1856, and 1860.

St. Pierre, Bernardin de. "Voyage à l'île de France" (orig. pub. 1773). In *Ile de France: Voyage et Controverses*. Mauritius: AlmA, 1996.

Samson, John. "Photographs from the High Rockies." *Harper's Monthly Magazine* 39 (September 1869), pp. 465–75.

Sauer, Carl O., et al. "The Relation of Man to Nature in the Southwest: A Conference." *Huntington Library Quarterly* 8 (February 1945), pp. 115–51.

Seaborn, Captain Adam. *Symzonia: Voyage of Discovery*. New York: J. Seymour, 1820.

———. *Symzonia: Voyage of Discovery*. Edited by J. O. Bailey. Gainesville, Fla.: Scholars' Facsimiles and Reprints, 1965.

Silliman, Benjamin. *A Visit to Europe in 1851*. 2 vols. New York: George P. Putnam, 1853.

Smith, Edmond Reuel. *The Araucanians; or, Notes of a Tour among the Indian Tribes of Southern Chili*. New York: Harper and Brothers, 1855.

Stillé, Alfred, M.D. *Humboldt's Life and Character: An Address before the Linnaean Society of Pennsylvania College*. Philadelphia: Linnaean Association, 1859.

Stoddard, Richard Henry. *The Life, Travels, and Books of Alexander von Humboldt*, with an introduction by Bayard Taylor. New York: Rudd and Carleton, 1859.

Symmes, Elmore. "John Cleves Symmes, The Theorist." *Southern Bivouac*, n.s., 2 (February, March, and April 1887), pp. 555–66, 621–31, and 682–93.

Thoreau, Henry David. *The Maine Woods*. 1864. Reprint, New York: W. W. Norton, 1950.

———. *Walden*. 1854. Reprint, Boston: Beacon Press, 2004.

———. "Walking." 1862. In *Nature/Walking*, edited by John Elder. Boston: Beacon Press, 1991.

Twain, Mark. *Roughing It.* 1872. Reprint, Berkeley: University of California Press, 1993.

Welpley, James Davenport. "Humboldt's Cosmos." *American Review: A Whig Journal of Politics, Literature, Art, and Science* 3 (June 1846), pp. 598–610.

Whitman, Walt. *Democratic Vistas*. In *The Portable Walt Whitman*, edited by Mark Van Doren. New York: Penguin, 1973.

———. *Leaves of Grass*. New York: New American Library, 1980.

Wilkes, Charles. *Narrative of the United States Exploring Expedition.* 5 vols. Philadelphia: Lea and Blanchard, 1845.

Williams, H. L. *History of the Adventurous Voyage and Terrible Shipwreck of the U.S. Steamer "Jeannette."* New York: A. T. B. De Witt, 1882.

Winthrop, Theodore. *The Canoe and the Saddle: Adventures among the Northwestern Rivers and Forests.* 1862. Reprint, Boston: Ticknor and Fields, 1864.

———. *Cecil Dreeme.* 1861. Reprint, New York: Henry Holt, 1876.

———. *A Companion to the "Heart of the Andes."* New York: Appleton, 1859.

———. *John Brent.* 9th ed. Boston: Ticknor and Fields, 1862.

———. *Life in the Open Air and Other Papers.* New York: John W. Lovell, 1862.

SECONDARY SOURCES

Humboldt

Alexander von Humboldt: Netzwerke des Wissens. Ostfildern-Ruit, Germany: Hatje Cantz Verlag, 1999.

Beck, Hanno. *Alexander von Humboldt.* 2 vols. Wiesbaden: F. Steiner, 1959 and 1961. In German.

Biermann, Kurt-R., Ilse Jahn, and Fritz G. Lange, eds. *Alexander von Humboldt: Chronologische Übersicht über wichtige Daten seines Lebens.* Berlin: Akademie Verlag, 1983.

Botting, Douglas. *Humboldt and the Cosmos.* London: Michael Joseph, 1973.

Bowen, Margarita. *Empiricism and Geographical Thought: From Francis Bacon to Alexander von Humboldt.* Cambridge: Cambridge University Press, 1981.

Brading, D. A. *The First America: The Spanish Monarchy, Creole Patriots, and the Liberal State, 1492–1867.* Cambridge: Cambridge University Press, 1991.

Brann, E. R. *Alexander von Humboldt, Patron of Science.* Madison, Wis.: Littel Printing, 1954.

———. *The Political Ideas of Alexander von Humboldt: A Brief Preliminary Study.* Madison, Wis.: Littel Printing, 1954.

Buffett, Jimmy. *A Salty Piece of Land.* New York: Little, Brown, 2004.

Bunske, Edmunds V. "Humboldt and an Aesthetic Tradition in Geography." *Geographical Review* 71 (April 1981), pp. 128–46.

Daum, Andreas. "Celebrating Humanism in St. Louis: The Origins of the Humboldt Statue in Tower Grove Park, 1859–1878." *Gateway Heritage* 15 (Fall 1994), pp. 48–58.

De Terra, Helmut. "Alexander von Humboldt's Correspondence with Jefferson, Madison, and Gallatin." *Proceedings of the American Philosophical Society* 103 (December 1959), pp. 783–806.

———. *Humboldt: The Life and Times of Alexander von Humboldt, 1769–1859.* New York: Knopf, 1955.

———. "Motives and Consequences of Alexander von Humboldt's Visit to the United States (1804)." *Proceedings of the American Philosophical Society* 104 (June 1960), pp. 314–16.

———. "Studies of the Documentation of Alexander von Humboldt" (two articles with the same title). *Proceedings of the American Philosophical Society* 102 (February and December 1958), pp. 136–41 and 560–89.

Dettelbach, Michael. "Humboldtian Science." In *Cultures of Natural History*, edited by N. Jardine, J. A. Secord, and E. C. Spary. Cambridge: Cambridge University Press, 1996, pp. 287–304.

———. Introduction to vol. 2 of *Cosmos*, by Alexander von Humboldt. Baltimore: Johns Hopkins University Press, 1997.

Ette, Ottmar, and Walther L. Bernecker, eds. *Ansichten Amerikas: Neuere Studien zu Alexander von Humboldt.* Frankfurt am Main: Vervuert Verlag, 2001.

Foner, Philip S. "Alexander von Humboldt on Slavery in America." *Science and Society* 47 (Fall 1983), pp. 330–42.

————, ed. *Alexander von Humboldt über die Sklaverei in den USA.* Berlin: Humboldt-Universität zu Berlin, 1981.

Friis, H. R. "Alexander von Humboldts Besuch in den Vereinigten Staaten von Amerika." In *Alexander von Humboldt: Studien zu Seiner Universalen Geisteshaltung,* edited by Joachim H. Schultze. Berlin: De Gruyter, 1959, pp. 142–95.

————. "Alexander von Humboldt's Visit to the United States of America." In *Records of the Columbia Historical Society of Washington, D.C., 1960–1962,* edited by Francis Coleman Rosenberger. Washington, D.C.: Columbia Historical Society, 1963, pp. 1–35.

Gascar, Pierre. *Humboldt L'Explorateur.* Paris: Gallimard, 1985.

Godlewska, Anne Marie Claire. "From Enlightenment Vision to Modern Science? Humboldt's Visual Thinking." In *Geography and Enlightenment,* edited by David N. Livingstone and Charles W. J. Withers. Chicago: University of Chicago Press, 1999, pp. 236–75.

Hartshorne, Richard. "The Concept of Geography as a Science of Space, from Kant and Humboldt to Hettner." *Annals of the Association of American Geographers* 48 (1958), pp. 97–108.

Harvey, David. "The Humboldt Connection." *Annals of the Association of American Geographers* 88 (December 1998), pp. 723–30.

Haverstock, Mary Sayre. "The Cosmos Captured." *Américas* 35 (January–February 1983), pp. 37–41.

Hein, Wolfgang-Hagen, ed. *Alexander von Humboldt: Life and Work.* Ingelheim am Rhein: C. H. Boehringer Sohn, 1987.

Helferich, Gerard. *Humboldt's Cosmos: Alexander von Humboldt and the Latin American Journey That Changed the Way We See the World.* New York: Gotham Books, 2004.

Humboldt Field Research Institute. "Alexander von Humboldt's Natural History Legacy and Its Relevance for Today." Special issue no. 1, *Northeastern Naturalist* (2001).

Kellner, L. *Alexander von Humboldt.* London: Oxford University Press, 1963.

Leitzmann, Albert, ed. *Georg und Therese Forster und die Brüder Humboldt.* Bonn: Verlag Ludwig Röhrscheid, 1936.

Macpherson, Anne M. "The Human Geography of Alexander von Humboldt." Ph.D. diss., University of California, Berkeley, 1972.

McCullough, David. "The Man Who Rediscovered America." *Audubon* 75 (September 1973), pp. 50–63.

McIntyre, Loren. "Humboldt's Way: Pioneer of Modern Geography." *National Geographic* 168 (September 1985), pp. 318–50.

Meyer-Abich, Adolf. *Alexander von Humboldt, 1769/1969.* Bonn: Inter Nationes, 1969.

Minguet, Charles. *Alexandre de Humboldt, Historien et Géographe de l'Amérique Espagnole, 1799–1804.* 1969. Reprint, Paris: L'Harmattan, 1997.

Nelken, Halina. *Alexander von Humboldt: His Portraits and Their Artists; A Documentary Iconography.* Berlin: D. Reimer, 1980.

Nicolson, Malcolm. "Alexander von Humboldt, Humboldtian Science and the Origins of the Study of Vegetation." *History of Science* 25 (June 1987), pp. 167–94.

————. "Historical Introduction" to *Personal Narrative of a Journey to the Equinoctial Regions of the New Continent,* by Alexander von Humboldt. Abridged ed. London: Penguin, 1995.

Oppitz, Ulrich-Dieter. "Der Name der Brüder Humboldt in aller Welt." In *Alexander von Humboldt: Werk und Weltgeltung,* edited by Heinrich Pfeiffer. Munich: R. Piper, 1969.

Pochmann, Henry. *German Culture in America: Philosophical and Literary Influences, 1600–1900.* Madison: University of Wisconsin Press, 1957.

Puig-Samper, Miguel Ángel. "Humboldt, un Pruisano en la Corte del Rey Carlos IV." *Revista de Indias* 59 (Mayo–Agosto 1999), pp. 337–55.

Rippy, J. Fred, and E. R. Brann. "Alexander von Humboldt and Simón Bolívar." *American Historical Review* 52 (July 1947), pp. 697–703.

Rupke, Nicolaas A. Introduction to vol. 1 of *Cosmos*, by Alexander von Humboldt. Baltimore: Johns Hopkns University Press, 1997.

Sachs, Aaron. "The Ultimate 'Other': Post-Colonialism and Alexander von Humboldt's Ecological Relationship with Nature." Theme issue on the environment, *History and Theory* 42 (December 2003), pp. 111–35.

———. "When Science Went Astray: Social Darwinism, Specialization, and the Forgotten Legacy of Alexander von Humboldt." *World Watch* 8 (March/April 1995), pp. 28–38.

Schwarz, Ingo. "Das Album der Humboldt-Lokalitäten in der Neuen Welt." *Magazin für Amerikanistik* 20 (nos. 2 and 3, 1996), pp. 64–66 (no. 2) and 52–54 (no. 3).

———. "The Second Discoverer of the New World and the First American Literary Ambassador to the Old World: Alexander von Humboldt and Washington Irving." *Acta Historica Leopoldina* 27 (1997), pp. 89–97.

———. " 'Shelter for a Reasonable Freedom' or Cartesian Vortex: Aspects of Alexander von Humboldt's Relation to the United States." *Debate y Perspectivas* 1 (2000), pp. 169–70.

Seminario de Historia de la Filosofía en México, de la Facultad de Filosofía y Letras. *Ensayos Sobre Humboldt*. México: Universidad Nacional Autónoma de México, 1962.

Stevens-Middleton, Rayfred Lionel. *La Obra de Alejandro de Humboldt en México: Fundamento de la Geografía Moderna*. México: Sociedad Mexicana de Geografía y Estadística, 1956.

Stoetzer, Carlos. "Humboldt: A Hundred Years After." *Américas* 11 (May 1959), pp. 2–8.

Van Dusen, Robert. *The Literary Ambitions and Achievements of Alexander von Humboldt*. European University Papers, vol. 52. Bern and Frankfurt/M.: Herbert Lang and Peter Lang, 1971.

Von Hagen, Victor Wolfgang. *South America Called Them: Explorations of the Great Naturalists La Condamine, Humboldt, Darwin, Spruce*. New York: Knopf, 1945.

Wadyko, Michael Anthony. "Alexander von Humboldt and 19th-Century Ideas on the Origin of American Indians." Ph.D. diss., West Virginia University, 2000.

Wasserman, Felix M. "Six Unpublished Letters of Alexander von Humboldt to Thomas Jefferson." *Germanic Review* 29 (October 1954), pp. 191–200.

Werner, Petra. *Himmel und Erde: Alexander von Humboldt und sein Kosmos*. Berlin: Akademie Verlag, 2004.

Whitaker, Arthur P. "Alexander von Humboldt and Spanish America." *Proceedings of the American Philosophical Society* 104 (June 1960), pp. 317–22.

Wilson, Jason. Introduction to *Personal Narrative of a Journey to the Equinoctial Regions of the New Continent*, by Alexander von Humboldt. Abridged ed. London: Penguin, 1995.

Reynolds

Almy, Robert F. "J. N. Reynolds: A Brief Biography with Particular Reference to Poe and Symmes." *Colophon*, n.s., 2 (Winter 1937), pp. 227–45.

Altschuler, Glenn C., and Stuart M. Blumin. *Rude Republic: Americans and Their Politics in the Nineteenth Century*. Princeton: Princeton University Press, 2000.

Bailey, J. O. "Sources for Poe's *Arthur Gordon Pym*, 'Hans Pfaal,' and Other Pieces." *PMLA* 57 (June 1942), pp. 513–35.

Balch, Edwin Swift. *Antarctica*. Philadelphia: Allen, Lane, and Scott, 1902.

Bandy, W. T. "Dr. Moran and the Poe-Reynolds Myth." In *Myths and Reality: The Mysterious Mr. Poe*, edited by Benjamin Franklin Fisher IV. Baltimore: Edgar Allan Poe Society, 1987, pp. 26–36.

Bertrand, Kenneth J. *Americans in Antarctica, 1775–1948*. New York: American Geographical Society, 1971.

Bradfield, Scott. *Dreaming Revolution: Transgression in the Development of American Romance*. Iowa City: University of Iowa Press, 1993.

Busch, Briton Cooper. *The War against the Seals: A History of the North American Seal Fishery*. Kingston, Ont.: McGill-Queens University Press, 1985.

Calman, W. T. "James Eights, a Pioneer Antarctic Naturalist." *Proceedings of the Linnean Society of London*, sess. 149, pt. 4 (1937), pp. 171–84.

Charvat, William. "American Romanticism and the Depression of 1837." *Science and Society* 2 (Winter 1937), pp. 67–82.

Clarke, John M. "The Reincarnation of James Eights, Antarctic Explorer." *Scientific Monthly* 2 (February 1916), pp. 189–202.

Collins, Paul. "Symmes Hole," in *Banvard's Folly: Thirteen Tales of People Who Didn't Change the World*. New York: Picador, 2001, pp. 54–70.

Coyle, William, ed. *Ohio Authors and Their Books*. Cleveland: World Publishing, 1962.

Daniels, George H. *American Science in the Age of Jackson*. New York: Columbia University Press, 1968.

Davis, Lance E., Robert E. Gallman, and Karen Gleiter. *In Pursuit of Leviathan: Technology, Institutions, Productivity, and Profits in American Whaling, 1816–1906*. Chicago: University of Chicago Press, 1997.

Garnett, R. S. "Moby-Dick and Mocha-Dick: A Literary Find." *Blackwood's Magazine* 226 (December 1929), pp. 841–58.

Gilmore, Michael T. *American Romanticism and the Marketplace*. Chicago: University of Chicago Press, 1985.

Gurney, Alan. *Below the Convergence: Voyages toward Antarctica, 1699–1839*. New York: W. W. Norton, 1997.

———. *The Race to the White Continent*. New York: W. W. Norton, 2000.

Helms, Randel. "Another Source for Poe's *Arthur Gordon Pym*." *American Literature* 41 (January 1970), pp. 572–75.

Holt, Michael F. *The Rise and Fall of the American Whig Party: Jacksonian Politics and the Onset of the Civil War*. New York: Oxford University Press, 1999.

Horsman, Reginald. "Captain Symmes's Journey to the Center of the Earth." *Timeline* 17 (September/October 2000), pp. 2–13.

Howe, Daniel Walker. *The Political Culture of American Whigs*. Chicago: University of Chicago Press, 1979.

Huntress, Keith. "Another Source for Poe's *Narrative of Arthur Gordon Pym*." *American Literature* 16 (March 1944), pp. 19–25.

Irwin, John T. *American Hieroglyphics: The Symbol of the Egyptian Hieroglyphics in the American Renaissance*. New Haven: Yale University Press, 1980.

Kohlstedt, Sally Gregory. "The Nineteenth-Century Amateur Tradition: The Case of the Boston Society of Natural History." In *Science and Its Public: The Changing Relationship*, edited by Gerald Holton and William A. Blanpied. Dordrecht: D. Reidel, 1976, pp. 173–90.

Kopley, Richard, ed. *Poe's Pym: Critical Explorations*. Durham, N.C.: Duke University Press, 1992.

Lenz, William E. *The Poetics of the Antarctic: A Study in Nineteenth-Century American Cultural Perceptions*. New York: Garland Publishing, 1995.

Martin, Lawrence. "James Eights' Pioneer Observations and Interpretations of Erratics in Antarctic Icebergs." *Bulletin of the Geological Society of America* 60 (January 1949), pp. 177–82.

Matthews, Jean V. *Toward a New Society: American Thought and Culture, 1800–1830*. Boston: Twayne Publishers, 1991.

Mawer, Granville Allen. *Ahab's Trade: The Saga of South Seas Whaling*. St. Leonard's, N.S.W.: Allen and Unwin, 1999.

McGrane, Charles. *The Panic of 1837*. 1924. Reprint, Chicago: University of Chicago Press, 1965.

Meisel, Max. *A Bibliography of American Natural History: The Pioneer Century, 1769–1865*. Vol. 2. New York: Premier Publishing, 1926.

Mitterling, Philip I. *America in the Antarctic to 1840*. Urbana: University of Illinois Press, 1959.

Nelson, Victoria. "Symmes Hole, or the South Polar Romance." *Raritan* 17 (Fall 1997), pp. 136–66.

Norton, Anne. *Alternative Americas: A Reading of Antebellum Political Culture*. Chicago: University of Chicago Press, 1986.

Peck, John Wells. "Symmes' Theory." *Ohio Archaeological and Historical Publications* 18 (1909), pp. 28–42.

Pessen, Edward. *Jacksonian America: Society, Personality, and Politics*. Homewood, Ill.: Dorsey Press, 1969.

Philbrick, Nathaniel. *In the Heart of the Sea: The Tragedy of the Whaleship* Essex. New York: Viking, 2000.

———. *Sea of Glory: America's Voyage of Discovery, The U.S. Exploring Expedition, 1838–1842.* New York: Viking, 2003.

Rhea, Robert Lee. "Some Observations on Poe's Origins." *University of Texas Bulletin, Studies in English* 10 (July 8, 1930), pp. 135–46.

Schlesinger, Arthur M. *The Age of Jackson.* Boston: Little, Brown, 1945.

Sellers, Charles. *The Market Revolution: Jacksonian America, 1815–1846.* New York: Oxford University Press, 1991.

Severin, Tim. *In Search of Moby Dick: Quest for the White Whale.* New York: Da Capo Press, 2001.

Silverman, Kenneth. *Edgar A. Poe: Mournful and Never-Ending Remembrance.* New York: Harper-Collins, 1991.

Simpson-Housley, Paul. *Antarctica: Exploration, Perception, and Metaphor.* London: Routledge, 1992.

Stackpole, Edouard. *The Sea-Hunters: The Great Age of Whaling.* Philadelphia: J. B. Lippincott, 1953.

Stanton, William. *The Great United States Exploring Expedition of 1838–1842.* Berkeley: University of California Press, 1975.

Starke, Aubrey. "Poe's Friend Reynolds." *American Literature* 11 (May 1939), pp. 152–59.

Van Doren, Carl. "Mr. Melville's Moby Dick." *Bookman* 59 (April 1924), pp. 154–57.

Viola, Herman J., and Carolyn Margolis, eds. *The U.S. Exploring Expedition, 1838–1842.* Washington, D.C.: Smithsonian Institution Press, 1985.

Wallace, Anthony F. C. *The Long Bitter Trail: Andrew Jackson and the Indians.* New York: Hill and Wang, 1993.

Watson, Harry L. *Liberty and Power: The Politics of Jacksonian America.* New York: Hill and Wang, 1990.

Whalen, Terence. *Edgar Allan Poe and the Masses: The Political Economy of Literature in Antebellum America.* Princeton: Princeton University Press, 1999.

Wilentz, Sean. *Chants Democratic: New York City and the Rise of the American Working Class, 1788–1850.* New York: Oxford University Press, 1984.

Willis, Thornton. "*Washingtonii Vita.*" *Imprimatur* 1 (April 1947), pp. 33–39.

Woodbridge, Richard G., III. "J.N. Reynolds: Father of American Exploration." *Princeton University Library Chronicle* 45 (Winter 1984), pp. 107–21.

King

Bartlett, Richard A. *Great Surveys of the American West.* Norman: University of Oklahoma Press, 1962.

Colby, Elbridge. *Bibliographical Notes on Theodore Winthrop.* New York: New York Public Library, 1917.

Cooper, John D., Richard H. Miller, and Jacqueline Patterson. *A Trip Through Time: Principles of Historical Geology.* Columbus: Merrill, 1986.

Crosby, Harry H. "The Great Diamond Fraud." *American Heritage* 7 (February 1956), pp. 58–63 and 100.

Dickason, David H. "Clarence King's First Westward Journey." *Huntington Library Quarterly* 7 (November 1943), pp. 71–87.

———. "Henry Adams and Clarence King, The Record of a Friendship." *New England Quarterly* 17 (June 1944), pp. 229–54.

Dingus, Rick. *The Photographic Artifacts of Timothy H. O'Sullivan.* Albuquerque: University of New Mexico Press, 1982.

Durham, Michael S. *Desert Between the Mountains: Mormons, Miners, Padres, Mountain Men, and the Opening of the Great Basin.* Norman: University of Oklahoma Press, 1997.

Eliot, Ellsworth. *Theodore Winthrop.* New Haven: Yale University Library, 1938.

Emmons, Samuel Franklin. "Biographical Memoir of Clarence King." *National Academy of Sciences Biographical Memoirs* 6 (1909), pp. 25–54.

Emmons, Samuel Franklin, et al. *Clarence King: A Memorial*. New York: The Engineering and Mining Journal, 1902.

Faragher, John Mack, ed. *Rereading Frederick Jackson Turner: "The Significance of the Frontier in American History" and Other Essays*. New York: Henry Holt, 1994.

Farquhar, Francis P. "The Whitney Survey on Mount Shasta, 1862: A Letter from William H. Brewer to Professor Brush." *California Historical Society Quarterly* 7 (June 1928), pp. 121–31.

Farquhar, Peter. "Site of the Diamond Swindle of 1872." *California Historical Society Quarterly* 42 (March 1963), pp. 49–53.

Fontana, Ernest L. "Cognition and Ordeal in Clarence King's *Mountaineering in the Sierra Nevada*." *Explorations* 4 (July 1977), pp. 25–30.

Goetzmann, William H. *Army Exploration in the American West, 1803–1863*. New Haven: Yale University Press, 1959.

———. *Exploration and Empire: The Explorer and the Scientist in the Winning of the American West*. New York: W. W. Norton, 1966.

———. "The Heroic Age of Western Geological Exploration: The U.S. Geological Survey and the Men and Events That Created It." *The American West* 16 (September/October 1979), pp. 4–13; 59–61.

Hague, James D. "Clarence King: An Appreciation." *New York Evening Post*, December 28, 1901, p. 5.

———, ed. *Clarence King Memoirs*. New York: G. P. Putnam's Sons, 1904.

Hodes, Martha, ed. *Sex, Love, Race: Crossing Boundaries in North American History*. New York: New York University Press, 1999.

Horan, James D. *Timothy O'Sullivan: America's Forgotten Photographer*. Garden City, N.Y.: Doubleday, 1966.

Huggett, Richard. *Catastrophism: Systems of Earth History*. London: Edward Arnold, 1990.

Kramer, Howard D. "The Scientist in the West, 1870–1880." *Pacific Historical Review* 12 (September 1943), pp. 239–51.

The Life and Poems of Theodore Winthrop, edited by his sister. New York: Henry Holt, 1884.

Lundberg, Ann E. " 'The Ruin of a Bygone Geological Empire': Clarence King and the Place of the Primitive in the Evolution of American Identity." *American Transcendental Quarterly* 18 (September 2004), pp. 179–203.

Manning, Thomas. *Government in Science: The USGS 1867–1894*. Lexington: University of Kentucky Press, 1967.

Morrison, Michael A. *Slavery and the American West: The Eclipse of Manifest Destiny and the Coming of the Civil War*. Chapel Hill: University of North Carolina Press, 1997.

Naef, Weston J., and James N. Wood. *Era of Exploration: The Rise of Landscape Photography in the American West, 1860–1885*. Buffalo: Albright-Knox Art Gallery and the Metropolitan Museum of Art, 1975.

Nash, Gerald D. "The Conflict between Pure and Applied Science in 19th-Century Public Policy: The California Geological Survey, 1860–1874." *Isis* 54 (June 1963), pp. 217–28.

O'Toole, Patricia. *The Five of Hearts: An Intimate Portrait of Henry Adams and His Friends, 1880–1918*. New York: Ballantine Books, 1990.

Paul, Rodman W. *Mining Frontiers of the Far West, 1848–1880*. 1963. Reprint, Albuquerque: University of New Mexico Press, 2001.

Shebl, James M. *King, of the Mountains*. Stockton, Calif.: Pacific Center for Western Historical Studies, 1974.

Skinner, Brian J., and Stephen C. Porter. *Physical Geology*. New York: John Wiley and Sons, 1987.

Smith, Henry Nash. "Clarence King, John Wesley Powell, and the Establishment of the United States Geological Survey." *Mississippi Valley Historical Review* 34 (June 1947), pp. 37–58.

Snyder, Joel. "Aesthetics and Documentation: Remarks Concerning Critical Approaches to the Photographs of Timothy H. O'Sullivan." In *Perspectives on Photography: Essays in Honor of Beaumont Newhall*, edited by Peter Walch and Thomas Barrow. Albuquerque: University of New Mexico Press, 1986.

———. *American Frontiers: The Photographs of Timothy H. O'Sullivan, 1867–74*. Millerton, N.Y.: Aperture, 1981.

Sollors, Werner, ed. *Interracialism: Black-White Intermarriage in American History, Literature, and Law.* New York: Oxford University Press, 2000.

Starr, Kevin. *Americans and the Californian Dream, 1850–1915.* New York: Oxford University Press, 1973.

Trimble, Stephen. *The Sagebrush Ocean: A Natural History of the Great Basin.* Reno: University of Nevada Press, 1989.

Utley, Robert M. *The Indian Frontier of the American West, 1846–1890.* Albuquerque: University of New Mexico Press, 1984.

Wilkins, Thurman. *Clarence King: A Biography.* Rev. and enlarged ed. Albuquerque: University of New Mexico Press, 1988.

Williams, John H., and Clarence B. Bagley. *Winthrop and Curtis.* Tacoma: J. H. Williams, 1914.

Wilson, Allen D. "The Great California Diamond Mines: A True Story." *Overland Monthly,* 2nd ser., 43 (April 1904), pp. 291–96.

Wilson, Richard. "American Vision and Landscape: The Western Images of Clarence King and Timothy H. O'Sullivan." Ph.D. diss., University of New Mexico, 1979.

Melville

Berger, Martin A. *Man Made: Thomas Eakins and the Construction of Gilded Age Manhood.* Berkeley: University of California Press, 2000.

Berton, Pierre. *The Arctic Grail: The Quest for the North West Passage and the North Pole, 1818–1909.* Toronto: McClelland and Stewart, 1988.

Bolles, Edmund Blair. *Ice Finders: How a Poet, a Professor, and a Politician Discovered the Ice Age.* Washington, D.C.: Counterpoint, 1999.

Cathcart, William Ledyard. "George Wallace Melville, Engineer-in-Chief of the United States Navy." *Cassier's Magazine* 11 (April 1897), pp. 459–80.

Chapin, David Alexander. "Exploring Other Worlds: Margaret Fox, Elisha Kane, and the Antebellum Culture of Curiosity." Ph.D. diss., University of New Hampshire, 2000.

Dillard, Annie. "An Expedition to the Pole." In *Teaching a Stone to Talk: Expeditions and Encounters.* New York: Harper and Row, 1982, pp. 17–52.

Ellsberg, Commander Edward. *Hell on Ice: The Saga of the Jeannette.* New York: Dodd, Mead, 1938.

Friis, Herman R., and Shelby G. Bale Jr., eds. *United States Polar Exploration.* Athens: Ohio University Press, 1970.

Guttridge, Leonard F. *Ghosts of Cape Sabine: The Harrowing True Story of the Greely Expedition.* New York: G. P. Putnam's Sons, 2000.

———. *Icebound: The Jeannette Expedition's Quest for the North Pole.* Annapolis: Naval Institute Press, 1986.

Henderson, Bruce. *Fatal North: Adventure and Survival Aboard USS Polaris, the First U.S. Expedition to the North Pole.* New York: New American Library, 2001.

Hoehling, A. A. *The Jeannette Expedition: An Ill-Fated Journey to the Arctic.* New York: Abelard-Schuman, 1967.

Holland, Clive, ed. *Farthest North: The Quest for the Pole.* London: Robinson, 1994.

Johnson, Clive. *The Devil's Labyrinth: Encounters with the Arctic.* Shrewsbury, U.K.: Swan Hill Press, 1995.

Loomis, Chauncey C. *Weird and Tragic Shores: The Story of Charles Francis Hall, Explorer.* New York: Knopf, 1971.

Lopez, Barry. *Arctic Dreams: Imagination and Desire in a Northern Landscape.* New York: Bantam, 1986.

Malaurie, Jean. *Ultima Thule: Explorers and Natives in the Polar North.* New York: W. W. Norton, 2003.

Maxtone-Graham, John. *Safe Return Doubtful: The Heroic Age of Polar Exploration.* 1988. Reprint, London: Constable, 2000.

Mirsky, Jeannette. *Elisha Kent Kane and the Seafaring Frontier*. Boston: Little, Brown, 1954.
————. *Northern Conquest: The Story of Arctic Exploration from Earliest Times to the Present*. London: H. Hamilton, 1934.
Neatby, L. H. *Discovery in Russian and Siberian Waters*. Athens: Ohio University Press, 1973.
Officer, Charles, and Jake Page. *A Fabulous Kingdom: The Exploration of the Arctic*. New York: Oxford University Press, 2001.
Oswalt, Wendell H. *Eskimos and Explorers*. 1979. Reprint, Lincoln: University of Nebraska Press, 1999.
Parry, Richard. *Trial by Ice: The True Story of Murder and Survival on the 1871 Polaris Expedition*. New York: Ballantine Books, 2001.
Robinson, Michael Frederick. "The Coldest Crucible: Arctic Exploration and American Culture, 1850–1910." Ph.D. diss., University of Wisconsin—Madison, 2002. (Book version due out in 2006 from the University of Chicago Press.)
Vaughan, Richard. *The Arctic: A History*. Phoenix Mill: Alan Sutton Publishing, 1994.
Wilmerding, John, ed. *Thomas Eakins (1844–1916) and the Heart of American Life*. London: National Portrait Gallery, 1993.
Ziff, Larzer. "Arctic Exploration and the Romance of Failure." *Raritan* 23 (Fall 2003), pp. 58–79.

Muir

Bockstoce, John R. *Whales, Ice, and Men: The History of Whaling in the Western Arctic*. Seattle: University of Washington Press, 1986.
Campbell, Robert Bruce. "Inside Passage: Alaskan Travel, American Culture, and the Nature of Empire, 1867–1898." Ph.D. diss., Yale University, 2003. (Book version forthcoming from University of Pennsylvania Press.)
Cohen, Michael P. *The Pathless Way: John Muir and American Wilderness*. Madison: University of Wisconsin Press, 1984.
Fleck, Richard F. *Henry Thoreau and John Muir among the Indians*. Hamden, Conn.: Archon Books, 1985.
Fox, Stephen. *John Muir and His Legacy: The American Conservation Movement*. Boston: Little, Brown, 1981.
Hinckley, Ted C. *The Canoe Rocks: Alaska's Tlingit and the Euramerican Frontier, 1800–1912*. Lanham, Md.: University Press of America, 1996.
Kollin, Susan. *Nature's State: Imagining Alaska as the Last Frontier*. Chapel Hill: University of North Carolina Press, 2001.
Limbaugh, Ronald H. *John Muir's "Stickeen" and the Lessons of Nature*. Fairbanks: University of Alaska Press, 1996.
Merriam, C. Hart. "To the Memory of John Muir." *Sierra Club Bulletin* 10 (January 1917); available on the Web at www.sierraclub.org/john_muir_exhibit/life/memory_jm_c_hart_merriam.html.
Miller, Sally M., ed. *John Muir in Historical Perspective*. New York: Peter Lang, 1999.
————. *John Muir: Life and Work*. Albuquerque: University of New Mexico Press, 1993.
Mitchell, Donald Craig. *Sold American: The Story of Alaska Natives and Their Land, 1867–1959*. Hanover, N.H.: Dartmouth College/ University Press of New England, 1997.
Turner, Frederick. *Rediscovering America: John Muir in His Time and Ours*. New York: Viking, 1985.
Wilkins, Thurman. *John Muir: Apostle of Nature*. Norman: University of Oklahoma Press, 1995.
Wolfe, Linnie Marsh. *Son of the Wilderness: The Life of John Muir*. New York: Knopf, 1945.
Worster, Donald. "John Muir and the Modern Passion for Nature." *Environmental History* 10 (January 2005), pp. 8–19.
Young, S. Hall. *Alaska Days with John Muir*. New York: Fleming H. Revell, 1915.

General

Allen, John L., ed. *North American Exploration.* 3 vols. Lincoln: University of Nebraska Press, 1997.

Arac, Jonathan, and Harriet Ritvo, eds. *Macropolitics of Nineteenth-Century Literature: Nationalism, Exoticism, Imperialism.* Philadelphia: University of Pennsylvania Press, 1991.

Avery, Kevin J. *Church's Great Picture: "The Heart of the Andes."* New York: Metropolitan Museum of Art, 1993.

Barrett, Andrea. *The Voyage of the Narwhal.* New York: W. W. Norton, 1998.

Bate, Jonathan. *Romantic Ecology: Wordsworth and the Environmental Tradition.* New York: Routledge, 1991.

———. *The Song of the Earth.* Cambridge, Mass.: Harvard University Press, 2000.

Bedell, Rebecca. *The Anatomy of Nature: Geology and American Landscape Painting, 1825–1875.* Princeton: Princeton University Press, 2001.

Bercovitch, Sacvan. *The American Jeremiad.* Madison: University of Wisconsin Press, 1978.

———. *The Rites of Assent: Transformations in the Symbolic Construction of America.* New York: Routledge, 1993.

Bercovitch, Sacvan, and Myra Jehlen, eds. *Ideology and Classic American Literature.* New York: Cambridge University Press, 1986.

Berger, John. *Ways of Seeing.* New York: Penguin, 1972.

Berkhofer, Robert F., Jr. *The White Man's Indian: Images of the American Indian from Columbus to the Present.* New York: Vintage, 1979.

Blaut, J. M. *The Colonizer's Model of the World: Geographical Diffusionism and Eurocentric History.* New York: Guilford Press, 1993.

Bledstein, Burton J. *The Culture of Professionalism: The Middle Class and the Development of Higher Education in America.* New York: W. W. Norton, 1976.

Bloom, Lisa. *Gender on Ice: American Ideologies of Polar Expeditions.* Minneapolis: University of Minnesota Press, 1993.

Botkin, Daniel. *Discordant Harmonies: A New Ecology for the Twenty-First Century.* New York: Oxford University Press, 1990.

Boyarin, Daniel, and Jonathan Boyarin. "Diaspora: Generation and the Ground of Jewish Identity." In *Identities,* edited by Kwame Anthony Appiah and Henry Louis Gates Jr., Chicago: University of Chicago Press, 1995, pp. 305–37.

Bramwell, Anna. *Ecology in the Twentieth Century: A History.* New Haven: Yale University Press, 1989.

———. *The Fading of the Greens: The Decline of Environmental Politics in the West.* New Haven: Yale University Press, 1994.

Brulle, Robert J. *Agency, Democracy, and Nature: The U.S. Environmental Movement from a Critical Theory Perspective.* Cambridge, Mass.: MIT Press, 2000.

Bryson, Michael A. *Visions of the Land: Science, Literature, and the American Environment from the Era of Exploration to the Age of Ecology.* Charlottesville: University Press of Virginia, 2002.

Buell, Lawrence. *The Environmental Imagination: Thoreau, Nature Writing, and the Formation of American Culture.* Cambridge, Mass.: Harvard University Press, 1995.

———. *Writing for an Endangered World: Literature, Culture, and Environment in the U.S. and Beyond.* Cambridge, Mass.: Harvard University Press, 2001.

Burnett, D. Graham. *Masters of All They Surveyed: Exploration, Geography, and a British El Dorado.* Chicago: University of Chicago Press, 2000.

Buttimer, Anne. *Geography and the Human Spirit.* Baltimore: Johns Hopkins University Press, 1993.

Caesar, Terry. *Forgiving the Boundaries: Home as Abroad in American Travel Writing.* Athens: University of Georgia Press, 1995.

Cahn, Robert. *American Photographers and the National Parks.* New York: Viking, 1981.

Callicott, J. Baird, and Michael P. Nelson, eds. *The Great New Wilderness Debate.* Athens: University of Georgia Press, 1998.

Cannon, Susan Faye. *Science in Culture: The Early Victorian Period.* New York: Dawson and Science History Publications, 1978.

Carnes, Mark C. *Secret Ritual and Manhood in Victorian America.* New Haven: Yale University Press, 1989.

Carnes, Mark C., and Clyde Griffen, eds. *Meanings for Manhood: Constructions of Masculinity in Victorian America.* Chicago: University of Chicago Press, 1990.

Carr, Gerald L. *Frederic Edwin Church: The Icebergs.* Dallas: Dallas Museum of Fine Arts, 1980.

Carter, Edward C., II, ed. *Surveying the Record: North American Scientific Exploration to 1930.* Philadelphia: American Philosophical Society, 1999.

Carter, Paul. *The Lie of the Land.* London: Faber and Faber, 1996.

———. *The Road to Botany Bay: An Exploration of Landscape and History.* New York: Knopf, 1987.

Chaffin, Tom. *Pathfinder: John Charles Frémont and the Course of American Empire.* New York: Hill and Wang, 2002.

Christie, John Aldrich. *Thoreau as World Traveler.* New York: Columbia University Press, 1965.

Clark, John P., ed. *Renewing the Earth: The Promise of Social Ecology.* London: Green Print, 1990.

———. "A Social Ecology." *Capitalism, Nature, Socialism* 8 (September 1997), pp. 3–33.

Clark, John P., and Camille Martin, eds. *Anarchy, Geography, Modernity: The Radical Social Thought of Elisée Reclus.* Lanham, Md.: Lexington Books, 2004.

Clifford, James. *The Predicament of Culture: Twentieth-Century Ethnography, Literature, and Art.* Cambridge, Mass.: Harvard University Press, 1988.

———. *Routes: Travel and Translation in the Late Twentieth Century.* Cambridge, Mass.: Harvard University Press, 1997.

Clifford, James, and George E. Marcus, eds. *Writing Culture: The Poetics and Politics of Ethnography.* Berkeley: University of California Press, 1986.

Cohn, Bernard S. *Colonialism and Its Forms of Knowledge: The British in India.* Princeton: Princeton University Press, 1996.

Connell, Evan S. *A Long Desire.* New York: Holt, Rinehart, and Winston, 1979.

Cosmos Club. *The Twenty-fifth Anniversary of the Founding of the Cosmos Club of Washington.* Washington, D.C.: Cosmos Club, 1904.

Crain, Caleb. *American Sympathy: Men, Friendship, and Literature in the New Nation.* New Haven: Yale University Press, 2001.

Cronon, William, ed. *Uncommon Ground: Rethinking the Human Place in Nature.* New York: W. W. Norton, 1996.

Cunningham, Andrew, and Nicholas Jardine, eds. *Romanticism and the Sciences.* Cambridge: Cambridge University Press, 1990.

Daniels, George H. *Science in American Society: A Social History.* New York: Knopf, 1971.

Davenport, Guy. *The Intelligence of Louis Agassiz.* Boston: Beacon Press, 1963.

De Botton, Alain. *The Art of Travel.* New York: Pantheon, 2002.

Dickason, David. *The Daring Young Men: The Story of the American Pre-Raphaelites.* Bloomington: Indiana University Press, 1953.

Dippie, Brian W. *The Vanishing American: White Attitudes and U.S. Indian Policy.* Middletown, Conn.: Wesleyan University Press, 1982.

Dirlik, Arif. "The Global in the Local." In *Global/Local: Cultural Production and the Transnational Imaginary,* edited by Rob Wilson and Wimal Dissayanke. Durham, N.C.: Duke University Press, 1996, pp. 21–45.

Dorman, Robert L. *A Word for Nature: Four Pioneering Environmental Advocates, 1845–1913.* Chapel Hill: University of North Carolina Press, 1998.

Dowie, Mark. *Losing Ground: American Environmentalism at the Close of the Twentieth Century.* Cambridge, Mass.: MIT Press, 1995.

Driver, Felix. *Geography Militant: Cultures of Exploration and Empire.* Oxford: Blackwell, 2001.

Dupree, A. Hunter. *Science in the Federal Government.* Cambridge, Mass.: Harvard University Press, 1957.

Edwards, Justin D. *Exotic Journeys: Exploring the Erotics of U.S. Travel Literature, 1840–1930.* Hanover, N.H.: University Press of New England, 2001.

Egan, Timothy. *The Good Rain: Across Time and Terrain in the Pacific Northwest.* New York: Vintage, 1991.

Elsner, Jás, and Joan-Pau Rubiés. *Voyages and Visions: Towards a Cultural History of Travel.* London: Reaktion Books, 1999.

Ette, Ottmar. *Literature on the Move.* Translated by Katharina Vester. Amsterdam: Rodopi, 2003.

Evernden, Neil. *The Social Creation of Nature.* Baltimore: Johns Hopkins University Press, 1992.

Faber, Daniel, ed. *The Struggle for Ecological Democracy: Environmental Justice Movements in the United States.* New York: Guilford Press, 1998.

Fabian, Johannes. *Out of Our Minds: Reason and Madness in the Exploration of Central Africa.* Berkeley: University of California Press, 2000.

Ferber, Linda, and William H. Gerdts. *The New Path: Ruskin and the American Pre-Raphaelites.* New York: Schocken, 1985.

Foerster, Norman. *Nature in American Literature.* New York: Macmillan, 1923.

Fritzell, Peter A. *Nature Writing and America: Essays upon a Cultural Type.* Ames: Iowa State University Press, 1990.

Fulford, Tim, Debbie Lee, and Peter J. Kitson. *Literature, Science, and Exploration in the Romantic Era: Bodies of Knowledge.* Cambridge: Cambridge University Press, 2004.

Fussell, Edwin. *Frontier: American Literature and the American West.* Princeton: Princeton University Press, 1965.

Gare, Arran E. *Postmodernism and the Environmental Crisis.* London: Routledge, 1995.

Gerbi, Antonello. *The Dispute of the New World: The History of a Polemic.* Translated by Jeremy Moyle. Pittsburgh: University of Pittsburgh Press, 1983.

Gilroy, Paul. *Against Race: Imagining Political Culture beyond the Color Line.* Cambridge, Mass.: Harvard University Press, 2000.

Gode-von Aesch, Alexander. *Natural Science in German Romanticism.* New York: Columbia University Press, 1941.

Goetzmann, William H. *New Lands, New Men: America and the Second Great Age of Discovery.* 1986. Reprint, Austin: Texas State Historical Association, 1995.

Goetzmann, William H., and Kay Sloan. *Looking Far North: The Harriman Expedition to Alaska, 1899.* New York: Viking, 1982.

Gould, Stephen Jay. "Art Meets Science in *The Heart of the Andes*: Church Paints, Humboldt Dies, Darwin Writes, and Nature Blinks in the Fateful Year of 1859." In *I Have Landed: The End of a Beginning in Natural History.* New York: Harmony Books, 2002.

———. *Ever Since Darwin: Reflections in Natural History.* New York: W. W. Norton, 1977.

———. *The Mismeasure of Man.* New York: W. W. Norton, 1981.

———. *Time's Arrow, Time's Cycle: Myth and Metaphor in the Discovery of Geological Time.* Cambridge, Mass.: Harvard University Press, 1987.

Greenblatt, Stephen. *Marvelous Possessions: The Wonder of the New World.* Chicago: University of Chicago Press, 1991.

Greene, John C. *American Science in the Age of Jefferson.* Ames: Iowa State University Press, 1984.

Greenfield, Bruce. *Narrating Discovery: The Romantic Explorer in American Literature, 1790–1855.* New York: Columbia University Press, 1992.

Grove, Richard. *Green Imperialism: Colonial Expansion, Tropical Island Edens and the Origins of Environmentalism, 1600–1860.* New York: Cambridge University Press, 1995.

Harrison, Robert Pogue. *Forests: The Shadow of Civilization.* Chicago: University of Chicago Press, 1992.

Harvey, Eleanor Jones. *The Voyage of the Icebergs: Frederic Church's Arctic Masterpiece.* New Haven: Dallas Museum of Art and Yale University Press, 2002.

Herron, John Paul. "'The Reality of Living': Science, Gender, and Nature in American Culture, 1865–1965." Ph.D. diss., University of New Mexico, 2001.

Hine, Robert V., and John Mack Faragher. *The American West: A New Interpretive History.* New Haven: Yale University Press, 2000.

Horwitz, Tony. *Blue Latitudes: Boldly Going Where Captain Cook Has Gone Before.* New York: Picador, 2002.

Huhndorf, Shari M. *Going Native: Indians in the American Cultural Imagination.* Ithaca, N.Y.: Cornell University Press, 2001.

Huntford, Roland. *The Last Place on Earth.* New York: Modern Library, 1999.

———. *Nansen: The Explorer as Hero.* London: Duckworth, 1997.

Huntington, David. "Landscape and Diaries: The South American Trips of F. E. Church." *Brooklyn Museum Annual* 5 (1963–64), pp. 65–99.

———. *The Landscapes of Frederic Edwin Church: Vision of an American Era.* New York: Braziller, 1966.

Huth, Hans. *Nature and the American: Three Centuries of Changing Attitudes.* Berkeley: University of California Press, 1957.

Hyde, Lewis. *Trickster Makes This World: Mischief, Myth, and Art.* New York: North Point Press, 1998.

Jackson, Donald. *Thomas Jefferson and the Stony Mountains: Exploring the West from Monticello.* Urbana: University of Illinois Press, 1981.

Johnston, Robert D. "Beyond 'The West.'" *Rethinking History* 2 (Summer 1998), pp. 239–77.

Jones, Howard Mumford. *O Strange New World: American Culture: The Formative Years.* New York: Viking, 1964.

Kammen, Michael. *People of Paradox: An Inquiry Concerning the Origins of American Civilization.* 1972. Reprint, Ithaca, N.Y.: Cornell University Press, 1990.

Karttunen, Frances. *Between Worlds: Interpreters, Guides, and Survivors.* New Brunswick, N.J.: Rutgers University Press, 1994.

Kasson, Joy. *Artistic Voyagers: Europe and the American Imagination in the Works of Irving, Allston, Cole, Cooper, and Hawthorne.* Westport, Conn.: Greenwood Press, 1982.

Katz, Jonathan Ned. *Love Stories: Sex between Men before Homosexuality.* Chicago: University of Chicago Press, 2001.

Keller, Robert H., and Michael F. Turek. *American Indians and National Parks.* Tucson: University of Arizona Press, 1998.

Kelly, Franklin. *Frederic Edwin Church.* Washington, D.C.: National Gallery of Art, 1989.

———. *Frederic Edwin Church and the National Landscape.* Washington, D.C.: Smithsonian Institution, 1988.

Kinsey, Joni Louise. *Thomas Moran and the Surveying of the American West.* Washington, D.C.: Smithsonian Institution Press, 1992.

Knoepflmacher, U. C., and G. B. Tennyson, eds. *Nature and the Victorian Imagination.* Berkeley: University of California Press, 1977.

Kolodny, Annette. *The Land Before Her: Fantasy and Experience of the American Frontiers, 1630–1860.* Chapel Hill: University of North Carolina Press, 1984.

———. *The Lay of the Land: Metaphor as Experience and History in American Life and Letters.* Chapel Hill: University of North Carolina Press, 1975.

Krech, Shepard, III. *The Ecological Indian: Myth and History.* New York: W. W. Norton, 1999.

Kroeber, Karl. *Ecological Literary Criticism: Romantic Imagining and the Biology of Mind.* New York: Columbia University Press, 1994.

LaFeber, Walter. *The New Empire: An Interpretation of American Expansion, 1860–1898.* 1963. Reprint, Ithaca, N.Y.: Cornell University Press, 1967.

Lasch, Christopher. *The True and Only Heaven: Progress and Its Critics.* New York: W. W. Norton, 1991.

Lears, T. J. Jackson. *No Place of Grace: Antimodernism and the Transformation of American Culture, 1880–1920.* New York: Pantheon, 1981.

Leed, Eric. *The Mind of the Traveler: From Gilgamesh to Global Tourism.* New York: Basic Books, 1991.

———. *Shores of Discovery: How Expeditionaries Have Constructed the World*. New York: Basic Books, 1995.

LeMenager, Stephanie. *Manifest and Other Destinies: Territorial Fictions of the Nineteenth-Century United States*. Lincoln: University of Nebraska Press, 2004.

Lewis, R. W. B. *The American Adam: Innocence, Tragedy, and Tradition in the Nineteenth Century*. Chicago: University of Chicago Press, 1955.

Light, Andrew, ed. *Social Ecology After Bookchin*. New York: Guilford Press, 1998.

Limerick, Patricia Nelson. *The Legacy of Conquest: The Unbroken Past of the American West*. New York: W. W. Norton, 1987.

Livingstone, David N. *The Geographical Tradition: Episodes in the History of a Contested Enterprise*. Oxford: Blackwell, 1992.

Livingstone, David N., and Charles W. J. Withers, eds. *Geography and Enlightenment*. Chicago: University of Chicago Press, 1999.

Lowenthal, David. *George Perkins Marsh: Prophet of Conservation*. 1958. Reprint, Seattle: University of Washington Press, 2000.

———. *The Past is a Foreign Country*. Cambridge: Cambridge University Press, 1985.

Lowenthal, David, and Martyn J. Bowden. *Geographies of the Mind: Essays in Historical Geosophy*. New York: Oxford University Press, 1976.

Lueck, Beth L. *American Writers and the Picturesque Tour: The Search for National Identity, 1790–1860*. New York: Garland, 1997.

Lundberg, Ann Elaine. "Mapping the Geologic Wilderness: Science, Nature Writing, and the American Self." Ph.D. diss., University of Notre Dame, 1999.

MacKenzie, John M., ed. *Imperialism and the Natural World*. Manchester, U.K.: Manchester University Press, 1990.

Manthorne, Katherine Emma. *Tropical Renaissance: North American Artists Exploring Latin America, 1839–1879*. Washington, D.C.: Smithsonian Institution, 1989.

Marx, Leo. *The Machine in the Garden: Technology and the Pastoral Ideal in America*. New York: Oxford University Press, 1964.

———. *The Pilot and the Passenger: Essays on Literature, Technology, and Culture in the United States*. New York: Oxford University Press, 1988.

Masur, Louis P. *1831: Year of Eclipse*. New York: Hill and Wang, 2001.

Mazel, David. *American Literary Environmentalism*. Athens: University of Georgia Press, 2000.

McCullough, David. *Brave Companions: Portraits in History*. New York: Touchstone, 1992.

McPhee, John. *Basin and Range*. New York: Noonday Press, 1981.

Meinig, D. W. *The Interpretation of Ordinary Landscape*. New York: Oxford University Press, 1979.

———. *The Shaping of America: A Geographical Perspective on 500 Years of History*. Vol. 3, *Transcontinental America, 1850–1915*. New Haven: Yale University Press, 1998.

Menand, Louis. *The Metaphysical Club: A Story of Ideas in America*. New York: Farrar, Straus, and Giroux, 2001.

Merideth, Robert. *The Environmentalist's Bookshelf: A Guide to the Best Books*. New York: G. K. Hall, 1993.

Merrill, George P. *The First One-Hundred Years of American Geology*. 1924. Reprint, New York: Hafner, 1964.

Miller, Angela. *The Empire of the Eye: Landscape Representation and American Cultural Politics, 1825–1875*. Ithaca, N.Y.: Cornell University Press, 1993.

Miller, David C. *Dark Eden: The Swamp in Nineteenth-Century American Culture*. New York: Cambridge University Press, 1989.

Miller, Perry. *Errand into the Wilderness*. Cambridge, Mass.: Harvard University Press, 1956.

———. *Nature's Nation*. Cambridge, Mass.: Harvard University Press, 1967.

———. *The Raven and the Whale: The War of Words and Wits in the Era of Poe and Melville*. New York: Harcourt, Brace, 1956.

Mitchell, Lee Clark. *Witnesses to a Vanishing America: The Nineteenth-Century Response*. Princeton: Princeton University Press, 1981.

Mogen, David, Mark Busby, and Paul Bryant, eds. *The Frontier Experience and the American Dream: Essays on American Literature*. College Station: Texas A&M University Press, 1989.

Mulvey, Christopher. *Anglo-American Landscapes: A Study of Nineteenth-Century Anglo-American Travel Literature*. New York: Cambridge University Press, 1983.

Nash, Roderick. *Wilderness and the American Mind*. 4th ed. 1967. Reprint, New Haven: Yale University Press, 2001.

Nicolson, Marjorie Hope. *Mountain Gloom and Mountain Glory: The Development of the Aesthetics of the Infinite*. 1959. Reprint, Seattle: University of Washington Press, 1997.

Novak, Barbara. *Nature and Culture: American Landscape and Painting, 1825–1875*. New York: Oxford University Press, 1980.

Oelschlaeger, Max. *Caring for Creation: An Ecumenical Approach to the Environmental Crisis*. New Haven: Yale University Press, 1994.

———. *The Idea of Wilderness: From Prehistory to the Age of Ecology*. New Haven: Yale University Press, 1991.

O'Grady, John P. *Pilgrims to the Wild: Everett Ruess, Henry David Thoreau, John Muir, Clarence King, Mary Austin*. Salt Lake City: University of Utah Press, 1993.

Olds, Elizabeth Fagg. *Women of the Four Winds: The Adventures of Four of America's First Women Explorers*. Boston: Houghton Mifflin, 1985.

Orr, David. *The Last Refuge: Patriotism, Politics, and the Environment in an Age of Terror*. Washington, D.C.: Island Press, 2004.

Pagden, Anthony. *European Encounters with the New World*. London: Yale University Press, 1993.

Perry, Adele. *On the Edge of Empire: Gender, Race, and the Making of British Columbia, 1849–1871*. Toronto: University of Toronto Press, 2001.

Phillips, Dana. *The Truth of Ecology: Nature, Culture, and Literature in America*. New York: Oxford University Press, 2003.

Pomeroy, Earl. *In Search of the Golden West: The Tourist in Western America*. Lincoln: University of Nebraska Press, 1957.

Porter, Charlotte M. *The Eagle's Nest: Natural History and American Ideas, 1812–1842*. Tuscaloosa: University of Alabama Press, 1986.

Pratt, Mary Louise. *Imperial Eyes: Travel Writing and Transculturation*. London: Routledge, 1992.

Price, Jennifer. *Flight Maps: Adventures with Nature in Modern America*. New York: Basic Books, 1999.

Quinn, D. Michael. *Early Mormonism and the Magic World View*. Rev. and enlarged ed. Salt Lake City: Signature Books, 1998.

Reingold, Nathan, ed. *Science in Nineteenth-Century America: A Documentary History*. New York: Hill and Wang, 1964.

Reynolds, David S. *Beneath the American Renaissance: The Subversive Imagination in the Age of Emerson and Melville*. Cambridge, Mass.: Harvard University Press, 1988.

———. *Walt Whitman's America*. New York: Knopf, 1995.

Richardson, Robert D. *Emerson: The Mind on Fire*. Berkeley: University of California Press, 1995.

Riffenburgh, Beau. *The Myth of the Explorer*. New York: Oxford University Press, 1994.

Robbins, Bruce. "Colonial Discourse: A Paradigm and Its Discontents." *Victorian Studies* 35 (Winter 1992), pp. 209–14.

Ronda, James P. *Finding the West: Explorations with Lewis and Clark*. Albuquerque: University of New Mexico Press, 2001.

Rosaldo, Renato. "Imperialist Nostalgia." In *Culture and Truth: The Remaking of Social Analysis*. Boston: Beacon Press, 1989, pp. 68–87.

Rotundo, E. Anthony. *American Manhood: Transformations in Masculinity from the Revolution to the Modern Era*. New York: Basic Books, 1993.

Runte, Alfred. *National Parks: The American Experience*. 1979. Reprint, 3rd ed., Lincoln: University of Nebraska Press, 1997.

————. *Yosemite: The Embattled Wilderness.* Lincoln: University of Nebraska Press, 1990.

Rupp, Leila J. *A Desired Past: A Short History of Same-Sex Love in America.* Chicago: University of Chicago Press, 1999.

Russett, Cynthia Eagle. *Darwin in America: The Intellectual Response, 1865–1912.* San Francisco: W. H. Freeman, 1976.

Ryden, Kent C. *Landscape with Figures: Nature and Culture in New England.* Iowa City: University of Iowa Press, 2001.

Said, Edward W. *Orientalism.* 1978. Reprint, New York: Vintage, 1994.

————. *Culture and Imperialism.* New York: Vintage, 1994.

Schama, Simon. *Landscape and Memory.* New York: Knopf, 1995.

Schnaiberg, Allan, and Kenneth Alan Gould. *Environment and Society: The Enduring Conflict.* New York: St. Martin's, 1994.

Schneider, Richard J., ed. *Thoreau's Sense of Place: Essays in American Environmental Writing.* Iowa City: University of Iowa Press, 2000.

Shi, David E. *Facing Facts: Realism in American Thought and Culture, 1850–1920.* New York: Oxford University Press, 1995.

————. *The Simple Life: Plain Living and High Thinking in American Culture.* New York: Oxford University Press, 1985.

Silverberg, Robert. *Mound Builders of Ancient America: The Archaeology of a Myth.* 1968. Reprint, Athens: Ohio University Press, 1986.

Simonson, Harold P. *Beyond the Frontier: Writers, Western Regionalism, and a Sense of Place.* Fort Worth: Texas Christian University Press, 1989.

Slaughter, Thomas P. *Exploring Lewis and Clark: Reflections on Men and Wilderness.* New York: Knopf, 2003.

Slotkin, Richard. *The Fatal Environment: The Myth of the Frontier in the Age of Industrialization, 1800–1890.* 1985. Reprint, New York: HarperPerennial, 1994.

————. *Regeneration through Violence: The Mythology of the American Frontier, 1600–1860.* 1973. Reprint, New York: HarperPerennial, 1996.

Smallwood, William M. *Natural History and the American Mind.* New York: Columbia University Press, 1941.

Smith, Henry Nash. *Virgin Land: The American West as Symbol and Myth.* 1950. Reprint, Cambridge, Mass.: Harvard University Press, 1978.

Smith, Michael. *Pacific Visions: California Scientists and the Environment, 1850–1915.* New Haven: Yale University Press, 1987.

Solnit, Rebecca. *As Eve Said to the Serpent: On Landscape, Gender, and Art.* Athens: University of Georgia Press, 2001.

————. *A Field Guide to Getting Lost.* New York: Viking, 2005.

————. *River of Shadows: Eadweard Muybridge and the Technological Wild West.* New York: Viking, 2003.

————. *Savage Dreams: A Journey into the Landscape Wars of the American West.* 1994. Reprint, Berkeley: University of California Press, 1999.

————. *Wanderlust: A History of Walking.* New York: Viking, 2000.

Spence, Mark. *Dispossessing the Wilderness: Indian Removal and the Making of the National Parks.* New York: Oxford University Press, 1999.

Sperry, Shelley Lynne. "Natural Relations: Women, Men, and Wilderness in California, 1872–1914." Ph.D. diss., University of Maryland—College Park, 1999.

Spirn, Anne Whitson. *The Language of Landscape.* New Haven: Yale University Press, 1998.

Spurr, David. *The Rhetoric of Empire: Colonial Discourse in Journalism, Travel Writing, and Imperial Administration.* Durham, N.C.: Duke University Press, 1993.

Stanton, William. *The Leopard's Spots: Scientific Attitudes toward Race in America, 1815–1859.* Chicago: University of Chicago Press, 1960.

Stegner, Wallace. *Beyond the Hundredth Meridian: John Wesley Powell and the Second Opening of the West.* 1953. Reprint, Lincoln: University of Nebraska Press, 1982.

————. *Where the Bluebird Sings to the Lemonade Springs: Living and Writing in the West.* New York: Penguin, 1992.

Stein, Roger. *John Ruskin and Aesthetic Thought in America, 1840–1900.* Cambridge, Mass.: Harvard University Press, 1967.

Sterling, Keir B. *Last of the Naturalists: The Career of C. Hart Merriam.* Rev. ed. New York: Arno Press, 1977.

Stewart, Frank. *A Natural History of Nature Writing.* Washington, D.C.: Island Press, 1995.

Stilgoe, John R. *Landscape and Images.* Charlottesville: University of Virginia Press, 2005.

Stocking, George W., Jr. *Race, Culture, and Evolution: Essays on the History of Evolution.* New York: Free Press, 1968.

Taft, Robert. *Photography and the American Scene: A Social History, 1839–1889.* 1938. Reprint, New York: Dover Publications, 1964.

Thomas, John L. *Alternative America: Henry George, Edward Bellamy, Henry Demarest Lloyd, and the Adversary Tradition.* Cambridge, Mass.: Harvard University Press, 1983.

Thomas, Keith. *Man and the Natural World: A History of the Modern Sensibility.* New York: Pantheon, 1983.

Thomas, William L., ed. *Man's Role in Changing the Face of the Earth.* Chicago: University of Chicago Press, 1955.

Tichi, Cecelia. *New World, New Earth: Environmental Reform in American Literature from the Puritans through Whitman.* New Haven: Yale University Press, 1979.

Todorov, Tzvetan. *The Conquest of America: The Question of the Other.* New York: HarperPerennial, 1992; orig. pub. in French in 1982.

Trachtenberg, Alan. *The Incorporation of America: Culture and Society in the Gilded Age.* New York: Hill and Wang, 1982.

————. *Reading American Photographs: Images as History, Mathew Brady to Walker Evans.* New York: Hill and Wang, 1989.

Tuan, Yi-Fu. *Cosmos and Hearth: A Cosmopolite's Viewpoint.* Minneapolis: University of Minnesota Press, 1996.

————. *Escapism.* Baltimore: Johns Hopkins University Press, 1998.

————. *Passing Strange and Wonderful: Aesthetics, Nature, and Culture.* Washington, D.C.: Island Press, 1993.

————. "Realism and Fantasy in Art, History, and Geography." *Annals of the Association of American Geographers* 80 (1990), pp. 435–46.

————. *Space and Place: The Perspective of Experience.* Minneapolis: University of Minnesota Press, 1977.

————. *Topophilia: A Study of Environmental Perception, Attitudes, and Values.* Englewood Cliffs, N.J.: Prentice-Hall, 1974.

Van Den Abbeele, Georges. *Travel as Metaphor from Montaigne to Rousseau.* Minneapolis: University of Minnesota Press, 1992.

Walls, Laura Dassow. *Seeing New Worlds: Henry David Thoreau and Nineteenth-Century Natural Science.* Madison: University of Wisconsin Press, 1995.

————. "*Walden* as Feminist Manifesto." *ISLE: Interdisciplinary Studies in Literature and Environment* 1 (Spring 1993), pp. 137–44.

Warner, Michael. "Thoreau's Bottom." *Raritan* 11 (Winter 1992), pp. 53–78.

Washburn, Wilcomb E. *The Cosmos Club of Washington: A Centennial History, 1878–1978.* Washington, D.C.: Cosmos Club, 1978.

Watson, David. *Beyond Bookchin: Preface for a Future Social Ecology.* Brooklyn: Autonomedia, 1996.

Wauchope, Robert. *Lost Tribes and Sunken Continents: Myth and Method in the Study of American Indians.* Chicago: University of Chicago Press, 1962.

White, David A., ed. *News of the Plains and Rockies, 1803–1865: Original Narratives of Overland Travel and Adventure Selected from the Wagner-Camp and Becker Bibliography of Western Americana.* Vols. 4 and 5. Spokane: Arthur H. Clark, 1998.

White, Richard. *The Organic Machine: The Remaking of the Columbia River.* New York: Hill and Wang, 1995.

Wiebe, Robert H. *The Search for Order, 1877–1920.* New York: Hill and Wang, 1967.

Williams, Raymond. *The Country and the City.* New York: Oxford University Press, 1973.

———. *Culture and Society: 1780–1950.* 1958. Reprint, New York: Columbia University Press, 1983.

Wilson, Alexander. *The Culture of Nature: North American Landscape from Disney to the Exxon Valdez.* Toronto: Between the Lines, 1991.

Wilson, Eric. *Romantic Turbulence: Chaos, Ecology, and American Space.* Houndmills, U.K.: Macmillan, 2000.

Wilson, R. Jackson. *Figures of Speech: American Writers and the Literary Marketplace, from Benjamin Franklin to Emily Dickinson.* New York: Knopf, 1989.

Wilton, Andrew, and Tim Barringer. *American Sublime: Landscape Painting in the United States, 1820–1880.* London: Tate Publishing, 2002.

Wolf, Bryan Jay. *Romantic Re-vision: Culture and Consciousness in 19th-Century American Painting and Literature.* Chicago: University of Chicago Press, 1982.

———. "A Grammar of the Sublime, or Intertextuality Triumphant in Church, Turner, and Cole." *New Literary History* 16 (Winter 1985), pp. 321–41.

Wolf, Daniel, ed. *The American Space: Meaning in Nineteenth-Century Landscape Photography.* Middletown, Conn.: Wesleyan University Press, 1983.

Worster, Donald, ed. *The Ends of the Earth: Perspectives on Modern Environmental History.* New York: Cambridge University Press, 1988.

———. *Nature's Economy: A History of Ecological Ideas.* New York: Cambridge University Press, 1977.

———. *A River Running West: The Life of John Wesley Powell.* New York: Oxford University Press, 2001.

———. *The Wealth of Nature: Environmental History and the Ecological Imagination.* New York: Oxford University Press, 1993.

Ziff, Larzer. *Literary Democracy: The Declaration of Cultural Independence in America.* New York: Penguin, 1981.

———. *Return Passages: Great American Travel Writing, 1780–1910.* New Haven: Yale University Press, 2000.

Zimmerman, Michael E. *Contesting Earth's Future: Radical Ecology and Postmodernity.* Berkeley: University of California Press, 1994.

Zweig, Paul. *The Adventurer: The Fate of Adventure in the Western World.* 1974. Reprint, Pleasantville, N.Y.: Akadine Press, 1999.

— IMAGE CREDITS —

1 Friedrich Georg Weitsch. Portrait of Alexander von Humboldt, 1806. Oil on canvas, 126 × 92.5 cm. Nationalgalerie, Staatliche Museen zu Berlin, Berlin. Digital image used by permission of Bildarchiv Preussischer Kulturbesitz / Art Resource, New York.

2 Detail of a map of South America, 1897. Single-sheet map apparently extracted from *Rand, McNally and Company's Indexed Atlas of the World* (Chicago: Rand, McNally, 1898).

3 Timothy H. O'Sullivan. "Clover Peak, East Humboldt Mountains, Nevada," 1868. Reproduction courtesy of the National Archives, Still Pictures Section, collection of the U.S. Geological Exploration of the 40th Parallel.

4 Timothy H. O'Sullivan. "Tufa domes, Pyramid Lake, Nevada," 1867. Reproduction courtesy of the National Archives, Still Pictures Section, collection of the U.S. Geological Exploration of the 40th Parallel.

5 Timothy H. O'Sullivan. "Trinity Mountains: Rhyolite Ridge, 1867." U.S. Geological Exploration of the 40th Parallel, Photos of Scenery of the American West. Banc Pic 1957.027-f. Courtesy of The Bancroft Library, University of California, Berkeley.

6 Alexander von Humboldt. Chart showing the climate zones of Chimborazo, Ecuador, 1805. Plate entitled *Géographie des Plantes Équinoxiales*, attached to Humboldt, *Essai sur la Géographie des Plantes* (Paris: F. Schoell, 1807). Photo courtesy of the Beinecke Library, Yale University.

7 Alexander von Humboldt. Chart comparing the height of mountain ranges, 1825. Humboldt, *Umrisse von Vulkanen aus den Cordilleran von Quito und Mexico* (Stuttgart und Tübingen: J. G. Cotta, 1853), plate 12.

8 Map showing the region of Humboldt's Orinoco journey, 1897. Single-sheet map apparently extracted from *Rand, McNally and Company's Indexed Atlas of the World* (Chicago: Rand, McNally, 1898).

9 Frederic Edwin Church. *The Heart of the Andes*, 1859. Oil on canvas, 66 × 119 in. 09.95. Metropolitan Museum of Art, New York, Bequest of Margaret E. Dows, 1909. Used by permission.

10 Map of Chile, 1823. In *A General Atlas Containing Distinct Maps of all the Known Countries in the World* (Baltimore: F. Lucas, Jr., 1823), plate 104.

11 James Eights. *Decolopoda australis*, 1834. Eights, "Description of a New Animal Belonging to the Arachnides of Latreille; Discovered in the Sea Along the Shores of the New South Shetland Islands," *Boston Journal of Natural History* 1 (Boston: Hilliard, Gray, 1837), plate 7.

12 Map showing southern Chile and the isle of Mocha, 1823. Detail from *A General Atlas Containing Distinct Maps of all the Known Countries in the World* (Baltimore: F. Lucas, Jr., 1823), plate 104.

13 Asa Gray. *Reynoldsia sandwicensis*, 1854. Gray, *United States Exploring Expedition: Botany: Phanerogamia*, vol. 1 (New York: George P. Putnam, 1854), plate 92.

14 Frederic Edwin Church. *The Icebergs*, 1861. Oil on canvas, 64 × 112 in. Dallas Museum of Art, anonymous gift, 1979.28. Used by permission.

15 Frederic Edwin Church. *Cotopaxi*, 1862. Oil on canvas, 48 × 85 in. Detroit Institute of Arts, Founders Society Purchase, Robert H. Tannahill Foundation Fund, Gibbs-Williams Fund, Dexter M. Ferry, Jr., Fund, Merrill Fund, Beatrice W. Rogers Fund, and Richard A. Manoogian Fund. Photograph © 1977 The Detroit Institute of Arts. Used by permission.

16 Timothy H. O'Sullivan. Clarence King—"Mountain Climbing," c. 1869. U.S. Geological Exploration of the 40th Parallel, Photos of Scenery of the American West. Banc Pic 1957.027-f. Courtesy of The Bancroft Library, University of California, Berkeley.

17 Clarence King and James T. Gardner. Map of the Yosemite Valley, 1865. California Geological Survey. G4362.Y62 1872.G3, Case C. Courtesy of The Bancroft Library, University of California, Berkeley.

18 Map showing the area covered by King's 40th Parallel Survey, 1870. "General Map of the United States," *Black's General Atlas of the World*, new and rev. ed. (Edinburgh: A. and C. Black, 1870), plate 41.

19 Timothy H. O'Sullivan. "Hot Sulphur Spring, Ruby Valley, Nevada," 1868. U.S. Geological Exploration of the 40th Parallel, Photos of Scenery of the American West. Banc Pic 1957.027-f. Courtesy of The Bancroft Library, University of California, Berkeley.

20 Timothy H. O'Sullivan. "The Falls of the Shoshone, Snake River, Idaho," 1868. Courtesy of The Bancroft Library, University of California, Berkeley.

21 Timothy H. O'Sullivan. "Geyser, Ruby Valley, Nevada," 1868. U.S. Geological Exploration of the 40th Parallel, Photos of Scenery of the American West. Banc Pic 1957.027-f. Courtesy of The Bancroft Library, University of California, Berkeley.

22 Timothy H. O'Sullivan. "Gould and Curry Works, Nevada," 1867–68. U.S. Geological Exploration of the 40th Parallel, Photos of Scenery of the American West. Banc Pic 1957.027-f. Courtesy of The Bancroft Library, University of California, Berkeley.

23 Timothy H. O'Sullivan. "Gold Hill, Nevada," 1867–68. U.S. Geological Exploration of the 40th Parallel, Photos of Scenery of the American West. Banc Pic 1957.027-f. Courtesy of The Bancroft Library, University of California, Berkeley.

24 Timothy H. O'Sullivan. "Gould and Curry Mine, Cave-in, Nevada," 1867–68. U.S. Geological Exploration of the 40th Parallel, Photos of Scenery of the American West. Banc Pic 1957.027-f. Courtesy of The Bancroft Library, University of California, Berkeley.

25 Timothy H. O'Sullivan. "Savage Shaft, Miners and Mine Car, Nevada," 1867–68. U.S. Geological Exploration of the 40th Parallel, Photos of Scenery of the American West. Banc Pic 1957.027-f. Courtesy of The Bancroft Library, University of California, Berkeley.

26 Timothy H. O'Sullivan. "Indians: Group (Shoshones), Nevada," 1868? U.S. Geological Exploration of the 40th Parallel, Photos of Scenery of the American West. Banc Pic 1957.027-f. Courtesy of The Bancroft Library, University of California, Berkeley.

27 Timothy H. O'Sullivan. "Shoshone Indians: Group (Buck and Staff), Nevada," 1868? U.S. Geological Exploration of the 40th Parallel, Photos of Scenery of the American West. Banc Pic 1957.027-f. Courtesy of The Bancroft Library, University of California, Berkeley.

28 Timothy H. O'Sullivan. "Summit, East Humboldt Mountains, Nevada," 1868. Courtesy of The Bancroft Library, University of California, Berkeley.

29 Carleton Watkins. "Mount Shasta, Siskiyou County, California," 1870. "Photographs of California from the U.S. Geological Exploration of the 40th Parallel (1870)." Reproduction courtesy of the Beinecke Library, Yale University.

30 Carleton Watkins. "Commencement of the Whitney Glacier, Mount Shasta, Siskiyou County, California," 1870. "Photographs of California from the U.S. Geological Exploration of the 40th Parallel (1870)." Reproduction courtesy of the Beinecke Library, Yale University.

31 Timothy H. O'Sullivan. "Survey Company: Group, 'The immortal few who were not born to die,'" 1867? U.S. Geological Exploration of the 40th Parallel, Photos of Scenery of the American West. Banc Pic 1957.027-f. Courtesy of The Bancroft Library, University of California, Berkeley.

32 Officers of the *Jeannette* (souvenir, 1879). "Composite photograph of ship and officers," USS *Jeannette* Collection, Naval Historical Foundation (NH #52007), courtesy of Capt. T. S. Wilkinson, USN, 1934. Reproduction courtesy of the Naval Historical Foundation.

33 "Map of the North Polar Region (Arctic Ocean), from the latest information, 1885." Engraved by Fisk and Co., New York, 1885. Single-sheet map, almost certainly extracted from an atlas. Author's collection.

34 Thomas Eakins. *Rear-Admiral George W. Melville*, 1905. Oil on canvas, 40 × 26¹⁵⁄₁₆ in. Na-

tional Gallery of Art, Washington, D.C. Gift (Partial and Promised) of Senator and Mrs. H. John Heinz III, in Honor of the 50th Anniversary of the National Gallery of Art. Image © 2005 Board of Trustees, National Gallery of Art. Used by permission.

35 *The Sinking of the* Jeannette, 1881. USS *Jeannette* Collection, Naval Historical Foundation (NH #52000), engraving taken from *The Voyage of the Jeannette*, 1883, vol. 1 (Emma De Long, ed.). Reproduction courtesy of the Naval Historical Foundation.

36 A page from George Wallace Melville's "Ice Journals," 1882. Reproduction courtesy of Dartmouth College Special Collections, Rauner Library.

37 The cairn George Wallace Melville erected for his dead companions, 1882. Engraving after a sketch by Melville, in his "Report of Chief Engineer Geo. W. Melville in connection with the Jeannette Expedition" (Washington, D.C.: Government Printing Office, 1882). Reproduction courtesy of Dartmouth College Special Collections, Rauner Library.

38 John Muir as a young man, 1860s? Photographer unknown. Used by permission of the Wisconsin Historical Society, Image ID: WHi-1946.

39 John Muir. Sketches of natives, 1881. Muir's journal of the summer of 1881, p. 37, John Muir Papers, microfilm reel 26, "Journals and Sketchbooks 1879–1881." Reproduction from the John Muir Papers, Holt-Atherton Special Collections, University of the Pacific Library. Copyright 1984 Muir-Hanna Trust. Used by permission.

40 John Muir. Sketch of a native village, Diomede Island, 1881. Muir's journal of the summer of 1881, p. 22, John Muir Papers, microfilm reel 26, "Journals and Sketchbooks 1879–1881." Reproduction from the John Muir Papers, Holt-Atherton Special Collections, University of the Pacific Library. Copyright 1984 Muir-Hanna Trust. Used by permission.

41 Edward S. Curtis. "House and Hearth—Plover Bay, Siberia," 1899. Harriman Alaska Expedition Souvenir Book, courtesy of Beinecke Library, Yale University.

— INDEX —

Page numbers in *italics* refer to picture captions.